Laboratory Experiments in the Social Sciences

Laboratory Experiments in the Social Sciences

Edited by
Murray Webster, Jr. and Jane Sell

AMSTERDAM • BOSTON • HEIDELBERG • LONDON
NEW YORK • OXFORD • PARIS • SAN DIEGO
SAN FRANCISCO • SINGAPORE • SYDNEY • TOKYO
Academic Press is an imprint of Elsevier

Academic Press is an imprint of Elsevier
30 Corporate Drive, Suite 400, Burlington, MA 01803, USA
525 B Street, Suite 1900, San Diego, California 92101-4495, USA
84 Theobald's Road, London WC1X 8RR, UK

This book is printed on acid-free paper. ∞

Library of Congress Cataloging-in-Publication Data
Laboratory experiments in the social sciences / edited by Murray Webster, Jr. and Jane Sell.
    p. cm.
    Includes bibliographical references and index.
    ISBN 978-0-12-369489-8
    1.   Social sciences—Experiments. 2.   Experimental design.
    I. Webster, Murray, 1941- II. Sell, Jane.
    H62.L23 2007
    300.72'4—dc22   2007017991

British Library Cataloguing-in-Publication Data
A catalogue record for this book is available from the British Library.

ISBN: 978-0-12-369489-8

For information on all Academic Press publications
visit our Web site at www.books.elsevier.com

Printed in the United States of America
07 08 09 10   9 8 7 6 5 4 3 2 1

Working together to grow
libraries in developing countries
www.elsevier.com | www.bookaid.org | www.sabre.org
ELSEVIER   BOOK AID
            International   Sabre Foundation

*We dedicate this book to the memory of
Elizabeth G. Cohen (1931–2005):
pioneer, scientist, and gentle humanitarian.*

CONTENTS

PART **II**

Designing and Conducting Experiments

PART **III**

Experiments Across the Social Sciences

18 Social Dilemma Experiments in Sociology, Psychology, Political Science, and Economics

Jane Sell

PART **IV**

Experiments in the Twenty-First Century

19 Experimental Political Science

Rose McDermott

20 Economic Games for Social Scientists

Catherine Eckel

21 Laboratory Experiments in Sociology

Morris Zelditch, Jr.

PREFACE

This book grew from our love of experimental methods and our frustration with the lack of published work on them. It seemed to us that experimental methods were routinely mentioned briefly and then ignored in most research methods books. In fact, it is not unusual for books that serve as overviews of social science methods to cite the experimental method as the "gold standard" for assessment of causality and then to assert without evidence that many, perhaps most, questions of interest to social scientists simply cannot be addressed through experiments. While there is impressive literature on statistical analysis of data from experiments, there is much less available on the design, development, and actual conduct of experiments. In fact, if someone were to ask where to go to learn about the philosophy, design, and operation of experiments, we would have been hard pressed to offer a practical suggestion.

Yet experiments have been studied, and there is a great deal of useful information on them and how to do them well; it is spread across a wide range of journal articles and unpublished papers presented at professional meetings, as well as countless "working papers" circulated internally at sites where experiments are conducted. That effectively makes it unavailable to anyone who doesn't already know the main experimental sites and journals. We hope this book will constitute a useful reference for anyone interested in beginning to use experimental methods, as well as offer information for experienced experimenters interested in learning new techniques or expanding their knowledge of what others are doing with this method.

What might lead to neglect of experimental methods in standard books on research in the social sciences? We think it may be because the method is quite new in most social sciences, and it is often misunderstood. Many social scientists misunderstand the philosophical foundations of experiments, and many others still believe that experiments are the purview of the natural sciences, not the social sciences. When a new methodology is subject to misunderstandings about its uses and value, it is little wonder that social scientists sometimes avoid it. Even those interested in taking a chance and trying out the method, as we noted above, cannot easily find a good place to learn about it. This book is for them.

Undergraduate courses, and even more so graduate courses in research methods, deal mostly with methods of analyzing data. Many such courses are aimed at someone who has already collected or acquired a large data set from a survey and wants to know how to answer questions from it. What's missing? First, there is nothing on the methodology of acquiring data, whether it be data from well-designed surveys, interviews, observations, or experimental methods. Students are unlikely to learn how to actually collect reliable social science data, and they certainly are not going to learn how to get good experimental data. Second, there is little on appropriate uses of different methods, and sadly, one could conclude from many contemporary courses that, while experiments might look good in principle, in actuality social scientists do not use them.

Most students learn that experiments offer a high degree of control over independent variables and control over the research setting to facilitate observations of dependent variables. They learn about control conditions and experimental conditions and the importance of random assignment of individuals to conditions. They learn that a well-designed experiment permits high confidence that changes observed in dependent variables are associated with (maybe even "caused by") changes in independent variables. So far, so good. At this point they may be excited about learning how to use this design because of its strengths.

Unfortunately, too often there is more, and some of it can be harmful. Students learn about confounding factors such as history, maturation, testing, and many others, which can mislead investigators who wish to impute causality. Those factors often are presented as if they are inevitable and make it difficult if not impossible to know for sure what happened in a study.[1] To make matters worse, some instructors also show a couple of horrible films depicting unethical social science experiments, and they may talk about the infamous Tuskegee medical experiments.[2] Students can conclude that experiments are complicated

---

[1] In 1963, Donald T. Campbell and Julian Stanley published a masterful consideration of potential confounding factors as one chapter in a methodology book. Their chapter is so useful that it was reprinted by itself 2 years later, and it is the standard reference on confounding factors. Oddly overlooked is the fact that Campbell and Stanley showed that a true experiment using random allocation, a control condition, and one or two experimental conditions, controls *all* the confounding factors they identified. Properly understood, their analysis is a strong argument in favor of using experiments whenever possible.

[2] Unethical research of any kind by anyone is unacceptable, as several chapters in this book emphasize.

and really do not show anything because of confounding factors, and that those who experiment on humans are often unethical.

Those are false conclusions, and we hope to clear up those and other misapprehensions about experiments with this book. The authors of these chapters are concerned with explaining how to design and conduct scientifically sound experiments and how to meet the highest ethical standards. It really is not that hard to meet those goals.

We learned to conduct experiments the same way as nearly all experimenters practicing today learned their skills: in mentoring and tutoring relationships. That is, we learned individually by working on research projects under close supervision of someone who watched to see when we needed guidance and was ready to step in if we got too far astray in the work. We and most experimenters remember those years of learning as hard work and immeasurably valuable; most of us enjoy mentoring and tutoring new investigators. While individual learning and mentoring are wonderful in many ways, they are inherently slow and limited to a few individuals who are fortunate enough to form an apprenticeship relation with an excellent laboratory and its practitioners. We believe that much of the knowledge required for experimental work can be taught in classes because some of the chapter authors here already teach superb courses in experimental methods. Someone who first studies experimentation before beginning work may still benefit from guidance and mentoring, but he or she will already have a large store of knowledge to build upon. This can be as useful to someone who has already used other research methods as it is to someone whose entire research career will involve experiments. Between us, we have over 70 years working with, writing about, and thinking about experiments. Still, we learned things from every single chapter in this volume. The chapters are wonderfully filled with knowledge.

In *Laboratory Experiments in the Social Sciences,* the chapter authors confront many of the misunderstandings about experimental methods, and they go beyond correcting misunderstandings to offer guidelines for how to do experiments well, and still more, they describe some exemplary research programs that have used these methods.

In the first section of the book, authors address the philosophical and methodological foundations for experiments. In this section, researchers address what kinds of questions might be answered through experiments, differences among different approaches, and how experiments relate to theory and evidence.

Section Two provides information to address the myriad of theoretical, methodological, and practical considerations of actually doing research. Initial considerations involved in ethics and the role of Institutional Review Boards, writing effective proposals, training experimenters, and developing the general parameters of the experiment are addressed.

Examples of different experimental programs, different strategies, and different outcomes are offered in the third section. These case studies provide illustrations of how the power of experimental designs directly feeds into theoretical and applied advancement. These examples reach across different disciplines and illustrate how very different kinds of questions at different levels of analysis can be investigated experimentally.

The fourth section considers common problems that might arise in experimental research and offers some solutions. In addition, researchers in three different disciplines—political science, economics, and sociology—consider how experimental research has affected the growth of theory in their disciplines and assess promising trajectories of research.

This book incorporates experiences and insights of some of the foremost researchers in experimental social science. Some of these researchers have been conducting experiments for many years and offer a wealth of information about what they have learned. Others came to experimentation much more recently. They offer insights into all the aspects of beginning an experimental research program and some of the unexpected things that "nobody told me." The authors bring different insights and kinds of experience to the chapters. All of them share a commitment to rigorous experimental methods.

This book would not have come into being without the vision of Executive Editor Scott Bentley from Elsevier, who initiated it and who demonstrated that he shares our commitment to it throughout its development. We thank our Development Editor Kathleen Paoni for her efficient responses to our many queries while keeping an eye on the big picture of one book in a heavy production schedule. We thank those who encouraged and nurtured our intellectual interest in experimental methods. In graduate school at Washington State University, Jane was challenged and encouraged by Lee Freese and Louis Gray; at Texas A&M, she was supported by Raymond Battalio who shared his intellectual excitement as well as his laboratory and its resources. Murray began as a research assistant for Bernard P. Cohen and then received extensive training from Joseph Berger. He is grateful to Peter H. Rossi, Ronald W. Maris, Margaret Zahn, and Schley Lyons, who supported him and his experimental research, sometimes mostly on faith that it would all work out somehow. Jane is grateful to her husband Philip Berke and their children LeeAnn and Timothy for sharing their ideas and their laughter; Murray leans on his friends every single day. Finally, we thank the researchers who annually meet at the Group Process meetings, a tradition created by Linda D. Molm. The interaction and intellectual excitement and new ideas at every Group Process meeting renew our energy for our own research and for spreading the word. Those scholars are our Significant Others (as Harry Stack Sullivan used that term), and if this book earns approval in the Looking Glass they embody, we will be deeply, respectfully grateful.

—*Murray Webster, Jr., Charlotte, North Carolina*
—*Jane Sell, College Station, Texas*

CONTRIBUTORS

Numbers in parentheses indicate the pages on which the authors' contributions begin.

JOSEPH BERGER (353), Department of Sociology, Stanford University, Stanford, CA 94305

GIOVANNA DEVETAG (407), Department of Law and Management, University of Perugia, Perugia 06123, Italy

JAMES E. DRISKELL (329), Florida Maxima Corporation, Winter Park, FL 32789

CATHERINE ECKEL (497), School of Social Sciences, University of Texas at Dallas, Richardson, TX 75083

MARTHA FOSCHI (113), Department of Sociology, University of British Columbia, Vancouver, BC, Canada V6T 1Z1

KAREN A. HEGTVEDT (141), Department of Sociology, Emory University, Atlanta, GA 30322

STUART J. HYSOM (289), Department of Sociology, Texas A&M University, College Station, TX 77843

WILL KALKHOFF (243), Department of Sociology, Kent State University, Kent, OH 44242

JENNIFER KING (329), Naval Research Laboratory, Washington, DC 20375

KATHY J. KUIPERS (289), Department of Sociology, University of Montana, Missoula, MT 59812

MICHAEL J. LOVAGLIA (243), Department of Sociology, University of Iowa, Iowa City, IA 52242

ROSE McDERMOTT (483), Department of Political Science, University of California–Santa Barbara, Santa Barbara, CA 93106

LINDA D. MOLM (379), Department of Sociology, University of Arizona, Tucson, AZ 85721

LEDA NATH (243), Department of Sociology, Anthropology, and Criminal Justice, University of Wisconsin–Whitewater, Whitewater, WI 53190

ANDREAS ORTMANN (407), Charles University and Academy of Sciences of the Czech Republic, 111 21 Prague, Czech Republic

LISA SLATTERY RASHOTTE (225), Department of Sociology and Anthropology, University of North Carolina–Charlotte, Charlotte, NC 28223

JANE SELL (5, 459), Department of Sociology, Texas A&M University, College Station, TX 77843

ROBERT K. SHELLY (267), Sociology Department, Ohio University, Athens, OH 45701

SHANE R. THYE (57), Department of Sociology, University of South Carolina, Columbia, SC 29208

LISA TROYER (173), Department of Sociology, University of Iowa, Iowa City, IA 52242

HENRY A. WALKER (25), Department of Sociology, University of Arizona, Tucson, AZ 85721

MURRAY WEBSTER, JR. (5, 193), Department of Sociology, University of North Carolina–Charlotte, Charlotte, NC 28223

DAVID WILLER (25), Department of Sociology, University of South Carolina, Columbia, SC 29208

RICK K. WILSON (433), Department of Political Science, Rice University, Houston, TX 77251

REEF YOUNGREEN (243), Department of Sociology, University of Massachusetts–Boston, Boston, MA 02125

MORRIS ZELDITCH, JR. (87, 517), Department of Sociology, Stanford University, Stanford, CA 94305

PART I

Introduction to the Philosophy of Experimentation

The book is divided into four parts or sections. Part 1 contains five chapters dealing with an overview of experimentation and topics related to the method, the philosophical foundations of the method, and how experiments can increase knowledge in social science.

Chapter 1, by the editors, outlines some of the historical points in the development of experiments in social science and some general questions and topics that arise in using this method. This chapter describes our understanding of experiments and their value.

Chapter 2, by Henry A. Walker and David Willer, distinguishes empirical from theoretical experiments and develops a view that the latter is the only type likely to yield enduring knowledge. Along the way, these authors show why abstractly conceptualizing problems is necessary for theoretical advancement and why naïve experimental realism can be harmful to advancing social science understanding.

Chapter 3, by Shane R. Thye, addresses the skeptical or cynical views some social scientists have of experimentation and shows why many of those are based on misunderstanding the purposes of experiments. He also describes several well-known threats to good experimental design and ways to avoid them or to compensate for the uncertainty they produce. This chapter explores the necessary conditions for inferring causality from different kinds of evidence and shows why properly designed experiments offer a strong likelihood of producing an increase in theoretical understanding.

Chapter 4, by Morris Zelditch, Jr., considers issues of external validity. A criticism of experiments that is as widespread as it is wrong is that, because they are artificial, experiments cannot tell anything about what critics like to label "the real world." This objection is one often raised and, while it has frequently been analyzed and shown to be incorrect, still is raised. Zelditch considers several possible meanings of "external validity," and he shows how they do not apply to theory-driven experiments, which, he maintains, are the only kind likely to produce enduring knowledge. It is the most detailed examination available from one of our senior scholars of a problem that he has returned to many times in the past 5 decades.

Chapter 5, by Martha Foschi, offers a close examination of several terms involved with experimentation that can be misunderstood. She considers the parts of a hypothesis and how those relate to empirical research, which she illustrates with an example from her own research program on double standards. Building on distinctions introduced by Walker and Willer, and by Thye, Foschi considers what operationalization means in the context of theory-testing experiments, making the important distinction of operationalizing theoretical variables from operationalizing accidental features of an experiment. She offers a set of design features that require manipulation checks, showing how to get satisfactory operations and measurements in experimental research.

CHAPTER 1

Why Do Experiments?

MURRAY WEBSTER, JR.
University of North Carolina–Charlotte

JANE SELL
Texas A&M University

ABSTRACT

Laboratory experiments in social science first appeared around 1900, but they developed most rapidly in the years since the end of the Second World War. Experiments offer powerful advantages for testing predictions because they offer the possibility both to control on theoretically relevant factors and to utilize the power of random assignment to eliminate spurious variables. Experiments are most useful when investigating predictions derived from explicit theories, and it is theories, rather than experimental results, that are properly applied to explain features of natural settings. Experimental research programs include issues of theoretical foundation, abstract design, operational concerns, and interpretation of outcomes. While not every research question lends itself well to experimental research, when questions are formulated abstractly, the range of experimental usefulness is much broader than many people appreciate.

I. A BRIEF HISTORY OF EXPERIMENTS

Many social scientists, and most physicists, chemists, and biologists, see experimental methods as one of the defining characteristics of scientific inquiry. While the experiment is far from being the only research method available to the social sciences, its usage has grown remarkably in the years since the Second World War. Many historical changes are associated with the growth in experimental methods, two of which are especially important: new topics and new technology.

In early decades of the twentieth century, sociologists were largely occupied with classifying types and growth of societies, or with development of different parts of cities. Following World War II, many social scientists became interested in phenomena that can be studied experimentally. In sociology and social psychology, for instance, topics such as interpersonal influence, distortions of judgment, and conformity processes seemed more pressing than they had seemed before authoritarian and repressive societies were common topics. With the new topics came new theories, many of them amenable to experimentation. Economists began to conceptualize strategic game playing and became interested in behavioral economics; political scientists developed rational choice theories of voting choice; communications scientists began to understand influence processes; sociologists had new theories of social exchanges; and psychologists, whose discipline had used experiments from its beginnings, expanded its study of effects of social factors that appear in the presence of one or more other individuals. New topics and new theory were both congenial to the development of experimental methods in social science.

The second factor is new technology. Starting at a few universities, experimental laboratories were built, followed by laboratories in government facilities and at private research firms. New laboratories required and facilitated development of many kinds of technological advance: coding schemes to record discussion groups; one-way mirrors and, later, television and computers to observe and control communication among researchers and experimental participants; sound and video recorders, and many other elements of contemporary experiments began developing in the years following World War II.

Even though experiments are a recognized part of today's social science research techniques, for many social scientists they are still not well understood. Training in laboratory experimentation is still not part of the graduate training of the majority of social scientists (psychology may be the exception). That is unfortunate for many reasons. Those who might wish to conduct experimental research may not feel confident enough in their skills to approach this method. Social scientists who use other methods—survey researchers in sociology, for instance—may misunderstand the goals and uses of experiments. As every science relies on peer review of research, misunderstandings can slow the accumulation of knowledge, good experiments may be criticized on inappropriate grounds, and real flaws in an experiment may be overlooked.

With the continuing growth and development of experimental methods in social science, it will not be long before understanding experiments is an important part of every social scientist's professional skills. We and the other authors in this book hope to contribute to that understanding. For new experimenters we offer suggestions to improve the quality of their work; for those who read and wish to assess experimental research, we describe techniques and offer guidelines. In these ways we hope to contribute to the growing quality of experimental research in social science. A poorly designed experiment will either produce no results or, worse, will produce results that are not what an experimenter thinks they are. This book brings together "best practices" of several of today's outstanding experimental researchers. The chapters can be read as "how to" manuals for developing one's own experiments, or as sources of criteria to judge and improve the quality of experimental research by practitioners and by the professional audience. All of the chapters contain background to their individual topics that explicitly addresses common and some uncommon points crucial for understanding this method.

Experimental research is one kind of intellectual activity. A good way to approach experiments, either those one plans to conduct or those conducted by others, is to ask what they contribute to knowledge. What do we know as a result of an experiment or what do we hope to learn from a contemplated experiment? As will be clear in several of the chapters, the central issue in experimental research, as well as in other kinds of research, is how the research can contribute to knowledge of social processes and social structures.

We begin with some terminology on research design. All research is about how things are related. In describing a research design, often it is convenient to distinguish *independent* and *dependent variables,* and we use those terms in describing what we mean by *experiments.* A *variable* is anything that takes on different values, something that can be measured. In research, variables are tied to measurement operations; for instance, the variable *socioeconomic status* (SES) may be measured by a person's or a family's income in dollars. Beginning students must become accustomed to the idea that so-called *independent variables* are usually controlled in some way by investigators, while *dependent variables* are left free to vary; they are controlled only by nature. Thus, a study might control educational level statistically by dividing a sample of individuals into groups of those whose education ended after eighth grade, after some high school, with a high school diploma, etc., to see education's effect on SES. Education is the independent variable in this study, and SES is the dependent variable. The design described uses survey methods.

As we use the term, a study is an *experiment* only when a particular ordering occurs: when an investigator controls the level of independent variables *before* measuring the level of dependent variables. In the preceding hypothetical survey design, presumably data on respondents' schooling and income were collected at the same time. The independent variable was partitioned afterwards, and the interest was in how SES divided after education was so divided. If the design had been an experiment, the investigator might have begun with a large group of children and placed them into different groups determining how many years of education they would receive. When members of the last group had completed their education, average income levels of the different groups might have been compared.

That hypothetical example illustrates two more points about research designs. First, not every research design is an experiment. Informal usage sometimes describes any study as an experiment, but if we are going to focus on creating good research designs, we need to be clear about what types of designs we work on. While many of the criteria of good scientific research apply to all kinds of research, there are specific criteria for good surveys that are quite different from the criteria of good experiments. For instance, sample representativeness is a large concern in many surveys, but less so (for reasons discussed in several chapters) for experiments. Random allocation of respondents to conditions is crucial for most experimental designs, but it is often impossible in surveys. Second, for many reasons, not every research question can be studied experimentally. Moral and practical considerations make the preceding hypothetical experimental design—deciding ahead of time how much education every person can attain—ridiculous. Knowing what we can study experimentally is important at the very earliest stages of research design, when an investigator selects a type of design to develop.

Experimental studies came into social sciences around 1900, beginning in psychology with studies of biological responses (in particular, saliva production in dogs) conducted by the Russian physiologist Ivan Petrovich Pavlov. Pavlov's training in the biological laboratory was demonstrated in his turning to an experimental design, as well as in his use of physiological terms such as "stimuli" to name independent variables in the work. American psychologists such as Edward L. Thorndike (1905) at Harvard University and J. B. Watson (1913) at Johns Hopkins University adopted and developed the method for many studies of individual differences and interpersonal influences, and the methods spread across the social sciences.

Social psychologists Solomon Asch (1951), Muzafer Sherif (1948), and Leon Festinger (Festinger & Carlsmith, 1957) developed experimental methods beginning in the 1940s, about the same time the economist E. H. Chamberlin (1948) began to study markets experimentally. Mathematically trained social scientists, including Siegel and Fouraker (1960) and Von Neumann and Morgenstern (1944), analyzed rational choices, negotiations, and games, providing the foundations for many contemporary theories in sociology, political science, communications, and economics. In the 1950s, Robert Freed Bales and his colleagues and students at Harvard University began studying discussion groups using techniques and technology of an experimental laboratory. Kurt Lewin and Dorwin Cartwright founded the Research Center for Group Dynamics at Massachusetts Institute of Technology in 1945, and Cartwright moved it to the University of Michigan at Lewin's death in 1947 (Cartwright & Zander, 1953). The experiment has been a significant part of all the social sciences for over half a century, and its particular advantages continue to attract researchers across the social sciences.

The history of experimental methods shows diffusion across disciplines. Starting with techniques Pavlov learned in biology, through widespread diffusion of Bales' studies of discussion groups (e.g., into contemporary "focus groups" in communications and business), imaginative researchers have built on what worked in other projects and other disciplines. This is all to the good. When investigators in one discipline come up with useful designs, other disciplines benefit from using those when appropriate. Designs and technology do not have to be reinvented constantly, and investigators can focus their attention more appropriately on research questions. Experimental methodologies are not tied to any particular discipline; it makes no sense to argue, for instance, that economic experiments are fundamentally different from experiments in psychology or sociology.

There are different theoretical concerns in different disciplines, and different disciplines have developed different typical designs and even display different "tastes" in design. For instance, some experiments in social psychology use deception to create independent variables, something most economists do not do and often do not approve. Importantly, however, there definitely are criteria for well-designed experiments and reliable experimental results, and those do not differ across disciplines. Quality experimentation is furthered by ecumenism.

II. COMPARISONS TO OTHER DESIGNS USED IN SOCIAL SCIENCE RESEARCH

Social scientists have a wide range of data collection methods. They include:

• Unstructured observation. This method is typically used when a social scientist is present at some unexpected event, such as an accident, natural disaster, or violent attack. The scientist can record relevant data on presumed causal factors and outcomes, also often identifying intervening factors and their effects. Other observers, such as news media or ordinary citizens, can also record unstructured observations, though in many cases the social scientist's training helps him or her to choose what to observe and record as being theoretically important.

• Structured observation. This method is used when a social scientist begins with a coding scheme of what to observe and, perhaps, also how to record it. In contrast to unstructured observation, here the observation site is predictable and well known enough so that observations can be chosen ahead of time. For instance, a sociologist might go to a courtroom to code jurors' facial expressions and other behaviors to see how those may be related to verdicts. The observations here are limited and focused, and it is possible to assess reliability if two observers record the same data from the situation.

• Historical archival research. This method relies upon documents as data for answering research questions. For instance, a researcher may compare recorded lynchings to the price of cotton in the old Confederacy states. Documents may be in written, video, or audio form. As with observational studies, archival documentary research does not exert control over factors, except through statistical techniques.

• Participant observation. This may be structured or unstructured; its defining characteristic is that the observer has privileged access to the setting by virtue of being a member of the group studied. The natural settings are extremely important for such studies; their explication is as high a priority as the interaction itself. So, for example, the rules of the sheltered workshop, the structure of the asylum, or the city in which a gang operates are important actors in their own right. In many ways, observation research of all three types is at the other end of a methodological continuum from experiments: The setting is natural, and the outgoing processes in actual settings are valued. Control and randomization, so highly prized in experimental research, are not usually possible in such studies.[1]

• Survey research. Surveys are usually defined as observations on or about individuals or groups. The United States Census is a famous survey widely used by researchers in many countries. It asks respondents about many behaviors or characteristics, and then analysis proceeds by controlling some factors to see how that affects outcome variables. Survey research is generally interested in generalizing from a sample to a particular empirical population. Interestingly, an incorrect presumption that generalization proceeds from

---

[1]There are exceptions. For example, Milton Rokeach's (1981) *The Three Christs of Ypsilanti* involved a kind of control because all three men claiming to be Christ were in the same setting and talking to the same psychologist (Rokeach).

experiments and from survey samples in the same way can lead to survey researchers criticizing experiments for being "artificial." We consider artificiality in greater detail later, as well as the differences between sample generalization in surveys and theoretical generalization from experiments.

- Combining different methods is a defining characteristic of case studies. Case studies often involve documentary historical data and may also include surveys and participant observation.[2]

III. ADVANTAGES AND DISADVANTAGES OF EXPERIMENTS

The benefits of any research method cannot be assessed independently of the questions the method is designed to answer. A beautiful research design cannot compensate for a flawed research question. This is especially true for experiments because they are designed to determine how specific kinds of independent variables and antecedent conditions affect dependent variables (or consequents). If a research hypothesis, for example, is derived from faulty assumptions or premises, even the most elegant research—experimental or otherwise—cannot save the study. What a good design can do, however, is help the investigator identify the part of the theory that is faulty. Disconfirmed predictions, when they have been derived from an explicit theory, are valuable because they can show which assumptions or conditions in the theory need improvement.

The greatest benefits of experiments reside in the fact that they are artificial. That is, experiments allow observation in a situation that has been designed and created by investigators rather than one that occurs in nature. Artificiality means that a well-designed experiment can incorporate all the theoretically presumed causes of certain phenomena while eliminating or minimizing factors that have not been theoretically identified as causal. In the language of research design, an experiment offers an opportunity to include the independent variables of theoretical interest while excluding irrelevant or confounding variables. An experiment (like a theory) is always simpler than a natural setting; because of that, it offers the possibility to incorporate factors of interest and limit extra-

---

[2]Robert K. Yin (1984) argues that case studies can share some of the characteristics of experiments. In particular he argues that multiple case studies can be chosen such that they represent variation on the independent variable. Then measurements can be taken on the dependent variables. So, for example, if the researcher had a theory concerning how land use compared between government-mandated programs and voluntary programs, she could randomly choose (or perhaps choose based upon specific criteria) mandated and voluntary programs, gather data, and examine the dependent variables.

neous factors so that the theoretical principle being examined is isolated. Experiments permit direct comparisons: most often the comparison is between conditions in which a factor is present (an experimental condition) or in which the factor is absent (a baseline condition).[3] In this way, the effects of a factor in the experimental condition can be gauged. Such direct testing of a theoretical prediction is much more difficult in other, less artificial settings. Because they are less artificial, they are more complicated and contain a myriad of factors that could interfere with, magnify, or dilute the effects of the particular factor being investigated.

Another technique that experiments often use to ensure comparison is random assignment. The power of randomization is the power assured by probability theory: if extraneous influences (errors) are distributed randomly, they sum to zero. When individuals are randomly assigned to different experimental conditions, different effects observed in different conditions are not due to uncontrolled factors, such as personal traits of the individuals studied, because those factors have been evenly distributed across conditions. So, for example, in drug studies there is often a control condition or treatment in which individuals receive only a sugar pill and an experimental treatment in which individuals receive only a treatment drug. When individuals have been randomly assigned to either the control or the treatment condition of the experiment, comparing outcomes in the control or baseline (sugar pill) condition with the treatment (drug) condition permits a gauge of the effect of the drug treatment. Random assignment of individuals to conditions assures a researcher that there is nothing unanticipated about the individuals that might have led to an effect. So, for example, if there is no random sampling there is the possibility that, by some chance, healthier people or people with better nutrition or taller people ended up in the same condition. With random assignment, we can eliminate the explanation that differences in conditions result from characteristics of the individuals or group treated.

Artificiality also allows experiments to provide settings that are difficult, if not impossible, to find. For instance, some studies of voting, such as the experiment of Wilson's (Chapter 17, this volume) allow individuals to record dozens of votes within a single experimental session to assess a theory of how individuals reach equilibrium points where each person's interests are represented as closely as possible. The situation is analogous to repeated voting on bills in Congress or to repeated voting among condominium residents. However, because Wilson studied it experimentally, he was able to assess the process not in months or years but within a couple of hours per group.

---

[3]Sometimes comparison may not be between two empirical conditions, but between an empirical condition and a theoretical prediction. Walker and Willer, Chapter 2, this volume, discuss making experimental comparisons.

Because experiments are artificial and controlled, they also invite and enable clear replication by other investigators and comparison across different settings. Findings from an experiment can be assessed by someone else who replicates the experiment. By contrast, findings that occurred in some natural setting can be impossible to replicate because it can be nearly impossible to find another natural setting enough like the first one so that the investigator can be confident it truly replicates the important features of the setting. Also, experiments are well suited to cumulative research programs that develop and test theories sequentially because they can be evaluated under consistent conditions. The researcher can be assured that differences do not arise over different settings or different operationalizations, but rather because of the different theoretical factors being tested. Such consistency can be crucial for theoretical cumulation—results for one study can be used for subsequent studies. For instance, the experimental or manipulated condition in a study at time one can become the baseline condition for study two.

Temporal or theoretical ordering also can be examined in experimental settings, brought by, again, artificiality. In interactions that occur in natural settings, it is often difficult to disentangle cause and effect, and antecedent and consequent conditions. So, for example, it is difficult to disentangle how experience within a group and the statuses of individuals within a group (often derived from their experience) affect individuals' behavior. In experiments, not only can experience and status be separated, but also they can be manipulated prior to the measurement of behavior. In this manner, the antecedent and the consequent can clearly be distinguished. In natural settings, this is much more difficult, even given longitudinal designs, because individuals possess many different characteristics and experiences, those factors may change dramatically over time, groups may be involved in different kinds of tasks, and group composition is constantly changing.

While enormous benefits accrue from artificiality, experiments receive a great deal of criticism for this defining property (see Babbie [1989], for example). The criticisms center around generalizability: because experiments are artificial, they do not mirror any real setting, and they are not representative of a particular empirical population. Those criticisms are correct, as far as they go. Further, they might be considered disadvantages of experiments. We argue that is true only insofar as experiments are not appropriate for certain kinds of questions. Basically, if the goal is to study properties of some natural setting itself, an experiment is not particularly appropriate. For studying theories that have abstracted some properties of natural settings, however, experiments can be ideally suited.

Experiments cannot attempt to simulate all the complexities of particular settings. They cannot reproduce, for example, elementary school classrooms in Texas or all features of the work environment of a particular corporation such as Apple. Experiments can, however, produce abstract features of school class-rooms (such as giving someone chances to perform and then evaluating per-formances), and they can produce authority structures that might also occur in a corporation like Apple. So, rather than reproduction, experiments are designed to match the characteristics of theories comprising precisely defined, abstract concepts. In this volume, Zelditch develops this argument in some detail in his analysis of external validity.

These concepts and, consequently, the theories comprising them are not defined by a particular time or place. Rather, they are abstracted from particular times and places. For instance, social scientists from several disciplines have studied "public goods," a line of research described by Sell, Chapter 18, this volume. Naturally occurring instances of public goods might include parks in a large city, listener-supported radio, or a free shuttle bus. Those instances have certain features in common. Individuals need not pay money in order to enjoy their benefits—anyone may be allowed to use a park or a playground in the city. However, there are always costs, such as taxes foregone by keeping businesses out of the park, and maintenance crews. Public goods dilemmas revolve around keeping enough people motivated to contribute a fair share for the things they can actually use without paying.

The term "public good" is an abstract concept meant to capture key features of concrete things like buses and parks. The term is independent of time and place; while a shuttle bus may be an instance of a public good at the present, the term public good will apply in any culture at any time where other concrete things may meet that definition. A theory using the term public good must include an abstract definition telling someone how to find instances of it. The definitions of abstract terms can be much more precise than definitions of con-crete terms. For instance, while someone might argue whether a particular piece of land is or is not a public park, if the theorist has done his or her job well, there will be no doubt whether something is a public good.[4] Chapters 2 and 3 in this volume by Walker and Willer and by Thye, respectively, contain more detail on abstract definitions and the parts of a theory. Our point here is that experiments are ideally suited to developing and testing abstract theories, and this is their most appropriate and most valuable use. If there is no good theory, it is too early to think about doing experimental research. The investigator would be better advised to concentrate on observation of natural settings for a while.

Since the theories are abstract and the experiments are artificial, the ques-tion arises as to how the results and general principles can be applied to the settings we are really interested in—natural settings like those elementary

---

[4]Logicians say such definitions are "exact class," and they obey the law of the excluded middle. That means that for an adequate definition, it must be clear whether something does or does not meet that definition; there is no middle ground where it "sort of" meets the definition.

school classrooms in Texas. The answer rests with the chain of activities in scientific research and application. At the first stage, an investigator decides upon a particular kind of research issue, problem, or theoretical dilemma. The issue could arise from observations of some phenomenon that, for whatever reason, she wishes to understand better.

However, observation is not a necessary prerequisite. The research issue could be suggested by elaborations of other theories or by purely deductive implications of a formal system. The investigator conceptualizes the phenomenon abstractly to develop propositions about how it functions under different conditions. In other words, she comes to develop a theory. Next, she may create an experiment to test her developing theory in exactly the kind of situation that illuminates the parts she is most concerned about. If the theory receives experimental confirmation, she comes to believe in her developing theory and is ready to apply it to understand other natural settings. She might even, if she is unusually energetic and skillful, decide to use her theory to devise interventions to produce desirable changes in some natural setting that fits the relevant conditions specified in the theory.

Note that the theory is the bridge between the laboratory experiment and natural settings. The experiment tests the theory, and the theory can be applied to natural settings as well as to the experiment.

So, for example, suppose that a theory predicted that norms for cooperation in dyads would strongly increase cooperation of a whole group (composed of many dyads) in a public good context. Further suppose that a scope condition for this theory was that the initial dyadic cooperation occurs in settings in which there are no authority figures endorsing the cooperation. Such a theory could be applied to classroom settings as long as no individual (such as a teacher) required dyadic cooperation.

IV. STEPS IN CONDUCTING EXPERIMENTAL RESEARCH

Experimental work takes place in four large blocks that we call *foundations, abstract design, operations,* and *interpretations. Foundations* include theory and hypotheses behind an experiment; they are the intellectual reasons to conduct the experiment. General questions at this stage are, "What do I want to learn?" and "How can an experiment answer those questions?" *Abstract design* refers to the "plan" of an experiment, including independent and dependent variables, measures, interaction conditions, and all other things that go into making up an experiment. The main question here is, "How can I design a situation to answer the previous research questions?" *Operations* are the way things actually look to someone watching or participating in an experiment. Variables become operational measures; interaction conditions become instructions or words used by participants to communicate with each other. An important question to consider here is, "How will this look in actual experience to someone in the experiment? Is that the same as I intended in my design?" *Interpretation* is what experimental results mean. The general question at this stage is, "What did we learn from the experiment?" The answer may be in terms of the theory or hypotheses stated at the foundations stage or in terms of some unrelated general issues such as methodological advance.

All four kinds of activity and all four questions must occur before an experiment is actually conducted. While that might be obvious for the first three blocks (*foundations, design, operations*), it is equally true for *interpretations*. That is, an experimenter should consider what the results of an experiment could mean before beginning the work. Experiments that work out—that is, that support research hypotheses—have fairly straightforward interpretations, but experiments resulting in disconfirmed hypotheses also mean something. We will expand this point later. Let us consider in greater detail what each of the four kinds of activities entails.

Foundations of an experiment are the reasons for doing it. An experimenter has some questions to answer or a set of ideas to assess. Ideally, the experimenter begins with a developed theory and rigorously derived hypotheses, though often theoretical and hypothesis developments are incomplete. However, by understanding the ideal case, often it is possible for an investigator to assess how damaging departures from that ideal may be—that is, whether the work is ready for experimental investigation or whether more time needs to be spent on the theoretical foundations of the experiment.

Theoretical foundations have different elements that affect experimental design. We consider scope and initial conditions, and derived consequences or hypotheses. (Other essential parts are the abstract propositions that constitute the actual theory, and the logical or mathematical calculus used to combine propositions to yield derivations or hypotheses. However, scope and initial conditions and hypotheses are crucially linked to experimental design, and we focus on those parts here.)

All theories have limited scope. Scope conditions describe classes of situations to which a particular theory claims to apply. Newton's laws of acceleration of falling bodies (the general propositions) famously apply only in the absence of factors such as air resistance or magnetic deflection; in other words, they take as scope conditions the absence of those factors. Einstein's special theory of relativity treats the speed of light as a constant; it does not use scope limitations of air resistance or deflection. Sociological theories of status processes treated in Chapter 13 take as scope conditions situations where individuals are task focused and collectively oriented; that is, in these situations, individuals see solving problems as their main reason for interacting (rather than, say, enjoying each other's company), and they believe it is important to let everyone contribute to problem solving (as opposed to taking tests without help).

Scope conditions tell the kinds of situations where a theory claims it can describe what happens. If a situation meets its scope conditions, the theory ought to be able to predict accurately. If a situation is outside that scope, the theory makes no claim to being able to make predictions. Given that scope conditions are met, confirmation of prediction increases confidence in the theory and disconfirmation indicates something wrong with the theory (or the test, if methodology could be a problem). Thus, confirmation and disconfirmation both have meaning *as long as a situation is within the theory's scope*. Outside the theory's scope, any results are irrelevant for judging the theory. This means that an experimenter must be careful to design and operationalize situations within a theory's scope.

It is unfortunately true that many social science theories have been offered without any explicit scope conditions. They are presented as if they applied to every conceivable kind of social situation. If you pressed a theorist about that, he or she might admit that the theory probably does not apply everywhere, but many theorists appear not to have considered just where their theories do and do not apply. In order to design an experiment to assess a theory that is not explicitly scope limited, an experimenter must infer (that is, guess) what kinds of scope limits the theorist had in mind. Put differently, suppose a particular natural situation or experiment produced results different from what the theory would expect. Would the theorist say, "Well, of course, I never intended the theory to apply to such cases"? If so, that is an implicit scope condition. All theories have scope limitations. While it is really a theorist's job to tell what those are, if a theorist does not do that, the experimenter must, for there is no way to design a relevant experiment until issues of scope have been settled.

Initial conditions are what prompt the theoretical process; they are the instigators of change in a situation. Presuming a situation meets a theory's scope conditions, then initial conditions describe the setup for what will happen. For example, in theories of status processes mentioned earlier, one situation of interest is what happens when two individuals occupying different status positions interact. Interactants having different statuses (however their society may define status) is an initial condition in this case. In that theory, if individuals have different status positions, that fact will become salient to their interaction; they will form differential performance expectations for each other, and they will treat each other differently in specific ways, all related to power and prestige inequality. From the theory, we could derive predictions or hypotheses that if a high-status person interacts with a low-status person, the former person will talk more and listen less, and the latter's discussion behavior will be the inverse. The initial condition to be created in an experimental test is status inequality.

Initial conditions often define the independent variables of experimental design. The preceding predictions might be tested in an experiment designed so that two individuals of different status positions will interact in a task-focused, collectively oriented situation. To be clear, the theory says that if a status difference exists, then certain behavioral outcomes will occur, given a situation meeting the scope conditions. An experiment to assess that prediction must, therefore, create a situation where individuals of different statuses encounter each other. The design is governed by initial conditions specified in the theory and by its predictions. Independent variables must reflect initial conditions, and dependent variables that reflect predicted theoretical outcomes must be possible.

Experimental operations are what actually will happen in an experiment. They are the instances an experiment offers of the abstractly defined independent and dependent variables. In the previous example, creating a situation where two individuals differing in status interact might be accomplished by creating a team consisting of a teenager and a 35-year-old woman. In order to use that operation, we would need to know that age constitutes a status characteristic for people in our society because the theory and the independent variable speak of "status difference," not "age difference." The theory predicts effects of status processes; it is definitely not a theory of age (or of gender or skin color or any other actual characteristic). The theory says that *if* age is a status characteristic for people in a particular society, then it will have certain interaction effects. The operational challenge is to instantiate status difference, which in our society can be done using age difference.

For a theorist, the distinction of the theoretical term "status" and the operational term "age" is important because societies change. Status characteristics are created and atrophy all the time, but the same theoretical processes apply to whatever may be the status distinctions existing at a particular time. The theoretical status process, not the historical fact of what happens to be status, is of primary interest here.

For an experimenter, the distinction of theoretical terms and operational terms also is crucial because a well-designed experiment must instantiate "status difference" in order to test that theory. In other words, many challenges at the stage of operational design involve creating situations that are instances of cases described theoretically. Partly that means understanding the culture in which the experimenter operates—for instance, what constitutes a status characteristic for this population.

However, operational challenges extend far beyond what an experimenter knows of society. He or she must also translate crucial concepts into terms meaningful for the experimental participants. For instance, if a situation must be task focused, as one of the scope conditions requires, that must be created in any experimental test of that theory. Yet experimental participants do not ordinarily use a term like "task focused," and it would not communicate much to tell them to try to simulate being task focused. The experimenter must create a situation that individuals will accept as task focused. In other words, experimenters must understand the world of participants well enough to instantiate situations they understand in the same ways as the experimenter does, even though the experimenter thinks in abstract, theoretical terms and participants think in concrete actual realities. Going back and forth between the theoretical world and the world of participants is one of the most difficult challenges facing experimenters.

Operations include introductions and instructions—whatever participants are told when they appear for an experiment. Of course, those elements should be as standard as possible, for in a well-designed experiment, participants in different conditions experience most of the same situation, with the exception of differences specified by independent variables and initial conditions. Reading instructions, for instance, is preferable to talking about a situation informally, for the former kind of administration is much more uniform.

The final part of operations is dependent measures—operational translations of dependent variables from the guiding hypotheses. Here again, translation is not straightforward. An experimenter has to think about what can be done technically, given the limits of human observers. Limits include technology, such as what can be seen, and human limits such as fatigue and inattention. In our status example, the hypothesis predicted more talking and less listening from high-status people. How can "more talking" be assessed? By recording their interaction and counting words? Maybe sentences would be a better measure, or maybe complete thoughts. How can an experimenter measure amount of listening? Is silence enough, or does the hypothesis entail some sort of paying attention, and if the latter, how can that be measured?

None of the steps in creating operational measures is likely to be simple, and yet the strength of any experimental design depends on doing all of them well. The general advice here is for the experimenter to think through absolutely every step in the operations as thoroughly as possible and then pretest everything to be sure it works as intended.

A crucial decision in experimental operations involves what data will be collected and how they will be used. In an experiment, it is possible to assess behavior of various sorts, to distribute questionnaires, and to interview participants. Selecting measures to use and carefully developing them are significant steps in experimental design. Just the simple case of testing a prediction about who will talk more, for instance, requires making decisions about how much difference matters. For instance, if an experimenter counted words, would a 10% difference in average words across high-status and low-status conditions be considered confirmation? Could we put the decision in the hands of statisticians so that any statistically significant difference counts as success? Besides mean differences across conditions, many other statistics are available that have potential value. For instance, variability across participants may give clues to both theoretical and methodological questions.

Presuming an experimenter has reached thoughtful decisions about what sorts of data to collect and what will constitute confirmation and disconfirmation of hypotheses, other questions arise. Usually, there are many hypotheses possible from a theory under test, and in the nature of things, not all of the test results will point in the same direction. That is, it is reasonable to expect some confirmations and some disconfirmations. An experimenter who gives some thought to what different possible patterns might mean before collecting the data is in a better position to collect additional information, such as other measures, when disconfirmations or indeterminate outcomes appear.

Even the purest cases—total confirmation and total disconfirmation—deserve some advance thought. Total confirmation of all hypotheses greatly strengthens confidence in the theoretical foundation, of course. This happy state should lead to the question, "What's next?" An experimenter who has even a sketch of an idea what questions to investigate following the present experiment is primed to move quickly once the data are in.

What is not often appreciated is that disconfirmation also is informative. It might, of course, mean that an entire theory or at least the parts of it under test were wrong. However, that outcome is superficial and it is never likely. Did the experimenter really begin with a theory so weak that he or she is not surprised by getting total disconfirmation? That seems unlikely. Most of the time, experimenters believe the theories they work with, and disconfirmation does not usually mean a theory ought to be abandoned. Abandoning a theory without a substitute theory really says, "I have no idea at all what's going on here." While that might be true in rare instances, for most social situations, we have at least vague intuitions about what is going to happen. (The work of an experimenter, as of a theorist, is to make vague intuitions explicit and testable.) Disconfirmation invites further thought. What parts of the theory were used in the disconfirmed prediction? How could those parts be modified to account for the outcomes of the experiment? What would an independent test of that reformulation look like? In a well-designed experiment, disconfirmation information is at least as useful as confirmation information.

Much of this section identifies challenges and potential problems with good experimental design. That reflects the world: good experiments are hard to design and conduct, and only a foolish person thinks otherwise. Fortunately, some of the best experimenters in social science have prepared guidance for all steps of the process in the succeeding chapters of this book.

V. THE PLACE OF EXPERIMENTS
IN SOCIAL SCIENCE

Experimental methods are one kind of data collection that may be used to assess theoretical knowledge. They are certainly not the only method, for many others are available: surveys, content analyses, structured and unstructured observation, and others. As noted earlier, all research looks for relations between concepts. All good research depends on being able to infer reliably that things are related in the ways a theory thought they would be or in understanding what parts of the predictive apparatus need revision.

The preceding discussion suggests that experiments are not well suited to study in the absence of any theoretical foundation—for instance, just "to see what will happen." Anytime someone sets up a situation and collects observations, we know that "something will happen." However, the real questions are what we learned from it. Did something happen for a reason we can specify, or was it just a chance occurrence? If the latter, it is not interesting scientifically. If there is a reason, then that is the beginning of a theoretical understanding. A scientist will usually want to work out the theoretical understanding before moving to empirical research. The reasons why are more interesting and important than the simple facts of what happened.

In fact, a reader may have realized another feature of experimentation from the discussions in this chapter: experimental results themselves are not really interesting except as they bear on a theory. To put it differently, who cares if two people in our status example talk different amounts? That becomes an interesting fact if it shows us something about how status operates, for we will encounter thousands of other situations of status inequality and, with the theory, we have some way to understand what is likely to occur in those new situations also. But the simple numbers from experimental data hold little interest without a theory. (For other kinds of research, say, unstructured observations of police interacting with drug dealers, the observations themselves are likely to be interesting; the cases may be described as "socially important.") For experimental research, the theory is the interesting part; the experimental results are interesting insofar as they tell us something about the theory.

At present, some phenomena do not lend themselves well to experimental research. Understanding how different patterns of parent–child interaction affect children's success in school, for instance, may require several years of observation in natural settings. However, sometimes the apparent inapplicability of experimental methods becomes less serious upon closer examination. Many years ago, Morris Zelditch, Jr. (one of the authors of this volume) wrote an article titled "Can you really study an army in the laboratory?" (Zelditch, 1969) and answered in the affirmative. While you cannot take an army into a laboratory, you certainly can study important theoretical features of armies, such as authority structures, power exercise, and organizational efficiency.

As the social sciences become more theoretical, experimental methods are likely to become increasingly important. Experiments generally offer the most convincing evidence for success or weakness of theoretical explanations. No other kind of research method produces data so directly relevant to a theory or suggests causality so conclusively as experiments do.

VI. HOW THIS BOOK CAN HELP

Most students first learn about experimental methods from coursework in college, but there is a long way to go from knowing experiments exist to being able actually to do one and get some useful results. It is unreasonable to expect a student who has studied chemistry to be able to go into a laboratory and create a drug to block the replication of HIV or build a safe cleaning solution for home use. Nobody would just give a biology student a preserved animal and ask her to figure out why it died. It is the same in the social sciences. There are some theoretical ideas about how social structures and social processes work, but those theories do not tell enough for someone to be able to study those things experimentally.

Laboratory experimentation is taught in all fields through apprenticeship programs. Students, from new undergraduates through postdoctoral trainees, work in laboratories with established scholars. At the early level, students learn some very basic facts about experimental methods, such as Mill's canons of difference and similarity, the importance of developing theoretical concepts and tests, and how to use and interpret statistical tests. At the most advanced level, a student learns advanced techniques, how to use and interpret very sensitive instrumentation, an extensive body of experiential knowledge about likely outcomes and what they mean, and other esoteric knowledge usually associated with a narrow specialization.

The odd fact is that for the vast area between the extremes of expertise, there is little in the way of reliable information. This means that someone who has never designed an experiment is unlikely to find a good source of the information needed to get useful outcomes from it. Even an experimenter in one specialty is unlikely to know how to devise and conduct useful experiments in another area. Because social science experimentation is newer and less widely known than natural science experimentation, many of us do not have a good understanding of why someone would want to do experiments, what they are good for, what problems can arise and how to deal with them, and what some exemplars of good experimentation look like in the different social sciences. This is the gap we address with the chapters in this book.

REFERENCES

Asch, S. (1951). Effects of group pressure upon the modification and distortion of judgment. In H. Guetzkow (Ed.), *Groups, leadership, and men* (pp. 177–190). Pittsburgh, PA: Carnegie Press.

Babbie, E. (1989). *The practice of social research*. Belmont, CA: Wadsworth.

Bales, R. F. (1999). *Social interaction systems: Theory and measurement*. New Brunswick, NJ: Transaction Publishers.

Berger, J. (1992). Expectations, theory, and group processes. *Social Psychology Quarterly, 55*, 3–11.

Campbell, D. T., & Stanley, J. C. (1966). *Experimental and quasi-experimental designs for research*. Chicago: Rand McNally.

Cartwright, D., & Zander, A. F. (1953). *Group dynamics: Research and theory*. New York: Harper & Row.

Chamberlin, E. H. (1948). An experimental imperfect market. *Journal of Political Economy, 56*, 95–108.

Festinger, L., & Carlsmith, J. M. (1959). Cognitive consequences of forced compliance. *Journal of Abnormal and Social Psychology, 58*, 203–210.

Rokeach, M. (1981). *The three Christs of Ypsilanti: A psychological study*. Columbia, NY: Columbia University Press.

Sell, J. (2007). Social dilemma experiments in sociology, psychology, political science, and economics. In M. Webster & J. Sell (Eds.), *Laboratory experiments in the social sciences* (pp. 459–479). Burlington, MA: Elsevier.

Sherif, M. (1948). *An outline of social psychology*. New York: Harper & Brothers.

Siegel, S., & Fouraker, L. E. (1960). *Bargaining and group decision making*. New York: McGraw–Hill.

Thorndike, E. L. (1905). *The elements of psychology*. New York: Seiler.

Thye, S. R. (2007). Logical and philosophical foundations of experimental research in the social sciences. In M. Webster & J. Sell (Eds.), *Laboratory experiments in the social sciences* (pp. 57–86). Burlington, MA: Elsevier.

Von Neumann, J., & Morgenstern, O. (1944). *Theory of games and economic behavior*. Princeton, NJ: Princeton University Press.

Walker, H. A., & Willer, D. (2007). Experiments and the science of sociology. In M. Webster & J. Sell (Eds.), *Laboratory experiments in the social sciences* (pp. 25–55). Burlington, MA: Elsevier.

Watson, J. B. (1913). Psychology as the behaviorist views it. *Psychological Review, 20*, 158–177.

Wilson, R. K. (2007). Voting and agenda setting in political science and economics. In M. Webster & J. Sell (Eds.), *Laboratory experiments in the social sciences* (pp. 433–457). Burlington, MA: Elsevier.

Yin, R. K. (1984). *Case study research: Design and methods*. Newbury Park, CA: Sage.

Zelditch, M., Jr. (1969). Can you really study an army in the laboratory? In A. Etzioni (Ed.), *Complex organizations* (2nd ed., pp. 528–539). New York: Holt, Rinehart, and Winston.

Zelditch, M., Jr. (2007). The external validity of experiments that test theories. In M. Webster & J. Sell (Eds.), *Laboratory experiments in the social sciences* (pp. 87–112). Burlington, MA: Elsevier.

CHAPTER 2

Experiments and the Science of Sociology

Henry A. Walker
University of Arizona

David Willer
University of South Carolina

I. INTRODUCTION

A laboratory experiment is "an inquiry for which the investigator plans, builds, or otherwise controls the conditions under which phenomena are observed and measured" (Willer & Walker, 2007). That broad specification encompasses two fundamentally different types of experiments—*empiricist* and *theory-driven*.[1] The empiricist experiment has the objective of discovering regular patterns of behavior. Unfortunately, contrary to the claims of John Stuart Mill, who systematized its logical foundations, the empiricist

---

[1]Sociology is plagued with many controversies—for example, theory versus method, qualitative versus quantitative research, and so on. The label "empiricist" is often hurled as an insult in our field. We use the term in its technical sense and not as a pejorative. An empiricist experiment is one that relies on generalizations from observations to make claims about phenomena outside the laboratory.

experiment cannot find causal laws. As a result, it does not lead to cumulative theoretical science. In contrast, theory-driven experiments are designed to test theories. Theories are the building blocks of science and theory-driven experiments are the engines that drive cumulative science programs in physics, chemistry, and biology and in sociology as well.

Only one of these types—the empiricist experiment—is described in most research methods texts in sociology and other social sciences. The theory-driven experiment is neglected almost entirely. Yet, the theory-driven experiment is the method best suited to advance sociological science (Willer & Walker, 2007; Willer, 1987). As we show in the following sections, theory-driven experimentation is occasionally employed in sociology, where it also supports the development of cumulative knowledge.

II. EXPERIMENTATION IN SOCIOLOGY

Sociology texts at the undergraduate and graduate levels describe the empiricist experiment as if it were the only experimental method. As two widely used undergraduate methodology texts describe it:

> *True experiments* must have at least three things: two comparison groups (in the simplest case, an experimental and a control group), variation in the independent variable before assessment of change in the dependent variable, [and] random assignment to the two (or more) comparison groups (Schutt, 2006, p. 201; emphasis in original).
>
> The most conventional type of experiment, in the natural as well as the social sciences, involves three major pairs of components: (1) independent and dependent variables, (2) experiments and control groups, and (3) pretesting and posttesting (Babbie, 1989, p. 213).

Schutt's and Babbie's assertions would surprise Galileo, physicist Ernest Rutherford, and pioneering physicist and physiologist, Hermann von Helmholtz. Each of them conducted important experiments, but they did not conduct empiricist experiments like those described by Schutt and Babbie.[2] They tested ground-breaking theories with experiments in which (1) there were no experimental or control groups, (2) there were no independent or dependent variables, and (3) nothing was randomly assigned.

Sadly, the poor and incomplete scholarship found in many methods texts has come to define an orthodoxy for sociology. As a result of that orthodoxy,

---

[2]The experimental-control group design is the hallmark of empiricist experiments, but theory-driven experiments do not use it. However, as we show in this chapter, the crucial difference between the two types is their logic of design. The logic of empiricist experiments results in the experimental-control group design. Because theory-driven experiments have a different logic, they do not have control groups.

theory-driven experiments have been forced all too frequently to masquerade as the other kind—further muddying the methodological waters. It is time to distinguish the two methods clearly and we will try our best to do so in the space available to us. (For a more detailed treatment, see Willer & Walker, 2007.)

We begin our discussion with very brief histories of the two experimental methods. Next, we examine both methods beginning with the logic of empiricist experimentation as described by John Stuart Mill. We analyze Mill's well-known method of difference and its refinement by the statistician Ronald A. Fisher. Solomon Asch's social psychological experiments are used to show that empiricist experiments are what the philosopher Hempel (1966) labeled *methods of discovery*. Our exposition continues with a discussion of the logic underlying theory-driven experiments, which are *methods of test* (Hempel). We use two examples of sociology experiments as examples of theory-driven experiments. We end this chapter by taking up the problem of artificiality and its consequences for the two experiment types.

III. TWO KINDS OF EXPERIMENTS

The history of modern theory-driven experimentation can be traced back at least four centuries to Galileo's tests of his theory of falling bodies. Galileo ([1636], 1954) completed his research in 1607 and in 1686 Isaac Newton published *Principia Mathematica*, which reconceptualized and extended Galilean mechanics (Newton [1686], 1966). Theory development, experimental tests of theory, and theory refinement proceeded rapidly in physics, later in chemistry, and more recently in biology, and they continue today.

The history of modern empiricist experimentation begins later, in the middle of the nineteenth century with the philosopher J. S. Mill's *A System of Logic* ([1843], 1967). Why did he propose a new method of inquiry that we are calling empiricist experimentation? Mill did not set out to build a new kind of science. Instead, he misunderstood how science works. He failed to recognize that cumulative, general theoretical knowledge is produced by the continuous interaction of theory and research. In spite of that misunderstanding, as we will show, empiricist experimentation has been refined as a useful tool for discovering patterned behaviors (or regularities) in sociology and elsewhere. It is useful for discovery precisely because its logic of investigation places high importance on the discovery of cause–effect relations between phenomena.

Nothing can be more important to an investigator than knowing what a method can and cannot do. Thye (Chapter 3, this volume) points out that sociologists have mixed opinions about the suitability of experiments for sociological research. Ironically, social scientists learn to judge the usefulness of nonexperimental methods (e.g., comparative historical analysis and most surveys) according to their capacity to emulate empiricist experimental designs.[3] It is to the logic of such methods that we turn.

A. THE LOGIC OF EMPIRICIST EXPERIMENTAL METHOD

Mill ([1843], 1967) set out the logical foundations of empiricist experiments in five canons: the methods of (1) agreement, (2) difference, (3) agreement and difference, (4) concomitant variation, and (5) residues. These methods are often described in social science research methods texts. Here we consider only the method of difference, which, for Mill, is *the* experimental method:

> If an instance in which the phenomenon under investigation occurs, and an instance in which it does not occur, have every circumstance in common save one, that one occurring only in the former; the circumstance in which alone the two instances differ is the effect, or cause, or an indispensable part of the cause, of the phenomenon (Mill, [1843], 1967).

Mill's logic is incorporated in the conception of experiments portrayed in the typical social science textbook. The "phenomenon under investigation" is the effect or dependent variable. The "circumstance in which alone the two instances differ" is the cause or independent variable. The cause occurs in the experimental group and does not occur in the control group.[4]

How are Mill's canons used to find general theoretical knowledge? Mill claimed that scientific knowledge is found by using his canons or methods to discover regularities in the world. The method assumes that such regular cause–effect patterns are laws.[5] For example, Mill asserted that the statements "fire burns" and "air has weight" are scientific laws. They are not. Here is a law: "Every body perseveres in its state of rest, or of uniform motion in a right line, unless it is compelled to change that state by forces impressed thereon" (Newton, [1686], 1966).

The preceding statement is Newton's first law and is often read, "an object at rest remains at rest and an object in motion continues to move in a straight line *unless force is applied to it.*"[6] The italicized phrase implies a scope

---

[3]Campbell and Stanley's 1966 influential book is exemplary of this perspective.

[4]Mill's description permits the "phenomenon" to be either cause or effect. Whether phenomena or circumstances are causes or effects is usually clear from their temporal ordering. Causes appear before their effects.

[5]An empiricist approach to knowledge "emphasizes the importance of observation and of creating knowledge by amassing observations and generalizing from these observations" (Cohen, 1989, p. 16; see Popper, 1963, p. 21 ff.). Mill's logic exemplifies an empiricist approach.

[6]Galileo ([1636], 1954) initially identified the idea as the law of inertia. Newton refined it and combined it with his second and third laws to create the Newtonian theory of motion.

statement that restricts the universe of phenomena to which the law applies. The law is general—it is not restricted to particular places or times—and has many practical implications when it is linked to other physical concepts. For example, the law implies that

$$v_0 = v_1$$

where v_0 and v_1 are an object's velocity at a given time (0) and a later time (1) and velocity is calculated as the distance an object travels from point A to point B divided by the time of travel.

To illustrate, if the first author's car is sitting in a parking garage in Tucson at 4:30 p.m. while he thinks about an upcoming Arizona Diamondbacks game, the law implies that it will be sitting there at 6:30 p.m. unless some force moves it. Similarly, the car will be at the ballpark 120 miles away in Phoenix when the umpire yells, "Play ball," at 6:40 p.m. if, from 4:30 to 6:30, it is traveling northwest toward Phoenix at an average velocity of 60 miles per hour. The law can also be used with other physical laws (e.g., Galileo's law of uniform acceleration) to explain how far the car moved between rest and attaining 60 miles per hour. Mill does not show that experiments found laws like Newton's first law or any other law in the 600+ pages that make up his *System of Logic*. We know. We looked.

Mill did not describe experiments that found laws because his method fails to find laws. It fails because it raises several problems that cannot be solved. One problem follows from the statement "every circumstance in common save one" in his description of the method of difference—a demand that has no practical solution. Mill asks the reader to imagine two circumstances:

$$ABCDE \rightarrow a$$

and

$$BCDE \rightarrow {\sim} a$$

Read the first as "when ABCDE are present, a is observed" and the second as "when BCDE are present, a is not observed." Because only A is missing in the second instance, the two instances of his hypothetical example have "every circumstance in common but one."

Can an experimenter create two *empirical* circumstances in the *real world* that are known to be exactly alike but for one and only one difference? While the goal seems reasonable, it is actually impossible to achieve. Let us say that an experimenter runs the first of the preceding conditions and then the second. But now there are two differences because the time in which the two were run is different. More generally, any two empirical instances differ from one another, not only in one way as in Mill's hypothetical example, but also in many, many ways, any one of which could be claimed as the cause or an indispensable part of the cause of the phenomenon under study. It follows that there is no practical way of ensuring that Mill's difference criterion is achieved.

A second problem is that Mill's criterion for identifying a lawful regularity cannot be satisfied. The criterion is that the relation of cause and effect is perfect. That is, when the cause occurs, as in the experimental group, the effect must always be present. When the cause does not occur, as in the control group, the effect must always be absent. Anything less than perfect regularity fails to satisfy Mill's criterion and thus is not a regularity as he uses the term. In fact, no perfect regularities have ever been found in sociology or in any other science. Measurement error alone assures that perfect regularities will never be found.

Mill argues that perfect regularities are laws, a claim that is the third problem. There is no system of logic that permits an analyst to treat perfect sets of observations as laws because subsequent observations may contradict previous ones. Perfect regularity requires all instances of the experimental and control conditions, whether past, present, or future, to satisfy the criterion. The possibility of uncovering negative findings in the future is an obstacle that cannot be overcome.[7]

Furthermore, scientific laws are not statements about empirical regularities. Laws relate theoretical constructs (Willer & Webster, 1970). Observations are described in concrete terms (i.e., variables) and there is no system of logic that permits the inference of relations between theoretic constructs from relations between variables. It follows that laws cannot be found by one or a thousand observations. Importantly, laws that find negative evidence within their scope are in danger of falsification. However, an isolated law cannot be falsified. Falsification can occur only when a set of laws embedded in theory finds consistently negative evidence *and* is replaced by better theory.

Finally, the history of science shows that observations often lead researchers *away from* rather than toward useful laws. Had he focused solely on observations, Newton would have observed that every object in motion eventually comes to a stop. Relying on observation alone, he would have come to the same conclusion as did Aristotle two millennia earlier: The natural state of objects is rest (i.e., without the continued application of force, there is a tendency for objects to come to zero velocity).

1. Fisher Rescues Mill

Statistician Ronald A. Fisher (1935, 1956) "solved" two problems confronting Mill's method. How are two circumstances to be made similar to each other but

---

[7]Philosophers of science have tried to work their way around this issue by substituting the idea of a lawlike statement (Goodman, 1947). Unlike laws that are axiomatically true, lawlike statements may or may not be true.

for one condition? For experiments with human subjects, that problem is solved, in part, by random assignment of people to experimental and control groups. Applying probability theory, two groups (or samples) randomly drawn from the same population will be similar with respect to population characteristics. According to the law of large numbers, that similarity will be greater for larger sample sizes.

Fisher resolved a second problem, the impossibility of finding perfect regularity, by abandoning Mill's criterion and replacing it with a new one: an observed difference between experimental and control groups is probably a regularity if it is unlikely to be due to chance. Because it is no longer necessary for results of experimental and control groups to be wholly distinct, Fisher's method, unlike Mill's, is workable. Researchers can find probable regularities. In practice, when n is large enough, results for experimental and control groups can overlap substantially and still satisfy statistical criteria for establishing probable regularities. As an example, the purely arbitrary probability values (or p-values) .05 and .01 have become standard values for determining acceptable statistical limits in sociology (Leahey, 2005).

How confident can one be of the results of Fisher-design experiments? When the null hypothesis of no differences is rejected, the modern analyst presumes that the observed difference is a probable regularity since it is unlikely to have occurred by chance. His or her confidence in the claim increases as the probability of chance occurrence declines (e.g., $p < .001$ inspires more confidence than $p < .05$). Such statements are not at all like descriptions of laws. As the philosopher and physicist Stephen Toulmin points out, we do not give odds of five to one on Snell's law (i.e., that the angle of incidence and the angle of reflection of light will be equal). Insofar as laws are concerned, "the words 'true,' 'probable' and the like seem to have no application" (Toulmin, 1953, p. 78).

The third problem confronting Mill—that experiments cannot find laws—is an insurmountable problem even for Fisher's revision of Mill. It embodies the same misunderstanding of science found earlier in Mill: that laws can be discovered by observation. Nevertheless, the empiricist experiment has proved to be an excellent method of discovery. Fisher's version of Mill's method can find, if not a perfect regularity, at least something that is likely to be nearly regular. That is a great achievement because Fisher's method identifies probable regularities that require theory to explain them. But all theory must be tested. We turn now to an examination of theory-driven experiments.

B. The Logic of Theory-Driven Experimental Method

Theory-driven experiments, unlike empiricist experiments, are not designed to uncover or establish regularities. What then is the purpose of theory-driven experimentation? The purpose is to test theory. We begin to explain our answer by observing that science has as its primary objective the explanation of phenomena through the development of general, abstract theory. Every theory must be tested by confronting evidence that can either support or falsify it. Theory is the method of the sciences. That is, *theory designs methods of test*. Once a theory has been formulated, the purpose of subsequent empirical investigations is either to test the theory or, once successfully tested, to apply it for prediction or explanation. Experiments are the best method for testing theory because they give researchers the greatest control over test conditions.

Experimental control serves different purposes in empiricist and theory-driven experiments. Empiricist experiments must control test conditions in order to create experimental and control conditions that are as similar as possible. By doing so, they satisfy criteria for making valid inferences about relations they find. Theory-driven experiments control test conditions in order to create conditions that are similar to those described by a theory. The objective is to permit investigators to claim that the experiment fits conditions described by theory or models drawn from theory. Then results can be credibly evaluated as supporting or disconfirming the theory.

Theory designs experiments. Therefore, theory-driven experiments will have designs that reproduce—in concrete form—the general relationships of the theories and models they test. It follows that the designs of theory-driven experiments will vary as widely as the theories that are tested. Experiments testing status characteristic theory (Berger, Fisek, Norman, & Zelditch, 1977) will not look at all like experiments testing elementary theory (Willer & Anderson, 1981) and both will be very different from experiments testing game theory or legitimacy theory. Nevertheless, as we show elsewhere, there are principles of good design common to all theory-driven experiments (Willer & Walker, 2007).

Theories explain relations between phenomena. Thus, a necessary step of all theory-driven experimentation is to produce the phenomena covered by the theory. As Freese and Sell (1980) discussed, an experiment is a replica of its theoretical model. Both replica and model will have initial conditions. A theory claims that a set of processes is set in motion once the initial conditions are established. Similarly, once initial conditions are set for replicas in the lab, researchers expect the processes described by the theory to occur in the lab. A theory is supported when there is a one-to-one correspondence between model and replica that extends from initial conditions through processes to final conditions.

But now an objection could be raised. Since theory builds the experiment, it creates the conditions of its own test. Does it not follow that negative results are precluded, that the test must support the theory? Not at all. It is the experience of both authors, and of science in general, that it is possible to use a theory to build experimental procedures, run the experiment, and find evidence that is inconsistent with the theory. There are several reasons for this.

First, experiments do not offer *direct tests* of theoretical arguments. Theories are built from theoretical concepts, which are not observable. Thus, the researcher's first task is to find or build measures for the concepts. Then the experiment is an indirect test because it is the relations among measures, not among concepts, that are observed. Assume that test results are negative. In that case, it may be that the theory is unsound or it may be that measures are poorly connected to concepts or both.

Second, theories have the property of falsifiability; that is why they must be tested. They are put forward as explanations for relations between phenomena, but explanations that do not fit the facts are of little use. Many theories have been shown to be false as a result of experiments built from the models they generate. Galileo's early research on falling bodies demonstrated fundamental flaws in Aristotle's explanation. Experiments built from models generated by Aristotle's arguments found data that were inconsistent with the theory and the theory was falsified when Galileo's theory was successfully tested. That a theory can produce disconfirming evidence is not a fault but an indication of its effectiveness.

Why are laboratory experiments preferred to natural settings for theory testing? There are two reasons: simplicity and scope. Simple models that lead to simple empirical conditions are preferred for theory testing because the reasons for the success or failure of the test can be pinpointed more easily. Test conditions are given in natural settings and an investigator cannot choose whether they fit a simple or a complex model. Instead, he or she must fit the model to what is present. If a complex model is needed and the simpler model fails, it may be difficult to determine precisely why it failed. Alternatively, when tests find evidence consistent with the model, it may be difficult to determine if it is due to processes described by the theory and its model or due to other, uncontrolled factors. By contrast, for experiments, theories can be used to build simple replicas of their simple models. Beyond making it easier to determine whether and why a model succeeded or failed, simple replicas are preferred because they are easier to build, and, once built, are easier to manage and outcomes are more easily measured.

The laboratory is also the ideal setting for testing theories across their scope. Many contrasting phenomena can satisfy the scope of a broadly applicable theory. As an example, strong coercive power structures fall within the scope of elementary theory (Willer & Anderson, 1981), a theory that explains relations between social structures and behavior within them. In natural settings, strong coercive power structures were found in slave systems in nineteenth-century Caribbean and Brazilian societies and in first-century Rome. However, we have only fragmentary evidence from those societies and in many instances it is difficult to determine if the conditions that existed satisfy elementary theory's scope restrictions.

Beyond being difficult or sometimes impossible to observe conditions suitable for theory testing, archival or field research may be prohibitively expensive. Therefore, testing a wide range of phenomena under the same scope restrictions may not be feasible and is always less precise in natural settings. On the other hand, given a powerful theory, widely contrasting conditions can be created easily in the laboratory and, when they can, they should be. Elementary theory's strong coercive power structures can be readily studied in the lab, and a variety of them have been (Willer, 1987).

At this point a reader might fairly ask, "Are not Mill's two problems also problems for theory-driven experimentation?" No, they are not and here is why. First, success or failure of theory-driven experiments does not depend on creating two conditions that differ in every circumstance "save one." In theory-driven experiments, theory designs the experiment. By doing so, theory determines what should be controlled and measured and what can be ignored. Furthermore, a theory-driven experiment may require only one condition to complete an adequate test. Of course, there are also uncountable numbers of things that could be controlled in theory-driven experiments, but "could be controlled" and "required by the theory to be controlled" are now distinct.

Second, the success or failure of theory-driven experiments does not depend on uncovering absolute regularities because their aim is to test theory, not to find regularities. Successful outcomes depend on the consistency between predictions derived from the theory and results of experimental tests. Here it could be fairly asked, "Isn't it true that theory-driven experiments find regularities?" Sort of. One measure of the power of a theory is its ability to produce its phenomena again and again—as in *replication*. After the first experiments on falling bodies satisfied his theory's predictions, Galileo said of replications of variants of the basic design that "in such experiments, repeated a full hundred times, we *always* found that the spaces traversed were to each other as the squares of the times, and this was true for *all* inclinations [angles] of the plane" (Galileo, [1636], 1954, p. 179; emphasis added). That the results of theory-driven experiments can be replicated is a statement about the effectiveness of the theory and the quality of experimental design. Replication of theory-driven results says little or nothing about whether observations reflect absolute or relative regularities in the absence of theory or whether such regularities exist outside the imagination of Mill and those who quote him. In the next section we use examples to compare theory-driven and empiricist designs.

IV. EXAMPLES OF EMPIRICIST AND THEORY-DRIVEN EXPERIMENTS

We begin with the experiments of the social psychologist Solomon Asch to show the logic of empiricist experiments. Asch's work heralded new understandings precisely because his studies were designed so well. His work is brilliantly creative and, while it did not produce theory, as we will show, it inspired a line of research that did lead to the development of general formulations.

A. THE ASCH EXPERIMENTS

Solomon Asch's studies of conformity and distortion of individuals' reports of judgments are still studied today. They are considered classic experiments—as they should be, given the excellence of their design. Asch's goal was to investigate social conditions that would "induce individuals to resist or to yield to group pressures when the latter are perceived to be *contrary to fact*" (Asch, 1958, p. 174; italics in original). Writing after the mid–twentieth-century defeat of Fascism and Nazism, Asch held that understanding why people submit to group pressure is of "obvious consequence for society" (p. 174).

Asch's first study sat eight men in a row (or semicircle); seven of the eight were confederates of the experimenter. Stimulus materials were 18 pairs of cards that research assistants showed in fixed sequence. On the first card of each pair was drawn a single vertical line called the "standard line." On the second card were drawn three vertical comparison lines. One comparison line was always exactly the same as the standard line. The other two lines were unambiguously shorter or longer than the standard. The task required subjects to judge which of the comparison lines was like the standard line and to report their judgments to the research staff.

Asch's study had a control group and an experimental group. In fact, Asch had a large number of experimental designs and we will comment on some others later. For the control group in one experimental-control group pair, the subject and all seven confederates made private judgments; they wrote them down. The lone subject was seated in the eighth or last position in the experimental group. In the experimental group, all reported their judgments in sequence and aloud. For 12 of the 18 stimulus pairs, every confederate gave the same erroneous answer. Then the subject gave his answer, also aloud. Otherwise, the experimental and control conditions were similar. The "errors" of the unanimous majority created social pressure, the independent variable. Asch's dependent measure was the subject's response, whether correct or in error. The control group permitted Asch to find a baseline error rate with which responses in experimental groups could differ. The difference measured the effect of the independent variable: social pressure.

In the face of a unanimous majority, the 50 subjects in the 50 experimental groups averaged 3.84 errors for an error rate of 32%. Thirty-seven subjects in the control groups averaged only 0.08 errors! The error rate of less than 1% (approximately 0.67%) in the control group contrasts sharply with the nearly one-third error rate in the experimental group.

Asch's study embodies the logic of the empiricist experiment. The control and experimental conditions were designed to differ on only one factor: the method of reporting judgments. Since some men (all participants were male) never conformed to the unanimous majority, Asch did not observe perfect regularity. Still, he did observe perfect regularity in one sense. All errors in the experimental group were in the direction of the majority's responses. In that limited way, Asch satisfied Mill's requirement of perfect regularity. On the other hand, Asch did not take advantage of Fisher's revision of Mill's criterion of perfect regularity. Today, we would use standard statistical tests to show that the findings are statistically significant, but Asch reported only differences. He did not report test statistics estimating the likelihood that the differences were due to chance.[8]

Asch was as interested in resistance to group pressure as in conformity to it. He ran an array of contrasting experimental group designs to investigate resistance. Some designs varied the number of confederates (1, 2, 3, 4, 8, or 16). One confederate had little effect on baseline error rates and the largest effects were found for three to four or more confederates. As the number of confederates increased beyond four, effects leveled out and declined slightly. Mill's canon of concomitant variation asserts that there is a regularity if two phenomena vary together. In this case, conformity rates varied with the size of the unanimous majority, and the relation between the two was curvilinear, not linear.

Asch employed two additional designs to study the effects of nonunanimous majorities. In one design, a second experimental subject was substituted for a confederate. In the other, one confederate was instructed to make correct judgments. The error rate, which had been over 30% for unanimous majorities, declined precipitously to 10.4 and 5.5%, respectively. Finally, in a little known variation, Asch reversed the design, placing a single erring confederate in a group of 16 subjects. Subjects' error rates were unaffected. Instead, the confederate was ridiculed.

Asch's experiments were well designed. The independent variable (social pressure) was easily varied while measurement of the dependent variable

---

[8]Today, beginning statistics students could "eyeball" these findings and identify them as "statistically significant." That is not true of the designs with nonunanimous majorities described later, where differences are not as large.

(the error rate) was straightforward. The control group differed (as it should following Mill) in one notable regard from the experimental group: subjects' and confederates' judgments were recorded privately rather than voiced aloud. The similarity of experimental and control groups strengthens confidence in inferences made about the link between the confederates' assertions and the subjects' voicing of judgments that they knew to be wrong.

Asch's studies are remembered for the very strong effects he was able to produce. His results were strong because he used a "sledgehammer approach." Using a sledgehammer approach, the strength of the independent variable is set as high as feasible—so high that it overwhelms much if not all error variation in the dependent variable. Asch's sledgehammer-like use of social pressure leaves no doubt about group pressure and its effect on what people say their judgments are. Beyond its sledgehammer approach, the effectiveness of the design was also due to its *experimental realism*. Experimental realism is the capacity of a design to involve the subject in the experimental task. In some designs, experimental realism is established by creating high *mundane realism*. Mundane realism refers to the similarity of experimental designs to some condition or conditions in the world outside the lab (Aronson & Carlsmith, 1969; Blascovich *et al.,* 2002). Many experiments achieve high experimental control but have low experimental and mundane realism. For Asch, the powerful psychological conflict induced in subjects seems to have ensured that high experimental realism (subject involvement) was not reduced by high experimental control. Yet, Asch's designs had low mundane realism.

Asch's design is not without flaws. Editors and ad hoc reviewers for contemporary journals would surely raise questions. Did he select subjects randomly? Did he assign them randomly to experimental and control conditions? Why did he fail to report tests of statistical significance? Yet, Asch's study is a classic that is taught to thousands of psychology, sociology, and social psychology students yearly. His experiments were designed to uncover conditions under which individuals resisted or yielded to group pressure and they accomplished that goal. Researchers learned that (1) group pressure can affect individuals' responses to *unambiguous stimuli*, (2) group size and degree of consensus combine with group pressure to affect subjects' responses, and (3) conformity is affected more by sociological than purely psychological factors. All were important discoveries but the last proved especially important.

Trait theories (Allport, 1937), which presume that human behavior is motivated largely by an individual's traits (or dispositional characteristics), were very popular at the time. Asch (1946) had previously done important work on traits, impression formation, and behavior. His conformity studies showed that situational factors like group size and unanimity of dissent could overwhelm trait effects on behavior. In that sense, Asch was on the leading edge of research that eventually weakened the influence of trait theories in psychology and social psychology. Yet, Asch's findings have been less useful outside the laboratory than they might have been had theory been developed and tested. Neither Asch's work nor hundreds of replications of it have culminated in a theory.

B. FROM PATTERN TO PROCESS: TOWARD A THEORY OF CONFORMITY

Bernard P. Cohen began the study of the phenomenon discovered by Asch in the mid-1950s. Eventually, he developed a representational model (Berger, Cohen, Snell, & Zelditch, 1962) of conformity processes and applied it to the Asch situation. Cohen (1958, 1963) recognized that subjects in the Asch situation were in a state of conflict. They could respond to internal pressures based on their perceptions or to group pressure but not to both. The task required them to resolve the conflict; they had to make a choice to conform with the majority or fail to conform on every trial.

Cohen reasoned that, on any trial t, a subject could be in one of four states. Call them (1) permanent nonconformity, (2) temporary nonconformity, (3) temporary conformity, and (4) permanent conformity. Subjects in State 1 fail to conform at t and on every subsequent trial. Subjects in State 2 fail to conform at t but may or may not conform on a subsequent trial. Subjects in State 3 conform at t but may or may not conform on a subsequent trial. Finally, subjects in State 4 conform at t and on every subsequent trial.

Cohen focused on the trial-by-trial pattern of responses—runs of consecutive conforming or nonconforming responses and alternations between conformity and nonconformity. He reasoned that the phenomenon could be treated as a stochastic process in which the likelihood of the next event or state (e.g., conformity or nonconformity) is affected by the current state. He devised the model shown in Figure 1, where the circles represent states and the arrows show transitions between states. The model was designed to represent general processes that produce majority influence on conformity, a phenomenon already discovered by Asch. Cohen's initial conflict model (1958, 1963) assumed

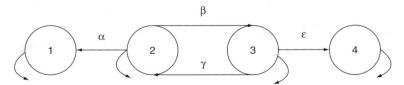

FIGURE 1   Cohen's conflict model.

that (1) people begin in State 2, (2) they cannot move out of States 1 and 4, and (3) they cannot move directly from State 2 to State 4 or from State 3 to State 1.

Cohen's model was not built from a theory, but the logic of his work is more consistent with theory-driven than empiricist approaches. A theory tries to explain the relationship between two phenomena (e.g., group pressure and conformity to it) by describing the *mechanisms* that link the two. For example, a theory linking group pressure and conformity might claim that, under given conditions, (1) disagreements create pressures on parties to disagreement, and (2) pressures motivate disagreeing parties to reduce pressure, etc. Cohen's representational model describes the sequences of possible behaviors in the situation and can be used to estimate the likelihood that people move from one behavior to another. Unlike a theory, the model neither identifies nor describes mechanisms that would cause them to make such moves.[9] A proper test of models like Cohen's generates estimates of the patterns expected under the model's initial conditions and compares the estimates to results from experiments built from the model. Similar to theory-based models, the conformity model describes *how* the relationship between unanimous majorities and conformity emerges. By contrast, empiricist experiments are concerned with identifying or more accurately specifying relations among phenomena.

Using the Asch setting as a starting point, Cohen designed two new experiments that differed from Asch's earliest studies in the following ways:

- Every experiment used six confederates and a lone subject.
- Subjects made a total of 38 judgments.
- Each set of three comparison lines was configured identically. The leftmost line (A) was extremely different from the standard line, the middle line (B) was identical to the standard line, and the rightmost line (C) was moderately different from the comparison line.
- Unlike Asch's design, "neutral" or nonerror trials were placed at the beginning of the series of 38 trials. No neutral trials were interspersed within the sequence of error trials.
- Confederates uniformly chose A in one set of experiments, called the extreme condition. They uniformly chose C in the moderate condition.

Cohen's experiments used the logic of theory-driven experiments and differed from Asch's because of his model's demands. A proper test of Cohen's model did not need—and he did not create—two experimental conditions that differed in only one way. Nor did he randomly assign subjects to experiments.

[9]Cohen initially considered a two-state model but opted for the four-state system after concluding that the two-state model could not reproduce the Asch result. (See Chapter 2 in Berger *et al.*, 1962.)

Cohen needed only to create conditions with unanimous group pressure as in the basic Asch situation, apply his model to those conditions, and compare the model's predictions to experimental results. As we said earlier, studies with theory-driven designs are successful when experimental results match a model's (or theory's) predictions.

Cohen used the experiment data to identify the proportions of subjects who were in states 1 and 4 on the first trial and to estimate the transition probabilities for movement into and out of states 2 and 3. With the parameter estimates in hand, he ran computer simulations and compared the trial-by-trial patterns produced by the simulations with the data from his experiments.

Cohen presumed that the size of the discrepancy between the standard line and the confederates' erroneous choices could affect the rates of conforming and nonconforming responses. Additionally, differences in the size of the discrepancy ought to affect the proportions of subjects who settled initially into the permanent states 1 and 4 and perhaps the transition probabilities for states 2 and 3. Thus, the conflict model generated different trial-by-trial patterns for the two experiments (extreme and moderate discrepancies). Cohen found that estimates from the simulations generated by the conflict model fit the experimental data quite well. Cohen also applied the model to some of Asch's original data, where the fit was not as good.[10] Later, the original model was refined and replaced with the Cohen–Lee model (Cohen & Lee, 1975).

Cohen's simulations were compared to data drawn from his experiments. As we said earlier, the purpose of the experiments was to produce data to test the logic of the four-state model, so Cohen did not need and did not use a control group. In fact, either the extreme or the moderate experiment would have been sufficient to test the model's validity. But replication is important and desirable. Cohen's initial model and the Cohen–Lee model developed 20 years later are representational models that have general applicability. They are not limited to the Asch situation. For example, the models can be used to analyze the conflicts partisan voters feel when they are pressured to cast ballots for opposition candidates. Similarly, juveniles experience conflict when one course of action (abstaining from drug use) is consistent with their values but an alternative (using illicit drugs) is favored or enacted by their peers. Finally, the logic of Cohen's experiments is identical to that underlying theory-driven experiments. It is time now to look more closely at experiments that use that kind of design.

---

[10]Generating simulations that reproduce the Asch results was complicated by Asch's original design, which mixed extreme and moderate responses in the same experiment. In that regard, Cohen's experiments created "cleaner" results. That is, they fit the model better than Asch's data because they eliminated unnecessary variation in the stimulus materials.

C. Two Theory-Driven Experiments

In this section, we analyze an experiment designed to test status characteristics theory (SCT) and another that tests elementary theory (ET). The two offer evidence of the range of contrasting designs of theory-driven experiments. An important limitation of the analysis offered here is that space does not allow us to present either theory. Whereas this limitation will present no problem for the many readers who are familiar with both theories, those who are not so familiar may wish to consult expositions of both. (See Berger *et al.,* 1977, for a presentation of SCT and Willer & Anderson, 1981, for a presentation of ET.)

We note that the two designs analyzed here have a single common feature. Both are computer mediated with experimental subjects seated in individual rooms interacting through PCs. However, the computer-mediated designs have little else in common and are used to achieve very different purposes. For example, the SCT setup is used to give subjects ambiguous information, whereas, for ET, the computer-mediated design is used to give subjects full and accurate information.

1. Status Characteristics Research[11]

Status characteristics theory explains how individuals' knowledge of their own and others' *status characteristics* affects their interaction patterns, including whether they or those with whom they interact will be more influential. Status characteristics are attributes (or features) of individuals who have status value and around which group membership is or can be organized. As examples, ethnic heritage and height are characteristics of individuals. Each has status value in the United States; being Anglo-American and tall garners higher prestige or status than being American Indian and short. Finally, both categories are used to name or classify groups of people (e.g., "tall people").

Tests of the theory require situations in which actors who differ on one or more status characteristics use status information to organize their interaction. The theory asserts that, in collective task settings, higher status people will influence lower status people more than the reverse. Each SCT session has only one real subject, who is led to believe that he or she is interacting through a computer interface with a subject in another room. The subject actually interacts with software that is on the computer or on a server to which the computer is connected.

To produce collective task orientation, subjects are told that (1) they will be working with a partner, (2) the objective is to give the best responses to the questions that will be posed, and (3) to that end, they will be given information

---

[11]See Joseph Berger (Chapter 14, this volume) for the historical development of this experimental design.

on the responses of their partner. In the computer-mediated setting we describe, a subject is seated at a PC that provides a complete set of task instructions.[12] The subject also receives information that permits comparison of self and purported other on some status characteristic. For example, if a 21-year-old undergraduate with a "C+" grade point average is to be placed in a low-status position, he or she could be told that the partner is a 30-year-old graduate student with an "A" average. For added realism, the subject could be shown a snapshot of the fictive partner chosen to represent a high- or low-status individual. Subjects are assured that they will not meet their partner after the study.

Standard tasks require subjects to view and make judgments about ambiguous stimulus material for which there is no objectively correct response. As an example, the contrast sensitivity task requires a subject to view a slide containing two rectangles placed one above the other. A typical screen as seen by a subject is given in Figure 2. As can be seen, the rectangles have nearly equal irregular patterns of black and white. Below the slides are three rectangular panels, each with a "light" and a button on each side. The first panel has the label "Your initial choice" in the center. The leftmost button is labeled "top" and the right button is labeled "bottom." The second panel is labeled "Your partner's initial choice" and the third panel is labeled "Your final choice." The left and right buttons on the last two panels are also labeled "top" and "bottom."

The task requires the subject to make a judgment as to whether there is more white (or black) on the top or bottom rectangle. The slide appears for approximately 5 seconds and, using mouse control, the subject clicks one of the two buttons on the initial choice panel. The appropriate light illuminates and there is a delay after which a light representing the choice supposedly made by the subject's fictive partner is illuminated. The delay after the subject's initial choice is computer mediated to vary somewhat, reflecting varying times needed for decision by the fictive other. Then the slide reappears and instructions on the screen ask the subject to make a final choice by clicking a button on the third or bottom panel.

Computer mediation ensures that the subject and the ostensible partner disagree on a number of criterion trials. The subject's final choice, whether consistent with his or her first choice or with the partner's first choice, is recorded as a measure of influence. The proportion of all disagreement trials on which the subject stays with his or her first choice on criterion trials, P(S), is called

---

[12]Several investigators, including Professors Martha Foschi (University of British Columbia), Lisa Troyer (University of Iowa), and Murray Webster (University of North Carolina, Charlotte), have built computer-mediated SCT designs. Our description uses Troyer's Iowa protocol.

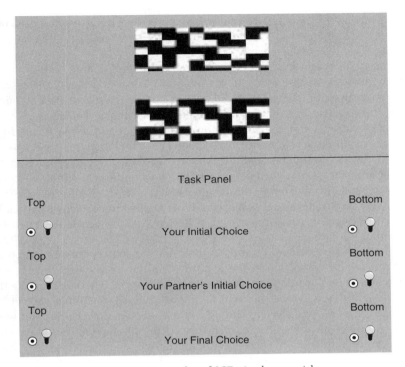

FIGURE 2   Screen shot of SCT stimulus materials.

the probability of a stay or self-response. High P(S) values reflect greater resistance to influence and the converse for low values of P(S). SCT implies that high-status actors will have higher P(S) values when interacting with low-status partners, who are expected to have lower values of P(S). The setting permits a researcher to investigate the effects of a wide range of diffuse (e.g., gender, sexual orientation, educational status) and specific (e.g., musical ability, athletic ability, or completely fictional abilities) status characteristics.

The SCT design has a number of good qualities. SCT designates conditions that must be initially present such as a task that is both collective and ambiguous. It then generates predictions of whether and to what degree the subjects' responses are affected by their partners' actions by linking initial conditions, including the status of the subject and (fictive) partner, to final conditions. A central focus is typically the effect of differences on one or more status characteristics. SCT experiments are typically paired. Two groups are drawn from the same pool of subjects with one acting from a high-status position and the other from low status. Recruiting both groups from the same pool avoids correlated biases.[13] The test of the theory is whether the predicted direction of effect is supported and, when status differences can be quantitatively estimated, the size of the effect.

Contrasts between the SCT experimental design and that of Asch draw out further qualities, qualities that flow from theory. Asch's design concentrated on finding effects of a majority. It did not control the relative status of confederates and subjects and did not physically separate them. The close physical contact of subjects and confederates opened the door to the use of interpersonal power through sanctioning (e.g., looks or verbal expressions of scorn). SCT is a theory of influence and the computer-mediated design ensured that it studied influence—not power—by setting two conditions. First, influence effects depend on some level of initial doubt that permits changes in beliefs or judgments. Asch eliminated subjects' doubts about the correct responses by using unambiguous stimuli. The SCT design established doubt by using ambiguous stimuli.

Second, SCT subjects did not meet their fictive partners before the experiment, got highly restricted feedback from them during the experiment, and were told that they would not meet after the experiment. Therefore, unlike the Asch design, the fictive partner had no opportunity to sanction a subject for his or her responses. Since sanctioning could not occur, power processes were ruled out. Given an ambiguous task and an absence of sanctions, it follows that SCT studies influence.[14] These important design qualities flow, not from empiricist rules of thumb, but from understandings built into SCT.

Like most theory-driven experiments, SCT experiments usually have no control groups. SCT experiments frequently have two or more conditions but they are contrasting status conditions. Neither condition is, as it would have to be for an empiricist method-of-difference design, an absence of status or status effects. Nor do the conditions represent variations required by Mill's

---

[13]SCT experiments rarely form two groups organized according to the status of subjects in the larger society because people from two contrasting status groups will also differ in many other ways. For instance, 18-year-old and 50-year-old persons differ on the status characteristic "age," and they also may be expected to differ on experience, workplace involvement, parenthood, and many other characteristics that will have unknown effects on some of the same behaviors as the status characteristic under study. Differences of that kind are called correlated biases. One of the great advantages of experimental over nonexperimental methods (e.g., survey research) is the opportunity to reduce correlated biases substantially.

[14]Asch (1958) describes his experiments as studies of influence. As we point out, influence requires changes in beliefs. Very few if any of Asch's subjects actually believed the erroneous answers they gave were correct. Their responses to questioning made clear that they were concerned about public ridicule and scorn if they defied the majority—responses that we take as evidence of power effects. Power effects depend on sanctions such as ridicule and scorn as negative sanctions. (See Chapter 3 in Willer & Walker, 2007 for extended discussion of this point.)

method of concomitant variation. Rather, they are required by theory. SCT predicts similarity or differences in the behavior of paired interactants. As we described previously, high-status actors paired with low-status actors are predicted to have higher P(S) values than their low-status partners. Therefore, the group (pair) is the proper unit of analysis and, given subjects' interaction with simulated others, two experimental treatments are required for a proper test of the theory.

An empiricist experiment designed to uncover status effects on behavior would create an experimental group with a status difference and a control group without a status difference. In contrast, SCT's theory-driven experiments focus on comparing data from contrasting status differences to theoretical predictions of their effects.[15] Furthermore, if the SCT experiment is successfully concluded, the theory can be brought to bear on instances outside the lab. This is not "generalization" of lab results and is very different from the common practice associated with the empiricist method.

Empiricist generalization uses laboratory observations to make claims or predictions about the universe of phenomena outside the laboratory. The practice follows from the application of probability theory and the inductivist (or inferential) approach advocated by Mill's and Fisher's methods. In contrast, theory that is supported by findings from theory-driven experiments can be applied with confidence outside the laboratory *if the applications satisfy the theory's requirements*. We stress that theory, not findings from tests of theory, is applied to phenomena outside the lab in theory-driven research.

2. Elementary Theory Experiments

Elementary theory (Willer & Anderson, 1981) is a theory of social structure and the behavior of actors within structures. The theory uses a set of laws and principles to explain how the arrangement of positions and the way that positions are connected affect behavior, including the distribution of benefits in social exchange. Since the late 1980s, much theoretical and research activity has focused on the related network exchange theory (NET) and the analysis of exchanges (Willer, 1999). However, ET and NET also model and explain coercive relations. Bargaining situations in which actors exchange goods or

---

[15]Some might argue that high-status and low-status conditions differ in only one respect and that the difference is like that of experimental and control groups. We respond by pointing out that SCT also can be used to make predictions about the behavior of paired actors who have equal status. That is, they ought to have equal or very similar P(S) values after allowing for subject idiosyncrasies and measurement error. Although equating status information has been used in some SCT experiments, it is not done often. The problem is that there is no consensus among status characteristics theorists about how individuals process equal status information to produce equivalent P(S) values. (See Walker & Simpson, 2000, for discussion of alternative theoretical ideas.)

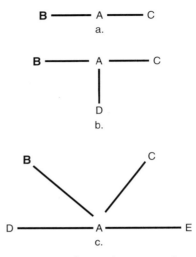

FIGURE 3   Three exchange networks.

services to their mutual benefit are typical of situations to which ET can be applied.

In contrast to SCT experiments, all ET experiments have multiple subjects ranging from as few as two to, perhaps, 10 or more. Each subject interacts with others using the PC as an interface to send and receive offers and complete agreements. At the University of South Carolina laboratory, the PCs are connected by the Internet to software now found at http://weblab.ship.edu. Each study is set up by the experimenter, who accesses and activates the software through the Web browsers on his or her PC and those used by the subjects.[16]

Subjects occupy positions in small social structures in ET studies. Shown in Figure 3 are three such structures, each centralized at position A. Each subject can actually see one of those structures displayed on the computer screen. As a part of the display, offers, counteroffers, and exchanges are shown as they occur on each line. A session investigating the Figure 3a structure requires three subjects: one placed initially in A, one in B, and one in C. The subject can easily locate his or her own position: the subject's letter is larger than the letters of the others. All three structures of Figure 3 are displayed from the point of view of a subject located in B. The screen of the A subject will, of course, show his or her letter largest and similarly for C's screen.

---

[16]The software is available to interested scholars without charge.

ET models subjects as rational decision makers. As a result, the subject–PC interface was designed to ensure that information given to the subject is clear and easily accessible. Additionally, considerable effort was expended to ensure that the subject's interaction across the PC interface is as natural as possible. Application of the theory requires subjects who have mastered the interface. Thus, each session is preceded by a tutorial explaining how to make offers, read offers from others, and complete exchanges. To assure expertise on the part of subjects, they practice on a structure, distinct from the one under investigation, after the tutorial and just before the experiment. Structures used for practice are different from those used for experimental rounds to eliminate the advantage that some subjects would gain from experience in a given position in a particular structure. (See following description of subject rotation.)

Tests of ET require specification of structural and relational conditions for which the theory predicts interactions and outcomes. One set of structural conditions, the network configurations, is given in Figure 3. Other structural conditions specify properties linking relations. For these example structures, two properties link the relations at A, either of which disadvantages A when exchanging with positions at the periphery. First, relations at A are *inclusively connected*. Inclusive connection requires A to complete exchanges in all relations before he or she can gain payoff from any of them. Second, relations are ordered. A must exchange first with B and only after that exchange is completed can A exchange with C—and similarly to the last peripheral. These conditions disadvantage A because each potential exchange has a high cost of failure. Because relations are ordered, if the *first* exchange is not completed, payoffs of all subsequent exchanges are lost—and similarly to the last exchange. For inclusive connection, if the *last* exchange is not completed, payoffs negotiated for the previous exchange are lost—and similarly back to the first exchange.

A is doubly disadvantaged and the properties of the structure that produce that disadvantage (but not that they are disadvantageous) are explained to all subjects. The software imposes the order such that A will have access to C only after exchanging with B, and similarly for the larger structures. ET predicts that the two structural properties taken together will result in highly unequal exchanges favoring the peripherals. Moreover, since the costs of failure increase with the number of peripherals, ET predicts that A of Figure 3a is less disadvantaged than the A of 3b, and that the A of 3c is the most disadvantaged of all.

ET makes precise predictions for each position in each relation of each structure. To facilitate comparisons across the three networks, exchange is simulated by placing a pool of resources between A and each peripheral. All pools in Figure 3 are the same size, 24 units, and each pool can be divided by mutual agreement. Thus, offers can range from 23 to 1 favoring self, through

12 to 12, to 1 to 23 favoring the other. Subjects are told (accurately) that they will be paid by points earned.

Many ET experiments rotate subjects to wipe out the effects of individual differences. Each experiment session consists of as many periods as there are subjects and each period consists of a number of rounds. Subjects are electronically rotated across positions such that each subject occupies every position in a structure for one period. Each round begins with new resource pools and concludes when all exchanges are completed or, if all exchanges are not completed, when a given amount of time has elapsed.

Experimenter control ensures that subjects in each session have uniform experiences. As each subject arrives, he or she is shown to a small room containing a PC. The tutorial is started, followed by interaction, first with the practice structure and then with the experimental structure. Subjects negotiate through rounds punctuated by periodic rotation. This computer-mediated setting is very high in subject involvement or experimental realism. Evidence of high subject involvement is frequently observed as subjects, though alone, vocalize. It is not unusual to hear, "Why wasn't my offer taken!" and "No, that is too much! I won't go for it." On rare occasions, very excitable subjects have been known to bang on walls. With the conclusion of the last round of the last period, subjects are paid by points earned and the session is completed.

The ET experiment has a number of good design qualities. Computer mediation allows excellent fit of replicas to theoretic models. The structures of Figure 3 are models drawn from the theory and also are replica structures as displayed on the subjects' screens. The two structural properties—ordering and inclusion—occur in the theoretic model and in the experiment's replica. Furthermore, no communication can occur in the experimental replica other than that modeled in the theory. That is, B cannot communicate with C, C cannot communicate with D, and similarly for all positions not immediately connected to each other. The content of communications is also controlled. While each subject is free to offer any resource division to anyone to whom he or she is connected, no threats or side payments can be sent. A side payment is an inducement or bribe a negotiator offers a partner in turn for an agreement. A would be offering a side payment to B if A told B that he or she would give B a date after the experiment if B would agree to a 12–12 resource split.

The software, called ExNet 3.0, can investigate all structural properties thus far discovered by ET in any structural configuration for up to 25 subject positions.[17]

---

[17]ET has discovered seven structural properties that affect exchange and other relations, and all may be investigated by ExNet 3.0. Since all other exchange theories recognize only one of the seven, ExNet 3.0 can cover their full scope as well (Willer & Emanuelson, 2006).

Furthermore, it can study those properties in quite pure form: qualities idiosyncratic to one or another subject have little opportunity to produce effects. For example, some subjects may be better negotiators than others. By producing new subject pairings, however, rotation across structural positions controls for differences in negotiation skills. Because subjects occupy both advantaged and disadvantaged positions, rotation has the added benefit of producing greater equality of subjects' earnings, thus dampening equity or fairness concerns.

Recent studies have shown that status differences can affect exchange (Thye, Willer, & Markovsky, 2006). In ExNet 3.0 experiments, subjects cannot identify the status or any other distinguishing quality of those with whom they are exchanging. Granted, subjects may see one or more others upon arriving at the lab, but no one can match persons to positions in the structure. Finally, ET models assume rational actors. If subjects are to make rational decisions, their attention to the task must be high so that alternatives are carefully weighed. ExNet 3.0's high experimental realism keeps attention high, as our descriptions of subjects' vocalizations make clear.

Like SCT experiments and most other theory-driven experiments, ET experiments neither need nor have control groups. The example study traced earlier investigates three contrasting structures, but none of the three is a control group. Empiricist methods require comparisons of differences between conditions, but theory-driven methods require a comparison of results from only a single structure with a theory's predictions. An investigator who studied all three structures in Figure 3 would have completed three tests.

ET can accurately predict effects for structures of different sizes and within the limits of exchange behavior; size is not a scope restriction on ET. Thus, studying groups of different sizes or with different types of connections is an example of scope sampling. With scope sampling (Willer, 1967; Willer & Walker, 2007), investigators test a theory under a variety of conditions that fall within its scope. The scope of ET includes structures that are inclusively connected like those described previously, as well as those that have other connection types. (See Willer, 1999, for a thorough discussion of connection types.) Scope sampling pushes a theory to demonstrate its applicability across a range of situations by opening it up to more opportunities to find falsifying instances. Finally, if ET experiments are successfully concluded, there will be no attempt to generalize from their findings. Instead, laboratory results that support theoretic predictions will encourage researchers to *apply the theory* to make predictions for structures in natural settings outside the lab.

V. ARTIFICIALITY, SIMPLICITY, AND IMPLICATIONS OUTSIDE THE LAB

An experimental situation is artificial if it fails to correspond to situations that exist in natural settings. Experimental designs are also simple situations when compared with the world outside the laboratory. A key question for students of research methods is whether artificial and simple designs impede or aid the development of science. There are two quite distinct answers for the two experimental methods.

Mill's imagery of the world as one chaos following another is an apt starting point (Mill, [1843], 1967, p. 248). The world must undergo analysis; it must be cut into smaller constituent pieces if science is to understand it. Experiments, whether empiricist or theory driven, are elements of analytical processes that simplify the world they seek to understand. Their simplicity ensures their artificiality; they fit situations in the world outside the lab less than perfectly. Empiricist and theory-driven designs are artificial and simple for different reasons, and artificiality and simplicity have different implications for their relationship to the world outside the laboratory.

A common complaint of social scientists is that, in contrast to some other forms of investigation, experiments are of limited use to science because their results are difficult to generalize to the world outside the laboratory. Results are difficult to generalize because the phenomena of experiments are artificial, simpler than and divorced from the world of natural phenomena. As this chapter shows, most social science references to experimentation are references to empiricist experiments and the reference to generalization is no exception.

The objective of empiricist experiments is to discover regularities that are also regularities outside the lab. Applying them outside the laboratory requires generalization and generalization of empiricist experiments requires point-by-point similarities between the lab and instances outside it. That is, empiricist experiments generalize from concrete cases in the laboratory to concrete cases of naturally occurring phenomena. Therefore, the more realistic and the less artificial an empiricist experiment is, the better it is. As an example, in a recent study, Lucas, Graif, and Lovaglia (2006) hypothesized that, as severity of crime increased, so did prosecutorial misconduct, including withholding information from the defense. Controlled conditions very much like those faced by prosecutors were devised and the hypothesis was supported. Given experimental–natural world prosecutorial similarities, generalization, although limited to the issue of prosecutorial misconduct, is straightforward.

By contrast, it is difficult to imagine finding conditions outside the lab like those Asch studied. People are seldom confronted by a sizable group of people who unanimously express the same obviously incorrect judgments. Even when that occurs in natural settings, people are permitted to question each other, to ask how they reached their conclusions, and so on. The Asch design did not permit such conversations. Why then did Asch and many other researchers design simple, artificial empiricist experiments?

Of course, we already know why Asch designed artificial experiments, but the answer points to a dilemma faced by anyone using the empiricist experimental method. They were designed to discover how group pressure affected levels of conformity (1) when the subject had sure knowledge that his conforming responses were objectively wrong, and (2) under such sparse conditions that the link between group pressure and subsequent conformity was clear. Now, here is the dilemma. The stronger the independent variable is and the simpler and more artificial are experimental conditions, the more certain a researcher can be of the link between independent and dependent variables. Identifying links between independent and dependent variables is the purpose of empiricist experimentation and the logic of the method designs the experiments. However, as we have just shown, it is difficult to apply (generalize) such results outside the lab, where conditions are messy and uncontrolled. Said somewhat differently, internal validity and external validity are in opposition. (See Campbell & Stanley, 1966, p. 5 ff., and Cohen, 1989, for a nontechnical discussion of the general idea of validity and of external and internal validity.)

Artificiality and tight experimental control increase a researcher's confidence that experimental effects are due to differences in the variable under study (i.e., they improve internal validity). On the other hand, such experiments have low external validity. That is, artificiality, simplicity, and tight experimental control reduce the fit between the experimental situation and naturally occurring situations in the world outside. Generalization is limited, as in the Lucas et al. research, to those few cases that correspond point by point to the experimental design. Artificiality and simplicity impede generalization of empiricist experiments to the world outside the lab.

Theory-driven designs are also artificial, but they are artificial because they are designed to test theory—not to discover phenomena. Theories are analytical simplifications of the world of phenomena, and theory-driven experiments are replicas of models derived from theory. Running such experiments requires activation of a theory's scope and initial conditions further simplifying the design under study. As a result, theory-driven experiments can have very high internal validity and low external validity. Fortunately, the opposition between internal and external validity that plagues empiricist experiments is not an issue for theory-driven experiments. We make that claim confidently because experiments in physics, chemistry, and biology are highly artificial; yet, their results are routinely brought to bear outside the lab.

Why is theory-driven research with high internal validity and low external validity easily applied outside the lab? Theory bridges between the concrete laboratory situation and concrete situations outside the lab. As said earlier, experiments are analytical tools. Mill and Fisher describe a method of analysis: the empiricist experiment that cuts the seemingly chaotic world into smaller and smaller parts to find regularities. The method fails to advance theoretic science because it does not have a method of *synthesis*—a way of reconstituting the world from the parts that have undergone analysis. On the other hand, theory is a method of analysis *and* synthesis.

A "good" theory-driven experiment is a faithful replica of its model and consistently produces its theory's phenomena under tightly controlled conditions. Tight control under the simplified and highly artificial conditions found in the lab enhances the design's internal validity. The fact that laboratory conditions are very dissimilar to conditions in the world outside is irrelevant. Theory-driven results are *not* generalized to the world outside. Instead, a theory that has found consistent support from laboratory tests can be applied to situations outside the lab as long as they meet the theory's scope and initial conditions. Although not discussed here, we show in our book on experimental method (Willer & Walker, 2007) that theory-driven experiments and the application of well-supported theory outside the lab are exactly the same in the physical and the social sciences.

Pure sodium is not found outside the chemistry laboratory, and chemical theories (e.g., of oxidation) explain why it is not. Yet controlled laboratory experiments that expose sodium to water or other elements and compounds under highly artificial conditions are useful tests of chemical theories that can then be applied to situations outside the laboratory. Similarly, it is very unlikely that a researcher studying natural settings will find two individuals who differ on a single status characteristic, are engaged at a collective task, and disagree on several key issues that must be resolved in order to complete the task. Status characteristics theory predicts what should occur under such conditions, and hundreds of artificial laboratory studies have supported the theory's predictions.

Armed with successful tests of SCT, Elizabeth G. Cohen and others who have used the intervention techniques that she developed have applied the theory to thousands of classrooms across the globe to improve educational outcomes for thousands of students (Cohen, 1998; Cohen, Lotan, Scarloss, & Arellano, 1999). Yet, classrooms are very different from the SCT designs we described previously. We emphasize what has been said earlier: one never generalizes from the results of theory-driven experiments to settings outside the laboratory. Instead, *one applies the experimentally supported theory to phenomena across its scope no matter where such phenomena are found.*

VI. CONCLUSIONS

This chapter has shown that there are two distinct experimental methods. The logic of the younger empiricist method was devised by John Stuart Mill and is based in his misunderstanding of science. Mill believed he was putting forward the logic of all scientific experiments. But, as explained earlier, his misunderstanding of science led him to develop the logic of empiricist experiments. Mill's method, as exemplified by the method-of-difference experiment, is unworkable. It is impossible to determine if an experiment satisfies his criteria of (1) creating two conditions with a single difference, and (2) finding perfect regularities. Without perfect regularities, Mill is left with no laws. However, as revised by R. A. Fisher, Mill's method has become a powerful and effective tool for discovering relations between empirical phenomena.

Our purpose in writing this chapter is not to discourage the use of the refined empiricist experimental method. Instead, our aim is to explain what it will and will not do and to encourage greater use of theory-driven experiments. Empiricist experiments can find evidence of relations between phenomena and, using refinements in statistical methods made since Fisher's earliest formulations, researchers can make informed statements about the likelihood that the patterns are "real." One thing that the empiricist method very definitely will not and cannot do is find laws. Consequently, it cannot produce, as Mill argued, cumulative scientific or theoretical knowledge.

To produce cumulative scientific knowledge, one needs first to develop theory. Observations of relations between phenomena like those uncovered by empiricist experiments can serve as the impetus for theory development (Willer & Walker, 2007). After developing theory, researchers can use it to design experiments that test the theory's precision and investigate its scope. As in other sciences, successful tests of theory-driven sociology experiments produce cumulative knowledge that can be readily brought to bear on phenomena outside the lab. For these reasons, it is important that sociologists and other social scientists make greater and more effective use of theory-driven experiments.

We conclude with a few words about complexity. Simplicity and complexity are not exclusive alternatives for theory-driven research. Simple replicas that correspond to simple theoretic models are the foundation of cumulative scientific work. Yet investigators may begin their work with complex models. At first glance, the world is a bloomin', buzzin' confusion and, to paraphrase Marx ([1845], 1972, p. 109), the point of science is to understand the world, to make sense of it. For that reason, science undertakes analysis; it cuts models for that world into smaller and smaller parts so that we can better understand it.

Empiricist and theory-driven experimentation share the conception of an analysis. But only theory-driven research has both a method of analysis and a method of synthesis. Theory-driven experiments bridge by synthesis from the simple world of the lab to the seemingly complex world that science ultimately wants to understand. Successful tests of theory open the door to improvements in precision and scope. Simple theories can also become more complex as theoretical knowledge cumulates.

Moreover, simple theories and their models can be combined as the sequence of (1) theory development, (2) experiment, and (3) theory refinement progresses with time. Our own work has combined ideas from network exchange theory, status characteristics theory, and legitimacy theory and applied them to the study of organizational behavior (Bell, Walker, & Willer, 2000). Others (Thye *et al.*, 2006) are doing similar work combining other theories. Readers familiar with such work know that simple models can be made more complex, combined with other complex models, and applied to explain events of the sometimes chaotic world that is our ultimate laboratory.

ACKNOWLEDGMENTS

The authors express their thanks to the National Science Foundation for supporting some of their work discussed in this chapter and to the editors of this volume for helpful suggestions.

REFERENCES

Allport, G. W. (1937). *Personality: A psychological interpretation*. New York: Henry Holt and Company.

Aronson, E., & Carlsmith, J. M. (1969). Experimentation in social psychology. In G. Lindzey & E. Aronson (Eds.), *Handbook of social psychology* (2nd ed., Vol. II, pp. 1–79). Reading, MA: Addison–Wesley.

Asch, S. E. (1946). Forming impressions of personality. *Journal of Abnormal and Social Psychology, 41*, 258–290.

Asch, S. E. (1958). Interpersonal influence. In E. E. Maccoby, T. Newcomb, & E. Hartley (Eds.), *Readings in social psychology* (3rd ed., pp. 174–83). New York: Holt, Reinhart and Winston.

Babbie, E. (1989). *The practice of social research*. Belmont, CA: Wadsworth.

Bell, R., Walker, H. A., & Willer, D. (2000). Power, influence and legitimacy in organizations: Implications of three theoretical research programs. In S. B. Bacharach & E. J. Lawler (Eds.), *Organizational politics* (pp. 131–178). Stamford, CT: JAI Press.

Berger, J. (2007). The standardized experimental situation in expectation states research: Notes on history, uses, and special features. In M. Webster & J. Sell (Eds.), *Laboratory experiments in the social sciences* (pp. 358–378). Burlington, MA: Elsevier.

Berger, J., Cohen, B. P., Snell, J. L., & Zelditch, M., Jr. (1962). *Types of formalization in small group research*. Boston: Houghton–Mifflin.

Berger, J., Fisek, M. H., Norman, R. Z., & Zelditch, M., Jr. (1977). *Status characteristics and social interaction*. New York: Elsevier.

Blascovich, J. *et al.* (2002). Immersive virtual environment technology as a methodological tool for social psychology. *Psychological Inquiry, 13*, 103–124.

Campbell, D. T., & Stanley, J. C. (1966). *Experimental and quasi-experimental designs for research*. Chicago: Rand McNally and Company.

Cohen, B. P. (1958). A probability model for conformity. *Sociometry, 21*, 69–81.

Cohen, B. P. (1963). *Conflict and conformity: A probability model and its application*. Cambridge, MA: M. I. T. Press.

Cohen, B. P. (1989). *Developing sociological knowledge* (2nd ed.). Chicago: Nelson–Hall.

Cohen, B. P., & Lee, H. (1975). *Conflict, conformity and social status*. New York: Elsevier.

Cohen, E. G. (1998). Complex instruction. *European Journal of Intercultural Studies, 9,* 127–131.

Cohen, E. G., Lotan, R. A., Scarloss, B. A., & Arellano, A. R. (1999). Complex instruction: Equity in cooperative learning classrooms. *Theory into Practice, 38,* 80–86.

Fisher, R. A. (1935). *The design of experiments.* London: Oliver and Boyd.

Fisher, R. A. 1956. *Statistical methods and scientific inference.* Edinburgh: Oliver and Boyd.

Freese, L., & Sell, J. (1980). Constructing axiomatic theories in sociology. In L. Freese (Ed.), *Theoretical methods in sociology: Seven essays* (pp. 263–368). Pittsburgh: University of Pittsburgh Press.

Galilei, G. ([1636], 1954). *Dialogues concerning two new sciences* (H. Crew & A. DeSalvio, Trans.). New York: Dover.

Goodman, N. (1947). The problem of counterfactual conditionals. *The Journal of Philosophy, 44,* 113–128.

Hempel, C. G. (1966). *Philosophy of natural science.* Englewood Cliffs, NJ: Prentice–Hall.

Leahey, E. (2005). Alphas and asterisks: The development of statistical significance testing standards in sociology. *Social Forces, 84,* 1–24.

Lucas, J., Graif, C., & Lovaglia, M. J. (2006). Misconduct in the prosecution of severe crimes: Theory and experimental test. *Social Psychology Quarterly, 69,* 97–107.

Marx, K. ([1845], 1972). Theses on Feurbach. In Robert C. Tucker (Ed.), *The Marx–Engels reader* (pp. 107–109). New York: Norton.

Mill, J. S. ([1843], 1967). *A system of logic.* London: Longmans, Green and Co.

Newton, I. ([1686], 1966). *Principia mathematica* (A. Motte, Trans.). Berkeley: University of California Press.

Popper, K. R. (1963). *Conjectures and refutations: The growth of scientific knowledge.* New York: Harper Torchbooks.

Schutt, R. K. (2006). *Investigating the social world: The process and practice of research* (5th ed.). Thousand Oaks, CA: Sage Publications.

Thye, S. (2007). Logical and philosophical foundations of experimental research in the social sciences. In M. Webster & J. Sell (Eds.), *Laboratory experiments in the social sciences* (pp. 57–86). Burlington, MA: Elsevier.

Thye, S., Willer, D., & Markovsky, B. (2006). From status to power: New models at the intersection of two theories. *Social Forces, 84,* 1471–95.

Toulmin, S. E. (1953). *The philosophy of social science.* London: Hutchinson's University Library.

Walker, H. A., & Simpson, B. T. (2000). Equating characteristics and status organizing processes. *Social Psychology Quarterly, 63,* 175–185.

Willer, D. (1967). *Scientific sociology.* Englewood Cliffs, NJ: Prentice–Hall.

Willer, D. (1987). *Theory and the experimental investigation of social structures.* New York: Gordon and Breach.

Willer, D. (1999). *Network exchange theory.* Westport, CT: Praeger.

Willer, D., & Anderson, B. (Eds.). (1981). *Networks, exchange and coercion: The elementary theory and its applications.* New York: Elsevier.

Willer, D., & Emanuelson, P. (2006). Testing ten theories. Paper presented at the annual meeting of the American Sociological Association, Montreal, August.

Willer, D., & Walker. H. A. (2007). *Building experiments: Testing social theory.* Palo Alto, CA: Stanford.

Willer, D., & Webster, M. (1970). Theoretical concepts and observables. *American Sociological Review, 35,* 748–57.

CHAPTER 3

Logical and Philosophical Foundations of Experimental Research in the Social Sciences

SHANE R. THYE

University of South Carolina

ABSTRACT

This chapter outlines the logical and philosophical underpinnings of contemporary experimental research in the social sciences. After explicating the fundamental principles of experimental design and analysis as established by John Stewart Mill and Ronald A. Fisher, I examine modern critiques of the experimental method and show how these are flawed. Along the way I consider how scientists think about causality, issues of internal and external validity, empirical versus theoretical experimentation, and the many varieties of experimental design.

I. INTRODUCTION

Social scientists of various academic stripes depict the experimental method in diverse ways—sometimes good, mostly bad. Lieberson (1985, p. 228) asserts that "the experimental simulation fails at present in social science research because we are continuously making counterfactual conditional statements that have outrageously weak grounds." Psychologists, who normally embrace experimental methods, have made even more radical claims: "The dissimilarity between the life situation and the laboratory situation is so marked that the laboratory experiment really tells us *nothing*..." (Harré & Secord, 1972, p. 51; italics in original). Even writers of popular methods textbooks make critical observations. Babbie (1989, p. 232) argues that "the greatest weakness of laboratory experiments lies in their artificiality." There are numerous costs and benefits associated with any research method, but for whatever reason, the experimental method seems to inspire its fair share of critics. More fairly, the disadvantages and advantages are really quite different from those just listed.

On the positive side, controlled experimental research gives one the ability to claim, with some degree of confidence, that two factors are causally linked. Social scientists are in the business of establishing causal laws that explain real-world events. This objective is realized through the painstaking building and systematic testing of scientific theory. As a sidebar, to say that one has an explanation for some phenomenon is to say that there is a well-supported scientific theory of that phenomenon. For example, if you "explain" why sky-divers descend to the earth by invoking the notion of gravity, what you really mean is that the observed descent conforms to the theory of gravitational forces. To date, controlled experimentation is the most widely embraced method for establishing scientific theory because it allows scientists to pinpoint cause–effect relations and eliminate alternative explanations. Laboratory experimentation is the gold standard for isolating causation because the logic of experimental research embodies the logic of scientific inquiry. *Thus, the advantage of the experimental method is that it allows one to see the world in terms of causal relations.*

Ironically, this also is a heavy burden carried by the experimental scientist. Consider a recent headline that read: "Bottled Water Linked to Healthier Babies." If it is true that expectant mothers who drink bottled water tend to have babies with fewer birth defects, higher birth weights, and other health benefits, this simple yet powerful intervention may carry major health consequences. But, my excitement over the power of water was short-lived as my newfound "parent" identity had to be reconciled with the "experimentalist" in me, who also had an opinion.

During that transformation a number of questions emerged. What is the *theoretical connection* between bottled water and healthy babies? Is there a biological mechanism or could the effect be due to other factors? The latter issue concerns whether or not the relationship is real. So-called spurious factors are unrecognized causes (such as socioeconomic status) that produce the illusion that two things (drinking bottled water and having healthy babies) are causally linked. Could it be that socioeconomic status causes both? Perhaps moms who can afford bottled water have healthier babies because they can afford to get better prenatal care, join gyms, eat healthier food, take vitamins, receive treatment at elite hospitals, and so on. For the same reason, I would lay heavy odds that moms who drive new BMWs also have healthier babies than moms who do not own a car, but that would be unlikely to grab the headlines.

In what follows I examine the underlying logic of experimental design and analysis. I consider how experimental research bears on establishing causation, explore recent critiques of Fisherian methods and show how these are flawed, and examine various forms of experimental design. The aim is not to provide a comprehensive discussion of all facets of experimentation. Instead, I hope to illuminate the logic and philosophical underpinnings of experimental research, dispel a number of myths and misconceptions, and generally excite the reader about the prospects of building scientific theory via experimentation. Along the way I consider a number of issues germane to all research, such as how evidence bears on theory, notions of causation, and the logic of applying or generalizing findings to other settings.

II. CLUES TO CAUSATION

The notion of "causality" has always been a challenge, in part, because causation is not directly observable. Rather, causation must be inferred from some manner of evidence. This section considers the various ways that scientists think about causality and the methods scientists use to infer that two phenomena are causally linked. The concept of causality has a twisting and convoluted history in the philosophy of science, and there are many ongoing discussions and debates. Most notions of causation are traced to Aristotle ([340 BC], 1947), who offered four different conceptions of causation. Of these, Aristotle's *efficient cause* captures the notion that one event (X) sets into motion, forces, creates, or makes another event (Y) occur. Although this kind of causation corresponds nicely with the everyday meaning of the term, it has become the focal point of controversy and debate.

Galileo developed an alternative causal ontology that equated causation with necessary and sufficient conditions. He argued that to say event *a* caused event *b* was to say that event *a* is a necessary and sufficient condition for event *b* (see Bunge, 1979, p. 33). A *necessary condition* exists when event *b never* occurs in the absence of event *a*. That is, event *b* follows event *a* with 100% regularity. A *sufficient condition* exists when event *b always* follows event *a* with perfect regularity. To illustrate, a lawnmower will only start if it has the correct kind of fuel in its fuel tank. Thus, the proper kind of fuel is a necessary condition for starting the mower, as it will *never* start without it. At the same time, fuel alone is not enough to start the mower; it requires a working engine and all requisite parts. This means that proper fuel is not a sufficient condition to make a mower start. The combinations of the proper fuel and a working engine are necessary and sufficient conditions; when both are in place, the mower *always* starts and, if either is missing, the mower *never* starts.

The philosopher David Hume ([1784], 1955) took the more radical position associated with British empiricism. Prior to Hume, popular notions of causation traced to Aristotle involved one event forcing, or setting into motion, another event. Hume offers a softer notion that robs causation of its force. He argued that we can never directly observe a causal force in operation, but, instead, all we can observe is the conjunction or correlation of two events that we presume are causally linked. Hume used the term "constant conjunction" to describe the situation where X always occurs in the presence of Y. Thus, for Hume and the empiricists, causation is an elusive thing—we can never be sure that one thing causes another or that events correlated today will be correlated tomorrow. More radical Humeans, such as Bertrand Russell, reject the idea of causation altogether. In his now infamous 1913 paper, Russell wrote: "The law of causality, I believe, like much that passes muster among philosophers, is a relic of a bygone age, surviving, like the monarchy, only because it is erroneously supposed to do no harm."

Despite the historical disagreements surrounding notions of causation, the majority of social scientists generally agree upon certain basic requirements that must be satisfied to support causal inference (Davis, 1985; Hage & Meeker, 1988). One can think of these requisites as "clues" to assess whether a relation is truly causal or not. Next, I consider six conditions that scientists use to assess causation. This discussion focuses not on points of disagreement, but rather on the general principles for which there is consensus.

Covariation. When a cause occurs, then so should its effect; when a cause does not occur, then neither should its effect. (This is one of the basic maxims underlying Mill's canons of inference discussed later.) In short, causes and their effects should covary or be correlated. At the same time, it is very important to remember that things that covary may not be causally related. For instance, the price of bourbon is correlated with the price of new cars in any given month. In this case, the prices of bourbon and new cars are not causally linked, but both are caused by the prevailing economic conditions. As Hume would agree, it is easy to focus on the correlation between car prices and bourbon while blindly missing the underlying causal force. As such, this leads to a very important principle: *correlation alone does not imply causation.* Sifting causation from correlation is a focal issue in virtually all social science research.[1]

---

[1]The basic experimental design provides an elegant solution to the problem of sorting causal relations from correlations. Because this problem has an even tighter stranglehold on those who deploy survey, historical, qualitative, or ethnographic methods, researchers in these domains often use the experiment as a template to design their own studies (Lieberson, 1991).

Contiguity. There is always some time lag between a cause and its effect. When the time lag between a cause (a paper cut) and its effect (bleeding) is short, we say the two events are contiguous. Some cause–effect relations are contiguous while others are not (such as conception and childbirth). In general, social scientists presume that noncontiguous causes set into motion other processes that, in turn, have effects at a later point in time. For example, greater parental education can lead to a variety of lifestyle benefits (such as greater income, more social capital, and advanced reading and verbal skills) that have effects on their children's educational attainment. Importantly, however, a causal claim between parent and child education levels is not warranted unless there is a theory that specifies the intermediary factors occurring between the two points in time. For example, more educated parents are more likely to read to and with their children than are uneducated parents, and the greater amount of reading by those children helps them succeed in school.

Time and asymmetry. One of the most basic principles of causation is that if X causes Y, then X must occur before Y in time. It is important to note that simply because X precedes Y does not mean X causes Y. Just because it rained before I failed my exam does not mean the rain caused me to fail (if I do make the connection between rain and failure, I am committing the *post hoc fallacy*). A related idea is that causation is assumed to run in one direction, such that causes have asymmetric effects. That is, we cannot simultaneously assert that X causes Y and Y causes X.

While the notion of *reciprocal causation* $X_1 \rightarrow Y \rightarrow X_2$ seems to violate the assumptions of time and asymmetry, it does not because the X at time two is not the same X as the X at time one. Instead, reciprocal causation implies that cause–effect sequences dynamically unfold. For instance, cybernetic feedback systems involve reciprocal causation in which the cause–effect relationships are reversed through time. The "cruise control" feature in a car operates on this principle. That is, an initial cause (the deceleration of the car) triggers an effect (an increase in engine RPM) that in turn feeds back on that initial cause (the acceleration of the car).

Nonspuriousness. When two things occur together but are truly caused by some third force, the original relationship is said to be spurious. There are many unusual spurious relations that illustrate the point. For instance, few people know that there is a strong positive correlation between ice cream sales and rape. That is, in months when ice cream sales are high, many rapes are reported. Does this mean that ice cream sales are somehow causally linked to the occurrence of rapes? The answer is of course not! The relationship is spurious because both factors are caused by a third factor: temperature. In warmer months more ice cream is sold and there are more sexual predators frequenting outdoor venues and social gatherings. In cold weather, both factors are attenuated. It would be a mistake to believe that rapes and ice cream sales are casually related without seeking alternative explanations. To establish causation, one must be able to rule out alternative explanations with some degree of confidence. Ruling out alternatives is a key activity in science and, as will be shown later, experimental research provides the best known method for doing so.

Consistency. Philosophers and scientists have long debated whether one should think of causality as *deterministic* (i.e., that a given cause X will always lead to effect Y) or *probabilistic* (i.e., that the presence of cause X will increase the likelihood of effect Y). Early writers such as Galileo ([1636], 1954) and Mill ([1872], 1973) leaned toward the deterministic end of the spectrum. However, with the advent of modern statistical tools, the majority of contemporary philosophers and social scientists evoke probabilistic notions of causality. To illustrate, there is abundant evidence that smoking causes a higher rate of lung cancer. Still, not *every* smoker develops lung cancer and not *every* lung cancer victim is a smoker. It seems more accurate to say that smoking increases the probability of lung cancer.

Hage and Meeker (1988) argue that probabilistic notions of cause are preferable for three reasons. First, there can be unrecognized countervailing causal forces (such as antibodies or gene combinations that make people resilient to cancer) that play a role. Second, most phenomena are affected by a multiplicity of causes (such as body chemistry or exposure to carcinogens) that can interact to obfuscate true causal relations. Third, chaos and complexity theories have shown that both natural and social phenomena can behave in unpredictable and nonlinear ways (Waldrop, 1992). To illustrate, those with advanced cancer can for some inexplicable reason go into remission. The upshot is that the deterministic views of causation may be overly simplistic.

Theoretical plausibility. Finally, scientists always view causal claims about the world with one eye trained on established scientific laws. When claims about the world violate or are inconsistent with those laws, the confirmation status of those claims is questionable without unequivocal evidence to the contrary. Simply stated, extraordinary claims require extraordinary evidence. For instance, Newton's law of inertia states that unless acted upon, a body at rest stays at rest and a body in motion stays in motion. Based on this thinking, many have attempted to build "perpetual motion" machines, ignoring the broader context of other physical laws. A true perpetual motion machine (if one could be built) would, in fact, violate several existing scientific laws.

For example, such a machine would need to consume no energy and run with perfect efficiency (thus violating the second law of thermodynamics) or produce energy without consuming energy as it runs (thus violating the first law of thermodynamics). Thus, while Newton's law of inertia suggests that perpetual motion machines are possible, in the context of the laws of thermodynamics, it seems that such machines are very unlikely ever to be produced.

It should not be surprising that, to date, dreams for perpetual motion machines have remained just dreams.

III. MILL'S CANONS AND INFERRING CAUSALITY

What is the logical connection between data and causation? Asked differently, how does one infer causation based on the outcome of some empirical test? Many, if not all, social scientists operate by approximating a model of evidence that can be traced to John Stuart Mill ([1872], 1973), who was influenced by Sir Francis Bacon and subsequently influenced Sir Ronald A. Fisher (1935, 1956). Mill presumed that nature is uniform and, as such, if a cause–effect relationship occurred once, it would occur again under similar circumstances. Mill developed five methods (or canons) to assess causation; these canons lie at the heart of contemporary inference and experimental design today. Here I briefly discuss two of the more important canons: the method of difference and the method of agreement. Whereas the method of difference aims to find a lone difference across two circumstances, the method of agreement seeks to find a lone similarity.

> *The method of difference:* If an instance in which the phenomenon under investigation occurs, and an instance in which it does not occur, have every circumstance in common save one, that one occurring only in the former; the circumstance in which alone the two instances differ, is the effect, or the cause, or an indisputable part of the cause, of the phenomenon ([1872], 1973, p. 391).
>
> *The method of agreement:* If two or more instances of the phenomenon under investigation have only one circumstance in common, the circumstance in which alone all the instances agree, is the cause (or effect) of the given phenomenon ([1872], 1973, p. 390).

Perhaps the best way to illustrate the methods is by example. Imagine that you and a friend attend a wild game cookout, sample a variety of dishes, and later that night you feel ill but your friend does not. How would you determine the cause of the illness? Imagine that you and your friend both sampled shrimp, pheasant, and duck, while you also enjoyed oysters but your friend did not.

The top panel of Figure 1 illustrates this scenario. Notice that there are two differences between you and your friend: you dined on oysters and then became ill while your friend did neither. Mill's method of difference suggests that oysters are the cause of the illness because it is the single condition that distinguishes illness from health. The same method can also be used as a method of elimination. Notice that both you and your friend ate shrimp, pheasant, and duck, but only you got ill. Thus, shrimp, pheasant, and duck can be eliminated from the list of possible causes. In this way the method can be used to eliminate alternative causes because *a deterministic cause that remains constant can never produce effects that are different.*

Panel A: The Method of Difference

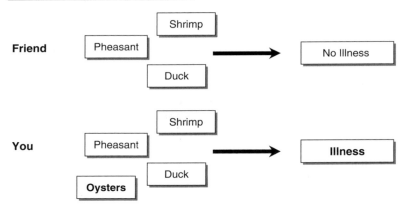

Panel B: The Method of Agreement

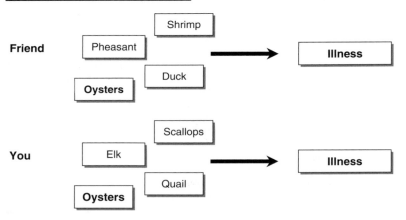

FIGURE 1   Mill's method of difference and method of agreement.

Now consider the method of agreement, which indicates that if a single factor (eating oysters) occurs in conjunction with a common effect, then that single factor is the likely cause. Panel B of Figure 1 illustrates this scenario. Assume that your friend enjoyed shrimp, pheasant, duck, and oysters, while you ate scallops, elk, quail, and oysters. Later that evening you both became ill. Since eating oysters is the common denominator preceding illness, you might infer that oysters are the cause. This illustrates another important principle: *A causal factor that differs across two circumstances can never generate precisely the same effect.* In this case, if eating shrimp were the true cause of illness, because your friend ate shrimp and you did not, we would expect your friend (but not you) to be ill. The method of agreement once again suggests that the oysters are suspect.

Limitations of Mill's canons. Mill's ([1872], 1973) method of difference and method of agreement provide useful guidelines for thinking about causation. Even so, it is now recognized that the canons are limited in a number of regards. A good critique of the methods can be found in Cohen and Nagel (1934; see also B. P. Cohen, 1989) and it may be useful to summarize their analyses here. Cohen and Nagel (p. 249) point out that Mill's methods are neither "methods of proof" nor "methods of discovery" (see also Hempel, 1965, and Walker & Willer, Chapter 2, this volume). In terms of proof, the canons generally presume that *all other possible causes* are contained in the factors examined.

In the preceding example, we presume that all possible causes of illness reside in the food that was eaten. But the methods cannot *prove* this definitively because there *could* be unrecognized causes in things besides the food. Imagine that you and your friend have an undetected mild shellfish allergy that only reacts when two or more varieties of shellfish are eaten simultaneously. Referring back to Figure 1, in every case of illness, both oysters and one other shellfish (shrimp or scallops) were consumed. As such, it might not be oysters per se causing the illness, but rather the unique combination of food and the allergy, what statisticians call an *interaction effect.* The method of difference and method of agreement are blind to this possibility.

Second, Mill's methods cannot be used as "methods of discovery" because they presuppose that one can identify, a priori, potential causes of the illness. In Figure 1, differences and similarities across food items are analyzed as potential causes for the illness. However, there are other potential causes that are unmeasured or unknown. These include personal factors (food consumed before the party), social factors (contact with other sick people), historical factors (flu or allergy season), or even genetic or biological factors (a weak immune system) that might fuel illness in a hard to detect way. In principle, there are an infinite number of additional causes that could contribute to illness, and the lion's share of these will be unmeasured and unknown. Again, the methods are not methods of discovery because they are oblivious to possible alternatives.

In summary, Mill's canons suffer two limitations: *proving* that a single factor is the unique cause and *discovering* causes beyond the immediate situation. Although it may not be immediately obvious, both limitations stem from a single issue: *Any phenomenon may have an infinite number of intertwined causes that researchers cannot measure or identify.* Given this seemingly insurmountable problem, one might guess that Mill's methods would have withered on the vine. Indeed, that might have occurred had it not been for the statistician R. A. Fisher,[2]

---

[2]Despite the tremendous vision Fisher possessed as a statistical prodigy and brilliant geneticist, he was, ironically, severely myopic. He claimed that this aided, rather than hindered, his creative insight because he was forced to rely more heavily on mental instead of physical representations.

who rescued the methods. Fisher (1935, 1956) did that by providing a logical and methodological basis for reducing or altogether eliminating the troublesome set of alternative explanations. Next, I illustrate Fisher's solution and other central features of experimental research through a detailed example. Following this, I reconsider Mill's methods in view of Fisher's solution and Cohen and Nagel's 1934 critique.

IV. FISHER'S SOLUTION AND HALLMARKS OF EXPERIMENTATION

The confluence of three features makes experimental research unique in scientific inquiry: random assignment, manipulation, and controlled measurement. To illustrate, presume that a researcher is intrigued by the relationship between viewing television violence and childhood aggression, a topic that captures much scrutiny. A typical hypothesis is that viewing television violence increases subsequent aggression by imitation. Correspondingly, let us assume that a researcher plans to study 100 fourth-grade children, their viewing habits, and their aggressive behaviors. The overall experimental strategy involves three distinct phases. First, the researcher must *randomly assign* each child to one of two groups. The first group will be exposed to violent TV (called the treatment group), and the second group will be exposed to non-violent TV (called the control group). Second, the researcher *manipulates* the content of the violent television such that one program contains violence while the other does not. Third, the researcher *measures* aggressive behavior across the groups in a controlled environment. Let us consider each feature in more detail.

Random assignment. R. A. Fisher (1935, 1956) offered the concept of random assignment as a way to hold constant spurious causes. Random assignment is defined as the placement of objects (people or things) into the conditions of an experiment such that each object has exactly the same probability of being exposed to each condition. Thus, in the two-condition experiment described previously, each child has exactly a 50% chance of being assigned to the experimental or control group. This procedure ensures that each group contains about 50 kids, half male and half female, to within the limits of random chance. The benefit of random assignment is that it *equates at the group level.* Said differently, random assignment ensures that the two groups are equal in terms of all *historical* factors (e.g., divorced parents, childhood poverty, being abused), *genetic* factors (eye color, chromosome distribution, blood type, etc.), *physical* factors (gender, height, weight, strength, etc.), *personality* factors (preferences, values, phobias, etc.), and *social* factors (role identities, dog ownership, etc.). The reason is that each factor has exactly the same probability of appearing in each group. The brilliance of random assignment is that it mathematically equates the groups on factors that are known, unknown, measured, or unmeasured.[3]

A common criticism is that experimental results are biased if the traits under the control of random assignment interact with the dependent variable. This idea, however, is slightly off the mark. For instance, assume that (1) both males and females respond aggressively, but differently, to watching violent television, and (2) our measure of aggression is sensitive to male, but not female, aggression. The result of the experiment would correctly show that males in the treatment group respond more aggressively than males in the control group *on this measure of aggression.* It would also correctly show no difference between the treatment and control females *on this measure of aggression.* Strictly speaking, the problem here is not one of random assignment or even one of incorrect inference. Instead, the problem is that one may not have a robust and valid measure of hostility that captures the kind of aggression expected to occur in both males and females. The relationship between exposure to violent television and aggressive behavior is properly guided by the theory that links the two phenomena. Such an interaction suggests a problem with the theory or a problem with the measurement procedures, but not the experiment per se.

Manipulation. Following random assignment, the independent variable is manipulated such that the treatment group is exposed to violent images while the control group is not. Ideally, the manipulation would be exactly the same in both conditions except for the factor hypothesized as causal. In our example, the researcher could ask the experimental group to watch a TV program of a couple engaged in a financial dispute that ends violently. The control group could watch the exact same couple end the exact same dispute in a nonviolent manner. Notice that in the context of random assignment, the basic experiment is comparable to Mill's method of difference in that the two groups are equated on virtually all factors except the independent variable. Apodictically, the Fisherian principle of random assignment is the cornerstone of experimental research, and in conjunction with controlled manipulation, these procedures render the method of difference workable.

Controlled measurement. The final step is to measure the dependent variable (aggressive behavior) in a controlled environment. Ideally, individuals from the treatment and control groups would have the opportunity to aggress in exactly the same manner toward the same target. In actual research, aggression

---

[3]There are deeply rooted statistical reasons to employ random assignment. For instance, violating random assignment can cause observations to be correlated, which can bias any ensuing statistical test, inflate the standard error of that test, or both. Thus, random assignment rests on logical and statistical foundations.

has been measured in a variety of ways, including the delivery of electric shock, hitting a doll with a mallet, or the slamming down of a telephone. Importantly, any difference between the treatment and control groups can only be attributed to the independent variable because, in principle, this is the only factor on which the two groups differ (remember that a cause that does not vary can have no effect). Overall, the experimental method is the method of causal inference because it equates two or more groups on all factors, manipulates a single presumed cause, and systematically records the unique effect.

How does the method measure up with respect to providing information on causation? Recall that causal inferences require information on three empirical cues (covariation, contiguity, and temporal ordering) and three other criteria (nonspuriousness, consistency, and theoretical plausibility). Overall, the basic experiment does an excellent job of attending to these matters. Evidence for covariance comes in the form of the effect appearing in the experimental group but not the control group. Also, the researcher has information on contiguity because he or she controls the time lag between the factors under investigation. Further, because the experimenter manipulates the cause before the effect, the temporal ordering is correctly instated. Of course, random assignment controls for potentially spurious factors. Finally, the experiment is a repeatable event and its data are always considered in the context of established theory.

V. FISHER'S PREMATURE BURIAL AND POSTHUMOUS RESURRECTION

The basic experiment approximates Mill's ([1872], 1973) method of difference in structure and design. However, recall that Mill's method was deemed intractable because there could be an infinite number of unknown and unknowable causes adding to (or interacting with) the presumed cause. A related problem is that Mill invokes a deterministic view of causation wherein empirical outcomes occur with perfect regularity, which of course, never occurs (Walker & Willer, Chapter 2, this volume; Willer & Walker, 2007).

Fisher (1935) developed the principle of random assignment to remedy the ailing method. When subjects are randomly assigned to conditions, any unknown and unknowable factors are distributed equally and thus (1) *the causal effects of those unknown and unknowable factors will be equated across experimental and control groups,* and (2) *factors that are the same can never cause a difference.* Thus, any differences can be reasonably attributed to variation from the independent variable. Although Fisher's principle of random assignment is widely recognized as the panacea that saved the method of difference, others still believe the method is grievously ill.

Bernard P. Cohen (1989, 1997) claims that random assignment does not remedy the problem. He argues that as the set of unknown or unknowable causes grows large, the probability that the experimental and control groups will differ on at least one of these causal factors approaches 1.0. This relationship is shown in Figure 2. Furthermore, Cohen notes that random assignment only equates experimental and control groups with infinite sample sizes and that, of course, real experiments always employ finite samples. He therefore concludes that there is always some unknown causal factor operating in any experiment, and, as such, the original problems that plagued Mill's ([1872], 1973) methods remain unresolved. In the end, Cohen (1997) does the only befitting thing and offers "a decent burial" for J. S. Mill and R. A. Fisher.

At first blush, Cohen (1989, 1997) appears to have brought the illness out of remission. However, he may have been premature in laying our progenitors and their methods to rest. His analysis centers on (1) the number of potential unknown causes, and (2) the number of subjects in the experimental and control groups—claiming that the former is too large and the latter is too small for random assignment to operate properly. However, in a mathematical sense, Cohen (1997) overestimates both the number of alternative causes that do exist and the sample size that *is* required for random assignment to work properly.

In terms of sample size, Cohen (1997) ignores the straightforward statistical relation between sample size and the nature of alternative causes. That relationship is described by the *law of large numbers*. This law dictates that as the size of the experimental and control groups increases, the average value of any factor differentiating those groups approaches a common value (i.e., the population value for that factor). This has implications for the number of

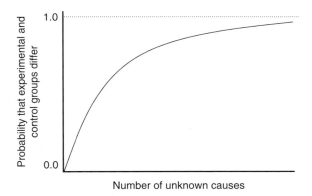

FIGURE 2   The relationship between unknown causes and probability that experimental and control groups differ.

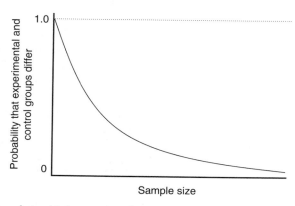

FIGURE 3   The relationship between sample size and probability that experimental and control groups differ.

alternative causes. The law implies that as group size increases and alternative causal factors become equated, the number of differentiating possible causes will diminish at an exponential rate. As this occurs the probability that the experimental group differs from the control group on any single factor approaches zero.[4] The relationship between sample size and probability of differentiating factors is shown in Figure 3. Thus, probability theory and the law of large numbers mitigate (but do not totally eliminate) the issue of unknown factors.

There are additional logical and empirical grounds that further salvage the method. Logically, it is important to distinguish factors that could make a difference from factors that do make a difference. Not all of the factors that make experimental groups different from control groups are relevant to the dependent variable; therefore, not all factors must necessarily be equated. Many differences simply do not matter. For instance, physical theories do not consider the color of falling objects for the same reason that bargaining theories do not consider the height of the negotiator: neither factor is relevant to the theoretical processes and phenomena of concern (Willer & Walker, 2007). Scientists specifically control for *theoretically relevant* factors in a given study, and in true experiments, they also control for *possibly relevant* factors using random assignment. Cohen heavily emphasizes those factors that are possibly relevant. However, given that laboratory research selectively focuses on specific theoretical problems, most of the infinite set of possibly relevant differences probably will be irrelevant.

---

[4]For instance, with just 23 randomly selected people, there is a 50% chance that two of them will share a birthday (Paulos, 1988)!

Now, let us play devil's advocate and presume that Cohen (1997) is correct in his assertion. That is, assume that a very large number of factors differentiate experimental and control groups and that these factors are relevant to the phenomena under consideration. Ironically, from a statistical point of view, even this situation does not pose a problem for the internal validity of the experiment. The *central limit theorem* explains how those factors would be distributed. This theorem suggests that, for a very large number of independent causal factors differentiating experimental and control groups, the aggregate effect of those causal factors would quickly converge and become normally distributed.

The shape of the distribution is important. Some causal factors would have positive effects on the dependent variable; others would have negative effects. When subjects are randomly assigned to conditions, the overall impact of these factors is to add random variance (or noise) to the dependent variable, and that noise would be normally distributed with an expected value of 0. *Thus, random assignment and the central limit theorem ensure that the errors are normally distributed and guarantee that there is no overall impact of these variables.* As such, even if the identified problems are real, they do not affect the basic logic of experimental inference.

In sum, the method of difference and the principle of random assignment provide the logical and statistical foundation for contemporary experimental design and analysis. Mill and Fisher left in their wake a powerful set of analytic and statistical tools that have become the method of choice in science. I have shown that the problems identified by Cohen (1989, 1997) and others are not problematic in a statistical or pragmatic sense. In practice, if the experimental method were flawed in the ways detailed above we would never expect to find consistent results produced by experimental research, nor would we expect to find cumulative theory growth in areas so informed. However, even Cohen (1989) acknowledges that cumulative research programs informed by experimental work abound. Despite the fact that both Mill and Fisher have long since passed away, the legacy of experimental testing they left behind continues to flourish.

VI. SIMPLE DESIGNS AND THREATS TO INTERNAL VALIDITY

At this juncture it seems that the logic of experimentation provides a fairly bulletproof way to establish causation. It would be a mistake, however, to imply that all experiments are cut from the same cloth. There are a number of well-understood threats to internal validity that can compromise results, but, fortunately, most if not all of the problems can be circumvented. *Internal validity* refers to the extent to which a method can establish a cause–effect relationship. The most severe threat to internal validity is a confound. A *confound* exists when more than one thing is unintentionally manipulated in an experiment. The problem is that when two factors are manipulated (say, X and Z), any change in the dependent measure (Y) could be caused by a change in X, a change in Z, or some combination of these two factors. Next, I detail nine important confounds that threaten internal validity. Again, I rely on a scenario to illustrate the issues.

Imagine that a school teacher is interested in determining whether or not a given reading program will improve reading aptitude in her elementary school classroom. Presume that the program is based on a psychometric theory of intelligence, and the theory makes clear predictions for which students will benefit the most from the program. One possibility would be to administer the program (X) and then measure reading ability (O) as such:

$$X \qquad O$$

This represents a "one-shot post hoc" experiment, a design recognized as having little (if any) utility in science. The design is severely limited because it does not enable the researcher to claim that X causes (or affects the probability of) O because the design does not eliminate key confounding factors. Sometimes, one-shot post hoc designs are the only possible designs (e.g., the effects of Hurricane Katrina on the Gulf Coast). The key problem here is that the design does not include some kind of control condition, so it cannot isolate the impact of other factors contributing to the outcome. The important point is that one-shot post hoc designs are problematic because the researcher can never be sure if X or some other factor is the mechanism responsible for producing O.

Being somewhat informed, our teacher opts for a better design. She decides to measure the students' reading abilities at the beginning of the semester (O_1), administer the reading program for 15 weeks (X), and then measure the students again (O_2) to determine if they improved. The overall design is a one-condition pretest/post-test design represented as follows:

$$O_1 \qquad X \qquad O_2$$

Let us assume that every student in the class performed better at time 2 than at time 1. The question becomes: can the teacher conclude that the reading program caused better performance? In short, she cannot make a causal claim because there could be a number of confounding factors masquerading as the treatment effect. Next, I review nine distinct threats to internal validity (see Campbell & Stanley, 1966, and Webster, 1994, for a good review).

- *History*. There are an unlimited number of factors that can occur between O_1 and O_2, in addition to the treatment X, that can create a change in the scores. For instance, there may be a national reading campaign that extols the virtues of reading, or the public library might offer incentives for parents to check out books with their children, and so forth. Any effect of the reading program is perfectly confounded with these factors, so the impact of that program cannot be separated from them.

- *Maturation.* Over the course of the semester the students will grow and mature at their own rate, both physically and cognitively. Students may improve simply because their brains become more fully developed and their cognitive skills have become sharper with experience. Because this maturation process is confounded with the reading program, again, the impact of the two cannot be distinguished.

- *Selection.* Individuals differ with respect to an array of learned and inherent characteristics. If the individuals in the study are somehow "selected" into groups on a nonrandom basis, then a selection bias exists because personal characteristics may cause a change in reading scores. To illustrate, suppose that a school administrator assigned kids to classrooms based on where they live such that kids from affluent neighborhoods share the same classroom. If so, then socioeconomic status, which may be the true cause, is confounded with the reading program.

- *Selection–maturation interaction.* There are sometimes unique effects of maturation processes (related to time) and selection biases (related to personal characteristics) that "interact" in producing an outcome. Suppose that a fair number of "gifted" children are assigned to our hypothetical classroom. The improvement in scores may be caused by the gifted children learning at an accelerated rate (compared to normal children). Thus, the change in scores may have nothing to do with the program per se, but instead could be caused by the interaction of aptitude and maturation.

- *Testing.* The prior measurement of a dependent variable can sometimes cause a change in the future measurement of that variable. For instance, the very act of measuring reading ability at the beginning of the semester could raise the students' levels of awareness with respect to their own reading ability. Being more aware, the students may read more often or learn to read more efficiently. Further, if the test is repeated, the students may learn some of the test items. As such, the change in reading ability may be caused by levels of awareness or prior exposure to the test, not the reading program itself.

- *Regression.* Whenever repeated measures are taken, extreme scores tend to become less extreme over time because they move (or regress) toward the group mean. The reason is simple: extreme scores are rare events and unlikely to occur in succession. In our example, if some kids scored very poorly on the first test by random chance alone, they should improve on the second test for the same reason. This change has nothing to do with the program but is a spurious factor fueled by chance fluctuations.[5]

- *Instrumentation.* It is well known that all measures contain some degree of unreliability that can cause the measurement of the dependent variable to change over the course of an experiment (Thye, 2000). Said differently, repeated measures may fluctuate due to random measurement error. When the response item is practiced or well rehearsed (such as responding to "What is your name?"), the random component is small. However, for novel or more difficult responses, the random component is larger. In our example, unreliability in the dependent measure would cause a change in scores unrelated to the true reading ability of the students.

- *Experimental mortality.* Over the course of an experiment, some individuals may drop out before the experiment is complete; this is not problematic if those individuals drop out on a random basis. But when individuals *selectively* exit an experiment in a nonrandom way, this can create an illusory effect. Imagine that at the beginning of the semester the children vary in terms of reading ability and, over time, those with less ability leave because the class it is too difficult or frustrating. At the end of the semester the average scores on the second test would improve—not because the program was effective, but because those children who lowered the original group mean are no longer present.

- *Experimenter bias.* Hundreds of studies now show that experimenter expectations can influence subjects' behaviors in subtle ways (Rosenthal & Rubin, 1978). In a classic study, Rosenthal and Fode (1963) led students to believe they were training either "bright" or "dull" rats to run a maze over the course of a semester. In reality, all students were randomly assigned five ordinary rats with which to work. At the end of the semester, those rats expected to be "bright" objectively learned more than those expected to be "dull." This occurred because students who expected bright rats actually handled them more frequently and were more patient with them during the training. Thus, student expectations had an impact on rat performance. Another kind of experimenter bias occurs when subjects try to be helpful in the experiment and act in ways to confirm what they believe to be the hypothesis (correctly or otherwise). I address both forms in more detail later.

---

[5]Strictly speaking, statistical regression to the mean can be caused by numerous factors, including chance fluctuations of subjects' true score values or random measurement error associated with the instrument.

Despite these threats to internal validity, the logic of experimental design provides a straightforward and powerful solution to eliminate confounds. Next, I turn to these solutions.

VII. USING EXPERIMENTAL DESIGN TO RESOLVE PROBLEMS OF INTERNAL VALIDITY

The last mentioned threat to internal validity, experimenter effects, is perhaps the easiest to prevent. In principle, experimenter effects occur because (1) subjects attempt to confirm or disconfirm the experimental hypothesis, or (2) experimenters unintentionally influence subject behavior. The solution for both problems is to use blinding techniques. A *single-blind experiment* is one in which the subject does not know the true hypothesis under investigation. If subjects do not know the hypothesis, they cannot act in ways to confirm or disconfirm that hypothesis. Researchers use a variety of techniques—from simply withholding information to outright deception—to prevent subjects from knowing the hypothesis. Single-blind techniques also prevent other kinds of subject-expectancy biases. For instance, medical research has shown that ingesting an inert substance (such as sugar or starch) can make some patients feel better simply because they believe they should feel better. Placebo effects are detected by keeping some subjects blind to the experimental treatment.

A more stringent way to prevent experimenter effects is to use a double-blind technique. A *double-blind experiment* is one in which neither the subject nor the experimenter knows the true hypothesis under investigation. Some double-blind studies involve two researchers. One researcher sets up the study knowing the experimental condition; the second interacts with the subjects and/or records their behavior without this knowledge. Other double-blind studies are computer mediated and require only one researcher. Here, the researcher interacts with the subject at the beginning of the study, but then a computer randomly assigns the subject to one of the experimental conditions. In either case, both researcher and subject are blind to the hypothesis. To further prevent experimenter bias, many experimenters use strict protocols that ensure the environment, written and verbal instructions, measures, and so forth are equated for all subjects. Some have said that the best experiment would use perfect clones as subjects (ensuring they are identical) and those clones would never come into contact with the experimenter (ensuring no experimenter effects).

Turning to the remaining threats to internal validity, all of these are circumvented through the use of a *true control* group. In our example the teacher could compare her classroom to another that did not receive the learning program.

In this case the second classroom is not a true control group but rather a *quasi-control* group. The design is:

Group 1	O_1	X	O_2
Group 2	O_1		O_2

The inclusion of a quasi-control group helps to eliminate some, but not all, threats to internal validity. For instance, the impact of history and regression to the mean should be equated across the groups because each group experiences these factors equally and simultaneously. However, because there may be true differences between these groups at the onset of the study, there could be differences in selection and maturation processes across the groups. Thus, while the design is better, there are still factors unrelated to the treatment that could cause changes in the dependent variable.

Without question, the best kind of experiment is the "true experiment," which involves the random assignment of subjects to conditions, blinding techniques, and the use of a true control group to eliminate threats to internal validity. Within this broad genre there are many different designs. The simplest is the two-condition "completely randomized design" illustrated next. This is similar to the previous example except subjects are now randomly assigned to conditions as follows:

Group 1	R	O_1	X	O_2
Group 2	R	O_1		O_2

Here, the individuals are randomly assigned (R) to one of the two groups before the study begins. Recall that random assignment ensures that subjects are equated on virtually all known and unknowable historic, genetic, physical, personality, and social factors. Thus, at the onset of the experiment, the groups are equated. Over the course of the experiment, the treatment group and control group are also equated in the extent to which they are affected by history, maturation, selection, selection–maturation interaction, testing, regression, instrumentation, and mortality. In this context, if the treatment group differs from the control group, it can be reasonably assumed that the difference is caused by exposure to the independent variable. Thus, the combination of random assignment and a carefully tailored control group provides the logical basis for making causal inferences.

More complicated designs involving multiple factors are also possible. Any design with two variables can be called a two-factor design; however, not all are factorial. A true factorial design is one where each individual subject only appears once in any experimental condition. That is, subjects are randomly assigned to one of several experimental conditions defined by the confluence

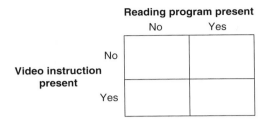

FIGURE 4   A 2 × 2 factorial design.

of two or more factors. Imagine that, in our example, students also can receive video instruction (or not) to complement their reading program. To study both factors, the teacher may use a "2 × 2" factorial design in which students are assigned to one of the four combinations. The 2 × 2 design is the most ubiquitous experimental design in the social sciences. The basic structure is shown in Figure 4.

Factorial designs offer numerous benefits. First, the design is more informative than the completely randomized design because the experimenter essentially gets two completely randomized designs in a single study. Second, the design is quite flexible. The design shown previously is an independent-groups design; that is, both factors are assignable and each individual is randomly assigned to one condition. However, one could also create a factorial design wherein one factor is not assignable (such as gender) or both factors are not assignable (such as child abuse by gender). Overall, the popularity of the design stems from its sheer strength and flexibility.

Finally, it is worth noting that although the better designs may seem costly in terms of time and money, this should not be a deterrent for those thinking about conducting experimental research. As a practical matter, experimenters often use variations of these basic designs that are more efficient and manageable. For instance, the two-group completely randomized design presented earlier involves four measurements—one for each group before and after the treatment. In practice, experimenters use a variation of this design in which the two measurements before the treatment (O_1) are omitted. The logic here is that if individuals are randomly assigned to conditions, the two observations at time 1 will almost assuredly be the same. Omitting O_1 does not alter the basic logic of the design, but it does make the design more elegant and accessible to those who have limited resources.[6]

---

[6]In fact, the very first "experiment" I ever conducted as an undergraduate student used this design to test pyramid power on the kitchen counter of my apartment.

VIII. VARIETIES OF EXPERIMENTS IN THE SOCIAL SCIENCES

Given the numerous experimental designs, it should come as no surprise that there are different kinds of experiments in the social sciences. For instance, some experiments are designed to test a theoretical prediction, and for these experiments, the measure of success is how well the experimental data conform to that prediction (Martin & Sell, 1979). Other experiments are designed to explore differences across settings and decipher empirical regularities. For these experiments, the metric of success is the ability to detect subtle differences if in fact those differences are real and sustainable. Both brands are informative, but in different ways and for different reasons. Here, I briefly consider these two kinds of experiments with specific emphasis on their purpose and design. This discussion is organized around a central idea: *regardless of the purpose of an experiment, the design of the experiment regulates its validity and overall utility.*

Theory-driven experiments. At one end of the spectrum lies experimental research devised to test scientific theory. The goal here is to compare the outcome of an experiment to a theoretical prediction. For instance, sociologists who attempt to predict social power often use theory to estimate how many points a person will earn during negotiation with another in bargaining games such as those Molm describes in Chapter 15 of this volume. Experiments are then designed that allow people to negotiate, as specified by the theory, and the outcomes of that negotiation are compared to the theoretical prediction. These kinds of experiments are critical in the process of developing and revising scientific theory because they provide a kind of "pure evidence" that either confirms or disconfirms the theory. However, it is important to note that the evidence is only "pure" to the degree that spurious factors cannot infiltrate and contaminate results. The issues of experimental design become critically important under these circumstances.

For theory-driven experiments to be useful they must be designed such that threats to internal validity are thwarted. At face value, the simplest kind of theory-testing experiment occurs when (1) a theory predicts that a given treatment will produce some effect, (2) this treatment is created in a laboratory situation, and (3) the observed outcome is compared to the prediction. Imagine that a new theory predicts that when males and females disagree, the males change their opinion exactly 42% of the time and stay with their opinion 58% of the time. Under this arrangement, we have a theoretical prediction (P), a treatment (X = mixed-gender disagreement), and an outcome (O = 42% opinion change). The design of the experiment is as follows:

$$P \qquad X \qquad O$$

Clearly, the reader should recognize that this design is simply a variant of the "one-shot post hoc" model outlined earlier and, as such, is vulnerable to all of that design's problems and pitfalls.

There are two issues that render this kind of experiment problematic. First, all of the spurious factors that threaten the internal validity associated with the one-shot post hoc design (history, maturation, etc.) also jeopardize the integrity of this design. Second, ironically, although this experiment is designed to "test theory," the outcome is equivocal with respect to theory. Assume that the experiment is conducted and the results correspond perfectly to the prediction that males change their opinion exactly 42% of the time. Does this mean the theory is confirmed? It is hard to say. Although the data confirm the prediction, the reason for this may have nothing to do with the processes asserted by the theory. It could be that males always change their opinion 42% of the time, regardless of the status of their partner. Said differently, the outcome may correspond to the prediction for reasons that have virtually nothing to do with the theory. The theory claims that X produces the outcome, but because the preceding design does not allow comparisons to situations without X, the confirmation status of the theory is obscure.

To conduct theory-driven research in its most instructive form necessitates certain kinds of controls. There are at least two varieties of control normally employed, alone or in conjunction, to sharpen the meaning of experimental data. The first and most straightforward type of control is the *control group*. Referring back to the previous example, the researcher can randomly assign some males to interact with females and some males to interact with males who are status equals. Assume the researcher finds that males interacting with females change their opinion 42% of the time as predicted, while males interacting with status equals change their opinion 53% of the time.[7] Such a finding speaks more to the veracity of the theory because it suggests that opinion change (O) is contingent on gender (X), as the theory predicts.

Second, researchers sometimes compare findings across studies as a kind of pseudo-control group in the absence of a true control group. For instance, countless studies in economics and sociology find that when two people negotiate in a dyad, neither has an advantage and they tend to profit equally. Further, this finding is predicted by numerous sociological (Emerson, 1981; Willer, 1999) and economic (Nash, 1950) theories. Given the breadth of these theories and their evidentiary basis, such findings are now taken for granted. Researchers can use these prior experimental sessions as a kind of "control" or "baseline" with which to compare new research findings. Often, this is done as a practical matter when a finding has been

---

[7]For this example, I presume the 11% difference between 42 and 53% has statistical and practical significance.

replicated to the point that yet another demonstration provides little or no new information.

Empirically driven experiments. At the other end of the spectrum are experimental studies targeting empirical problems. For instance, the most common use of experimentation in medicine, biology, agriculture, and education is to detect differences across two or more treatments. The experimental food scientist may attempt to understand how different levels of refrigeration accelerate bacterial growth on Atlantic shrimp; the experimental educator may want to understand whether computer-mediated instruction improves student retention. In both cases, the purpose of the experiment is to identify possible differences across or between treatments. These experiments require the same kinds of controls over spurious factors as do theory-driven experiments for the same reasons discussed previously, so those points will not be duplicated here. However, because the purpose of the experiment is to make inferences to a larger empirical population, the empirically oriented experimenter must confront another set of problems.

In empirically driven experiments, the researcher is almost never interested only in the properties or behavior of the specific experimental units under investigation—that is, these shrimp or these students. Rather, the goal of the experiment is to make inferences about the population from which these units originate—something akin to all shrimp stored at this temperature or all students exposed to this style of teaching. Said differently, the experiment is a device used to make population inferences from sample data. In this business, there are two kinds of mistakes that can be made, and these have different consequences for the conclusions that can be drawn.

The first mistake occurs when the researcher claims a false positive—that is, when the study finds a difference in the sample data but there is no true difference in the population. Statisticians refer to this as *Type I* statistical error. Type I error is regulated by the experimenter because the experimenter sets the alpha level (or Type I error rate) for the statistical test. In a statistical sense, this controls how frequently one rejects the null hypothesis (i.e., how often one finds differences across treatments) when that hypothesis should be accepted (i.e., there are no differences across treatments).

There are certain principles that guide how scientists set the Type I error rate. For instance, when an experiment bears on the benefits of a treatment, most empirically oriented experimenters adopt a more stringent Type I error rate such as 5% or 1%. The idea here is to be skeptical about the claimed benefit of a treatment and demand fairly rigorous proof of it. On the other hand, when the experiment bears on the negative or harmful impact of a treatment, most researchers use a more liberal Type I error rate such as the 10% level. The corresponding logic is to be more sensitive to potentially harmful effects.

The second kind of mistake is to declare a negative result when that conclusion is unwarranted. A *Type II* error occurs when one finds no difference between treatment groups but true differences exist in the population. The inverse of Type II error, *statistical power,* is the probability of detecting a difference in your sample when that difference is real in the population. Power is affected by a number of factors, including the size of the experimental groups, the Type I error rate, the variability of the phenomenon, and magnitude of the differences between treatments.[8] In many ways, the lack of statistical power (Type II error rate) is the most serious problem confronting all experimenters because, in principle, virtually any experimental treatment will have some effect on the dependent variable. That is, following a manipulation, the probability that a treatment and control group are exactly the same on the outcome measure is 0. The question is: does the experiment have adequate power for the researcher to detect that effect correctly? When designed with adequate power, empirically driven experiments are refined tools for understanding the empirical world.[9]

IX. EXTERNAL VALIDITY AND ARTIFICIALITY

Because only a relatively small number of social scientists employ experimental methods, it may be worthwhile to address some general misconceptions regarding the role of experimentation in the process of producing general knowledge. Perhaps the most misunderstood issue surrounding experimentation is that of external validity. *External validity* refers to the degree that experimental findings hold for other persons, in other locations, at other times. Recall from the introduction that Babbie (1989) and other writers of methods texts (Silverman, 1977) frequently claim that because laboratory studies are conducted in "artificial" environments, their findings do not generalize to other circumstances.

Overall, the two kinds of experiments detailed earlier (theory driven versus empirically driven) have different purposes and deal with the issue of external validity in different ways. To foreshadow, the purpose of theory-driven experiments is to provide one kind of test for a scientific theory. For these experiments, *theory,* not empirical *generalization,* connects the experiment to naturally occurring phenomena (see also Lucas, 2003; Mook, 1983; Thye, 2000; Webster & Kervin, 1971). For empirically driven experiments, *statistical inference*

---

[8]Robert Becker created the "OJ" page (http://www.socialresearchmethods.net/OJtrial/ojhome.htm), which introduces the notion of statistical power as you pass judgment on O. J. Simpson's guilt or innocence.

[9]Type I and Type II errors can also be made in theory-driven research, but with more severe consequences. Importantly, these errors will ensure that true theoretical assertions die and false ones survive.

provides the link between the experimental outcome and the larger population. Next, I address each kind of generalization.

Theory-driven experiments. The most frequent argument against the use of theory-driven experiments is that they have low external validity. A related argument is that the laboratory is an artificial environment, so it has limited utility to understand events in the real world. It is true that laboratory results cannot directly generalize beyond the laboratory. However, they can empirically document theoretical principles and rule out alternative explanations in ways that are useful for understanding aspects of "real-world" events with appropriate qualification (see Zelditch, 1969). It is the case that some who conduct experiments in the social and behavioral sciences engage in rhetorical practices that suggest their results are directly generalizable without theory. This only weakens the apparent value of experimental research in the eyes of those who employ other research methods.

Nevertheless, discounting or altogether rejecting laboratory methods because they do not directly generalize indicates a misunderstanding of the role that these kinds of experiments play in the research process. Issues of theory testing and external validity are no different for sociology than for any other field employing laboratory methods. Yet, (1) all laboratories in all sciences create artificial conditions, and (2) all sciences try to explain real-world phenomena, but (3) experimental methods are embraced proudly, unquestioningly, and highly successfully by other sciences.

Used properly, theory-driven experiments do not produce phenomena whose descriptions may be generalized to natural settings. *These kinds of experiments test theories.* Unfortunately, whenever experimentalists in psychology, social psychology, and sociology operate with less than explicit, a priori theories—which sometimes they do—their experiments disengage from their proper theory-testing function. When one is not actually testing hypotheses derived from theories, it may be tempting to infer that experiments are intended to serve as microcosms of real-world contexts, permitting the generalization of experimental findings to those natural settings. Even while some experimentalists encourage such interpretations, there is no basis for doing so (see Thye, 2000). As Webster and Kervin (1971, p. 269) explain:

> The proper use of the laboratory permits no direct generalization of laboratory results to the outside world; the only permissible connection between the two is the theory. Therefore the artificiality of the laboratory setting is an irrelevant issue when one is speaking about results in the natural environment. What is relevant is whether the natural environment ever contains instances which approximate situations described in the scope conditions of the theory.

When a laboratory experiment is used to confirm or disconfirm a theory, findings in that setting are as relevant to that theory's truth or falsity as findings gathered from any other setting. Moreover, given the degree of control over extraneous factors that the laboratory affords, the relevance of a given experimental test to a specified theory usually is even greater than a parallel test in a natural setting. Indeed, Webster and Kervin (1971) argue that what may be needed in sociology is more artificiality in the form of experimental research, for the history of science has demonstrated it to be the most efficient way to promote theoretical development. The advantage of the experiment lies, to a great extent, in its ability to filter the noise that may interfere with drawing sound inferences.

To summarize, the purpose of theory-driven experimentation is to create a testing ground that falls within the scope of the theory under consideration. Perhaps Mook (1983, p. 80) said it best when he stated: "We are not *making* generalizations, but *testing* them." Experimental conditions are only relevant as they relate to the highly stylized abstract theories they inform. In turn, those theories may be useful for understanding a range of phenomena. In physics, for example, experimentally informed theories make claims under conditions known never to occur in nature, such as "in a perfect vacuum." Yet, these theories and their experimental foundations provide key insights for resolving real-world problems. Using theory to understand various problems across diverse domains is a process of *theoretical inference.*

Empirically driven experiments. The issue of generalization is somewhat different for experiments designed to detect empirical differences. Such experiments are not guided by theory; instead, they tackle practical problems and issues (e.g., which drug reduces pain, which therapy increases flexibility, which grain is most resistant to drought, and so on). Here, it is not theoretical inference but *statistical inference* that allows one to forge connections between sample data and the larger population from which that sample came. R. A. Fisher (1935) was the first to formalize statistical models that allowed such inference.[10] These models enable one to claim, with some probability, that differences between experimental treatments are real and exist in the population. The issue of statistical generalization is no different for experimental work than for any other quantitative area. In all cases, correct inference is governed by probability theory and the statistical power associated with the hypothesis test.

Although critics assert that experimental findings do not directly generalize to other settings, only a few researchers have actually taken the time to check. For instance, Dipboye and Flanagan (1979) examined the content of empirical articles from major psychological and organizational journals over several years to see if field research is broader than the typical laboratory study. Contrary to popular belief that field research is more representative, they

---

[10]In fact, the statistical "F-test" which accompanies modern ANOVA models is named in honor of Fisher.

found studies in the field to be just as narrow as those in the laboratory in terms of the subjects, behaviors, and situations under investigation.

More directly, Locke (1980) examined research findings from industrial organizational psychology, organizational behavior, and human resource management. These are fertile testing grounds because the prevailing theories in these areas (1) are heavily informed by laboratory research, and (2) ultimately guide working conditions in businesses and organizational environments. The question is whether or not the laboratory findings reproduce themselves in the business world. It turns out they do. Time and again Locke and colleagues found remarkable consistency between the field and the lab. Locke writes, "Both college students and employees appear to respond similarly to goals, feedback, incentives, participation, and so forth, perhaps because the similarities among these subjects (such as in values) are more crucial than their differences" (1986, p. 6). Thus, despite the prevalence of the claim, the data suggest that many laboratory findings generalize to field settings.[11]

X. CONCLUSION

I began this chapter with the goal of outlining the philosophical and logical foundations of experimental methods in the social sciences. For those of us engaged in the business of designing and executing experiments, the advantages this method affords in terms of promulgating scientific theory and aiding empirical exploration are unmistakable. However, we are still a small (though growing) segment of practicing social scientists. Critics are decidedly more abundant, and vocal, as they galvanize around issues of random assignment, artificiality, external validity, and experimental utility. In some ways it is hard to conceive that the experimental strategy—the touchstone of scientific inquiry in physics, medicine, biology, chemistry, and so on—was argued to be in critical condition less than a decade ago. In truth, experimental methods in the social sciences serve the same function as they do in other scientific arenas. *Experimentation is the best known way to examine theoretical hypotheses, eliminate alternative explanations, and provide clues to causal inference.*

In closing, the experimental method is the sine qua non of scientific inquiry, spanning disciplines from particle physics to aerospace exploration and everything in between. The ubiquity of the experimental method likely stems from

---

[11]It may not be obvious, but the structure of generalization in empirically driven experiments is the same as in theory-driven experiments. In both cases, research findings from one setting must be thought about more abstractly before those findings can be mapped onto and empirically checked in another setting.

its multifaceted nature and utility. The method provides a powerful yet elegant way to (1) build, revise, and sometimes dismantle scientific theory; and (2) explore the empirical world and seek practical solutions to problems not well understood. The two processes are intertwined; theories are used to understand practical problems and such problems are the inspiration for theory development. The interactions of these two domains fuel and propel the scientific process. As long as both are present—theories to test and real-world problems to solve—the experimental method will be alive and well.

ACKNOWLEDGMENTS

I thank the National Science Foundation (SES-0216804) and the University of South Carolina for supporting my research. I thank the editors, Murray Webster and Jane Sell, for their insightful comments and suggestions. I am also grateful to Lisa Dilks, Kyle Irwin, Edward J. Lawler, Tucker McGrimmon, Barry Markovsky, Lala Steelman, and Jennifer Triplett, who helped me identify a humbling number of errors in this manuscript. Regretfully, none of us can be sure that we got them all.

REFERENCES

Aristotle. ([340 BC], 1947). *Metaphysics*. Cambridge, MA: Harvard University Press.
Babbie, E. (1989). *The practice of social research*. Belmont, CA: Wadsworth.
Bunge, M. (1979). *Causality and modern science*. New York: Dover.
Campbell, D. T., & Stanley, J. C. (1966). *Experimental and quasi-experimental designs for research*. Chicago: Rand McNally.
Cohen, B. P. (1989). *Developing sociological knowledge: Theory and method*. Chicago: Nelson–Hall.
Cohen, B. P. (1997). Beyond experimental inference: A decent burial for J. S. Mill and R. A. Fisher. In J. Szmatka, J. Skvoretz, & J. Berger (Eds.), *Status network and structure: Theory development in group processes* (pp. 71–86). Stanford, CA: Stanford University Press.
Cohen, M. R., & Nagel, E. (1934). *An introduction to logic and scientific method*. New York: Harcourt Brace and World.
Davis, J. A. (1985). *The logic of causal order*. Newbury Park, CA: Sage.
Dipboye, R. L., & Flanagan, M. F. (1979). Research settings in industrial and organizational psychology: Are findings in the field more generalizable than in the laboratory? *American Psychologist, 34,* 141–151.
Emerson, R. (1981) Social exchange theory. In M. Rosenberg and R. H. Turner (Eds.), *Social psychology: Sociological perspectives* (pp. 30–65). London: Transaction.
Fisher, R. A. (1935). *The design of experiments*. London: Oliver and Boyd.
Fisher, R. A. (1956). *Statistical methods and scientific inference*. Edinburgh: Oliver and Boyd.
Galilei, G. ([1636], 1954). *Dialogues concerning two new sciences*. New York: Dover.
Hage, J., & Meeker, B. F. (1988). *Social causality*. Boston: Unwin Hynman.
Harré, R., & Secord, P. F. (1972). *The explanation of social behavior*. Oxford, UK: Oxford University Press.
Hempel, C. (1965). *Aspects of scientific explanation*. New York: Free Press.
Hume, D. ([1748], 1955). *An inquiry concerning human understanding*. New York: Bobbs–Merrill.
Lieberson, S. (1985). *Making it count: The improvement of social theory and research*. Berkeley: University of California Press.

Lieberson, S. (1991). Small N's and big conclusions: An examination of the reasoning in comparative studies based on a small number of cases. *Social Forces, 70,* 307–20.

Locke, E. A. (Ed.). (1986). *Generalizing from laboratory to field: Ecological validity or abstraction of essential elements, in generalizing from laboratory to field settings* (pp. 3–9). Lexington, MA: D. C. Health and Company.

Lucas, J. (2003). Theory testing, generalization, and the problem of external validity. *Sociological Theory, 21,* 236–253.

Martin, M., & Sell, J. (1979). The role of the experiment in the social sciences. *The Sociological Quarterly, 20,* 581–590.

Mill, J. S. ([1872], 1973). *A system of logic ratiocinactive and inductive.* Toronto: University of Toronto Press.

Molm, L. D. (2007). Experiments on exchange relations and exchange networks in sociology. In M. Webster & J. Sell (Eds.), *Laboratory experiments in the social sciences* (pp. 379–406). Burlington, MA: Elsevier.

Mook, D. (1983). In defense of external invalidity. *American Psychologist, 38,* 379–387.

Nash, J. (1950). Equilibrium points in n-person games. *Proceedings of the National Academy of Science, USA, 36,* 48–49.

Paulos, J. (1988). *Innumeracy: Mathematical illiteracy and its consequences.* New York: Vintage Books.

Rosenthal, R., & Fode, K. L. (1963). The effect of experimenter bias on the performance of the albino rat. *Behavioral Science, 8,* 183–189.

Rosenthal, R., & Rubin, D. B. (1978). Interpersonal expectancy effects: The first 345 studies. *Behavioral and Brain Sciences, 3,* 377–86.

Russell, B. (1913). On the notion of cause. *Proceedings of the Aristotelian Society, 13,* 1–26.

Silverman, I. (1977). *The human subject in the psychological laboratory.* New York: Pergamon Press.

Thye, S. (2000). Reliability in experimental sociology. *Social Forces, 74,* 1277–1309.

Waldrop, M. M. (1992). *Complexity: The emerging science at the edge of order and chaos.* New York: Simon & Schuster.

Walker, H. A., & Willer, D. (2007). Experiments and the science of sociology. In M. Webster & J. Sell (Eds.), *Laboratory experiments in the social sciences* (pp. 25–55). Burlington, MA: Elsevier.

Webster, M. (1994). Experimental methods. In M. Foschi & E. Lawler (Eds.), *Group processes: sociological analyses* (pp. 43–69). Chicago: Nelson–Hall.

Webster, M., & Kervin, J. (1971). Artificiality in experimental sociology. *Canadian Review of Sociology and Anthropology, 8,* 263–72.

Willer, D. (1999). *Network exchange theory.* Westport, CT: Praeger.

Willer, D., & Walker, H. A. (2007). *Building experiments: Testing social theory.* Palo Alto, CA: Stanford Press.

Zelditch, M., Jr. (1969). Can you really study and army in the laboratory? In A. Etzioni (Ed.), *A sociological reader on complex organizations* (pp. 531–39). New York: Holt, Rinehart & Winston.

CHAPTER 4

The External Validity of Experiments That Test Theories

MORRIS ZELDITCH, JR.
Stanford University

ABSTRACT

External Validity refers to the validity of generalization from experiments. There is a problem with saying exactly what that means. The purpose of the present chapter is to explicate the concept of external validity (i.e., to say precisely what is generalized from what, to what, and under what conditions). The remainder of the paper studies what the explication implies for the design of experiments: sources of external invalidity, how to detect them, how to test for them, and how to remedy them.

I. PROBLEM

Because they randomize the allocation of subjects (Ss) to treatments, experiments solve the problem of internal validity better than any other method. But there are many features of an experiment that it does not randomize. Although the allocation of Ss to treatments is randomized, the particular treatments (x), the particular measures of their effects (y), the particular population of Ss, the particular period during which it is run, and the particular setting of the experiment—everything from its task and interaction conditions to the physical and social space in which it is carried out—are fixed. If any of these features of the experiment interacts with the process that it is the purpose of the experiment to investigate, the inference that x causes y may be internally valid but it may not be valid to generalize it. Because there is no way to randomize a constant, randomization does nothing to solve the problem of external validity.

But what exactly *is* the problem of external validity? As soon as we ask exactly what it is, we run into difficulties. Campbell and Stanley (1963), the earliest systematic investigators of it, say that

Criterion [1][1]: External validity asks the question of *generalizability* (italics in the original): To what population, settings, treatment variables, and measurement variables can this effect be generalized? (Campbell & Stanley, 1963, p. 5).

By this they sometimes meant generalization from a particular sample of Ss to the population from which the Ss were drawn, for example, from college students in an experiment to college students who were in the same pool of Ss but not in the experiment; sometimes from a particular population, for example, students to other populations, for example, nonstudents; sometimes to all populations, to all persons, periods, or settings. But note that these examples are all concrete entities: students, nonstudents, persons, periods, settings. Only in exceptional cases did they think in terms of variables. In the case of treatments and, sometimes, measures, they did define the problem of external validity as the problem of generalizing to other, nonidentical representations of the same treatment or measure, implying variables. But even in the case of treatments and measures, their examples were often almost as concrete as population, period, and setting. For example, two teachers using the same teaching method were nonidentical treatments (Campbell & Stanley, 1963, p. 32).

But there is no way one can generalize from concrete treatments, measures, populations, periods, or settings to any other concrete treatment, measure, population, period, or setting. Every concrete entity is unique; everything about it is relevant, and there are an infinite number of things about it. Absent variables, there is no generalization (cf. Blalock, 1968, p. 181). Taken literally, therefore, [1] is self-contradictory. No concrete experiment is generalizable. But, no matter how internally valid it is, there is no point to an experiment

---

[1]For ease of reference (there will be more than one criterion proposed before we are through), I will number each proposed criterion and enclose its number in brackets.

if it cannot be generalized. Generalization is its whole purpose (cf. Berger, Zelditch, & Anderson, 1972). A concept of external validity that contradicts the whole purpose of experiments is profoundly unsatisfactory.

Campbell and Stanley did make a well-argued point that not everything is relevant to everything; otherwise no generalization would be possible (pp. 17–18). For example, they were willing to assume that orientation of Ss in the magnetic field of the laboratory is irrelevant to teaching experiments (p. 17). But that merely makes a point of the fact that the problem of external validity is, in the first instance, a problem of what is or is not relevant. Equivocating between analytic and concrete concepts of external validity, Campbell and Stanley often had wise things to say about it, but even if [1] is not taken literally, it still provides no guide to what is or is not relevant. In their more analytic mode, they conceded that identifying the potential sources of external invalidity, if it is defined by [1], is therefore largely a matter of guesswork (p. 17). Thus, [1] not only is self-contradictory, but also leaves external validity indeterminate: it provides no guide to what can be generalized, under what conditions, to what. The names of concrete things—teacher, student, classroom—offer no answers.

The indeterminacy of [1] is not as profoundly unsatisfactory as its self-contradiction, but it is still unsatisfactory. Because it is the whole purpose of experiments to generalize, an experiment that is externally invalid is fatally flawed. It is obviously desirable to know precisely what about it can be generalized, under precisely what conditions, to precisely what.

At first sight, the obvious guide is theory.[2] That is,

Criterion [2]: What one generalizes is theory; what one generalizes to is any instance that satisfies the theory; and the conditions under which an experiment generalizes have to do with whether in fact both the experiment and the situation to which it is generalized satisfy the instantiation and scope conditions of the theory (Lucas, 2003; Zelditch, 1969).

But [2] opens Pandora's box. Certainly the problem posed by [1] lies in its concreteness. Certainly the solution to it lies in a more abstract approach. And

[2]Sufficiently obvious, in fact, that Campbell (1986) argued that all external validity (not only of treatment and measurement variables, but also of persons, periods, and settings) should be reconceptualized as the validity of the generalization of a theory (p. 71). Nevertheless, he continued to equivocate, in the end defining external validity as the proximal (i.e., detailed) concrete similarity of experimental to natural setting (p. 75). Campbell had credited Cook with influencing him to take a more abstract approach, but the five principles of external validity offered by Cook (1990) were equally equivocal. Shadish, Cook, and Campbell (2002) even more fully reconceptualized persons, periods, and settings in terms of the variables that describe them, but again the definition of external validity was equivocal, as was much of the rest of the literature. Influential papers by Messick (1989) and Brewer (2000) were thoughtful and systematic, but, perhaps because they were also comprehensive (hence eclectic), just as equivocal. For two exceptions, see John and Benet-Martinez (2000) and Haslam and McGarty (2004).

certainly [2] does not contradict itself. Furthermore, given a theory sufficiently developed that it is possible to state its instantiation and scope conditions, [2] is determinate. Its only problem is that it vastly overreaches. If the theory supported by an experiment did not generalize to a situation to which it was applicable, [2] would count as external invalidity: not only bad design, but also bad theory, bad measurement, and bad application. (See [5] for more precise specification of [2] and the paragraphs that follow it for a more detailed critique of it.)

The objectives of the present chapter are, therefore, first, to explicate the concept of external validity more precisely (i.e., to say more precisely what is generalized, from what, to what, under what conditions) and, second, to ponder what the answers imply for the design of experiments (i.e., for the sources of external invalidity) and how to detect them, test for them, and remedy them.

II. WHAT IS EXTERNAL VALIDITY?

If [1] is both self-contradictory and indeterminate, it might seem natural to propose instead that

Criterion [3]: What one generalizes from are the variables x' and y' in the experiment; what one generalizes to is the variables x^* and y^* in some natural setting, s^*; and what one has generalized is externally valid if and only if, given the same initial conditions, the effect of x^* on y^* in s^* is the same as the effect of x' and y' in the experimental setting.

This proposal does have an important consequence that certainly is valid: one never generalizes the effect of an experiment absent its conditions. For example, one cannot and should not generalize from the Asch (1951) experiment that a third of behavior or three-quarters of the population is conformist.[3] Nevertheless, the proposal encounters two immediate difficulties. Admittedly, one is easily remedied. "The same initial conditions" is open to the

---

[3]In Asch's experimental situation, a naïve S must make a sequence of choices between correct and incorrect stimuli in the face of unanimous and incorrect responses by seven of S's peers. The peers are confederates of the experimenter. An experimental trial requires S to match a standard line with the one of three comparison lines that is the same length as the standard. S is instructed to give his or her response orally and the oral response of six of the seven confederates precedes S's response. The instructions lead S to believe that he or she is participating in an experiment in visual perception so that there is motivation for responding correctly. On the other hand, the unanimity of the confederates in responding incorrectly exerts pressure on S to conform. The typical experimental procedure consists of a sequence of trials on each of which the S may respond correctly or conform to the group of confederates. Individuals in a control group where there are no instructed role players judge the lines with almost complete accuracy. A unanimous majority deflects a third of Ss' choices in its direction and 75% of Ss make at least one error in the presence of the majority (Asch, 1951).

interpretation that external validity requires maximal similarity of the experimental to the natural setting (e.g., Campbell & Stanley, 1963, p. 18). "Maximal similarity" refers to variables, not names, but it is no better than names at distinguishing what is relevant from what is not (Locke, 1986, chap. 1).

But even if one requires only relevant similarities, there is still a difficulty if, as is the case in many social psychological experiments (cf. Blalock, 1989, p. 456), the treatments, measures, and conditions of the experiment are described in the preanalytic, common-sense categories of natural English—for example, the unanimous majority, incorrect responses, or voluntary Ss in the Asch experiment (see note 3), all of which are, if not concrete, then nearly so. Natural language tends to package variables in complex clusters that include many causally irrelevant specificities of the particular situation they describe. For example, the near-concreteness of "conformity" in the Asch experiment has left open what exactly conformity *is* in the experiment (Deutsch & Gerard, 1955). Such experiments typically give rise to "effect" programs in which subsequent experiments refine the causes and conditions of an effect (cf. Asch, 1951 and 1956; also Cohen, 1963, and Sell, Chapter 18, this volume).

Even if the causes and conditions are precisely specified, there is still a difficulty: the more similar the conditions of a concrete natural setting are to those of a concrete experimental setting, the greater is the external validity of the experiment but the less its explanatory power. If the initial conditions of the experiment are commonly found in natural settings, why exactly did one do the experiment? It would seem to vitiate the whole purpose of experiments. But if the initial conditions of the experiment are not commonly found in natural settings, then generalization to just those natural settings that do in fact satisfy them severely limits the explanatory power of the experiment.

An obvious solution would be to elaborate the proposed criterion a little by raising its level of abstraction even further. It is not x' or y' that we care about. What we care about is the latent, unobserved, unmeasured variables X and Y, of which x' and y' are merely particular observable indications (cf. Blalock, 1971; Costner, 1971). Nor do we particularly care about the particular factors, say z', that condition the effect of x' on y' in the experiment. What we care about is the latent, unobserved, unmeasured Z of which z' is a particular observable indication. Furthermore, there is no reason why x', y', or z' in an experiment must be identical to x^*, y^*, or z^* in a natural setting to which x', y', and z' are generalized, providing each reflects X, Y, and Z. This holds not only for different values of a particular instance of X, Y, or Z—for example, levels of education different from the levels of it in a particular status characteristics experiment—but also even for concretely different instances of X, Y, and Z—for example, other status characteristics, such as race, ethnicity, gender, or occupation.

Thus, one might well propose:

Criterion [4]: What one generalizes from are the variables x' and y' in the experiment. What one generalizes to is any x^* and y^* that are instances of the same unobserved, unmeasured variables X and Y in some natural setting. What one has generalized is externally valid if, and only if: given that z^* in the natural setting is an instance of the same conditions Z as z' is in the experimental setting, the effect of x^* on y^* is the same in the natural setting as the effect of x' on y' in the experimental setting.

For example, Moore (1968) investigated the effect of the educational level of a partner (high school, junior college, university) on the partner's rate of influence. The participants were female junior college Ss performing a team-oriented spatial judgment task, where the partner's rate of influence was measured by the probability that, if they disagreed, S resisted accepting the partner's initial choice of a response in making her own final choice. Moore found that Ss whose educational level was higher than their partners' had a greater probability of resisting influence than Ss whose educational level was lower than their partners'.[4] This result should extend not only to other levels of education but also to other status characteristics, such as race, ethnicity, gender, or occupation; to other team-oriented tasks; and to other measures of influence.

Whether, for example, a characteristic *is* a status characteristic or not is an empirical question and the answer must be independent of any test of the hypothesis that, given a collective task, if it *is* a status characteristic, it determines rates of influence. But if, say, occupation *is* a status characteristic, then the effect found in Moore (1968) should generalize to the effect of occupation in Strodtbeck's jury studies (Strodtbeck, James, & Hawkins, 1957), despite the fact that in those jury studies, the task was to decide compensation unanimously for an injury. This was a team-oriented task, but one that in no way involved spatial judgment; the measure of influence was choice of a foreman,

---

[4]Moore (1968) led pairs of junior college females to believe they were working as a team with either a pupil at a nearby high school or a student at a nearby university, whom they could not actually see because they were separated by an opaque partition. Their task, said to involve "spatial judgment ability," was to decide whether each rectangle in a sequence of rectangles, each made up of 100 smaller black and white rectangles, was more black or white—a task that was made ambiguous by the fact that each rectangle, although configured differently, was almost exactly half black. Each decision was made in three stages, each S making an initial choice, exchanging choices, and then making a final choice. The exchange was made by means of circuitry controlled by the experimenter in such a way that initial choices disagreed on 28 out of 44 trials. Because the choice was binary, a final choice was either a change-response (indicating that S was influenced by the other) or a stay-response (indicating that influence by the other was resisted). How successful a team was at the task was defined by a team score, with no record kept of the relative contributions of the partners. The result was that, if S was team oriented, the probability that S made a stay-response was inversely proportional to the rank of S's partner.

which is informational social influence (cf. Deutsch & Gerard, 1955), but not the probability of resisting it.

But [4] actually leaves us only a little better off than we were before. A more abstract criterion of external validity does have an important consequence that is certainly valid: in order to determine the external validity of an experiment, it must be established that one is generalizing from and to concrete instances of the same abstract variables. Furthermore, the test of the experiment's instantiation of the abstract variables X, Y, and Z must be independent of the test of the experiment's hypothesis. But however high the criterion climbs the ladder of abstraction, the paradox of similarity still afflicts it. If the initial conditions of the experiment are conditions commonly found in natural settings, the experiment vitiates the whole purpose of experiments. If they are not, the explanatory power of the experiment is severely limited.

Part of the trouble lies in too narrow a view of it. Criteria [1], [3], and [4] all assume that a single experiment is directly extrapolated to some natural setting. If the purpose of the experiment is to test a theory, there is no reason to limit generalization of the theory to generalization of the particular experiment. Not all experiments test theories. But if they do, the hypothesis tested by any one experiment is typically embedded in a larger theoretical structure. The more complex the theoretical structure is, the less likely it is that any one experiment tests the whole of it.

Theoretical research—research oriented to testing, refining, and extending theories—typically simplifies before it complicates (cf. Berger, Wagner, & Zelditch, 1985, pp. 40–43; Berger et al., 1972; Meeker & Leik, 1995; Zelditch, 1969, 1974). It separates the effects of colinear factors—for example, studying the effect of a single status characteristic before studying multiple characteristics. It separates actors—for example, manipulating the behavior of one to study the behavior of a second, while controlling the behavior of any others, before studying open interaction. It separates the links in a chain of effects—for example, separating how performance expectations emerge from how they affect influence once they have emerged. It separates moderating conditions, for example studying legitimation separately from sentiment structures, either of which modulates the effect of performance expectations on resistance to influence. (For these examples, and others, see Wagner & Berger, 2002. But similar examples are found in almost any other theory program. For anthologies of other programs, see Berger & Zelditch, 1993, 2002.)

Without simplification it would be difficult to understand any complex process, but almost no natural situation satisfies the fixed initial conditions of any one experiment in a program. But if not, one almost never directly extrapolates from any one experimental test of a theory to any natural situation. What one extrapolates is its consequence for a theory. Its relation to any natural situation is indirect, linked to it by the theory. Hence, if an experiment in a setting s' tests a theory[5]:

Criterion [5]: What is generalized is the theory; what it is generalized to is any situation to which the theory is applicable (i.e., any situation that satisfies the instantiation and scope conditions of the theory); and an experimental setting is externally valid if application of the theory predicts or explains the behavior of the process the theory describes in any situation to which the theory is applicable (Lucas, 2003; Zelditch, 1969).

But, almost immediately, new difficulties appear: criterion [5] is much too weak. Certainly an experiment is externally valid if the theory supported by the experiment predicts and explains the behavior of the process described by the theory in situations to which the theory applies. Furthermore, the criterion does have an important consequence that certainly is valid: it makes no sense to generalize an experiment to a situation that does not satisfy the instantiation and scope conditions of the theory it tests. But the criterion counts as externally invalid much that is invalid for reasons that have nothing to do with the experiment.

For example, bad theory: Bales thought he had discovered role differentiation because the emergence of task leadership did not correlate with the emergence of socioemotional leadership in initially undifferentiated task-oriented groups (Bales & Slater, 1955). But role differentiation did not generalize to heterogeneous groups (Lewis, 1972). It is a by-product of the legitimacy problems of the emergence of inequality in homogeneous groups (Burke, 1967). But is that external invalidity or just bad theory?

But [5] conflates external invalidity with bad measurement as well as bad theory. For example, Deutsch and Gerard (1955) argued that the Asch experiment (note 3) confounded informational social influence—the use of others as evidence of what there is—with "normative social influence"—the effect of what others expect of S. Taking errors in the direction of the majority as an indicator of conformity to the group conflates the two in one observable indicator. An observable indicator that points to two distinct underlying concepts is normally taken to be bad measurement (e.g., Blalock, 1968) because the two dimensions (by definition) behave differently. Deutsch and Gerard showed that the Asch effect could be largely accounted for by informational social influence. That mattered because, for example, whether Ss were a group or not, or anonymous or not increased the effect of normative but not informational social

---

[5]Unlike [1], [3], or [4], [5] applies only to experiments that test, refine, or extend theories or test their applications. Also note that, because such experiments often give rise to programs, [5] refers to the experimental setting s' rather than any particular experiment in it. It is the setting that matters, not any one experiment, because a moderating condition (say z') may be constant in any one experiment but vary across experiments. It makes no sense to count omission of the interaction $x'z'$ as a methodological flaw of either a particular experiment or its setting if z' varies across experiments in the same setting.

influence. All this makes generalization of the Asch effect hazardous. But should such bad measurement be counted as external invalidity?

The proposed criterion not only conflates external invalidity with bad theory and bad measurement, but also conflates it with bad application. This follows from the fact that the proposed criterion has a second important consequence that is also certainly valid: if generalization of an experiment depends on a theory's applicability, then it depends as much on research that tests its applicability to the situation to which it is generalized as it does on the experiment that tests the theory (Zelditch, 1969). "Applied" research tests the hypothesis that the situation to which a theory is applied is an instance of the theory and satisfies its scope conditions. It is independent of the body of theoretical research that tests the theory. It also typically complicates what theoretical research simplifies. Theoretical research isolates unit processes and controls extraneities (e.g., friction, air resistance). But almost no natural situation is unitary; a unit process is likely to be conjoined with other processes in natural situations, though particular conjunctures of processes often differ from one situation to another (otherwise, they would be one unit process). Nor are many natural instances free of extraneities, though particular extraneities often differ from one situation to another.

Like instantiation and scope, therefore, the conjunctures and extraneities of a particular situation have to be determined for each application of a theory. Application of a theory will therefore entail empirical models of particular natural situations (Berger & Zelditch, 1997) that not only instantiate it and define its scope conditions but also specify the particular initial conditions of the particular situation to which it is applied and how the theory relates to other processes and extraneities in the situation. (Application of a particular hypothesis implied by a theory of a particular process—as opposed to predicting or explaining the joint effects of two or more interrelated processes—may require no more than knowing how to control for the extraneities and other processes in the natural situation, but it does require at least the capacity to control for them.)

But suppose a bad application: for example, in equity experiments, Ss who are unfairly under-rewarded, when given the opportunity to reallocate rewards, redress inequity, as do Ss who are unfairly over-rewarded (Leventhal, Allen, & Kemelgor, 1969; Leventhal, Weiss, & Long, 1969; Messick & Sentis, 1979; see also Hegtvedt & Markovsky, 1995). Neither is found in applications of equity theory to natural situations (Bachrach & Baratz, 1970; Moore, 1978). But the problem is not the external invalidity of equity experiments. The problem is that in many concrete natural situations (communities in Bachrach and Baratz's case; societies in Moore's), processes that distribute rewards are embedded in processes that distribute resources. Resources give rise to power as well as rewards. Power not only offers fewer opportunities to redress inequity, but also, because it has other sources of legitimacy, dampens actual pressures to redress it (Zelditch, Harris, Thomas, & Walker, 1983). But is that external invalidity or just a bad application of the theory?

Thus, the proposed criterion counts as external invalidity much that is not intuitively meant by it. Here, I think Campbell and Stanley (1963) had it right: intuitively, external invalidity is flawed design, not flawed theory, measurement, or application. Furthermore, the *kind* of flawed design is reasonably well understood.

If T is a theory of a process **P** occurring in a class of situations {s} and m' is an empirical model of T that instantiates it in a particular instance (say, an experimental setting s'), then, intuitively, the external invalidity of an experiment designed to test m' in s' is, in the first instance, a specificity introduced into s' by its method, such as a reactive pretest (Solomon, 1949) or a demand characteristic (Orne, 1962), that makes the behavior of T(**P**) in s' differ from the behavior of T(**P**) in {s}. Furthermore, whatever the source of the irrelevant specificity, it is something that is a constant in s' because it would otherwise have been randomized and therefore uncorrelated with x'. Finally, whatever the constant is, it is something that interacts with x', the experimental treatment in s' (cf. Campbell & Stanley, 1963). Thus, as in [5], if an experiment tests, refines, extends, or tests the application of a theory,

Criterion [6.1]: What is generalized is a theory; what it is generalized to is any situation to which the theory is applicable; and an experimental setting is externally valid if application of the theory predicts or explains the behavior of the process the theory describes in any situation to which the theory is applicable.

However, unlike [5],

Criterion [6.2]: If application of the theory does not predict or explain the behavior of the process that the theory describes in a situation to which it is applicable, then an experimental setting is externally valid only if the theory tested in it is applicable to the experimental setting and no constant specific to its methods interacts with its treatments.

Put another way, an experimental setting is externally valid if it is an instance of the theory it tests and neither adds nor subtracts anything from the theory.

III. IMPLICATIONS

The fundamental test of the external validity of an experiment lies in the application of the theory it tests. But if the theory does not predict and explain in a situation to which it is applicable, the fault may lie not in the experiment, but rather in the theory, the model of it tested in the experiment, the model of it applied in the particular instance, or the applied research testing the applicability of the theory to that instance. The fundamental test is necessary but not sufficient. Theory growth depends on ruling out alternative explanations of the failure of the theory to apply to a particular situation, say s^*. To rule out the external invalidity of s', [6.2] implies that two conditions are sufficient:

Criterion [7]: If a theory is applicable to s^* but application to s^* fails to predict or explain the behavior of process p^* in it, then the experimental setting s' in which the theory was tested is externally valid only if

(1) the process p' in s' satisfies the instantiation and scope conditions of $T(P\{s\})$, and

(2) if m' is a model of $T(P\{s\})$ in s', then,

1. No constant z' that interacts with p' is introduced by its methods into s' that is omitted from m'.
2. No constant z' that interacts with p' in m' is omitted by its methods from s'.

Criteria [6.1] and [6.2] also imply that applied research in s^* is as important to the external validity of s' as theoretical research in s' and they have something to say about the methods of this research. Applied research—say, an experiment that tests the applicability of $T(P\{s\})$ to s^*—has its own extraneities, independent of any introduced by s'. But [6.1] and [6.2] imply that what it has to say about extraneities in s' is also true of the methods of applied research in s^*.

In the rest of this chapter, I simply elaborate these implications.

A. Threats to the External Validity of Experiments That Test Theories

Criteria [6.1], [6.2], and [7] refine and extend the sources of threats to external validity. They refine them because [6.1], [6.2], and [7] exclude concrete ways of thinking about them. Both the purpose and nature of experiments that test theories are the sources of many concrete differences between s' and s^*. Their purpose is to test theories under controlled conditions. Because this is their purpose, s' analytically simplifies s^* (Zelditch, 1969; Walker & Willer, Chapter 2, this volume). Many concrete features of s^* are irrelevant to the theory tested in s', which incorporates only the features relevant to the theory it tests. Because the purpose is to manipulate and control the conditions under which they are tested and to measure their effects precisely, they are also, by definition, artifice—the source of even more concrete differences between s' and s^*.

Thus, a test of expectation states theory constructs a fictitious ability ("meaning insight ability") in order to control the effect of any other characteristics, such as race and gender, on informational social influence; it constructs fictitious interaction conditions, controlled by an "interaction control machine," in order to control the number of disagreements in the interaction process (hence the number of influence attempts). It constructs a fictitious task that repeatedly requires a choice between two and only two alternatives;

because it is a binary choice, it precisely measures the acceptance or rejection of influence. All these features are different from any concrete instance of the effect of status on informational social influence. Furthermore, the experiment continually repeats a condition—precisely the same performance by actors who differ in status—that is rarely seen in any natural setting. Finally, because the experiment isolates the status process from any other, s' taken as a whole omits many features of most concrete instances, such as s^* taken as a whole.[6]

But the many concrete dissimilarities between s' and s^* matter only if they are relevant to the theory being tested by the experiment. Not all features of s^* matter to the influence process; what matters to s' is its theoretical, not its maximal, similarity to s^*. Naturalism of the features of s', their surface similarity to the variables in s^*, also does not matter; what matters is only their psychological realism—that the variables created by E in s', whether fictitious or not, have the same meaning to S in s' that they would have were S to experience their naturalistic counterparts in s^*. If s' is made up of all and only the variables relevant to the theory it tests, and all have the same meaning in s' that

---

[6]Berger and Conner (1969) led pairs of 162 university students who volunteered from various university classes to believe they were working as a team with a partner who was less competent than, as competent as, or more competent than they were at a purely fictitious task requiring a purely fictitious ability. The experiment had two parts, Phases 1 and 2. In Phase 1, the two Ss were publicly given fictitious scores on a test that purported to measure their ability at the Phase 2 task. The task in Phase 1 consisted of repeated trials presenting sets of three words, one in English and phonetic spellings of two in a fictitious language; the instructions informed Ss that one of the two non-English words had the same meaning as the word in English. They were told that by comparing the sounds of the non-English words they could decide which meant the same as the English word. The ability to do this was called "meaning insight ability." Scores were interpreted to them as either exceptionally superior or exceptionally poor. The conditions created by the test in Phase 1 were (1) high self, low other [+ −]; (2) high self, high other [+ +]; (3) low self, low other [− −]; or (4) low self, high other [− +]. The task in Phase 2 also consisted of repeated trials presenting sets of three words, except that one was in the same fictitious language as that in Phase 1 and two were in English. The task was to decide which of the two English words had the same meaning as the non-English word. But in Phase 2, selection of a correct answer had three stages. Every time an S was presented with a set of alternatives, S first made a preliminary selection, exchanged information with his or her partner as to which alternative each initially had selected, and then made a final choice. The Ss could not verbally communicate or even see each other; they indicated their choices to E and to each other using a system of lights and push-button switches. Except for 3 of a total of 25 trials, Ss were led to believe that their initial choices disagreed. The final choice was private. The purpose of exchanging information was defined as seeing how well they worked together as a team. They were told, moreover, that their final decision would be evaluated in terms only of a team score, which was simply the sum of the number of correct final choices each made and would not record their relative contributions to the score. Because the choices were binary, the final choice in the 22 disagreement trials indicated either acceptance of or resistance to the influence of the other. The result was that the probability of acceptance of the influence of the other was greatest in Condition (4) (i.e., in the [− +] condition), least in Condition (1) (the [+ −] condition), and about equal in Conditions 2 and 3 (the [+ +] and [− −] conditions).

they have in s^* (however concretely different they are), then s' has "experimental realism" (Berkowitz & Donnerstein, 1982) even if it does not have "mundane realism" (Aronson & Carlsmith, 1968). It is experimental, not mundane, realism that experiments that test theories require.

Furthermore, that the values the variables take in s' are sometimes rare and therefore statistically unrepresentative of the values they take in s^* also does not matter. What matters is theoretical, not statistical, generalization. The universe of theoretical generalization is not the universe from which its population, period, or setting is drawn; it is a population of instances that satisfy the instantiation and scope conditions of the theory that it tests. Representative design (Brunswick, 1955) serves no purpose at all in an experiment that tests theory because it is not the purpose of an experiment that tests a theory to describe the initial conditions likely to be found in instances to which the theory is to be applied.

If neither concrete similarities nor dissimilarities matter to tests of a theory, they do not matter to external validity. They neither promise it nor threaten it. This matters in two ways: Requiring concrete similarity does not guarantee it and concrete dissimilarity does not necessarily threaten it. Also, expressing threats in terms of concrete entities, like population, period, and setting, rather than abstract variables does not allow one to say what is or is not relevant. Thus, [6.1], [6.2], and [7] exclude any form of misplaced concreteness from any list of threats to external validity.

But [6.1], [6.2], and [7] extend as well as refine sources of threat to the external validity of experiments that test theories. This is partly because [7] notices some sources that typically go unnoticed and partly because it unites some sources that, though noticed, tend to be treated in separate literatures. Criterion [7] implies that the chief threats to the external validity of experiments that test theories are errors of instantiation or specification. Most of the literature, following Campbell and Stanley (1963), focuses only on specification error, neglecting instantiation error. Criterion [7] also implies that specification error is a matter both of what E did do in s' that introduced interactions into it that were not in m' and what E did not do in s' that omitted interactions from it that were in m'. Again, like Campbell and Stanley, most of the literature that focuses on specification error focuses only on errors of commission, neglecting errors of omission. But let me start where most of the literature starts.

1. Specification Errors I: Introducing Interactions Not in E's Test Model

Specification error refers to errors in the specification of the causal model m' that justifies causal inference from an experiment in s'. While experimentalists have typically been hostile to causal models, Blalock (1971), Costner (1971), and Smith (1990) have shown that no experiment that generalizes is model free, and causal models of experiments are as prone to specification error as causal models of nonexperiments. They are not model free because x', y', and z' in s' presuppose latent, unmeasured, unobserved constructs X, Y, Z or they do not generalize. If they presuppose X, Y, Z, causal inference from observation of x', y', z' depends on predetermined assumptions about causal order and the relations between measured and unmeasured variables, unmeasured variables included in the model and unmeasured variables omitted from it, and among the unmeasured, omitted variables.

It is causal models that link theories to experiments. But they are as prone to error as causal models of nonexperiments because, while randomization may be a panacea for omitted disturbances, it is no panacea for confounds of x' introduced by the methods of the experiment itself (Blalock, 1971; Costner, 1971) or for omitted interactions (Smith, 1990). Furthermore, despite their hostility to causal models, most of what experimentalists have had to say about external validity has been, except for its language, about specification errors of one kind or another.

When I say they have been errors of commission, I mean that they are interactions not in m' that are introduced into s' by its methods. Virtually all the threats to external validity in Campbell and Stanley (1963) were errors of this kind. This kind of errors includes, for example, the effects of: the diffusion of information among Ss between treatments; treatment complexity on treatment integrity; measurement complexity on construct validity; the voluntarism of Ss; the reactivity of before-tests of y'; and the procedures of the experiment (signaling E's intentions) and the experiment itself (introducing Hawthorne effects). All these effects introduce irrelevant specificities into s'. Much of the subsequent literature, especially on the social psychology of the experiment, has also dealt with interactions present in s' but absent from m'. They have included, for example, the effects of: E's expectancies, hypotheses, and status characteristics; S's own expectancies about E's purposes, the processes being investigated by the experiment, about science, and about experiments; evaluation apprehension by S and of the social desirability cues about which S is apprehensive; and the demand characteristics inferred from the procedures of the experiment.[7]

Artifice is one thing, artifact another. The idea of an experiment is to isolate a process from the many extraneities that obscure it in natural settings— not to introduce extraneities of its own, especially extraneities that impair its generalizability.

---

[7]See especially Alexander and Knight (1971); Foschi (2006); Gustavson and Orne (1965); Lana (1969); Orne (1962, 1969); Orne and Evans (1965); Orne and Scheibe (1964); Rosenberg (1965, 1969); Rosenthal (1963, 1966, 1969, 1976); Silverman and Shulman (1970); and Weber and Cook (1972).

2. Specification Errors II: Omitting Interactions That Are in E's Test Model

Not all specifications errors are errors of commission. Some are errors of omission. By that, I mean omitting an interaction present in m' but not, because of its methods,[8] realized in s'[9]: ineffective manipulation of x'; range restriction of its manipulation (a problem if it is nonlinear); ineffective manipulation of z' (especially a problem if z' conditions the scope of m', e.g., Asch's failure to create a collectively oriented group task); range restriction of the manipulation of z'—for example, homogeneity of Ss not only within one experiment but also across all experiments in s' (a problem if a status characteristic of S, e.g. gender, is a moderating condition in m'); or restriction of the time horizon of an experiment (a problem if the time horizon of the Ss is a moderating condition in m').

What E does not do in s' can be as much a threat to the external validity of s' as what E does do. Most errors of omission go unnoticed in the literature on external validity, although a few are mentioned by Campbell and Stanley (1963); indeed, a few, such as ineffective manipulations, are on every experimenter's mind while pretesting an experiment. They are noticed more—in fact, oversold—in the extensive literature objecting to experiments in the social sciences, especially the restricted range of variables such as the size of the group, its history, its social relations, its incentives, and the time horizon of the Ss in and the duration of the experiment. Often the objections are to factors that in fact are irrelevant because they are not variables in m' or, if they are, are present in other forms in s' (Zelditch, 1969; Zelditch & Hopkins, 1961). But the defense against them can also be oversold. What you did not do in s' can be as fatal a threat to its external validity as what you did do.

3. Errors of Instantiation

Criterion [7] implies that instantiation error is one of the chief threats to the external validity of s'. Instantiation error has often been at issue, but discussion of it tends to be found in a running debate, largely in a literature of its own, over just how "real" s' must be. When allusion is made to this debate, however, classic sources, such as Campbell and Stanley (1963), and even its many, more sophisticated reconceptualizations (see note 2) have required maximal similarity of s' to s^*—also known as "mundane" as opposed to "experimental" realism of s' (Aronson & Carlsmith, 1968).[10]

---

[8]A procedure that systematically omits a moderating condition of a model (e.g., race or gender) is a matter of method, not theory.

[9]It is not logically possible to say that a variable is omitted from s'. Even at $z' = 0$, z' is a variable in s'. What is omitted, because z' is a constant, is its interaction with x'.

[10]The debate over mundane versus experimental realism, as Aronson and Carlsmith (1968) put it, begins more or less with Brunswick (1955), who argued that experimental designs ought to be

Criteria [6.1] and [6.2] make it clear that maximal similarity takes the wrong side in this debate. But it would be a mistake to dismiss too easily what one *can* learn from the debate. What is right about it is that, if s' is not an instance of $T(P\{s\})$, then, assuming that s^* *is* an instance of it, no test of T in s' sustains generalization of T to s^*. If T does not generalize to s^*, then a test of T in s' is externally invalid. Maximal similarity is not required, but experimental realism is. The problem is that experimental realism is easier said than done and it is simply hubris for an experimenter to assume that anything he or she does in an experiment, ipso facto, has it.

Both the purpose and the artifice of experiments are potential sources of instantiation error. Its purpose requires simplifying s', neglecting features of s^* that do not matter to **P**. But it is easy to neglect features that do matter. Recall Deutsch and Gerard's critique of Asch's failure to create the conditions in his experiment required to say anything about conformity to group influence. The analytic nature of theory is a justification for the simplification of s', but not at the expense of its instantiation.

Artifice, too, is a potential source of instantiation error: experiments create fictitious tasks, abilities, and interactions in the laboratory, like expectation-states theory's meaning insight and spatial judgment ability or its exchange of manipulated disagreements. They are important features of the experiment, controlling natural extraneities and not a problem if they are psychologically real to S. But experimental realism is not at all easy; it is one reason why it takes a considerable investment of time to create a fruitful experimental setting. (It took years to create the standardized experimental setting used to investigate expectation-state processes. A substantial part of that time was given to establishing its experimental realism. See Berger, Chapter 14, this volume.)

Even taking advantage of natural characteristics brought by Ss into the laboratory, which should ease the task, does not guarantee instantiation. Early status-characteristics experiments (such as Moore, 1968, and Berger, Cohen, & Zelditch, 1972) took it for granted that race, gender, and education were status characteristics in the population from which its Ss were drawn. No attempt was made to create fictitious status characteristics in these experiments; they merely activated the characteristics already possessed by their Ss. But whether a characteristic is a status characteristic or not is a question of history, not theory. Foschi and LaPointe (2002) have shown that the attitudes of college students towards gender have sufficiently changed over time that one can no longer assume that gender instantiates the theoretically relevant concepts of status characteristics

---

"representative" of natural situations. Berkowitz and Donnerstein (1982) make one of the more cogent arguments for experimental realism. For various forms of maximal—detailed, proximal—and surface similarity, see Campbell and Stanley (1963, p. 18), and also Campbell (1986), Cook (1990), and Shadish, Cook, and Campbell (2002).

theory. But if it is not an instance of the theory, it does not test it and the outcome of the experiment does not generalize to any instance that does satisfy its instantiation and scope conditions.

4. Application Errors

Criteria [6.1] and [6.2] also imply that applied research in s^* is as critical to the external validity of s' as theoretical research in s'. If $T(P\{s\})$ does not apply to s^*, failure of its application to s^* says nothing about the external validity of s'. It is therefore a crucial function of applied research to test the applicability of $T(P\{s\})$ to s^*. But applied research may introduce extraneities of its own into tests of a theory's applicability. The extraneities of applied research are entirely independent of any introduced by the methods of s'. Theoretical research in s' and applied research in s^* may both be externally valid; theoretical research in s' may be, but applied research in s^* may not be externally valid; applied research in s^* may be, but theoretical research in s' may not be externally valid; or both may be externally invalid.

The discovery of extraneities in applied research in s^* may explain why a theory tested in s' does not generalize to s^*, but it does nothing to rule out the external invalidity of s'. Nevertheless, it is something to take into account. It helps that [6.2] implies that the extraneities of applied research parallel those of theoretical research; though the conditions of the two are independent of each other, the same kinds of conditions that are sufficient for the external validity of s' are also the kinds of conditions sufficient for the external validity of applied research in s^*.

B. DETECTING, TESTING, AND REMEDYING EXTERNAL INVALIDITY

The fundamental test of the external validity of an experiment in s' is the application of the theory it tested to another instance of the theory, say s^*. If the theory does not generalize to s^*, and one suspected cause is the external invalidity of s', much effort goes into detecting its sources. But diagnosis of a potential source of threat is not, by itself, sufficient. If a potential threat is detected, one still has to test the hypothesis that the potential threat actually accounted for the external invalidity of s'. Furthermore, if, say, the threat z' did actually interact with x' in s', it does not follow that, absent z', x' would have had an effect on y'. All that a test for the effect of z' in s' shows is that the experiment in s' did not provide adequate proof that it did. One must still retest the theory, either ridding s' of its flaws, controlling for them, or estimating the magnitude of their effect on y'.

Criteria [6.1], [6.2], and [7] refine the methods available for diagnosing, testing, and remedying external invalidity. They do this in the same way they refine the sources of it, excluding those that are concrete, descriptive, or atheoretical.

1. Refinement of Methods of Detecting, Testing, and Remedying External Invalidities

In the first instance, because they exclude concrete, descriptive threats, [6.1], [6.2], and [7] refine where one looks for their sources. It is tempting, for example, to look for the sources of external invalidity in their peculiar, often unrepresentative populations of Ss (like college students), periods (like strikes or wars), or concrete settings (like laboratories). But concrete populations, periods, or settings, even if they matter, say nothing about why they matter. Appeal to populations, periods, or settings is useless unless one can say what particular aspect of them explains *why* they matter. Nor is it at all useful to look at the fixed initial conditions of the experiment, even if one looks only at variables. They will typically not describe s^*, but whether or not they are representative makes no difference to the theory that is tested in s'. One does not say an experiment is externally invalid because it controls air resistance or friction or legitimates female leaders, even if none of them is common outside the experiment.

But [6.1], [6.2], and [7] also exclude some popular methods of looking for threats. Some diagnostic tools are used theoretically by some experimenters, but atheoretically by others. It makes no sense to exclude the tools themselves, because they are sometimes useful, sometimes not. It all depends on how they are used. But used atheoretically, they say nothing about external validity simply because, absent a theory, the external validity of an experiment is indeterminate. Used atheoretically, therefore, some popular ways of looking for threats to external validity simply cannot say whether an experiment is or is not externally valid.

For example, a particularly useful diagnostic is meta-analysis (the statistical analysis of comparable statistical analyses, such as aggregation of separate estimates of an effect obtained by multiple replications of an experiment) (Glass, 1976, 2000). But meta-analysis is frequently mindless. Its logic is impeccable: if two experiments study the same hypothesis by the same methods, it makes sense to pool their separate estimates of the size of their effect. It also makes sense to aggregate tests of the same hypothesis by different methods and to use heterogeneity tests to determine whether the differences in method moderate the size of the effect (in other words, to use meta-analysis to detect methods that impair generalizing from experiments) (Hedges & Olkin, 1985). But often the selection of experiments is concrete, atheoretical; sometimes it involves nothing more than a similarity in the titles of the publications of the experiments. Theory-guided meta-analysis is a valuable tool for detecting potential sources of external invalidity; atheoretical meta-analysis is at best useless, at worst misleading.

In the same way, [6.1], [6.2], and [7] exclude concrete, descriptive, and/or atheoretical tests of the sources detected (e.g., the recourse to field studies as a reaction to the findings of theory and research on the social psychology of the experiment). On the one hand, the fundamental test of the external validity of experiments that test theories obviously lies in the field. Although logically s^* could be an experimental setting, most applications of theories are to natural settings and the fundamental test must apply to any occurrence of the process described by $T(P\{s\})$. Locke (1986) successfully exploited this test to justify generalization from industrial psychology in the laboratory to industrial psychology in the field. But the success of his test lay in generalizing scope-defined theories supported by experiments to scope-defined applications of them to organizations in the field. Absent instantiation and scope conditions, the test would probably have failed, but even if it had not, it would not have been an appropriate test of external validity.

On the other hand, the idea that there even *is* an other hand—that recourse to the field, though it gives up much internal validity, buys external validity—simply misunderstands the problem. Field studies do not generalize any more than laboratory experiments do (Dipboye & Flanagan, 1979). Excluding all recourse to the field would be "throwing out the baby with the bath water." But atheoretical field studies are no more generalizable than atheoretical experiments.

The same is true of an otherwise potentially useful test of external invalidity, cross-validation. It is frequently mentioned in the literature on external validity, and "robustness" of a cross-validated finding is frequently offered as a criterion of it. Furthermore, cross-validation of an experiment against other experiments using different methods to study the same theory is a valuable method, and robustness of the theory across methods is a sound criterion of external validity. But frequently what is advocated is cross-validation across multiple, concrete populations, periods, and settings. Multiple populations, periods, or settings—concrete and not scope defined—do not cross-validate the results of an experiment that tests a theory. All robustness in this sense means is that, whatever the theory that lies behind the experiment, it apparently is—or is not—a very general theory.

On the other hand, [6.1], [6.2], and [7] unequivocally exclude statistical generalization as a test of external validity. I am not suggesting that experiments can do without statistical inference or that statistical inference has nothing to do with the logic of sampling from a population. But statistical generalization has nothing at all to do with theoretical generalization, which is what external validity is about. There is no difficulty saying that there is a universe, a population in the technical sense, to which theories generalize, just as there is for statistical generalization. The universe to which a theory tested by an experiment generalizes is the population of instances that satisfy the instantiation and scope conditions of the theory. One can even say that the population, because its instances are concrete, is just as concrete as any other population to which any other statistic is generalized. But what matters is that the population is scope defined—that concreteness does not *define* the population to which a theory tested by an experiment is generalized—and, of course, that the purpose of the experiment is to test, refine, or extend the theory, not to describe the population.

All this obviously also holds for remedies: no concrete, descriptive, or atheoretical remedy remedies the external invalidity of an experiment that tests a theory (not representative design; not surface similarity; especially not maximal, detailed, proximal, or mundane similarity). Representative design does not matter if the experiment successfully creates all the theoretically relevant features of the process described by the theory it tests. Surface similarity does not matter if psychologically real; mundane realism does not matter if experimentally real. What matters is that the experiment create all and only the theoretically relevant features of the phenomenon and that they be psychologically real to, and the setting experimentally real for, the Ss in the experiment.

2. Extension of Methods of Detecting, Testing, and Remedying External Invalidities of Experiments

Criteria [6.1], [6.2], and [7] do not extend diagnoses, tests, or remedies for external validity in the way they do its sources. Nevertheless, developments in the last half-century in the design, modeling, measurement, and analysis of experiments have. Developments in modeling, measurement, and analysis have extended methods of detecting them. Developments in experimental design have extended methods of testing and remedying them.

Methods of detecting specification error, in particular, owe much to the extension of structural equation modeling from nonexperimental to experimental methods (Blalock, 1971; Costner, 1971; Smith, 1990). Even earlier, as a by-product of concern for construct validity, Campbell and Fiske (1959) had developed the multitrait, multimethod matrix method of detecting methodological specificities in the manipulation of x' and measurement of y'. But structural equation modeling has been especially influential in multiplying indicators of x' and y' to detect potential sources of external invalidities such as experimenter effects, demand characteristics, and evaluation apprehension, a method generalized and made widely available by Joreskog and Sorbom (1983). In addition, confirmatory meta-analysis has extended structural equation modeling between as well as within experiments, allowing detection of methodological moderators by comparing experiments using different methods to test the same theories. The method is easily abused (see preceding discussion), but when the experiments satisfy the same instantiation and scope conditions of the same theories, it becomes a powerful tool. Its test for heterogeneity due to method is a valuable test for specification error (Hedges & Olkin, 1985).

Methods of testing the threats detected owe themselves mostly to developments in experimental design. Perhaps a better way of putting it is that developments in experimental design have extended the function of experiments. Not only do they test theories, but they also increasingly test and remedy the external validity of experiments that test theories. The extension of the control group to the control of reactivity has allowed experimental test of the hypothesis that some interaction (say, $x'z'$), has impaired the generalizability of the theory tested by an experiment. The method depends on being able to separate the effect of z' from x' and either set it to 0 or vary it (hence, test either for the effect of x' absent z' or for the effect of z'). It has been used to test for the effect of reactive measurement (Solomon, 1949), demand characteristics (Gustafson & Orne, 1965; Orne & Evans, 1965; Orne & Scheibe, 1964), experimenter expectancies (Rosenthal, 1963, 1966, 1976), evaluation apprehension (Rosenberg, 1965), the voluntarism of Ss (Rosenthal & Rosnow, 1969), social desirability cues (Alexander & Knight, 1971; Rosenberg, 1969), subject expectancies (Rosenthal, 1976), subject awareness of the experiment (Foschi, 2006), and even instantiation (Foschi & LaPointe, 2002).

The extension of simulation methods to reactivity has facilitated the separation of x' from z'. While sometimes treated as an alternative to experiments, simulation is simply one way of manipulating the variables of an experiment. The use of an observer-S who is exposed to the reactive arrangements of an experiment but not its treatment (because the observer-S only observes another experiencing x' without actually experiencing it himself or herself) provides a way to separate the effect of a treatment from the effect of any reactive arrangement confounded with it. It has been useful in separating treatment effects from the effects of demand characteristics (Orne, 1962) and social desirability cues (Alexander & Knight, 1971).

Reactivity control groups have, in turn, led directly to altered replication as a remedy for the external invalidities of experiments. Solomon (1949) pioneered a design that crossed a treatment with a measurement effect, allowing separate tests of the effect of before-measurement of a dependent variable, independent of the treatment effect, but also of the treatment effect independent of the measurement effect. Solomon's four-group design was generalized to *any* pretreatment procedure by Lana (1969) and further generalized by Rosenthal (1976) to an eight-group design multiplying the number of reactivities controlled.

It may appear illogical to argue that a flawed method is, after all, the best method of testing for and remedying the flaws in a flawed method. But an argument due to Rosenberg makes sense: it has been experiments that have demonstrated the errors in experiments. The fact that they are capable of demonstrating the sources of systematic bias in experiments and of investigating the mechanisms of the process through which they operate also suggests how to control for and/or eliminate them (Rosenberg, 1969, p. 347).

IV. SUMMARY AND CONCLUSION

The goal of experiments is to test, refine, or extend theories under rigorously controlled, reliably observed conditions. If that is their purpose, then their external validity is a matter of generalizing the theories supported by them to other situations that satisfy their theories' instantiation and scope conditions. If the external validity of experiments is a matter of theoretical generalization, what their settings require is all of and only the theoretically relevant aspects of the situations to which their theories apply. Analytic simplification of their settings isolates the processes described by their theories from the many extraneities that obscure how the processes operate in natural settings. All the artifice of experiments facilitates manipulation and control of the causes and conditions of the isolated processes and rigorous methods of observing them.

But this should not be at the expense of the experiment's external validity. The purpose of an experiment is to control the extraneities that muddle observation of a process in its nonexperimental settings—not to introduce extraneities of its own into its observation in its experimental setting. If its whole purpose is generalization, external invalidity is a fatal flaw in an experiment.

But the fundamental test of an experiment's external validity is as abstract and as analytic as the experiment. An experiment is externally valid if the theory it supports predicts and explains the behavior of the process it describes in any situation to which the theory is applicable—that is, any situation that satisfies its instantiation and scope conditions. The story line is more complicated if the experiment fails the test, but the strategy that informs detecting, testing, and remedying external invalidity is just as theory driven and[11] just as abstractly analytic as its fundamental test.

The problem of the fundamental test is that, if the experiment fails it, its external invalidity is not the only possible explanation. The explanation may be the theory itself, its measurement, its application, or the applied research required to apply it. If the setting of the experiment has been successfully used

---

[11]But "theory" in this case means not only T(P{s}) and, with it, all the theoretical research oriented to testing, refining, and extending it, but also all the available theory and research bearing on its measurement and all the available theory and research on the social psychology of the experiment.

before, E will tend to look elsewhere for the explanation. But, *ab initio,* the experiment is the first place to look because the goal of a theory-oriented experiment is theory growth: discovering a substantive moderator that conditions a theory advances its growth. A methodological moderator that conditions it only impedes theory growth. But the methods of detecting, testing, and remedying it, and their criteria of assessment, admit no concrete sources, no concrete methods of detecting, testing, or remedying them, no concrete criteria of assessing them: not surface, maximal, detailed, proximal, or mundane similarity to the natural situation analytically simplified by the theory; not atheoretical meta-analysis; not atheoretical cross-validation on ill-defined, new, but concrete populations, periods, or settings; not equally atheoretical statistical generalization to them; not representative design; and least of all, not by abandoning the experiment for atheoretical, complex, under-analyzed field observation.

Application, and therefore the field, is the fundamental test of external validity, but if a theory supported by an experiment fails the test, an appeal to naturalistic, atheoretical field observation is no solution. The problem of external validity *has* no atheoretical solution. On the other hand, although it may look like a paradox, the lesson of theory and research on the social psychology of the experiment is that, if there *is* a theory, the solution to the problem of the experiment is the experiment.

REFERENCES

Alexander, C. N., & Knight, G. W. (1971). Situated identities and social psychological experimentation. *Sociometry, 34,* 65–82.

Aronson, E., & Carlsmith, M. (1968). Experiments in social psychology. In G. Lindzey & E. Aronson (Eds.), *Handbook of social psychology* (2nd ed.). New York: McGraw–Hill.

Asch, S. E. (1951). Effects of group pressure upon the modification and distortion of judgments. In H. Guetzkow (Ed.), *Groups, leadership, and men* (pp. 177–190). Pittsburgh: Carnegie Press.

Asch, S. E. (1956). Studies of independence and submission to group pressure: I. A minority of one against a unanimous majority. *Psychological Monographs, 70*(9), No. 416.

Bachrach, P., & Baratz, M. S. (1970). *Power and poverty.* Oxford: Oxford University Press.

Bales, R. F., & Slater, P. (1955). Role differentiation in small decision-making groups. In T. Parsons & R. F. Bales (Eds.), *Family, socialization and interaction process* (pp. 259–306). Glencoe, IL: Free Press.

Berger, J. (2007). The standardized experimental situation in expectation states research: Notes on history, uses, and special features. In M. Webster & J. Sell (Eds.), *Laboratory experiments in the social sciences* (pp. 353–378). Burlington, MA: Elsevier.

Berger, J., Cohen, B. P., & Zelditch, M. (1972). Status characteristics and social interaction. *American Sociological Review, 37,* 241–55.

Berger, J., & Conner, T. L. (1969). Performance expectations and behavior in small groups. *Acta Sociologica, 12,* 186–98.

Berger, J., Wagner, D. G., & Zelditch, M. (1985). Introduction: Expectation states theory: Review and assessment. In J. Berger & M. Zelditch (Eds.), *Status, rewards, and influence: How expectations organize behavior* (pp. 1–72). San Francisco: Jossey–Bass.

Berger, J., & Zelditch, M. (Eds.). (1993). *Theoretical research programs: Studies in the growth of theory*. Stanford, CA: Stanford University Press.

Berger, J., & Zelditch, M. (1997). Theoretical research programs: A reformulation. In J. Szmatka, J. Skvoretz, & J. Berger (Eds.), *Status, network, and structure: Theory development in group processes* (pp. 29–46). Stanford, CA: Stanford University Press.

Berger, J., & Zelditch, M. (Eds). (2002). *New directions in contemporary sociological theory*. New York: Rowman and Littlefield.

Berger, J., Zelditch, M., & Anderson, B. (1972). Introduction. In J. Berger, M. Zelditch, & B. Anderson (Eds.), *Sociological theories in progress* (Vol. ii, pp. ix–xxii). Boston: Houghton Mifflin.

Berkowitz, L., & Donnerstein, E. (1982). External validity is more than skin deep: Some answers to criticisms of laboratory experiments. *American Psychologist, 37*, 245–57.

Blalock, H. M., Jr. (1968). Theory building and causal inferences. In H. M. Blalock, Jr., & A. Blalock (Eds.), *Methodology in social research* (pp. 155–198). New York: McGraw–Hill.

Blalock, H. M., Jr. (1971). Causal models involving unmeasured variables in stimulus–response situations. In H. M. Blalock, Jr. (Ed.), *Causal models in the social sciences* (pp. 335–347). Chicago: Aldine.

Blalock, H. M., Jr. (1989). The real and unrealized contributions of quantitative sociology. *American Sociological Review, 54*, 447–460.

Brewer, M. B. (2000). Research design and issues of validity. In H. T. Reis & C. M. Judd (Eds.), *Handbook of research methods in social and personality psychology* (chap. 1). New York: Cambridge University Press.

Brunswick, E. (1955). Representative design and probabilistic theory in a functional psychology. *Psychological Review, 62*, 193–217.

Burke. P. (1967). The development of task and social–emotional role differentiation. *Sociometry, 30*, 379–392.

Campbell, D. T. (1986). Relabeling internal and external validity for applied social scientists. In W. M. K. Trochim (Ed.), *Advances in quasi-experimental design and analysis: New directions for program evaluation* (pp. 67–77), no. 31. San Francisco: Jossey–Bass.

Campbell, D. T., & Fiske, D. W. (1959). Convergent and discriminant validation by the multitrait multimethod matrix. *Psychological Bulletin, 56*, 81–105.

Campbell, D. T., & Stanley, J. C. (1963). *Experimental and quasi-experimental designs for research*. Chicago: Rand McNally.

Cohen, B. P. (1963). *Conflict and conformity: A probability model and its application*. Cambridge, MA: M.I.T. Press.

Cook, T. D. (1990). The generalization of causal connections: Multiple theories in search of clear practice. In L. Secrist, E. Perrin, & J. Bunker (Eds.), *Research methodology strengthening causal interpretation in nonexperimental data* (pp. 9–31). DHHS Publication No. PHS 90-3454. Rockville, MD: Department of Health and Human Services.

Costner, H. (1971). Utilizing causal models to discover flaws in experiments. *Sociometry, 34*, 398–410.

Deutsch, M., & Gerard, H. B. (1955). A study of normative and informational social influences upon individual judgment. *Journal of Abnormal and Social Psychology, 51*, 629–36.

Dipboye, R. L., & Flanagan, M. F. (1979). Research settings in industrial and organizational psychology: Are findings in the field more generalizable than in the laboratory? *American Psychologist, 34*, 141–50.

Foschi, M. (2006). On the application files design for the study of competence and double standards. *Sociological Focus, 39*, 115–132.

Foschi, M., & Lapointe, V. (2002). On conditional hypotheses and gender as a status characteristic. *Social Psychology Quarterly, 65*, 146–162.

Glass, G. V. (1976). Primary, secondary, and meta-analysis of research. *Educational Researcher, 5,* 3–8.

Glass, G. V. (2000). Meta-analyis at 25. http://glass.ed.asu.edu/gene/papers/meta25.html.

Gustafson, L. A., & Orne, M. T. (1965). Effects of perceived role and role success on the detection of deception. *Journal of Applied Psychology, 49,* 412–417.

Haslam, S. A., & McGarty, C. (2004). Experimental design and causality in social psychological research. In C. Sansone, C. C. Morf, & A. T. Panter (Eds.), *Sage handbook of methods in social psychology* (chap. 11). Thousand Oaks, CA: Sage.

Hedges, L. V., & Olkin, I. (1985). *Statistical methods for meta-analysis.* New York: Academic Press.

Hegtvedt, K. A., & Markovsky, B. (1995). Justice and injustice. In K. S. Cook, G. A. Fine, & J. S. House (Eds.), *Perspectives on social psychology* (pp. 257–280). Boston: Allyn and Bacon.

John, O. P., & Benet-Martinez, V. (2000). Measurement: Reliability, construct validation, and scale construction. In H. T. Reis & C. M. Judd (Eds.), *Handbook of research methods in social and personality psychology* (chap. 13). New York: Cambridge University Press.

Joreskog, K. G., & Soerbom, D. (1983). *LISREL V user's guide analysis of structural relationships by maximum likelihood and least square methods.* Chicago: International Educational Services.

Lana, R. E. (1969). Pretest sensitization. In R. Rosenthal & R. L. Rosnow (Eds.), *Artifact in behavioral research* (chap. 4). New York: Academic Press.

Leventhal, G. S., Allen, J., & Kemelgor, B. (1969). Reducing inequity by reallocating rewards. *Psychonomic Science, 14,* 295–296.

Leventhal, G. S., Weiss, T., & Long, G. (1969). Equity, reciprocity, and reallocating rewards in the dyad. *Journal of Personality and Social Psychology, 13,* 300–305.

Lewis, G. H. (1972). Role differentiation. *American Sociological Review, 37,* 424–434.

Locke, E. A. (1986). *Generalizing from laboratory to field settings: Research findings from industrial-organizational psychology, organizational behavior, and human resource management.* Lexington, MA: Heath.

Lucas, J. W. (2003). Theory-testing, generalization, and the problem of external validity. *Sociological Theory, 21,* 236–253.

Meeker, B. F., & Leik, R. K. (1995). Experimentation in sociological social psychology. In K. S. Cook, G. A. Fine, & J. S. House (Eds.), *Sociological perspectives on social psychology* (pp. 629–649). Needham Heights, MA: Allyn and Bacon.

Messick, D. M., & Sentis, K. P. (1979). Fairness and preference. *Journal of Experimental Social Psychology, 15,* 418–434.

Messick, S. (1989). Validity. In R. L. Lynn (Ed.), *Educational measurement* (3rd ed., pp. 13–103). New York: Macmillan.

Moore, B. (1978). *Injustice: The social bases of obedience and revolt.* New York: M. E. Sharpe.

Moore, J. (1968). Status and influence in small group interactions. *Sociometry, 31,* 47–63.

Orne, M. T. (1962). On the social psychology of the psychological experiment. *American Psychologist, 17,* 776–83.

Orne, M. T. (1969). Demand characteristics and the concept of quasi-controls. In R. Rosenthal & R. L. Rosnow (Eds.), *Artifact in behavioral research* (chap. 5). New York: Academic Press.

Orne, M. T., & Evans, F. J. (1965). Social control in the psychological experiment. *Journal of Personality and Social Psychology, 1,* 189–200.

Orne, M. T., & Scheibe, K. E. (1964). The contribution of nondeprivation factors in the production of sensory deprivation effects: The psychology of the panic button. *Journal of Abnormal and Social Psychology, 68,* 3–12.

Rosenberg, M. J. (1965). When dissonance fails: On eliminating evaluation apprehension from attitude measurement. *Journal of Personality and Social Psychology, 1,* 28–42.

Rosenberg, M. J. (1969). The conditions and consequences of evaluation apprehension. In R. Rosenthal & R. Rosnow (Eds.), *Artifact in behavioral research* (pp. 280–349). New York: Academic Press.

Rosenthal, R. (1963). On the social psychology of the psychological experiment: The experimenter's hypothesis as unintended determinant of experimental results. *American Scientist, 51,* 268–367.

Rosenthal, R. (1966). *Experimenter effects in behavioral research.* Cambridge, MA: Harvard University Press.

Rosenthal, R. (1969). Interpersonal expectations: Effects of the experimenter's hypothesis. In R. Rosenthal & R. L. Rosnow (Eds.), *Artifact in behavioral research* (chap. 6). New York: Academic Press.

Rosenthal, R. 1976. *Experimenter effects in behavioral research* (enlarged ed.). Cambridge, MA: Harvard University Press.

Rosenthal, R., & Rosnow, R. L. (1969). The volunteer subject. In R. Rosenthal & R. L. Rosnow (Eds.), *Artifact in behavioral research* (chap. 3). New York: Academic Press.

Sell, J. (2007). Social dilemma experiments in sociology, psychology, political science, and economics. In M. Webster & J. Sell (Eds.), *Laboratory experiments in the social sciences* (pp. 459–479). Burlington, MA: Elsevier.

Shadish, W. R., Cook, T. D., & Campbell, D. T. (2002). *Experimental and quasi-experimental designs for generalized causal inference.* Boston: Houghton Mifflin.

Silverman, I., & Shulman, A. D. (1970). A conceptual model of artifact in attitude change studies. *Sociometry, 33,* 97–107.

Smith, H. L. (1990). Specification problems in experimental and nonexperimental social research. *Sociological Methodology, 20,* 59–91.

Solomon, R. L. (1949). An extension of control group design. *Psychological Bulletin, 46,* 137–150.

Strodtbeck, F. L., James, R., & Hawkins, C. (1957). Social status in jury deliberations. *American Sociological Review, 22,* 713–719.

Wagner, D. G., & Berger, J. (2002). Expectation states theory: An evolving research program. In J. Berger & M. Zelditch (Eds.), *New directions in contemporary sociological theory* (pp. 41–76). New York: Rowman and Littlefield.

Walker, H. A., & Willer, D. (2007). Experiments and the science of sociology. In M. Webster & J. Sell (Eds.), *Laboratory experiments in the social sciences* (pp. 25–55). Burlington, MA: Elsevier.

Weber, S. J., & Cook, T. D. (1972). Subject effects in laboratory research: An examination of subject roles, demand characteristics, and valid inference. *Psychological Bulletin, 77,* 273–295.

Zelditch, M. (1969). Can you really study an army in the laboratory? In A. Etzioni (Ed.), *Complex organizations* (2nd ed., pp. 528–539). New York: Holt, Rinehart, and Winston.

Zelditch, M. (1974). Forward. In M. Webster & B. Sobieszek (Eds.), *Sources of self evaluation: A formal theory of significant others and social influence* (pp. vii–xiv). New York: John Wiley & Sons.

Zelditch, M., Harris, W., Thomas, G. M., & Walker, H. A. (1983). Decisions, nondecisions, and meta-decisions. *Research in Social Movements, Conflicts and Change, 5,* 1–32.

Zelditch, M., & Hopkins, T. (1961). Laboratory experiments with organizations. In A. Etzioni (Ed.), *Complex organizations* (1st ed., pp. 464–478). New York: Holt, Rinehart, and Winston.

CHAPTER 5

Hypotheses, Operationalizations, and Manipulation Checks

MARTHA FOSCHI
University of British Columbia

ABSTRACT

This chapter concerns the formulation of theoretical hypotheses and their empirical test in laboratory contexts. I examine the different types of variables that make up such hypotheses and discuss several issues related to operationalizing their terms and carrying out manipulation checks and replications. Throughout the chapter, I illustrate these topics with an experiment on level and type of received performance evaluations, inferred task competence, and exerted social influence.

I. INTRODUCTION

The formulation and empirical test of hypotheses are two fundamental activities in scientific research. In this chapter, I examine some aspects of these activities within the context of experimental research in the social sciences

113

and, in particular, laboratory experiments in group interaction. I focus on how the empirical testing of theoretical hypotheses requires both that their abstract terms be translated into concrete operations and that the effectiveness of these operations be verified.

Because the term "hypothesis" has many meanings, it is useful to begin by presenting a definition. By a hypothesis I mean a statement, intended to be tested empirically, that proposes how a set of variables (or factors) are related. An example would be "the more knowledgeable a person appears to be, the more he or she will be liked by others." The hypothesis states a positive or direct relationship between an independent variable, "perceived level of knowledge," and a dependent (or outcome) variable, "level of liking." It is, however, a very simplistic hypothesis, as it ignores other factors that most probably also affect that relationship.

The hypothesis could become both more complex and more likely to be true if one were to reformulate it as follows: "In groups where members have no history of interaction with each other and where they share a strong motivation to solve a common task, the more knowledgeable a group member appears to be about that task, the more he or she will be liked by the other group members." The hypothesis has now acquired four *scope conditions* (some stated more explicitly than others). These conditions indicate circumstances that define the applicability of the hypothesis and are as follows: a group with a common goal, participants with no history of interaction with each other, a strong motivation to achieve that goal, and communicated knowledge that is perceived to be related to the task (rather than knowledge about any matter). Other scope conditions could also be added.

Moreover, the hypothesis could be made more complex by including other types of variables. Thus, an additional *independent* variable could specify that the relationship will be affected by whether the person communicates his or her task-related knowledge either directly or indirectly. The more direct the expression of that knowledge is, the more pronounced its effect on liking will be. An added *dependent* variable could stipulate that not only liking but also degree of influence exerted by that person will be simultaneously affected. Finally, an *intervening* variable such as "degree of respect towards that group member's opinions" could mediate the relationship between the independent and the dependent variables.[1] A hypothesis can, and often should, have

---

[1]This hypothesis is meant only as an example of how these variables could be related. Thus, rather than considering "communication (either direct or indirect) of knowledge about the group task" as an additional independent variable, "a direct communication of that knowledge" could be treated as a scope condition. In turn, a strong motivation to solve the common task could be a value of an independent variable spanning from "very strong" to "very weak" in that motivation. Several other models could also be proposed.

a number of each of these components. (On the different types of variables that make up a hypothesis, see Baron & Kenny, 1986; Foschi, 1997.)

The variables in a hypothesis may range from notions that summarize everyday experiences (such as "book" and "food") to constructs that are part of a theory (such as "cognitive dissonance" or "social attribution"). The hypothesis itself, in turn, may be proposed as a stand-alone generalization or as a derivation from a scientific theory (here referred to as "theory").

A theory is a system of ideas consisting of (1) defined and undefined concepts that incorporate selected variables, (2) logically interconnected statements (assumptions and derived hypotheses) that relate these concepts to each other, and (3) rules for the derivation of statements (see, e.g., Cohen, 1989, chap. 10; Jasso, 1988; Wagner, 1984, chap. 2). Such a theory constitutes both an explanation of and a prediction about the occurrence of a class (or classes) of phenomena; a successful theory fulfills both explanation and prediction functions while exhibiting firm empirical support for its hypotheses.

A related, useful distinction is to classify research as either "exploratory" or "hypothesis testing." In the former case, a set of variables is identified but no relationship between them is proposed; in the latter, such a relationship is advanced. Hypothesis-testing studies that are embedded in a theory have a much larger long-term payoff than exploratory work that is not part of a theory. Whereas such exploratory work may or may not produce interesting findings, theoretical hypothesis-testing studies always yield useful outcomes; in the latter studies, there is a much narrower context within which the results can be interpreted—that is, they either give full, partial, or no support to the hypothesis. If full support is not obtained, the findings can point to flaws in either the hypothesis or the design. Moreover, assuming an appropriate design, results of any level of support are informative not only about the hypothesis but also about the theory. Results from atheoretical exploratory studies, on the other hand, stand by themselves. The absence of a theory means that there are no guidelines to sort out the various (often many) possible interpretations of those results.

Next I propose a set of theoretical hypotheses and construct an experimental test for them. Through this example, I put forth in concrete terms my thoughts about hypotheses and design issues. The points I make, however, are intended to apply beyond the specific details I present in this chapter.

II. HYPOTHESES

Suppose that a person had received results from a test of logical skills and that these results were very poor. Would the person quickly conclude that he or she does not possess those skills? The answer would depend on several factors, such as the extent to which the results are accepted to be objective evaluations, and how tired the person was at the time of the exam. Here I consider the perceived diagnosticity of the test and formulate my example within the context of expectation states theory, a system of ideas that meets all the requirements for a scientific theory listed previously.

The central interest of expectation states research is on how members of task groups assign competence[2] to each other (Berger, Fisek, Norman, & Zelditch, 1977; for a recent review, see Wagner & Berger, 2002). Two of the core concepts are "status characteristics" and "performance expectations." The former is defined as any socially valued attribute seen by an actor as implying task competence. Such characteristics consist of at least two levels or "states" (e.g., either high or low mechanical ability), one of which is viewed as having more worth than the other. "Performance expectations" are beliefs about the likely quality of group members' future performances on the task at hand and reflect levels of assigned competence. These beliefs need not be conscious or have an objective validity.

Status characteristics are classified as varying from specific to diffuse, depending on the perceived range of their applicability. A specific characteristic is associated with performance expectations in a limited domain; a diffuse characteristic carries expectations about performance on a wide, indeterminate set of tasks. For instance, in many societies, gender, ethnicity, nationality, formal education, and socioeconomic class constitute diffuse status characteristics for large numbers of individuals and in a variety of settings. Expectation states are said to develop for "self" (the focal actor) relative to each of the other members of the group; all propositions are formulated from self's point of view.

The theory specifies how, and under what conditions, expectations are formed. For example, these may be based on the group members' status characteristics, or on the evaluations they receive on their task performances by a "source" outside the group, or on both. Two central scope conditions are that members are "task oriented" (i.e., they value the ability required to do the task well and are motivated to achieve that result) and "collectively oriented" (i.e., they are prepared both to consider and to use the other group members' opinions for the solution of the task).

Expectation states theory is, in fact, a set of interrelated theories or "branches." In this chapter I focus on one of the main branches—namely, evaluations-and-expectations theory. This branch concerns the processes through which evaluations of units of performance generalize to become stable beliefs about levels of competence and result in performance expectations. Within this branch, research has considered properties of the performers and of the evaluations, and has tested for the effects of these variables on expectation

[2]I treat "competence," "skill," and "ability" as synonyms. I also use "participant," "subject," "actor," "self," and "respondent" as equivalent notions.

formation (see, e.g., Webster & Sobieszek, 1974). My interest here is in one feature of the evaluations: specifically, the extent to which they are seen as representing a diagnostic test of the ability in question. By "diagnosticity" I mean the degree to which a sample of a person's performances on a given task is considered to be a valid indication of his or her overall ability for that task. My general prediction is that, for good as well as poor levels of performance, a test believed to be highly diagnostic results in more conclusive (i.e., certain) inferences of either ability or lack of ability than does a less diagnostic test. In turn, I relate these inferences to levels of influence rejection, as I describe later.

This research topic is closely related to the work presented in Foschi, Warriner, and Hart (1985). In that study, we investigated the role that standards for both competence and lack of competence play in the formation of expectations. Here I formulate my hypotheses for the same setting as in that study (which is also the setting investigated in a large proportion of expectation states research, as discussed in Berger, Chapter 14, this volume). That situation involves (1) two persons, self and a partner ("other"), who perform a task, first individually and then as a team, and (2) a source of performance evaluations. The following scope conditions apply:

- The task consists of a series of trials, each having the same level of difficulty. Self has no prior expectations about the ability required for this task and believes this ability to be both valuable and specific.
- Self is motivated to perform the task well (i.e., self is task oriented).
- Performance evaluations originate in a source—namely, a person or procedure considered by self to be more capable of evaluating performances than he or she is. The source is the only basis of evaluations available to self to judge his or her task ability relative to that of the partner (and to form corresponding expectations).
- During the team phase of the task, self is prepared both to take into account and to use the partner's ideas for the task solution (i.e., self is collectively oriented).[3]

I propose the following hypotheses:

- If self has received *better* evaluations than the partner in the individual phase, then he or she will form *higher* expectations for self than for the

---

[3]Note that some scope conditions are required at the beginning of the interaction as well as throughout (e.g., that the participant be task oriented), whereas others are required only at the beginning (e.g. that, initially, the participant have no other bases for self–other expectations). Sometimes, the latter conditions are called "initial." I, however, prefer Cohen's terminology (1989, p. 80) and reserve the expression "initial conditions" for singular statements describing a particular system at a particular time and place. An initial condition is thus a statement that a scope condition has been operationalized as intended (see also Foschi, 1997, and Section V in this chapter).

partner and, in the team phase, will reject *more* influence from that person than if self had received *no* performance evaluations.

- If self has received *worse* evaluations than the partner in the individual phase, then he or she will form *lower* expectations for self than for the partner and, in the team phase, will reject *less* influence from that person than if self had received *no* performance evaluations.
- *Higher expectations* for self than for other will be *more conclusive* and result in *more* rejection of influence when the task has been perceived to be high rather than low in diagnosticity.
- *Lower expectations* for self than for other will be *more conclusive* and result in *less* rejection of influence when the task has been perceived to be high rather than low in diagnosticity.

In addition to the scope conditions, the hypotheses contain four variables: two independent (task diagnosticity and level of performance evaluations), one intervening (performance expectations), and one dependent (influence rejection). To my knowledge, these hypotheses have not yet been investigated within the expectation states tradition. It is also worth noticing that they implicitly contain what I call *irrelevant variables of theoretical interest*. These are factors that could reasonably be proposed to have a theoretical impact on the dependent variables, but that nevertheless have not been included (i.e., are treated as irrelevant) (Foschi, 1980, p. 93). For example, relative to self, the source could be of various levels of perceived superiority in capacity to evaluate performances (such as "marginally superior" and "clearly superior"). I have assumed that those levels do not affect the results. Similarly, I am treating as theoretically irrelevant whether or not, across the trials, the distribution of evaluations favoring one performer over the other follows a pattern.

Every researcher has to make decisions about identifying irrelevant factors. In some cases the factors *are known* to be irrelevant to the process under study (as supported by previous research). At other times, there is insufficient knowledge about a topic; in that case, one takes educated risks as to what factors can be left out. Finally, in still other cases, a factor can be assumed to be irrelevant for the time being (e.g., on the basis of what would be feasible to test). Regardless of the reasons, irrelevant factors are seldom explicitly listed; this is only done if one wants to call special attention to the factor. Rather, it is generally understood that a theory (or a set of hypotheses within a theory) need not include all factors of possible interest; it is sufficient that only the key ones be listed.[4] If, however, a factor is deemed to be irrelevant, though of

---

[4]For simplicity, here I list only a few factors in my hypotheses. Thus, I am not relating the hypotheses to other expectations states work that identifies variables (such as the strictness of performance standards and the number of evaluations) that elaborate and expand on how evaluations are interpreted.

theoretical interest, then it is included in the hypotheses through a clause specifying that they hold "regardless of the level of this variable." The same as scope conditions, such theoretically irrelevant variables help define the class (or classes) of phenomena to which a hypothesis applies, and are thus an instrument in generalizing it. I discuss irrelevant variables in more detail later in this chapter.

Let us now design an experimental test of the hypotheses.

III. THE EXPERIMENT

For research cumulativeness, work on expectation states has used a standardized experimental setting for hypothesis testing. The setting was created by Joseph Berger (Berger *et al.*, 1977, chap. 5; Camilleri & Berger, 1967; for a review of its history, see Berger, Chapter 14, this volume) and it has been used in several dozens of studies. In my test, I propose to use the computerized version that I introduced in Foschi (1996). This version enables the researcher to recreate the same theoretical variables as does the original setting, and has been an effective instrument for the study of expectation formation (for examples of other work using this version, see Foschi, Enns, & Lapointe, 2001; Lovaglia & Houser, 1996). The following summarizes the main features of the procedures I propose for my test.

The experiment will be conducted over a series of sessions; two previously unacquainted persons (either two men or two women) will take part in each. Subjects will be either first- or second-year undergraduates at my school, the University of British Columbia, and their ages will range between 18 and 21 years. They will be volunteers for a "study in visual perception," and will be recruited from large classes at my institution. Each person will be paid $16 for his or her participation, which will last approximately 1 hour and 20 minutes. The sessions will take place in a specially equipped research facility at the university. The two persons will be seated individually at adjacent computer stations said to be linked to each other. The stations will be separated by a fixed partition and subjects will be precluded from both seeing and talking with each other before or during the session; the only information they will be given about self and partner is that they are of the same sex category and year at the university, and of similar age.

The experimenter will be a female research assistant who will introduce herself as a graduate student in sociology (a statement that will be supported by her professional and confident demeanor); she will also be slightly older than the subjects. She will read the instructions from a position that will enable her to make eye contact with the two persons; a summary of her statements will also appear on their computer monitors as the instructions are being read.

The experimenter will remain in the room throughout the session, and her presence will be visible to the two participants at all times.

She will inform the participants that they will be asked to solve "contrast sensitivity" problems. This task involves viewing a series of rectangles made up of smaller ones (either red or white) that form an overall abstract pattern. Subjects have to decide which color is predominant in each case. Although the proportion of the two colors is almost the same in all patterns, their configurations (different for each trial) create the impression that discriminating between the two colors is a possible yet difficult task. The high ambiguity in the proportion of the two colors and the limited time (a few seconds) that the subjects have for each decision make it impossible for them to complete their task with certainty (e.g., by counting the colored rectangles). The experimenter will, however, state that "reliable research has established contrast sensitivity to be a newly discovered, important, and mainly innate ability." She will also mention that that research has so far determined it to be relatively specific—that is, not related to attributes such as sex category, intelligence, or artistic skills.

An overview of the design and predictions is shown in Table 1. Participants will *first* work individually on a series of 20 patterns, and then will receive the

TABLE 1 Overview of the Experiment

	Phase I			Phase II
	Test score (out of 20) received by		Diagnosticity of test	Rejection of influence measured*
	Self	Other		
Condition				
(1) Conclusive *higher* expectations for self than for other	17	3	High	a
(2) Inconclusive *higher* expectations for self than for other	17	3	Low	b
(3) Unformed expectations	No information		No information	c
(4) Inconclusive *lower* expectations for self than for other	3	17	Low	d
(5) Conclusive *lower* expectations for self than for other	3	17	High	e

*Hypotheses: a > b > c > d > e.

"scores" obtained by self and partner. These will be communicated through each person's monitor as well as printouts. The scores will be either 17 for self and 3 for the partner or the opposite combination. Subjects will also be informed of the trial-by-trial results for self and partner; these results will show one person consistently (though not on all the trials) outscoring the other. In addition, one group of control subjects will perform the task but will receive no scores.[5] Pairs of participants will be assigned at random to one of these three conditions. In turn, those receiving scores will be assigned, also at random, to hear a description of the task as either high or low in ability to diagnose contrast sensitivity reliably. The study thus involves five conditions or groups: one control (or baseline) and four experimental (or treatment).

Next, participants in all the conditions will be instructed to work as a team on 25 patterns of a similar task said to require the same ability. This phase will be presented to the subjects as an additional test of contrast sensitivity. On each of these trials, they will first make an initial choice between the two colors, receive the other person's "choice" via the monitor, and then make a final decision (either agreeing or disagreeing with the partner).[6] At the end, each person will individually complete a written questionnaire and will then be interviewed and debriefed. The questionnaire and interview will provide key manipulation checks.

IV. OPERATIONALIZATIONS AND MANIPULATIONS

An operationalization is the translation of a theoretical variable into procedures designed to give information about its levels. I treat "operationalization," "operational definition," "measure," "indicator," and "observable" as close notions. Operationalizations tie theoretical ideas to evidence. For example, in the experiment proposed here, the case of "better evaluations for self than for the partner" becomes a score of 17 for one person and 3 for the other. My interest here is in the logic of the relationship between theoretical variables and their operationalizations; issues such as how to construct the measures,

---

[5]Note that other control groups are also possible (and indeed desirable). For example, in one such group the two performers could receive equally average (or good, or poor) evaluations; in another, the first phase could be omitted altogether. In my view, the control group that I have included will provide the most basic information needed in this case.

[6]The prearranged communications to the subjects through their monitors enables the experimenter to assign two persons to the same condition in each session. For example, both can be informed simultaneously that self has received a higher score than the partner and that the task is high in diagnosticity. Moreover, in all five conditions, each person is informed *on the same prearranged trials* that the partner either agrees or disagrees with self's choices.

how to pretest them to ensure their validity and reliability, and what sample size to select are outside the scope of my discussion.

The notion of "manipulating a variable" overlaps in some respects with that of an operationalization. In an experiment, the researcher has a high degree of control over the variables that are either predicted or suspected to affect the participants' responses. Ideally, all except the outcome variables (namely, the intervening and the dependent variables) are controlled through manipulation. Although the term "manipulation" is often used to refer only to the creation of levels of the independent variables at the operational level, I interpret the term more broadly, as follows. *Direct* manipulation involves both establishing specific levels of different types of variables (e.g., the belief that contrast sensitivity is a valuable ability, or that the test has either high or low diagnosticity) and keeping factors such as the subjects' ages either constant or within a narrow range.

With respect to other factors that are not under study—and often not even identified as having potential relevance to the research (e.g., personality attributes of the participants, or their moods during the experimental session)—control is exerted *indirectly* through the use of random assignment either to experimental or to baseline conditions. Random assignment tends to distribute these other factors evenly across those groups, and thus ensures that they are not a variable affecting the responses.[7] It is useful to think of an experiment as a stage (or framework, or set of constraints) within which several factors are controlled and a few (the intervening and dependent variables) are allowed to vary freely.

At this point it is also helpful to make the following distinction. The independent variables of a hypothesis may be factors that are established by the researcher, such as the levels of performance evaluations and test diagnosticity described earlier, and for which random assignment can be implemented. These are *truly experimental* variables. On the other hand, sometimes naturally occurring differences in characteristics of either the participants or the experimental situation (or both) are of interest to the researcher, as when he or she considers subjects' sex category or location of the study as possible independent variables. Since random assignment cannot take place in these cases, the variables are known as *quasi-experimental* (or "organismic," since they are an inherent part of the entity under study). Many statistical tests of significance assume random assignment to conditions by the experimenter; it is only by accepting a relaxation of this requirement that those tests are used with quasi-experimental factors.

---

[7]Even if the distribution of such a factor is the same across conditions, the mean and/or the shape of the distribution could still affect the results. This issue can only be addressed through further work in which the factor is singled out and its levels are purposely varied across studies.

V. ON PARTICULAR FEATURES

Setting the stage in order to test a theoretical hypothesis experimentally involves many design decisions. These reflect a combination of (1) decisions about how to operationalize the theoretical variables of interest, (2) requirements dictated by good experimental design (such as random assignment to conditions), and (3) common-sense, practical matters (such as the length of the experimental session). All of these decisions contribute to narrowing down the theoretical ideas to specific situations. While all theoretical hypotheses are abstract, all evidence is concrete—that is, bound by particular circumstances, including those of time and place. Thus, such hypotheses always contain a larger number of cases than does the evidence about them. It is also important to consider that, even in a carefully constructed study, design decisions other than the ones taken could always have been made. Both in theoretical formulation and test design, research is a continuously changing and self-improving activity.

In what follows, I examine the design decisions in my proposed experiment in relation to three types of particular features: (1) those that operationalize theoretical variables, (2) those that are truly irrelevant, and (3) those that represent experimental limitations. Note that the intervening variable of performance expectations is viewed in this theory as an unobservable construct, and therefore is not measured directly in the standardized experimental setting (Berger *et al.*, 1977, p. 19).[8]

A. OPERATIONALIZATION OF THEORETICAL VARIABLES

Scope conditions. Scope condition (a) is implemented through the features of the contrast sensitivity task. All patterns making up this task have the same high level of ambiguity, which participants are expected to perceive as high difficulty. The number of patterns in each phase of the session reflects the following considerations: a much smaller number would probably be seen as inadequate for a test; a much larger number would likely make participants lose interest. In addition, the ambiguity of the task and its description as involving a specific and newly discovered ability make it very unlikely that participants would hold prior self–other expectations. The ability is also described as valuable. This statement is supported by the fact that the session

---

[8]Expectation states theory makes predictions about participants' behaviors, not about their self-reports. Notice, however, that some authors (see, e.g., Foddy & Smithson, 1999; Foschi, 1996) have included such reports of self–other expectations in their designs, as either auxiliary measures or manipulation checks. See also my comments, later in this chapter, on behaviors and self-reports.

takes place at a university facility, the reference by the experimenter to "reliable research on contrast sensitivity," the perception that the current project is part of such research, and the computerized nature of the task. It is predicted that the latter feature will serve to associate the project with up-to-date technological advances, and thus add interest and prestige to it. (For a further discussion of these points, see Foschi, 1997, pp. 541–543).

The operationalization of scope condition (b) ("task orientation") is closely related to the creation of the belief, specified in scope condition (a), that the task is valuable. Assuming that belief, the instructions will encourage the participants to try to do their best to achieve a correct solution to the patterns on which they will be working.

The source of evaluations identified in scope condition (c) is operationalized through the combination of the perceived scientific status of the project and the graduate-student status of the experimenter. Her age and demeanor also enhance her overall superior status relative to the subjects. She will administer a "test" of contrast sensitivity in the first phase of the session and will inform the participants of the results. Note that the experimenter does not generate the scores; rather, these derive from the test. Since subjects have no basis for evaluating the performances, and the setting is associated with university research, it is highly likely that they will not only accept the scores as a better indicator of ability than any judgment that they can make by themselves, but also will consider the scores accurate. Other possible bases for performance evaluation are blocked by informing each pair of participants that they are peers in several respects (such as sex category, age, and year at the university).

Scope condition (d) ("collective orientation") is implemented by the task changes that occur in the second phase: after making an initial choice, each person will receive information on the partner's selection, and will then be asked to make his or her own final choice. The instructions will emphasize that the pair should be working as a team, and that it is not only useful but also appropriate to consider and follow the other person's choice when making a final decision.

Independent variables. Levels of performance evaluation are operationalized by two clearly different scores for self and partner, one first-rate and the other poor. In the control group, no scores will be given. Levels of task diagnosticity will be created by statements by the experimenter advising participants that the 20 patterns of the test either have or have not been proven to be a very reliable indication of a person's contrast sensitivity.

Dependent variable. During the 25 trials of the second phase, the partner's initial choice will be programmed to show on the monitor that he or she disagrees with self on 20 trials and agrees on 5. These agreements are included because the occurrence of disagreements on all trials would arouse suspicions; instead, the agreements will take place every four to six trials in a way that does not suggest a prearranged sequence. The disagreements provide the opportunity for rejection of influence from the partner to be assessed. Thus, the high ambiguity of the task creates uncertainty in self about the correctness of a response; this, in turn, increases self's reliance on the only information that can help him or her decide on the "right" answer—namely, the scores (or their absence) during the first phase. A final choice in which self stays with his or her initial choice after a disagreement is an indicator of influence rejection. The level of this variable is measured by the proportion of "self" or "stay" responses over these trials.

Irrelevant variables of theoretical interest. As mentioned earlier, I highlight two factors in this category: level of the source's perceived superiority over self in ability to evaluate performances, and level of consistency, across the evaluations, of one performer's superiority over the other. It follows from the status of "theoretically irrelevant" that I assign to these variables that they can be operationalized at any level. However, to avoid the possibility of introducing additional factors (particularly if the sample size were to be small), it is a good design feature to keep them constant *within* a given experiment. Thus, the procedures to create the difference between self and source in ability to evaluate will be the same throughout the experiment, and so too will be the sequence of correct and incorrect scores given to self and other.

Let us now turn to the other two types of particular features.

B. TRULY IRRELEVANT VARIABLES

My design also contains the implicit decision that a very large variety of factors (of different levels of abstractness) are truly irrelevant to the process under study. Examples of particular features of the experiment that reflect these factors are the time of the day the experimental sessions are carried out, and the type of music the experimenter likes. These features are never listed; if a researcher were to be asked explicitly about them, the answer would be that they have no actual or presumed theoretical linkage to the hypotheses.

I also count in this group some of the changes I have made in my computerized version of the standardized setting, such as the manner in which agreements and disagreements are conveyed to the subjects at each trial. In this version, the monitor shows both the color of self's choice and of the partner's "choice," and a statement indicating either "partner agrees" or "partner disagrees." In the original version, self can see whether the two have either agreed or disagreed from a display of lights showing self's initial answer and the partner's "response." That is, I consider the impact of the statements on the monitor to be a nonsignificant addition to what is conveyed by the lights in the original version. To my knowledge, there is no research evidence that disconfirms this idea conclusively.

C. EXPERIMENTAL LIMITATIONS

There are also variables that have an in-between status. They are neither theoretically irrelevant nor truly irrelevant; that is, they have not been made a part of the hypotheses. However, they could be of relevance, and it is for that reason they are listed in the procedures and either kept constant or within a narrow range. They are also often mentioned in the interpretation of the results as part of ad hoc explanations. I call these factors "test (or experimental) limitations" (Foschi, 1980). In the present design, examples are the educational status of the participants, their age, the amount they are being paid, the sex category of the experimenter, and the country where the experiment is being conducted. Since these factors are formulated at a low level of generality, they could not have been incorporated into the hypotheses in their current form; for that, they would have to be treated as instances of more abstract notions (Foschi, 1997, pp. 544–545). (For an interesting compatible proposal on test limitations, see Poortinga & van de Vijver, 1987.)

Let us consider, for example, payment level. In the study I propose here, that level is consistent with the amount currently offered at my university for taking part in social science experiments. Personal experience indicates that paying less would not be sufficient to motivate participation, while paying more would not significantly increase the incentive. Many students volunteer out of curiosity and because it is a common undergraduate experience to participate in experiments. Still, since level of payment could have an effect, it is important to mention that amount in the report on the study, so the reader can arrive at his or her own conclusions about any possible effects from this factor.

In my view, the test limitation that would be most worthwhile to investigate in the present example is the sex category of the experimenter relative to that of the participants. This factor could have a role in the extent to which the experimenter and the project are accepted as a source; results from Foschi and Freeman (1991) and Foschi *et al.* (2001) point in that direction. My approach would be to view "sex category" as an instance of the theoretical variable "status characteristic." I would then not only replicate the study with a male experimenter (while not changing any of the other manipulations), but also investigate other possible status factors, such as this person's age and ethnicity. (For work on the related topic of the status characteristics of the evaluators, see Webster & Sobieszek, 1974, chap. 6.)

VI. FURTHER COMMENTS ON OPERATIONALIZATIONS

What cannot and what should not be done. It is important to realize that some operationalizations are either difficult to implement or simply cannot be carried out.

Thus, a theoretical hypothesis may remain experimentally untested because some or all of its variables/values cannot be manipulated in the laboratory. In some cases, this is due to practical reasons. For instance, a researcher may not be able to secure the funds (for laboratory facilities, equipment, or payment to subjects) that his or her experiment requires. There are also ethical reasons why some operationalizations should not be done. For example, in my proposed experiment, it would be unethical to try to increase the perceived value of contrast sensitivity by informing the participants that their scores are a very reliable indication of either high or low levels of intelligence. Intelligence carries strong emotional connotations for many people, and using it as an experimental manipulation might subject such people to unacceptable levels of anxiety.

Let us focus on problems that may emerge about the feasibility of the operationalizations in relation to the choice of subject population. First, one should ensure that there are no language or other comprehension barriers to the understanding and acceptance of the instructions. For example, this becomes a significant design concern if the subject population contains individuals with varying levels of proficiency in the language used in those instructions. Moreover, the researcher should be aware that the subjects' attitudes and beliefs could play an important role in the acceptance of the instructions. In all cases, it is essential to pretest the procedures with participants from the same population as those who will be taking part in the experiment proper, to determine whether or not the experiment is viable in all of these respects (see Section VII of this chapter, and Rashotte, Webster, & Whitmeyer, 2005). The next three examples illustrate some of these issues in relation to the standardized design.

- It could be that, in spite of the university laboratory setting and what the researcher considers to be convincing arguments presented in the instructions, the subjects are not prepared to accept that contrast sensitivity is a valuable ability. Participants may also be skeptical about the value of scientific research in general. If more than a few subjects fall into these categories even after strengthening the relevant instructions, it would be necessary to plan the test with either a different population or with a different ability.
- The creation of a collective orientation in the laboratory may involve more than simply conveying instructions to that effect. Cross-cultural researchers have identified "cooperation–competition" as a useful dimension for the classification of societies and cultures (including segments within them) in terms of their central norms. "Collectivism–individualism" is a closely related concept. Although the number of studies investigating whether these dimensions affect group-level interaction is still rather limited, there is research showing that differences at that level are in line with these dimensions (Mann, 1988, pp. 192–193; Miller-Loessi & Parker, 2003, pp. 539–542). It should then not be surprising to anticipate

that an experiment that requires collective orientation would be more difficult to implement in societies that stress competitiveness than in those that place more value on cooperation. The following should also be considered: some societies may emphasize cooperation to such a high degree that task orientation is compromised; that would be the case if an answer from a higher status person is almost always accepted regardless of its merit. (On effects in the standardized setting from a collective versus an individual orientation, see Wilke, Young, Mulders, & de Gilder, 1995.)

- A number of expectation states experiments have investigated the linkage between status and task, and contrast sensitivity has proven to be a highly versatile instrument for this purpose. For example, this task has been credibly presented to the subjects as (1) "masculine" (Foddy & Smithson, 1999; Foschi, 1996; Wagner, Ford, & Ford, 1986), (2) either "masculine" or "feminine" in different experimental conditions (Foddy & Smithson, 1989; Rashotte, 2006), or (3) "of no known relationship to gender" (Foschi & Lapointe, 2002). Researchers have also been success-ful in introducing contrast sensitivity as either explicitly or implicitly associated to characteristics such as education level, military rank, and age (Berger *et al.*, 1972; Freese & Cohen, 1973; Moore, 1968).

In the case of gender, however, results from two of the studies linking it directly to the task (Foddy & Smithson, 1999, p. 317; Foschi, 1996, p. 246) indicate that although subjects behaved as if they had accepted the experi-mental instruction that the task was masculine, they were reluctant to admit this in written questionnaires. (On this point, see also Section VIII of this chapter). The lack of correspondence between behaviors and self-reports sug-gests that gender may have now lost some of its status value (both as a source of information and as a norm) for the participants in those studies.

It is also worth noting that the operationalization of the task's sex linkage consisted of informing the subjects that research had found men to be better than women at solving contrast sensitivity problems. Suppose that I wanted to replicate the experiment proposed here, but this time defining the ability as masculine in some experimental conditions and feminine in others. Although the possibility of reactivity should always be taken into account, the opera-tionalization could be made stronger by, for instance, including charts and graphs supporting the sex-linkage claim. One should also consider that it is likely (as well as hoped for) that in a not too distant future, participants will not readily accept that there are skills associated with ascribed attributes such as sex category and skin color.

Replications and multiple operationalizations. It is always useful to plan the test of a theoretical hypothesis so that the evidence originates in more than a single study. Thus, experiments need to be replicated. Following Aronson,

Ellsworth, Carlsmith, and Gonzales (1990), I distinguish between two types of replications: "direct" and "systematic." In a direct replication, a study is carried out in such a way that the researcher attempts to duplicate, as closely as possible, the procedures of the original design. This type of replication has also been called "exact" (Campbell & Stanley, 1966, pp. 32–33; Hendrick & Jones, 1972, pp. 356–357). As well, this term is applied to the repetition of a session or "run" of an experiment that is done to obtain the desired number of participants per condition. It is worth noting that even in direct replications one should expect variation (across either sessions or studies) in the factors that are considered to be truly irrelevant.

In a systematic replication, a study is repeated either (1) varying the operationalizations of theoretical concepts or (2) letting the experimental limitations take on other values. For instance, a systematic replication of the experiment proposed here could involve using a task other than contrast sensitivity, provided it has the features specified in scope condition (a) of the hypotheses. Examples of studies incorporating such different tasks are Berger and Conner (1969), Freese and Cohen (1973), Foschi *et al.* (1985), and Martin and Sell (1980). As to the experimental limitations, I would focus on varying the characteristics of the experimenter, as I proposed in a previous section. Systematic replications strengthen the evidence by reducing the possibility that results are confounded with particular operationalizations. Furthermore, this type of replication can contribute to theoretical growth when experimental limitations are recast in theoretical terms. (I discuss this point in reference to cross-cultural replications in Foschi, 1980.)

VII. MANIPULATION CHECKS

Earlier in this chapter I referred to an experiment as a stage or framework within which several factors are controlled and a few others vary freely. A manipulation check is a procedure designed to verify that the controlled factors have indeed been implemented as expected. Without this check, the stability of the framework is not certain. Thus, it is an essential part of an experiment to include manipulation checks of scope conditions, independent variables, and irrelevant variables of theoretical interest, as well as of test limitations. It is equally important to report the results of these checks.

In the present experiment, the checks will be similar to those used in comparable expectation states studies. Some of these procedures will involve verifying that there has not been any experimental error (e.g., in the reading of the instructions and in the assignment of scores to the participants); this can be done by inspecting the logs that the experimenter will keep for each session. Other checks will be as follows. Levels of rejection of influence obtained from the control group can serve to establish that participants had no prior self-other expectations. Moreover, the pattern of self-responses can be examined to determine if participants show lack of interest (e.g., by always choosing the same color as the initial response, or by unvaryingly alternating between accepting and rejecting influence). Most of the checks that I plan, however, will be accomplished through the postexperimental questionnaire and interview.

It is generally understood that manipulation checks should be carried out on the scope conditions and the independent variables. Still, as mentioned earlier, such checks should also be done on the irrelevant variables of theoretical interest that have been identified and controlled, since these procedures imply that those variables have been recognized as part of the hypotheses. Similarly, levels of test limitations should be checked, to verify that each of these factors had indeed been kept as a constant or varied within a narrow range as intended. In this way, their inclusion in the interpretation of the results will be on firm ground.

The rest of this section outlines some of the questions I would use in the postexperimental instruments.

Independent variables. Regarding level of performance, I would ask the subjects to recall the scores obtained by self and partner and to assess the level achieved by each person on bipolar five-point scales ranging from "my partner's performance was much better than mine" to "my partner's performance was much worse than mine." I would evaluate the perceived diagnosticity of the test through several questions about self's confidence that the results from that test can be reliably generalized to assessments of task ability.

Scope conditions. I would use a variety of questions in this area. I would investigate the perceived level of difficulty of the trials, and the extent to which subjects were convinced that contrast sensitivity was both valuable and specific, and had no prior expectations regarding self's and partner's task performances. I will also assess the perceptions of the project and the experimenter as the source of evaluations, and the degree to which subjects understood and accepted the scores they received. Finally, I would use several questions formulated as Likert scales with five categories of agreement each (from "strongly agree" to "strongly disagree") to check on task and collective orientation. Two examples of these questions are: "It was really too difficult to try to figure out which pattern had more red, so I just guessed" (for task orientation) and "Agreeing as a team regarding the correct decision was more important to me than my own choice" (for collective orientation). In turn, for each scope condition, my exclusion rule for the data analysis would require that a person give extreme values in the wrong direction over several questions.

Irrelevant variables of theoretical interest. In an earlier section, I suggested that variables that have been explicitly identified as belonging to this category should be kept constant throughout the experiment. I would thus check that the source's perceived ability to evaluate performances (relative to self's) is similar across participants, and that there are no major discrepancies between the received and perceived sequences of "correct" and "incorrect" scores.

Test limitations. Although these factors are not theoretical variables, researchers often highlight them because they *could* be in that category. It is therefore important to verify that their levels have been created as intended. As discussed earlier, I would focus on two test limitations: level of payment and sex category of the experimenter relative to the participants. I would therefore probe about the perceived fairness of the payment and the extent to which the fact that the experimenter was a woman affected the acceptance of the scores. Since the latter item concerns gender as status, it is important that questions about it be as unobtrusive as possible (see following discussion). One possibility would be carefully to mask items about the experimenter's sex category with items about other characteristics, both status and nonstatus, of this person.

VIII. FURTHER COMMENTS ON MANIPULATION CHECKS

Exclusion of subjects. An important issue related to manipulation checks concerns the exclusion of subjects from the data analysis. The topic is often misunderstood. Since the test of a hypothesis assumes that the scope conditions and the independent variables have been implemented as intended, it is legitimate to exclude from the data analysis those subjects for whom that implementation has failed.[9] Rejection rules, however, should be explicitly formulated beforehand, to avoid the possibility of the exclusions being misunderstood as an attempt to keep only those subjects who support the researcher's hypotheses. The rules should also be as conservative as possible because a substantial proportion of exclusions (often taken to be 20% or higher of the total number of participants) opens the possibility that random assignment has been compromised. If an experiment results in that level of excluded subjects, it is wise to rethink the design and the procedures, and redo the study.

Behaviors and self-reports. As described previously, most manipulation checks in the standardized setting utilize self-reports obtained through the postexperimental questionnaire and interview. It is important to remember

---

[9]Because, as mentioned earlier, it is only for design reasons that irrelevant variables of theoretical importance are kept constant, the consequences of failing manipulation checks in their case does not warrant exclusions. Similarly, if a test limitation fails a manipulation check, this does not affect the test of the theoretical hypotheses—only the post hoc interpretation of the results with respect to this factor.

that self-reports are a subject's reflection of what has occurred during the session and, as such, may be affected by factors such as memory, self-presentation norms regarding one's task ability, level of support for the perceived goals of the research project, and even suspicion. Self-reports also increase a respondent's awareness of his or her answers, and this is particularly a concern if the topic is of a sensitive nature.

The behavioral measure of rejection of influence, on the other hand, is likely to be less obtrusive and thus to avoid these effects. Yet, self-reports are valuable tools in assessing manipulation checks, as often it is not feasible to obtain behavioral data on all variables of interest. Every effort should be made to control for factors that may affect these checks; at the very least, one should be apprised of the possible effects of these factors when interpreting the results. (The relationship between self-reports and behaviors is a classic topic in social science methodology and has been examined by many authors; for a useful discussion of this topic in the context of expectation states research, see Driskell & Mullen, 1988.)

As an example of the different responses often obtained from the two types of measures, let us consider the results from Foschi et al. (2001). Subjects in the standardized expectation states setting participated in same-sex pairs and received scores indicating that one of the two persons clearly had more ability than the other. Rejection of influence showed no effects from sex category of dyad (as predicted, since participants were peers in this respect). There were, however, significant effects from sex category of subject in the self-reports: at every level of scores received (in each case, the same for male and female dyads), men reported higher levels of ability relative to the partner than did the women. The authors interpreted this finding in terms of the operation of different, gender-based self-presentation norms that emphasize modesty in women but not in men.

The following is a useful procedure that minimizes the effects of extraneous variables on self-reports. It consists of dividing prospective subjects at random into two groups, one to participate in the experiment proper, and the other to be assessed on those variables that require manipulation checks. An often cited example of this procedure is found in Goldberg's 1968 study. In the experiment proper, the extent to which the subjects were biased by the sex category of a (fictitious) performer was measured; the other group of subjects provided data on their perceived sex linkage of various occupations. In this way, this second group supplied information that was not affected by possible bias in the task of assessing performances. Moreover, because of random assignment, results on sex linkage of occupation from these subjects can be assumed to apply to a similar extent to the group participating in the experiment proper.

Finally, the inclusion of control groups as part of a standardized expectation-states experiment provides an effective way of doing manipulation checks that yield behavioral data. Wagner *et al.* (1986) used such control groups to assess the combined effects of gender as status and the instructions that contrast sensitivity was a valuable, masculine task. Similarly, in Foschi and Lapointe (2002), we used such control groups to assess the extent to which gender functioned as a status characteristic for the subjects in our study. In both cases, the control groups avoid asking subjects directly about possibly sensitive issues.

IX. SUMMARY AND CONCLUSIONS

In this chapter I examined several issues concerning the experimental test of theoretical hypotheses. I began by identifying the different types of factors involved in such a test: independent, intervening, and dependent variables; scope conditions; test limitations; irrelevant variables of theoretical interest; and truly irrelevant variables. I focused on operationalizations and manipulation checks in the context of laboratory experiments. Operationalizations are a researcher's translation of theoretical variables into procedures that render them observable. In turn, the effectiveness of these procedures must be checked. I proposed an example involving a set of hypotheses within the expectation states tradition and an experimental design to test them, and used this example to illustrate a variety of methodological topics. Thus, I discussed limits on operationalizations in terms of what cannot and should not be done, and emphasized the value of replications and multiple indicators.

Regarding manipulation checks, I examined the reasons why the exclusion of some subjects may be necessary, and compared behaviors and self-reports in terms of the type of information that they yield. Carefully designed and executed laboratory experiments are an invaluable instrument in the construction and test of theories. Throughout the chapter, I offered ideas on how to enhance the design and interpretation of such experiments.

ACKNOWLEDGMENTS

Preparation of this chapter was supported by Standard Research Grant #410-2002-0038 from the Social Sciences and Humanities Research Council of Canada. I gratefully acknowledge this support. I also thank Lok See Loretta Ho, Vanessa Lapointe, and Maria Zeldis for valuable comments.

REFERENCES

Aronson, E., Ellsworth, P. C., Carlsmith, J. M., & Gonzales, M. H. (1990). *Methods of research in social psychology* (2nd ed.). New York: McGraw–Hill.

134   Laboratory Experiments in the Social Sciences

Baron, R. M., & Kenny, D. A. (1986). The moderator–mediator variable distinction in social psychological research: Conceptual, strategic, and statistical considerations. *Journal of Personality
and Social Psychology, 51,* 1173–1182.

Berger, J. (2007). The standardized experimental situation in expectation states research: Notes on
history, uses, and special features. In M. Webster & J. Sell (Eds.), *Laboratory experiments in the
social sciences* (pp. 353–378). Burlington, MA: Elsevier.

Berger, J., & Conner, T. L. (1969). Performance expectations and behavior in small groups. *Acta
Sociologica, 4,* 186–198.

Berger, J., Cohen, B. P., & Zelditch, M., Jr. (1972). Status characteristics and social interaction.
American Sociological Review, 37, 241–255.

Berger, J., Fisek, M. H., Norman, R. Z., & Zelditch, M., Jr. (1977). *Status characteristics and social
interaction: An expectation states approach.* New York: Elsevier.

Camilleri, S. F., & Berger, J. (1967). Decision-making and social influence: A model and an experimental test. *Sociometry, 30,* 365–378.

Campbell, D. T., & Stanley, J. C. (1966). *Experimental and quasi-experimental designs for research.*
Chicago: Rand McNally.

Cohen, B. P. (1989). *Developing sociological knowledge: Theory and method* (2nd ed.). Chicago:
Nelson–Hall.

Driskell, J. E., Jr., & Mullen, B. (1988). Expectations and actions. In M. Webster, Jr., & M. Foschi
(Eds.), *Status generalization: New theory and research* (pp. 399–429; 516–519). Stanford, CA:
Stanford University Press.

Foddy, M., & Smithson, M. (1989). Fuzzy set and double standards: Modeling the process of ability
inference. In J. Berger, M. Zelditch, Jr., & B. Anderson (Eds.), *Sociological theories in progress:
New formulations* (pp. 73–99). Newberry Park, CA: Sage.

Foddy, M., & Smithson, M. (1999). Can gender inequalities be eliminated? *Social Psychology
Quarterly, 62,* 307–324.

Foschi, M. (1980). Theory, experimentation, and cross-cultural comparisons in social psychology.
Canadian Journal of Sociology, 5, 91–102.

Foschi, M. (1996). Double standards in the evaluation of men and women. *Social Psychology
Quarterly, 59,* 237–254.

Foschi, M. (1997). On scope conditions. *Small Group Research, 28,* 535–555.

Foschi, M., Enns, S., & Lapointe, V. (2001). Processing performance evaluations in homogeneous
task groups. *Advances in Group Processes: A Research Annual, 18,* 185–216.

Foschi, M., & Freeman, S. (1991). Inferior performance, standards, and influence in same-sex
dyads. *Canadian Journal of Behavioural Science, 23,* 99–113.

Foschi, M., & Lapointe, V. (2002). On conditional hypotheses and gender as a status characteristic.
Social Psychology Quarterly, 65, 146–162.

Foschi, M., Warriner, G. K., & Hart, S. D. (1985). Standards, expectations, and interpersonal
influence. *Social Psychology Quarterly, 48,* 108–117.

Freese, L., & Cohen, B. P. (1973). Eliminating status generalization. *Sociometry, 36,*
177–193.

Goldberg, P. (1968). Are women prejudiced against women? *Transaction, 5,* 28–30.

Hendrick, C., & Jones, R. A. (1972). *The nature of theory and research in social psychology.* New York:
Academic Press.

Jasso, G. (1988). Principles of theoretical analysis. *Sociological Theory, 6,* 1–20.

Lovaglia, M. J., & Houser, J. A. (1996). Emotional reactions to status in groups. *American
Sociological Review, 61,* 867–883.

Mann, L. (1988). Cultural influences on group processes. In M. H. Bond (Ed.), *The cross-cultural
challenge to social psychology* (pp. 182–195). Newbury Park, CA: Sage.

Martin, M. W., & Sell, J. (1980). The marginal utility of information: Its effects upon decision-
making. *The Sociological Quarterly, 21,* 233–242.

Miller-Loessi, K., & Parker, J. N. (2003). Cross-cultural social psychology. In J. Delamater (Ed.), *Handbook of social psychology* (pp. 529–553). New York: Kluwer Academic/Plenum.

Moore, J. C., Jr. (1968). Status and influence in small group interaction. *Sociometry, 31,* 47–63.

Poortinga, Y. H., & van de Vijver, F. J. R. (1987). Explaining cross-cultural differences: Bias analysis and beyond. *Journal of Cross-Cultural Psychology, 18,* 259–282.

Rashotte, L. S. (2006). Controlling and transferring status effects of gender. Paper presented at the annual meeting of the International Society of Political Psychology, Barcelona, July.

Rashotte, L. S., Webster, M., Jr., & Whitmeyer, J. M. (2005). Pretesting experimental instructions. *Sociological Methodology, 35,* 163–187.

Wagner, D. G. (1984). *The growth of sociological theories.* Beverly Hills, CA: Sage.

Wagner, D. G., & Berger, J. (2002). Expectation states theory: An evolving research program. In J. Berger & M. Zelditch, Jr. (Eds.), *New directions in contemporary sociological theory* (pp. 41–76). Lanham, MD: Rowman & Littlefield.

Wagner, D. G., Ford, R. S., & Ford, T. W. (1986). Can gender inequalities be reduced? *American Journal of Sociology, 51,* 47–61.

Webster, M., Jr., & Sobieszek, B. (1974). *Sources of self-evaluation: A formal theory of significant others and social influence.* New York: John Wiley & Sons.

Wilke, H., Young, H., Mulders, I., & de Gilder, D. (1995). Acceptance of influence in task groups. *Social Psychology Quarterly, 58,* 312–320.

PART II

Designing and Conducting Experiments

Chapters in Part 2 are designed to offer essential information and guidelines for social science experimenters. These chapters describe issues faced by everyone intending to conduct experimental research in the social sciences, and they offer invaluable insights and information for successfully meeting those challenges. These chapters are based on the accumulated experience of these experimental researchers.

Chapter 6, by Karen A. Hegtvedt, analyzes issues regarding the rights of experimental subjects. In so doing, she considers the functioning of institutional review boards (IRBs), and how experimenters can successfully interact with them. Nobody may (and nobody should) conduct research involving human subjects without IRB review and approval. Drawing on her extensive experience as both an experimental social scientist and a head of an IRB,

Hegtvedt describes ethical requirements and principles as they apply to this type of work. She describes four broad types of potential mistreatment—objectification of research subjects, harming them, exploiting them, and breaches of confidentiality—with criteria for designing experimental research without such problems. She also describes for social scientists how an IRB is set up and how it functions, and provides advice for interacting with them. The chapter includes an extensive set of suggestions for promoting productive, sensitive, ethical, and methodologically sound experimental research.

Chapter 7, by Lisa Troyer, is a contemporary digest of technology available for social science experimentation. Most social scientists use computers, but the latest techniques for using them in experimental design are not widely known. Troyer shows how technology is always a part of experiments, and how technological advances can pace development of better experimental methods. Using a wide range of different experimental designs for illustration, Troyer shows how technology can be part of overall design, independent variables, and dependent variables. Describing some of her own research on how technology changes parameters of experiments, she identifies key areas experimenters will want to compare when changing technologies. She also describes some of the latest technologies for measuring physiological changes accompanying social processes, such as changes in brain waves and hormones, and notes how this technology may foster greater interdisciplinary links of social and natural sciences.

Chapter 8, by Murray Webster, Jr., describes the growth of external funding, with particular application to experimental research. He lists some of the sources of funding, and describes the proper role relations between their officers and researchers. The chapter includes an extended discussion of writing successful research proposals.

Chapter 9, by Lisa Slattery Rashotte, is designed to help a researcher move from theoretical ideas to experimental research. It includes a wide range of suggestions and topics that a researcher must address in developing experiments. They include deriving the most useful empirical hypotheses, thinking about independent and dependent variables, pretesting designs and procedures, and analyzing experimental data. Power analyses to decide the needed sample size and experimenter effects on the data receive extended treatments here.

Chapter 10, by Will Kalkhoff, Reef Youngreen, Leda Nath, and Michael J. Lovaglia, addresses a neglected topic in experimental design: who the participants will be, how to recruit them, and what happens after they volunteer for the research. They consider a variety of benefits and costs to participants, and offer suggestions for maximizing the former as part of good research design. The authors describe the many steps in finding and contacting reliable sources of participants, deciding which of them are most suitable for an experimental project, record keeping, and interpersonal relations with the participants.

Chapter 11, by Robert K. Shelly, describes how he and other skilled experimenters select, train, and manage members of research teams who conduct the experiments. Anyone intending to conduct an experiment needs to give careful thought to the participants (Chapter 10) and the experimenters, for those are the two central roles in any experiment. Shelly likens an experiment to a theatrical performance in which an alternate reality is created and sustained for a particular purpose, though with certain key differences. The research team is not only the face of the research organization for participants, but also the source of information an experimenter uses in deciding success or failure of an experiment and of its underlying theoretical structure. Thus, it is crucial to devote detailed planning to helping the team do its work as effectively as possible.

Chapter 12, by Kathy J. Kuipers and Stuart J. Hysom, describes a number of practical problems that can occur in setting up experimental research and provides solutions for them. The authors speak from recent experience, being themselves young researchers who have faced problems getting their own research careers started, and they share with readers the expected and unexpected problems they faced, and how they and others have overcome them. Challenges faced include everything from issues of experimental design, to dealing with others in the organizational environment, to what to do when scheduled subjects fail to show up, to managing the funds to pay for their participation.

CHAPTER 6

Ethics and Experiments

KAREN A. HEGTVEDT

Emory University

ABSTRACT

The pursuit of knowledge through social scientific research requires consideration of the many ways to protect the human dignity of study participants. This chapter reviews issues inherent in conducting ethical research, both generally and specifically with regard to experimental studies. To the extent that researchers remain mindful of these ethical concerns, they are more prepared to navigate the (federal) regulations pertaining to the protection of the rights and welfare of research subjects. The chapter concludes with suggestions on how to successfully meet the demands of institutional review boards, which oversee these regulations.

I. INTRODUCTION

The role of ethics in research is no longer simply a matter of concern for individual researchers, but rather a matter of public discourse. Regulations governing the protection of the rights and welfare of human research participants shape that discourse, but emphasis on underlying ethical concerns fade while concerns with regulatory compliance gain ascendancy. This chapter contextualizes federal requirements in terms of ethical issues and principles. Discussion focuses on four general issues regarding the ethics of research in the social sciences (objectification, potential harms, exploitive practices, and confidentiality) and two issues specific to experiments (deception and subject pools). Ways in which researchers may deal with these issues in their attempts to meet federal regulatory requirements are outlined. Increased awareness of ethical issues and knowledge of regulations is likely to ensure the safety and well-being of individuals who participate in experiments as well as maintain the public's trust in the scientific endeavor.

Social scientists investigate a variety of social processes (e.g., bargaining, power, status, justice, cooperation, conflict, attitudes, decision making) using laboratory experiments. While some may contend that such methods fail to capture the "real world," as noted elsewhere (Martin & Sell, 1979; Zelditch, 1969), it is not the intent of experimenters to recreate the real world in the laboratory. Undoubtedly, however, many elements of the real world affect the development and execution of laboratory experiments. One very important element—beyond the vagaries of funding and the challenges of operationaliaztions—is ethics.

The post-World War II era saw the transformation of ethics in research from a matter of concern for individual researchers to a matter of public discourse (McBurney & White, 2004). The emergence of modern ethical codes stems from the set of standards resulting from the Nuremberg trial of Nazi doctors who conducted cruel medical experiments on concentration camp prisoners (see Dunn & Chadwick, 2002). The Nuremberg Code introduced the requirement of informed consent for nontherapeutic research. Subsequent documents (e.g., the 1964 Declaration of Helskinki by the World Medical Association) and U.S. federal regulations (e.g., the 1974 National Research Act and subsequent policy, codified in 45 CFR § 46) extend the premises and safeguards of the Nuremberg Code. Concern with ethics in research is also evident in other countries, though often less formalized than in the United States and typically focused primarily on biomedical studies.

In addition, by the turn of the twenty-first century, nearly all U.S. professional associations within the social sciences had established their own codes of ethics.[1] Such codes represent contemporary consensus on what a profession believes to be acceptable practices. The associations, however, have no means

---

[1]Professional associations' ethics codes may be found as follows: American Psychological Association (2002), http://www.apa.org/ethics/homepage.html; American Sociological Association (1999), http://www.asanet.org/page.ww?section=Ethics&name=Ethics); American Political Science Association (http://www.apsanet.org/imgtest/ethicsguideweb.pdf); American Anthropological Association (1998), http://www.aaanet.org/committees/ethics/ethcode.htm. One exception is the American Economic Association, which appears to have no ethics code (personal communication with Edda Leithner, staff member at AEA, June 6, 2006).

of routinely regulating the behavior of individual researchers (Rosnow, 1997). In the United States, regulation falls under the rubric of institutional review boards (IRBs) established by universities and other organizations as specified by federal regulations.[2]

Although critics of IRBs may rightly claim that guidelines for the protection of human research participants are more suitable to medical research than to social and behavioral studies (Citro, Ilgen, & Marrett, 2003; DeVries, DeBruin, & Goodgame, 2004; Israel & Hay, 2006), the history of social science research includes landmark studies that highlight the harm that can occur in both field and laboratory research on processes such as conformity and influence. Most frequently cited is Milgram's 1974 experimental study of obedience to authority in which adult volunteers in the role of "teacher" were led to believe that they administered electrical shocks when another ostensible volunteer in the role of "learner" failed to give a correct answer. Such intentional deception has been harshly criticized (see Baumrind, 1985).

Yet, ethical concerns emerge with regard to studies even in the absence of deception. Zimbardo's examination of the behavior of randomly assigned volunteers to the roles of "prisoner" and "guard" in a mock prison setting who endured psychological and physical abuse or became sadistic, respectively, exemplifies such concerns (Zimbardo, 1973; Zimbardo, Banks, Haney, & Jaffee, 1973). The debriefing that Milgram gave his subjects and the early halt to the mock prison study that Zimbardo enacted indicate researchers' recognition of the potential and actual harms stimulated in their studies. There was, however, little subsequent attention paid to the lingering effects of these harms.

Thus, concern for welfare of research participants permeates the sphere of social science research as well as that of biomedical research. This chapter first addresses the meanings of ethics in various aspects of research. Attention then shifts to general ethical issues that arise in research as well as specific issues characteristic of laboratory experiments. The chapter also considers how IRBs operate with regard to the review of social and behavioral research. Such consideration provides the basis for outlining strategies that experimenters may employ to meet the federal regulatory requirements regarding protection of the rights and welfare of human research participants. Insofar as such protection captures the essence of ethical research, the strategies may also generalize to meeting the demands of ethics committees in other countries. Conclusions focus on the mesh between researcher aims and values and the demands of the broader environment to ensure ethical research.

[2]Federal Policy for the Protection of Human Subjects, 45 CFR § 46 (2001) may be found at http://www.hhs.gov/ohrp/humansubjects/guidance/45cfr46.htm. Elsewhere on the OHRP (Office of Human Research Protections) Web site are postings of guidance for IRBs and researchers on a variety of issues.

II. DEFINING ETHICS IN RESEARCH

Most broadly, ethics refer to the moral principles governing behavior—the rules that dictate what is right or wrong. Yet, as the preceding examples illustrate, what constitutes right or wrong is subjective, defined by groups with particular aims. Such aims underlie the fundamental conflict between (social) scientists who pursue knowledge that they hope may benefit society and the rights of research participants themselves (Neuman, 2007; McBurney & White, 2004). In the absence of moral absolutes, professional associations and others craft rules for what is proper and improper with regard to scientific inquiry to try to ameliorate this conflict. The resulting ethics codes reflect philosophical ideas and, as evidenced more recently, attempt to bridge to regulatory requirements. The ethical conduct of research pertains to more than a phase of data collection involving human participants and it is more encompassing than simply complying with specific federal regulations pertaining to protections for such participants.

Discussions of scientific misconduct (e.g., Altman & Hernon, 1997; Neuman, 2007) focus on unethical behavior often stemming from the pressures researchers feel to make their arguments and build their careers. Failure to identify the shortcomings of one's research or to suppress findings of "no difference" may be mildly unethical practices. Taking shortcuts that involve falsifying or distorting data or research methods or actually hiding negative findings, however, is a more egregious violation. A recent, much broadcast example of such fraud is the case of the Korean scientist, Hwang Woo-Suk, who faked data about what he was able to achieve with regard to cloning processes.[3] Not only does such fraud delay scientific developments, it also undermines the public's trust in scientific endeavors.

Another form of research misconduct is plagiarism, which occurs when a researcher claims work completed or written by others (e.g., colleagues, students) as his or her own without adequate citation. Plagiarism, while not technically illegal if the "stolen" materials are not copyrighted (e.g., presenting Ibn Khaldun's words as one's own), violates what the American Sociological Association's code of ethics labels "integrity." The principle of integrity charges scholars to be honest, fair, and respectful of others and to act in ways that do not jeopardize their own or others' professional welfare.

To some extent, the classification of these behaviors as forms of scientific misconduct derives from philosophical principles similar to those underlying the concern for the protection of the welfare of human research participants. Israel and Hay (2006) analyze philosophical approaches to determining how people might decide what is morally right—what should be done—under

---

[3]See "Cloning Scientist Is Indicted in South Korea," *New York Times*, May 12, 2006.

certain circumstances. One approach is to focus on the consequences of a behavior. Such an approach, characteristic of the utilitarian philosopher John Stuart Mill, essentially invokes a cost-benefit analysis. In simplified form, if the benefits that arise from a behavior outweigh the risks or harm associated with that behavior, then it is morally acceptable. This approach, however, begs the question of what constitutes a benefit or harm. In contrast, nonconsequential approaches stemming from the works of Immanuel Kant suggest that what is right is consistent with human dignity and worth. This perspective also emphasizes duties, irrespective of the consequences per se.

Social psychologist Herbert Kelman (1982) discusses this principle of consistency with human dignity in his evaluation of ethical issues in different social science methods. Kelman notes two components of human dignity: identity and community. The former refers to individuals' capacity to take autonomous actions and to distinguish themselves from others, while the latter emphasizes the interconnections among individuals to care for each other and to protect each other's interests. Thus, to promote human dignity requires people to accord respect to others, to foster their autonomy, and to care actively for their well-being.

In so conceptualizing human dignity, however, Kelman also draws attention to consequences, but not in a utilitarian sense. He argues that "respect for others' dignity is important precisely because it has consequences for their capacity and opportunity to fulfill their potentialities" (Kelman, 1982, p. 43). For example, lying to colleagues about scientific results or deceiving subjects about the purpose or procedures of an experiment violates human dignity by creating distrust within a community or by depriving individuals of information to meet their needs or to protect their interests. Using the principle of human dignity as a "master rule," Kelman indicates that it may be useful in resolving conflicts that arise in the development of a research project by weighing the costs and benefits of taking various courses of action, and choosing the actions that are most consistent with the preservation of human dignity.

Kelman's abstract approach to human dignity substantively undergirds the more accessible principles promulgated in the Belmont Report (National Commission, 1979), which exists as the cornerstone for the federal requirements for the protection of human research participants. The report describes three main principles. First, respect for persons captures the notion that individuals are autonomous agents. The principle also allows for the protection of those with diminished capacity (i.e., members of vulnerable populations who have limited autonomy owing to legal status, age, health, subordination, or the like). Second, beneficence refers to an obligation to maximize possible benefits and to avoid or minimize potential harms. This principle is consistent with Kelman's emphasis on the means to resolve conflicts between rules by opting for the best means to preserve human dignity. Third, the principle of justice pertains to who ought to receive the benefits of research and bear its burdens. In this sense, justice pertains to the selection of research participants insofar as those who bear the burden of research should also be the ones to benefit from it. In addition, the justice principle requires reasonable, nonexploitative procedures. These elements of justice highlight Kelman's emphasis on the element of the community in protecting human dignity.

The principles of the Belmont Report reflect the work of bioethicists Beauchamp and Childress (2001), who offer "calculability and simplicity in ethical decision making" (Israel & Hay, 2006, p. 18), especially in comparison to the lofty abstraction of other philosophical traditions. Although such principles seem to translate more readily into guidelines for ethical research, the abstract moral principles provide the larger framework in which to consider what is right and what is wrong in the pursuit of a scientific understanding of social behavior. In other words, it is important not to lose the fundamental concern with protecting human dignity—both for the individual and for the community—when designing a study, interacting with study participants, and communicating the study's results.[4] Indeed, researchers must consider the ethics of their research and take steps to protect study participants even when they are not strictly required to do so by federal regulations.

III. ETHICAL ISSUES IN LABORATORY EXPERIMENTS

Ethical concerns arise with the use of any social scientific method, both in general terms and in terms of particular characteristics about each method. Laboratory experiments challenge the protection of human dignity in ways like those of other methods and also in ways unique to the experimental method.

A. GENERAL ETHICAL CONCERNS

Methodology chapters pertaining to research ethics include lists of concerns (e.g., Babbie, 1998; McBurney & White, 2004; Neuman, 2007). Although the lists are of varying lengths and sometimes employ different labels, four interrelated

---

[4]Although Kelman (1982) draws attention to identity and community as complementary aspects of the principle of human dignity, some (e.g., Hoeyer, Dahlager, & Lynöe, 2005) argue that the ethics shaping biomedical research tend to privilege the individual and his or her autonomy over the political implications of the research. The latter concern is more characteristic of views in anthropological research traditions that share a collective memory of the complicity of some researchers in the execution of power over indigenous peoples and that depend upon the development of rapport, reciprocity, and trust with their research informants.

topics generally emerge: objectification of research participants; potential harms; coercive, exploitative, or intrusive practices; and maintenance of privacy or confidentiality. Each concern implicitly or explicitly raises issues of power and trust between the researcher and research participant. As Neuman (2007) notes, "A researcher's authority to conduct social research and to earn the trust of others is accompanied always by an unyielding ethical responsibility to guide, protect, and oversee the interests of the people being studied" (p. 49).

1. Objectification

As a consequence of the traditional view of science that separates the observer (the scientist) from the observed, researchers often deem the individuals that they invite to participate in their studies as their "subjects." The term "subject" transforms an autonomous actor or moral agent into an object of study. The researcher controls—manipulates in some instances—the context in which the subject is allowed to behave. The extent of control is variable, ranging from very little (e.g., observation studies occurring in public places, mail surveys) to a great deal (e.g., laboratory experiments). Use of the term "subject" compared to the more neutral term "study participant" may also raise a graver consequence than the potential loss of autonomy.

The transformation of individuals into subjects carries the danger of treating participants as merely "research material" (Veatch, 1987) and separating the researcher's humanity from that of his or her study participants (Neuman, 2007). Conceiving of subjects as objects rather than people, researchers distinguish themselves as members of a more powerful group and subjects as members of a subordinate group. In doing so, researchers may grow callous with regard to the potential harms that their studies inflict on their subjects.[5]

An alternative view of the role of research participants, embraced by feminist and humanistic social scientists, is that of collaborator (McBurney & White, 2004). Some researchers take this conception to an extreme by encouraging participants to contribute to the study design. Others consider the more general prospective participant perspective that suggests seeking ethical advice from members of groups who will serve as study participants (Fisher & Fyrberg, 1994). The critical aspect of these ideas is the reminder that experimental subjects or participants are human beings and deserve esteem and respect, in line with Kelman's (1982) principle of protecting human dignity.

---

[5]The development of such callousness is much like what emerges in in-group versus out-group interactions. To the extent that members of a group perceive themselves as distinct from—perhaps even superior to—members of another group, they are more inclined to discriminate against out-group members (see Hogg, 2003).

2. Potential Harms

A study participant may or may not realize that he or she is being treated as an object. If such a realization occurs, the individual may feel distressed or resentful. Whether the level of those negative feelings exceeds similar feelings experienced in daily life, however, may depend upon other elements of the situation, such as how much information the researcher provides and the researcher's reactions to any display of distress. More generally, in assessing the potential harms that may occur in any research study, consideration must be given to (1) the nature of the harm; (2) the degree or intensity of the harm, often assessed by comparing with harms associated with activities of everyday life; and (3) the actual likelihood or risk that a harm will occur, distinguishing between the possibility of a harm occurring and its probability (i.e., many harms may be possible but the probability of any particular harm may be extremely low).

Experiences that detract from well-being or create an ill effect may constitute harms (Kelman, 1982). In social science research, a variety of harms may occur. Sieber (2003) identifies various types of harms as well as degrees of each type. She also offers means to ameliorate each type of harm.

The most extreme type of harm, physical harm, is very rare in social and behavioral research (Neuman, 2007) compared to the risk of it in biomedical studies. Sieber (2003) describes physical harm as minimal (transitory or a minor injury) or major (involving assault or creating a life-threatening situation). It is incumbent upon the researcher to take appropriate safety measures, including anticipating the possibility of risk, screening out vulnerable populations, and monitoring participation.

More common in social science research studies is inconvenience, psychological harm, or social harm. Inconvenience may stem from asking participants to engage in boring, repetitive tasks or asking for the time of individuals with many existing demands. Generally, forewarning of and consenting to what a study involves ameliorates such harm.

Psychological harm is, perhaps, the most likely risk in social and behavioral investigations. Researchers sometimes create situations that lead to embarrassment, worry or anxiety, depression, shame or guilt, or even loss of self-confidence or self-esteem. These discomforts may stem from requesting that participants reveal private facts or traumatic experiences, or act in ways they did not anticipate. In doing so, individuals may come to question aspects of their own identities. The degree to which subjects experience these harms is highly variable, depending upon the characteristics of the research study itself (i.e., questions asked, conditions manipulated, tasks required, employment of deception) as well as the degree of respect the investigator affords the study participants. Properly informing subjects about a study, including the possibility of these psychological harms, may reduce their impact. Also, it is incumbent upon the researcher to sense distress in participants and to assure them by giving them the opportunity to skip questions or discontinue participation.

Social harms generally refer to threats to an individual's reputation or his or her relationships with others. Revealing private facts about a study participant (e.g., HIV/AIDS status, drug use) may compromise his or her standing in a work place or community. Subjecting a person to an embarrassing situation may decrease the level of esteem accorded to him or her by friends, family members, or coworkers. Careful consideration of the likelihood that study procedures may cause these harms and how they may be circumvented is the task of the researcher. In addition, as discussed further later, to avoid compromising a study participant's standing, the researcher must be prepared to ensure the confidentiality of the data collected.

Sieber (2003) also describes economic and legal harms. Economic harms may range from the loss of a few dollars to the loss of financial opportunities, credit, jobs, and so forth. Again, it is important for the researcher to ensure confidentiality and potentially to compensate the participant should loss occur. The potential for legal harm is most likely in studies of illegal behavior and includes involvement with lawyers or law enforcement and the potential for convictions. A key strategy to protect participants from such harms is the assurance of anonymity, as noted later, and the a priori recognition of the risks both for the subject and for the researcher (should he or she witness illegal activities and/or be called upon to testify).

As Babbie (1998) notes, just about any study might potentially harm someone in some way. The strongest scientific grounds must exist for pursuing any study that causes participants to suffer ill effects. Moreover, it is the researcher's responsibility to recognize and minimize those risks of harm.

3. Coercive, Exploitative, and Intrusive Practices

Potential harms are typically consequences of the researcher's actions. Equally harmful, however, are coercive, exploitative, or intrusive practices that investigators may employ knowingly or unknowingly in their quest for knowledge. The ethicality of this sort of practices highlights the power relationship between the investigator and potential participants.

To the extent that a researcher has something (e.g., a treatment, money, a grade, other special rewards) that a potential subject might desire, he or she is in a power-advantaged position vis-à-vis those who might enroll in the researcher's study. Using the desired resource to entice participation constitutes a form of coercion. To avoid the appearance of or actual coercion, emphasis on such enticements must be minimal; that is, the researcher must limit the amount of money to be paid to the subject and must never make course grades dependent upon study participation. Moreover, it is important for researchers to understand that participants are volunteers who are—and who must know that they are—free to decline participation.

Even once individuals have agreed to participate, researchers cannot exploit this agreement by asking them to answer types of questions or perform behaviors that had not been previously described or by prohibiting their withdrawal from the study. Although people may be apprehensive about participating in studies, once committed they may feel obligated as "research subjects" to stick with it, despite experiencing discomfort (McBurney & White, 2004). To take advantage of adherence to such role demands is a form of power use, of exploitation. To minimize the potential for exploitation, researchers must be mindful of the voluntary nature of subjects' participation as well as recognize their obligation to respect their participants.

Such respect may also temper how intrusive the requirements of a study may seem to a participant. Babbie (1998) points out that participation in any study takes time and energy. In addition, some studies ask participants to reveal information about which even friends or coworkers know nothing. Providing such information, moreover, may not be associated with any direct benefits to the individual participants. Researchers, however, may recognize the need to obtain certain types of information to advance knowledge or to develop policies to address social problems.

In other words, because social research has the potential to generate social benefits, intrusive practices may be seen as justified. The researcher who wields expert power must explain to participants why particular types of information are being solicited or types of behaviors observed. In other words, the researcher fully informs the participant. With such explanations, it is then up to the individual to decide whether or not to participate. The voluntariness of subjects' participation is one means to equalize the power relationship.

4. Maintenance of Privacy and Confidentiality

Insofar as researchers implicitly or explicitly ask their study participants to disclose information about themselves or their behavior, they are obtaining information that people might want to keep private. Ethical concerns arise regarding how those revelations will be held. Promises of anonymity or confidentiality are a basis for protecting individuals' privacy (Babbie, 1998).

Normative rights to privacy assure individuals that they need not reveal information about themselves. In other words, people control who has access to such information, which in turn affects their willingness to participate in research (Sieber, 1992). Moreover, privacy allows individuals to establish personal boundaries. When individuals agree to participate in a study, they provide the researcher with access to certain types of information. In some instances, it is possible to assure research participants that their revelations will be anonymous—that no names or other unique identifiers are associated with the information given.

More typically, however, researchers know their subjects' names and may be able to link those names to their revelations. Such links allow researchers to track study participants for follow-ups or triangulation of data sources (e.g., official records with interview or survey responses). To protect the privacy of study volunteers, researchers may promise that the data will be held confidentially. Specifically, confidentiality refers to how records about the study participant will be handled (Sieber, 1992), including identification of who may access the information, how the information will be securely stored, and how it will be disseminated. The potential for breaches of confidentiality is a major source of potential harms in social and behavioral research studies (Citro *et al.*, 2003). Technological developments in data collection, processing, dissemination, and analysis may compound the likelihood of social, legal, and economic harms. For example, the responses of participants in online experiments or surveys may be traceable to individuals and stored data accessible to unauthorized personnel.

To prevent breaches of confidentiality and thus protect private information, researchers may enact several interrelated strategies (Babbie, 1998). These strategies include the ethical training of researchers and their assistants, de-identification of study documents by removing names, separate storage of and limited access to master files linking identities with study data, password-protected computerized data files, and presentation of data only in aggregate form or only with pseudonyms. Although researchers usually stress that they will keep data confidential and do all that they can to avoid public disclosure of identified information, they can only do so to the extent allowed by law. Law enforcement officials may, under certain conditions, subpoena research documents and failure of researchers to provide them may result in contempt charges.

A certificate of confidentiality issued by the U.S. Department of Health and Human Services (via the National Institutes of Health or the National Institute for Justice) protects researcher–participant privilege and may be granted for funded or unfunded studies for which protection of confidentiality is necessary to achieve study purposes (Sieber, 1992). Researchers may obtain such certificates for studies on sensitive topics that might result in social, economic, or legal harm to the study participants should identifiable information be released. Studies on mental health, use of addictive products (e.g., drugs, alcohol), sexual attitudes and practices, genetic makeup, illegal behavior, or the like may qualify for certificates of confidentiality.[6]

---

[6]For more information, consult http://grants2.nih.gov/grants/policy/coc/.

Most experimental investigations are likely to fall outside the range of studies that qualify for certificates of confidentiality. Thus, it is imperative that experimenters protect study participants' identities in other ways. By sincerely informing research participants that the information that they provide will be held confidentially, the researcher indicates respect, potentially guards against various types of harms, and minimizes perceptions of objectification and power-based practices. Researchers who are cognizant of these ethical issues and act upon them in all stages of research, from recruitment of participants to data collection to presentation of results, are likely to reinforce the public's trust in the research endeavor in general. Experimentalists must also attend to several specific issues.

B. Specific Ethical Concerns

Experimental research is often unique from other methodological approaches insofar as researchers' emphasis on the isolation of key factors (and thereby the control of extraneous factors) leads them to use deception to enhance experimental control. Even when deception is not used, there is a tendency not to inform study participants fully about the study in order to develop more control over experimental treatments. In such cases, the completeness of informed consent becomes an issue. Also, because laboratory studies may involve special equipment located in dedicated spaces, experimenters have come to rely upon participants who are readily available, such as those who are enrolled in courses within departments.[7] In some instances, students become part of highly organized "subject pools," the members of which take part in research studies as part of the curriculum. Concerns about deception and subject pools embody issues of objectification, potential harms, coercion, and confidentiality in various ways.

1. Deception

Research involving deception has a long history, especially in psychology (see Korn, 1997). Deception refers to acts of providing false information or withholding information intended deliberately to mislead others into believing something that is untrue (Bordens & Abbott, 2005; Sieber, 1992; Sieber, Iannuzzo, & Rodriguez, 1995). Such acts are distinct from the typical practice in experimental research of not acquainting subjects fully in advance with all

---

[7]For experimental research, the use of nonrandom samples, such as a pool of college students, is not problematic so long as a study is theoretically based and the college students meet the theory's scope conditions (see Walker & Willer, Chapter 2, this volume).

aspects of the research being conducted, such as the hypotheses to be tested and the nature of all of the experimental conditions (Baumrind, 1985; Ortmann & Hertwig, 2002). Use of deception research is hotly debated, largely owing to its ethical implications (see, for example, Baumrind, 1985, who opposes deliberate deception, and Bonetti, 1998, who argues its benefits). Yet insofar as the nature of deception and the nature of the research in which it is embedded vary, the practice continues, albeit with precautions to prevent injury to human dignity.

A number of scholars have noted the potential harms of deception research at all levels. Elms (1982) indicates that subjects, researchers, researchers' professions, and society may be harmed by the use of deception in research. At the individual level (Baumrind, 1985; Korn, 1997; Seiber, 1992), deception undermines research participants' autonomy and right to self-determination because the lack of full disclosure of experimental purpose and procedures impairs participants' decision-making ability (Kelman, 1982). In addition, individuals may feel embarrassed or even suffer a loss of self-esteem upon realization that they have been duped. Kelman argues that such consequences deprive research subjects of the respect that ensures human dignity in interpersonal relationships. The absence of that respect may enhance the objectification of the subject. In addition, the failure of researchers to demonstrate respect and veracity in their dealings with subjects may undermine their own reputations in the long run.

The cumulated effects of the use of deception may negatively affect a profession as well. Researchers within a discipline who rely heavily on deception methodologies may be seen to abuse the power associated with their expertise (Korn, 1997). In addition, deception may stimulate suspicion and negative attitudes about research among potential subjects that undermine the work of their colleagues using nondeceptive methods (Baumrind, 1985; Bordens & Abbott, 2005; Sieber, 1992). Of particular importance to economists is a related practical concern: if study participants are aware that deception may be used, their behavior may be shaped more by psychological reactions to suspected manipulations rather than by actual situational circumstances (e.g., induced monetary rewards) (Davis & Holt, 1993). Economists (Davis & Holt; Hertwig & Ortmann, 2001) also emphasize the importance of maintaining a reputation for honesty and claim that the use of deception may undermine that reputation, even if practiced by only a few.

At a more general level the concern is that deception in research has the potential to erode the public's trust in social scientific endeavors (Baumrind, 1985; Kelman, 1982; Korn, 1997) and lead to cynicism (Neuman, 2007). This erosion in trust may play out in a university by making students and others unwilling to volunteer for research of all types. At the societal level, experience with deception in research may create a willingness in others to use deception themselves and contribute to a lack of trust in general (Elms, 1982; Sieber, 1992), thereby eroding a fundamental social value that binds relationships (Kelman, 1982).

Ortmann and Hertwig (2002) attempt to assess empirically some of these potential harms. Drawing largely from psychological studies, they show that subjects' direct experience with deception generates suspicion, which in turn may alter their judgment and behavior in subsequent studies. In this way, use of deception may invalidate the results of future investigations, perhaps through the operation of demand characteristics that lead subjects to guess at the objectives of the study and to alter their behavior accordingly. Empirical evidence, however, is lacking for the spillover effect of indirect experience— that is, knowledge that deception is used in laboratory studies—on emotional or behavioral responses to research.

Whether deception causes direct harm to subjects is ambiguous, given the conflicting evidence (Ortmann & Hertwig, 2002). For example, Christensen (1988) notes that his review of the literature demonstrates that subjects in deception studies do not perceive that they have been harmed. Likewise, although Epley and Huff (1998) detect the direct effect of deception on suspicion, they show that participants had few negative reactions to being in a deception experiment. However, when researchers directly ask college students how they are likely to feel upon learning that they participated in a study involving deception, students indicate anticipating negative feelings such as discomfort, sadness, or embarrassment (Fisher & Fyrberg, 1994).

Little evidence exists for the potential threats to the credibility of a profession or threats to human dignity in society (Ortmann & Hertwig, 2002). Student evaluators of deceptive research (Fisher & Fyrberg, 1994) largely believe that studies are scientifically valuable and valid and that a study's social benefits outweigh the costs. The potential harms signal that deception should be used cautiously and only with substantial scientific and ethical justification.

Typically, the use of deception in research is defended in terms of its benefit to scientific ends, which in turn benefit society (Elms, 1982; Korn, 1997). These benefits are weighed in view of the costs or potential harms identified previously. Deception studies should not cause discomfort at a level that would have prevented people from participating had they known fully in advance what the study would involve (Fisher & Fyrberg, 1994). Sieber (1992; Sieber et al., 1995) offers four defensible justifications for deception research:

- Deception may be employed to achieve stimulus control when no other alternative procedures are feasible. In such cases, the use of deception presumably ensures the execution of valid research with scientific, education, or applied value and not simple trickery.
- Deceptive conditions may allow the study of responses to low-frequency events.

- Designs involving deception must not present any serious risk of harm to study participants.
- Deception may be an appropriate means to obtain information that would otherwise be unobtainable because of subjects' anxiety, fears, embarrassment, defensiveness, and so forth.

These considerations diminish a researcher's ability to coerce subjects into a study involving unjustified deception. Decisions about the ethics of deceptive research extend beyond these justifications. As Ortmann and Hertwig (2002) suggest, "whether deception ... is considered acceptable by a participant, is a function of ... the nature and severity of deception, the methods of debriefing, and the recruitment mode" (p. 117).

Some deceptions are relatively innocuous, involving false expectations about experimental procedures, while others are more serious, such as providing subjects with false feedback about their performances on tasks that bear on their self-evaluations outside the laboratory (McBurney & White, 2004). Deception may be active (e.g., misrepresenting the purpose of the research, misidentifying the researcher, providing misleading information on equipment or procedures, using confederates of the experimenter) or passive (e.g., concealed observation, unrecognized conditioning) (see Bordens & Abbott, 2005).

Sieber *et al.* (1995) elaborately codify the harmfulness of deception in terms of (1) the kinds of failures to inform (e.g., false informing about a study or its procedures, no informing, consent to deception, waiver of the right to be informed); (2) the nature of the research in terms of the harmfulness of the behavior studied, the privacy of the behavior, the confidentiality of the data collected, and the power of the means of inducing the behavior; (3) the topic of the deception (e.g., the study's purpose, stimulus materials, feedback about self, feedback about others, actual participation in a study); and (4) the nature of the "debriefing" or explanations given after the study. Deception studies focusing on harmful behaviors that subjects would most likely prefer to keep private, induced by powerful means, involving false information about oneself without suitable debriefing are most likely to (rightfully) prompt ethical objections. Although Sieber *et al.* document that published deception research declined from 1969 to 1986 and then rose again by 1992, they conclude by cautioning that "to be acceptable, deception research should not involve people in ways that members of the subject population would find unacceptable" (p. 83).

The process of debriefing pretest subjects is a way researchers can determine what potential study participants may find to be objectionable or harmful. Researchers may use responses of pretest subjects to modify their procedures to reduce or at least ameliorate potential harms and to further clarify important information conveyed by the debriefing. For both pretest and study participants, debriefing typically involves two processes: dehoaxing and desensitizing

(Holmes, 1976a, b). Dehoaxing is the process of informing subjects after the session of the experiment's true purpose and of revealing the nature of the deception. Indicating how equipment actually worked or providing actual performance scores on tests is part of this process. The researcher needs to dehoax without increasing subjects' embarrassment over being duped and with the intent of eliminating any mistrust engendered by the study with regard to the scientific endeavor. A majority of Fisher and Fyrberg's subjects note that if people felt embarrassed after learning of the deception, they would be unlikely to reveal that emotion to the experimenter but might rather reveal annoyance or anger (1994).

Desensitizing focuses on attempting to remove any emotional harm that the study and specifically the deception may have caused.[8] The intent is to restore a sense of positive well-being, presuming that any emotional distress is temporary; studies that damage self-esteem or involve lasting harms should be avoided in the first place (Elms, 1982; Sieber, 1992). If necessary, the researcher should draw upon additional resources to handle the stress created by the study itself as well as, potentially, by the debriefing. By providing the subjects with the opportunity to ask questions, receive adequate responses, and withdraw from the study, the researcher may further minimize negative consequences of the study. These actions demonstrate respect for study participants and may engender continued trust in scientific research.

Deception research continues in some disciplines (e.g., psychology, sociology, political science) and is de facto prohibited in others (e.g., economics). Geller (1982) suggests that forewarning of the possibility of deception, but not giving the actual description of it, may be an alternative, more ethical approach. Associated methodological issues (e.g., demand characteristics, threats to random sampling), however, may compromise the validity of the data. Fisher and Fyrberg (1994) find that potential study participants think that forewarning would discourage participation owing to the discomfort people may feel in being "controlled."

Social scientists continue to debate the use of deception, often invoking different philosophical approaches to considering what is ethical. At a minimum, a focus on protecting human dignity may circumvent the emergence of potential harms to individuals, to professions, and to society. Moreover, careful pretesting may alert researchers to unexpected harms and lead them to devise more ethical procedures. Ironically, to the extent that experimental situations avoid mundane realism yet appear realistic enough to be taken seriously, the potential harms may be less likely to arise because subjects may attribute deception to

---

[8]Awareness of the possibility of discomfort, anxiety, distress, and the like should be indicated during consent procedures and monitored throughout the study. The researcher should halt any study session in which such significant suffering become obvious.

the artificiality of the laboratory rather than to the specter of having been treated without dignity.

2. Subject Pools

In order to provide a ready stream of research subjects, many (psychology) departments require students in introductory or other lower division courses to participate in studies conducted by faculty or other (graduate and under-graduate) students.[9] The number of hours of participation or number of studies of which a student must be a part varies, but the participation counts toward the course grade. In other words, students get credit for becoming research subjects or they are offered alternative activities (e.g., writing a paper, doing extra homework, taking a quiz, and viewing a movie or demonstration) to earn the same credit.

The creation and use of such subject pools produces conflict between the development of knowledge and the ethical treatment of research participants (Sieber & Saks, 1989). To the extent that students feel pressured into partici-pation, they suffer from both objectification (i.e., existing as fodder or "raw material" for someone's study) and coercion. In effect, researchers violate subjects' autonomy and demonstrate disrespect, thereby compromising human dignity.

As noted before, voluntariness in the decision to participate in a research study is a key aspect of ethical treatment. In psychology, investigators support the long-standing and pervasive practice (Landrum & Chastain, 1999; Sieber & Saks, 1989) by emphasizing the benefit of subject pools: the development of valid scientific knowledge while providing students with an important hands-on educational experience. The educational value of participation in research via the subject pool is students' direct exposure to the research domain and the process of scientific inquiry. In addition, some argue that students may learn how to conduct research ethically by participating in presumably ethical experiments.

Although Waite and Browman (1999) show that students who participate in research hold more favorable attitudes toward it than those who do not participate, other empirical work questions whether the promised educational value of participation materializes. Landrum and Chastain (1999) note that, in largely undergraduate departments, students do not typically assess the

---

[9]Sieber and Saks (1989) report that 74% of psychology departments have subject pools. Of these, only 11% indicate that their subject pool is entirely voluntary (i.e., there are no require-ments for participation in studies or for extra credit). In 4-year colleges, public institutions are more likely than private institutions to have subject pools (49.9% and 35.1%, respectively). Undergraduate institutions are more likely than those with graduate programs to make their sub-ject pools totally voluntary (34.7%).

educational value. Sieber and Saks (1989) discuss the difficulty of assessing the educational value of participation in studies in departments hosting graduate programs. They indicate that departments vary widely in what they consider as mechanisms to ensure a valuable educational experience (e.g., from providing an abstract of results months later to a 5-minute or more discussion of the study coupled with the subjects' own evaluations of the research experience). In order to contend that required participation in research has educational value, Sieber (1999) urges that researchers should at least debrief participants verbally and in writing on the background of the study and the current purposes, in language understandable to the students. Typically, such debriefing should occur immediately following the research session.

In the absence of an educational benefit, it is more difficult to justify subject pool practices as ethical largely owing to the extent to which they violate the principle of voluntary participation. Emphasis on alternative activities for the same credit is one means by which departments ameliorate the appearance of coercion. Yet the typical alternative of writing a paper may be more onerous (Sieber & Saks, 1989). In addition, coercion may arise in the debriefing process. Fisher and Fyrberg (1994) note that students may feel restrained in the questions they ask or what they say once they learn about the study for fear that saying the wrong thing might jeopardize their credit. In effect, the students may not feel totally free to withdraw from the study because they may fear—rationally or irrationally—losing the required credit. Thus, it is important for researchers to assure subjects that they will receive credit, regardless of their responses during the debriefing or even if they withdraw partway through the study.[10]

Sieber (1999) and Landrum and Chastain (1999) suggest procedures to ensure the ethical treatment of subject pool participants. The procedures emphasize ways in which researchers can demonstrate respect toward the members of the participant pool. For example, information on subject pool requirements should be widely available to students (via catalogs, course Web sites and descriptions, specific handouts phrased clearly and respectfully, etc.). Students should also be made aware of the process by which they may file complaints about their experience in research studies. Consideration should also be given to offering equivalent (in terms of educational experience, time, effort, and so forth) alternatives. In addition, departments may devise means to monitor the behavior of researchers with regard to their awareness of subject pool protocols and their proper treatment of participants in their studies. Although subject pools have long existed, practices in many departments have or must evolve in keeping with changes in professional codes and the federal regulatory requirements of research.

---

[10]Some IRBs require explicit language in the consent document to indicate the conditions under which credit will or will not be awarded.

IV. MEETING REGULATORY REQUIREMENTS

In the United States, the federal regulations for the protection of human research participants (45 CFR § 46) derive from the three ethical principles developed in the Belmont Report (1979): justice, beneficence, and respect. Adopted by 17 federal departments and agencies, these regulations constitute the "common rule" governing human subject research sponsored by the federal government. The regulations specify mechanisms to ensure IRB review of research, informed consent of study participants, and institutional assurances of compliance (Dunn & Chadwick, 2002). Most universities extend the review to all funded and unfunded research conducted by faculty, staff, and students and instigate additional criteria to be met prior to the approval of a project. Debate centers on whether the federal regulations fulfill the promise of protecting human dignity, especially in reference to social and behavioral research. Given the moral mandate for ethical research, researchers must learn to navigate the regulatory system.

A. IRB REVIEW OF SOCIAL SCIENCE RESEARCH

Presumably, IRBs protect the human dignity of individuals who volunteer to participate in research and ensure compliance with federal regulations. Yet IRBs may lose sight of the ethical foundations of human research protections. Dubois (2004) notes that many education programs focusing on human protections offer inadequate reasons for compliance with regulatory demands. Ironically, promoting compliance for its own sake or to avoid potential penalties is contrary to moral notions of promoting human dignity. He also notes that to the extent that IRBs unnecessarily impede research, they are acting unethically. Indeed, researchers frequently view IRB review as an unnecessary bureaucratic hurdle (DeVries et al., 2004). They focus on IRB actions that circumvent their presumed rights to conduct research with human subjects, without recognition of the IRB's ethical mission.

Researchers, however, do not have an inalienable right to pursue research with human subjects (Oakes, 2002). As with access to grants and publishing outlets, researchers face different types of reviews. In addition to their responsibility to ensure human well-being, IRBs also have a responsibility to foster research (Dubois, 2004) and not simply to enforce regulations mindlessly without attention to the diversity of research questions, subject populations, and methods (Eckenwiler, 2001).

Social scientists have long been among the most vocal critics of IRBs (see Citro et al., 2003). Designed largely for biomedical research, the regulations give short shrift to social and behavioral research (Oakes, 2002; Singer & Levine, 2003).

DeVries *et al.* (2004) note that social scientists fault IRBs for being "overprotective, unreasonable in their demands for consent, impractical in their directives for the protection of confidentiality, and excessive in the time required for review" (p. 352). These failings, they contend, stem from three interrelated sources.

First, although some institutions do have boards consisting wholly of social scientists, in general, few social scientists serve on IRBs. In the absence of social scientific knowledge on the boards, the second source of problems is the application of a biomedical model of research to social and behavioral investigations. Such an application results in unreasonable requirements or in overestimating risks of nonphysical harm and underestimating the benefits of a project because the study is unlikely to produce direct benefits (like improved health) for the subjects. For example, it is certainly unreasonable to obtain written documentation of informed consent from illiterate study participants and may be incongruous to demand parental consent for adolescents' participation in a minimal risk study that concerns behavior that parents often forbid or about which they are unaware (e.g., smoking) (see Diviak *et al.*, 2004).

The unintended consequences stemming from the inappropriate application of the biomedical model are compounded by a third problem, documented through empirical observation of actual board meetings by a qualitatively trained sociologist (see DeVries *et al.*, 2004): IRBs are likely to scrutinize social and behavioral research protocols more intensely than they do biomedical ones. As noted earlier, however, the risks of harm are likely to be far lower in social and behavioral research than in biomedical research.

Thus, a human subjects review of social and behavioral research suffers from the failure of IRBs to understand the nature of the research and to use the flexibility inherent in the federal regulations. As a result, regulatory compliance supplants the execution of a reasonable approach to protecting the human dignity of individuals involved in social scientific research. At its worst, in some cases, "by following the letter of the regulations one may act unethically" (Sieber, 2004, p. 403) by creating unintended consequences that increase rather than minimize harms.

To rectify these problems, a number of scholars emphasize the need for social scientists to grow familiar with federal regulations, to serve on IRBs, and to secure ethics training, and the need for IRBs to learn more about issues associated with social and behavioral research (Citro *et al.*, 2003; DeVries *et al.*, 2004; Dubois, 2004; Oakes, 2002). Of particular importance is the maintenance of emphasis on the underlying ethical reasons for regulations that protect the well-being and the rights of people who participate in research and applying those ethical premises to the interactions between researchers and IRBs as well. To move in that direction, social scientists must learn how to navigate the IRB review process at their institutions and might also consider how their knowledge of attitudes, group processes, and the like may contribute to an empirical understanding of issues relevant to human research protections (see Sieber, 2004).

B. Navigating the IRB

Although institutions have different procedures for obtaining IRB review, all are guided by the federal regulations and advisory postings of the Office of Human Research Protections (see note 2). Researchers and IRB members and staff have distinct sets of responsibilities (see also Dunn & Chadwick, 2002).

Most importantly, researchers bear the ultimate responsibility for the welfare of their study participants (see Table 1). Consent procedures and IRB review help to ensure human subject protections, but it is incumbent upon the researcher, who is the expert and the contact for study participants, to anticipate threats to well-being, to monitor the study for the emergence of harms, to cease data collection until changes can be made to minimize the risk of harm, and to notify the IRB of any adverse events. To ensure the welfare of study participants, researchers must use their professional judgment to recognize the benefits of their research in view of the risk of harm or violation of rights.

In addition, researchers must familiarize themselves with federal regulations and institutional policies. They may demonstrate compliance with the requirements by clearly communicating to the IRB the substantive questions of their work and the methodology to be used, outlining the benefits of the research in view of the likelihood of various potential harms, and indicating how they will obtain the consent of study volunteers and maintain the confidentiality of their responses. Moreover, researchers must be able to convey

---

TABLE 1 Researchers' Responsibilities in Protecting Research Participants

- Ensure the safety and welfare of research participants
- Use good professional judgment to determine research benefits in view of potential harms of varying types
- Demonstrate knowledge about federal regulations and institutional policies
- Comply with federal regulations and institutional policies
- Convey information to research participants in a way that they can understand

similar information to study participants in a way that they can readily understand. In effect, researchers mesh the ethical concerns embedded in their work with the demands of the regulations and institutional policies.

The IRB has the responsibility of guiding researchers through the review process. Boards must clearly indicate what information must be submitted. In addition, IRB members and staff must readily answer investigators' questions. Those answers, moreover, should take into account the nature of the research and the subject population in order to avoid a cookie-cutter approach, which may create unintended negative consequences. IRB reviewers must have sufficient expertise to address the wide range of substantive issues addressed in the studies that they review. The boards themselves include community representatives who recognize harms to which professional researchers may be blind. IRBs should also conduct reviews in a reasonable amount of time and, when changes are requested, explain the concerns underlying the requests. By doing so, IRBs further educate investigators.

The federal regulations identify three types of IRB review (see Citro et al., 2003; Oakes, 2002).[11] Regulatory definitions of research and of human subjects determine which studies are excluded from IRB review. Exclusions include studies not involving direct (like interviews) or indirect (like surveys) interaction. Most social scientific research falls into the category of "minimal risk," meaning that the probability and magnitude of harms are no greater than those ordinarily encountered in daily life and also may be ameliorated by the actions of the researchers (e.g., stopping a procedure that seems to produce stress, ensuring confidentiality of data). Institutions rather than researchers judge both the level of risk and designate the nature of the review: exempt, expedited, or full board.

When a study involves public observation of behavior, anonymous surveys, or interviews with adults who are not vulnerable in any way (e.g., owing to pregnancy, diminished capacity, or incarceration), it is likely to involve only minimal risk and to fall into the "exempt" category of review. Although, technically, such studies fall outside the purview of the federal regulations, the IRB must designate the study as "exempt." Minimal risk studies that involve written documentation of consent, possible risk of loss of confidentiality, collection of voice, video, and/or image recording for research purposes, or collection of small biological specimens through noninvasive clinical procedures are likely to be subject to "expedited" review. Institutions often execute exempt and expedited reviews in a similar fashion, employing rolling reviews by a designated committee or staff member.

---

[11]These descriptions of the different types of reviews are simplified. Please consult the Web site specified in note 2 for detailed explanations as well as the regulatory definitions of research and human subjects.

Full board review is typically reserved for studies involving greater than minimal risk, vulnerable populations (e.g., children, prisoners, pregnant women when there is a threat to the fetus, individuals with diminished capacity), deception, and/or changes in or waivers of informed consent elements. Although researchers sometimes request a particular type of review for their studies, ultimately it is the IRB that determines the review level. Attempts to thwart higher levels of review are typically unsuccessful. Thus, it is in the investigator's best interest to prepare a thorough, explicit, and honest IRB proposal.[12]

C. Preparing an IRB Proposal

Although most researchers deem the IRB proposal a bureaucratic hurdle to be overcome in order to secure grant monies and/or collect data, the moral mandate for ethical research should implicitly facilitate the preparation of materials. Moreover, to the extent that investigators cooperate with their IRBs, mutually shared goals of protecting human research participants are likely to be achieved. To ensure success in the IRB review process, investigators must prepare clear, consistent, and complete documents that convey their concern with designing procedures to protect the well-being of study participants.[13]

In addition to the (paper or electronic) application form designed for use within a specific institution, an IRB proposal typically includes the research protocol, consent documents, recruitment materials, and instruments (see Table 2). For experimental research, additional items may include the debriefing statement, details of experimental procedures, and specific rationale for any deceptive practices. Institutions may also require signatures from department chairs and/or faculty mentors, statements of researchers' qualifications, documents supporting the research from those with authority over the research context (e.g., clinic administrators, school superintendents or principals). In the case of international studies, researchers may also have to provide information about a local contact or in-country ethics review procedures. Described more fully later and listed in Table 2 are the documents to include in any IRB application, as well as those especially relevant to experimental studies.

---

[12]To omit relevant information intentionally in order to secure, for example, an exempt review would constitute unethical behavior, which would reinforce the views of some IRB staff that investigators are not to be trusted.

[13]Sieber (1992) has a guide for students, relevant also to faculty and to IRBs, for planning ethically responsible research. She includes a detailed discussion of what to include in an IRB proposal. Oakes (2002) provides definitions of regulatory terms and advice regarding the preparation of IRB proposals. Gillespie (1999) outlines IRBs' concerns and productive measures for researchers.

TABLE 2 IRB Application Materials

- IRB application (electronic or paper)
- Research protocol: describe or identify
- Theoretical background of substantive issue to be addressed
  - Benefits of the research
  - Research design
  - Characteristics of study participants
  - Recruitment procedures, including incentives
  - Potential types of harms and their likelihood of occurring
  - Deceptive practices, if any, and how the study meets the requirements for waiver of fully informed consent
- General study procedures
- Data handling procedures, especially with regard to confidentiality
- Data analysis plan
- Consent documents
  - Required elements
  - Optional elements
  - Institutional requirements
- Recruitment materials (e.g., flyers, ads, verbal pitches)
- Data collection instruments (e.g., surveys, questionnaires, measures)
- Experimental procedures (e.g., verbal or written instructions to participants, computer screens)
- Debriefing materials

The research protocol describes the substantive question to be investigated, using terminology for an educated lay audience. This description situates the project in the larger literature, demonstrating how it fills a void and thus will provide social benefits through augmenting knowledge about particular issues or processes. By doing so, the researcher begins to convey concern with the ethical principle of beneficence: maximizing good outcomes for science, humanity, and study participants (Seiber, 1992).

Within the protocol, the researcher should provide detailed information on who will be involved in the study. Characteristics of the participants, their selection and recruitment, and ultimately how they are treated in the course of the study provide the means to assess the principle of justice. Sieber (1992) suggests that justice pertains both to the fair distribution of costs and benefits among persons so that those who bear the risks of the research might also benefit from it and to ensuring reasonable and nonexploitive procedures. Materials indicating how subjects are to be recruited (e.g., flyers, advertisements, Web sites, scripts) must provide a brief description of what is involved in the study and who the investigators are without overstating the benefits of the study or suggesting excessive inducements. The latter raise concerns with exploitation or coercion, blatantly unjust strategies. The application of the principle of justice in experimental work also requires that researchers explain why they plan to use particular sets of participants and how they will make sure that the conditions to which subjects are randomized carry similar levels of risks of harm and of benefits.

The researcher's analysis of the risks inherent in the research should accompany the description of the actual study procedures (e.g., experimental manipulations, explanations of the means to measure dependent variables). Although physical harms are generally unlikely, researchers should assess the extent to which psychological, social, or other harms may occur in the course of a study. An outline of the means by which to maintain the confidentiality of the records should convince the IRB that the researcher has minimized the potential for a breach in confidentiality, which is a major risk in social and behavioral research. More generally, discussion of strategies to ameliorate potential harms is part of the risk-benefit analysis underlying the principle of beneficence. Such analysis is particularly important in research involving deception.

The process of debriefing and the concerns that may arise in the course of debriefing must be detailed in an IRB proposal. Moreover, the researcher should anticipate any negative effects from the debriefing itself, such as inducing a feeling that the participant is gullible or easily swayed by others' opinions. In all cases, researchers should assure the IRB that subjects, once fully informed of the nature of the research, will have the opportunity to drop out of the study with no loss of benefits (i.e., credit or compensation).

A key element of any IRB proposal is the description of the informed consent process. This process is the fundamental means by which the researcher conveys respect to potential subjects. Respect signals the investigators' concerns with the autonomy of the participants—their ability to make decisions for themselves.[14] In this process, the investigator communicates sufficient information to and answers questions from potential study participants so that they can choose whether or not to volunteer to participate in the study (Dunn & Chadwick, 2002; Sieber, 1992). In effect, the researcher and subject engage in an exchange of information before, during, and sometimes after the study. It is the researcher's responsibility to make sure that the information provided is at a level that the subject may comprehend. The consent process may be fully oral or involve written documents that may or may not require a signature. The default method of informed consent involves a written document that participants review and dis-

---

[14]The federal regulations recognize that some individuals have limited autonomy and may constitute traditionally vulnerable populations (e.g., prisoners, children, the mentally disabled). The additional federal safeguards and consent procedures for these populations may also apply to equally vulnerable populations of individuals in subordinate positions or resource-deprived environments (De Vries *et al.*, 2004; Dunn & Chadwick, 2002).

cuss with the investigator and then sign to signal agreement to participate (i.e., "written consent with written documentation").

What form the consent process takes depends upon the nature of the research question, the subject population, and the data collection procedures.[15] Flexibility inherent in the federal regulations allows IRBs to grant waivers of written documentation of consent if the consent form is the only document linking the participant to the study and the study involves no more than minimal risk. Under certain conditions, investigators may request waivers of other elements of consent. In the case of deception research, an investigator is essentially requesting a waiver of informed consent because the deception makes it impossible to convey all of the details of the study without jeopardizing its scientific validity. To justify the waiver, researchers must explain in their IRB proposal that (1) the research could not be otherwise carried out without the deception, (2) the study involves no more than minimal risk of harm, (3) omissions in the consent process do not adversely affect participants' rights or welfare, and (4) subjects will be provided with pertinent information whenever appropriate and be given the opportunity to withdraw. In other words, investigators must justify the use of deception in terms set forth by the federal regulations and with recognition of the ethical issues that deception engenders.

The federal regulations also stipulate required elements of consent. The core elements relevant for social and behavioral research include:

- introduction to the study that indicates it is a research project, by whom it is conducted, and the source of funding (if any)
- description of the study purpose, procedures, and duration
- disclosure of risks/benefits
- statements regarding confidentiality of records
- specification of the voluntariness of participation (which, for example, includes freedom to withdraw from the study or simply to skip questions)
- means by which to contact both the investigator, IRB officials, and, if relevant, a local contact in international contexts

Other elements of information should be included if they are relevant to the study and to what the participants may need to know to make an informed decision. For example, the number of subjects in a study may affect an individual's decision making if he or she has concerns about being identified.

---

[15]Variation in the means of obtaining consent is more common in qualitative or ethnographic research than in survey or experimental research, though it may also occur in the latter. Citro *et al.* (2003) identify issues associated with the default method of obtaining consent, and recommend that alternative methods may be more ethical in certain circumstances (e.g., when research involves special populations whose language skills or cultural attitudes make the default method inappropriate or when research is on illegal behavior or highly sensitive topics and privacy and confidentiality trump the need for written documentation of consent).

That situation may be very likely to arise in studies involving subject pools or participation of class members in their instructor's research. Also, if subjects are to be compensated, they should know in advance what to expect (e.g., a specific hourly amount; a range of possible outcomes, depending upon bargaining or exchange behavior; a specific amount of credit). Compensation per se, however, is not considered a direct benefit to the participant and thus does not enter into the risk-benefit analysis. Institutions also may specify reading level (e.g., eighth grade), structure of written consent document (e.g., with or without headings for subsections, font size, version dates), phrasing of particular elements, or additional elements of information. Researchers should consult with their local IRBs to learn of additional requirements.

In order to assess the consistency between what is specified in the IRB proposal and consent documents as well as to gauge the reliability of the researcher's own descriptions of potential harms, investigators must supply copies of their data collection instruments to the IRB. These instruments may be the actual questionnaires, interview schedules, or standard measures to which participants will respond. What may seem like harmless questions to an investigator may raise an IRB reviewer's concerns for the well-being of study participants. The IRB may also consider the explicit and detailed experimental instructions (e.g., what researchers say to subjects to manipulate variables, what information appears on computer screens) to constitute data collection instruments. With such information, reviewers may more readily imagine what a subject would experience, which may aid them in determining whether or not the investigator has anticipated potential risks of harms and/or created procedures to minimize such risks. Ultimately, the detailed description of the study significance and methods allows reviewers to assess the scientific merit of the study in view of its risks and benefits.

The IRB proposal is, at one level, much like writing the methods section to a journal article: investigators describe what they intend to do and how they intend to do it. In contrast to simply documenting what was done in order for others to replicate a study, the IRB proposal demands that researchers consider the implications of their procedures in terms of the rights and well-being of the people they invite to participate in their study. By ensuring justice, minimizing risks, and demonstrating respect in their interactions with subjects, researchers are likely to meet the moral goal of maintaining human dignity. That goal may also characterize interactions with IRB staff and committee members.

D. ADVANCING PRODUCTIVE INTERACTIONS

An overarching goal of both researchers and IRBs is the production of ethically responsible research that contributes useful knowledge about humans and their societies (Citro *et al.*, 2003). At the abstract level, their respective responsibilities are complementary. IRBs should assist researchers in meeting ethical standards for the treatment of human research participants. By doing so, IRBs ensure public accountability (Adair, 2000), researchers uphold the ethical codes of their disciplines, and volunteers are protected and likely to continue to trust in the scientific endeavor. Yet, at extremes, researchers condemn IRBs as obstructionists and IRBs perceive investigators as ethically bankrupt. Instead of stressing the complementarity of their responsibilities to create a cooperative relationship, researchers and IRBs often find themselves embroiled in an adversarial relationship. In addition to the call for mutual education noted previously (Citro *et al.*, 2003; DeVries *et al.*, 2004; Dubois, 2004; Oakes, 2002), facilitating the extent to which the ethical principles of respect and justice pervade researcher–IRB interactions may attenuate the negative tenor of the relationship.

Mutual education provides a basis for learning about the goals, roles, and behaviors of each group. Beyond gaining information, such education enhances the likelihood that researchers will be likely to imagine what IRBs want and IRBs may be able to understand researchers' goals and frustrations more clearly. The symbolic interactionist concept of role taking (Mead, 1934) captures this idea of imagining the interests and behaviors of others, which may prevent misunderstandings and facilitate interaction. Likewise, Eckenwiler (2001) emphasizes how incorporating the perspectives of others into decision making enhances moral thinking, and Dubois (2004) notes that anticipating the consequences of an action for others and taking them into consideration provides a sounder basis for a moral judgment.

While such processes are fundamental to determining the potential harms or benefits in an empirical study, they apply similarly to directions for (and assessments of) interactions between researchers and the IRB. Dubois (2004) also implies that by adopting a role-taking perspective, researchers are more likely to internalize the norms underlying regulatory demands. Generally, to the extent that disciplines nurture cultures of research ethics independently of IRBs, future generations of scholars may feel less overwhelmed and frustrated by the IRB process (Adair, 2000).

The inclination toward such role taking may emerge from the communication and interactions that routinely occur between investigators and IRB staff and committee members. Oakes (2002) emphasizes how important it is for researchers to feel free to ask the IRB questions about IRB proposals and procedures as well as federal regulations. Citro *et al.* (2003) contend that miscommunication between IRBs and researchers is a source of frustration for both sides. Those authors explicitly reinforce the notion that mutual education will create a better understanding of the functions of the IRB and the concerns of researchers that will pave the way to open, clear communications. At a minimum, all communications should be civil as a means to convey the courtesy that a general principle of respect entails (Sieber, 1992).

Drawing from procedural and interactional justice (Tyler & Lind, 1992; Bies, 2001), respect is key to ensuring that individuals in general feel fairly treated. Demonstrating respect involves treating people with sincerity, politeness, and dignity while refraining from deliberately being rude or attacking them. Accepting or esteeming other people's rights is also a way to show respect. Research indicates that authorities who demonstrate respect toward subordinates are likely to be perceived as more legitimate and to elicit greater compliance with their mandates (see Tyler, Boeckmann, Smith, & Huo, 1997). In addition, individuals are more likely to see decisions as fair if justifications are provided (e.g., Bies & Shapiro, 1988; Greenberg, 1993).

These patterns of findings provide a basis for understanding faulty and promising IRB–researcher interactions. To the extent that "IRB demands are perceived as unjust, researchers react by ignoring IRB demands and bypassing IRBs—and feel justified rather than unethical" (DuBois, 2004, p. 389). The dynamic of interaction requires that IRBs and researchers alike act justly and respectfully, just as the federal regulations mandate with regard to the treatment of study participants. Despite the increasing strictness of IRBs with regard to their review of social and behavioral research (Oakes, 2002), maintenance of the human dignity of individual actors in their roles as researchers, IRB staff, or IRB members may enhance cooperation that ultimately facilitates the creation of IRB proposals, their review, and compliance with federal regulations.

V. CONCLUSIONS

The production of knowledge is a venerable pursuit. It is, however, not an untethered one. While multiple definitions of what constitutes ethical research exist, at a minimum, producers of knowledge must be mindful of the implications of their study procedures for the rights and welfare of the people who volunteer for their investigations.

This chapter has outlined ethical concerns regarding social science research in general as well as issues more specific to experimental methodology. In addition, because the outside world intrudes upon research in terms of federal regulations enacted for the protection of human research participants, the chapter has attempted to reinforce the ethics underlying the regulations and to provide advice with regard to navigating the review process. Ironically, social scientists could also provide advice to IRBs by bringing their substantive and methodological expertise to bear upon determining the best ways to ensure the protections of study volunteers (Citro *et al.*, 2003; Seiber, 2004).

Other chapters in this volume provide more explicit and, perhaps, concrete advice with regard to the purpose and construction of laboratory experiments in the social sciences. Such advice is critical because poorly defined studies cannot yield meaningful results and thus cannot justify the demands on study participants (Rosenthal, 1994). In other words, invalid research is also unethical (Sieber, 1992). Also, it is within the purview of IRBs to decline approval of such studies. In contemplating the design of valid studies, ethical directives may constitute methodological challenges (Rosnow, 1997) that, when successfully met, have implications beyond the particular substantive issue under study. They protect human dignity and reinforce trust in the scientific endeavor.

ACKNOWLEDGMENTS

I would like to thank Melanie Clark and the editors of this volume for their helpful comments on earlier drafts. I would also like to express my appreciation to Felice Levine and Joan Sieber for augmenting my understanding of and dedication to human research protections. Please direct comments to the author at the Department of Sociology, Emory University, Atlanta, GA 30322 or khegtv@emory.edu.

REFERENCES

Adair, J. G. (2001). Ethics of psychological research: New policies; continuing issues; new concerns. *Canadian Psychology, 42,* 25–37.

Altman, E., & Hernon, P. (Eds.) (1997). *Research misconduct: Issues, implications, and strategies.* Greenwich, CT: Ablex Publishing Corporation.

Babbie, E. (1998). *The practice of social research.* Belmont, CA: Wadsworth Publishing.

Baumrind, D. (1985). Research using intentional deception. *American Psychologist, 40,* 165–174.

Beauchamp, T. L., & Childress, J. F. (2001). *Principles of biomedical ethics.* New York: Oxford University Press.

Bies, R. J. (2001). Interactional (in)justice: The sacred and the profane. In J. Greenberg & R. Cropanzano (Eds.), *Advances in organizational justice* (pp. 85–108). Stanford, CA: Stanford University Press.

Bies, R. J., & Shapiro, D. L. (1988). Voice and justification: Their influence on procedural fairness. *Academy of Management Journal, 31,* 676–685.

Bonetti, S. (1998). Experimental economics and deception. *Journal of Economic Psychology, 19,* 377–395.

Bordens, K. S. & Abbott, B. B. (2005). *Research and design methods: A process approach.* Boston: McGraw–Hill.

Christensen, L. (1988). The negative subject: Myth, reality or a prior experimental experience effect? *Personality and Social Psychology Bulletin, 14,* 664–675.

Citro, C. F., Ilgen, D. R., & Marrett, C. B. (2003). *Protecting participants and facilitating social and behavioral sciences research.* Washington, DC: National Academy Press.

Davis, D. D., & Holt, C. A. (1993). *Experimental economics*. Princeton, NJ: Princeton University Press.

DeVries, R., DeBruin, D. A., & Goodgame, A. (2004). Ethics review of social, behavioral, and economic research: Where should we go from here? *Ethics and Behavior, 14,* 351–368.

Diviak, K. R., Curry, S. J., Emery, S. L., & Mermelstein, R. J. (2004). Human participants challenges in youth tobacco cessation research: Researchers' perspectives. *Ethics and Behavior, 14,* 321–334.

DuBois, J. M. (2004). Is compliance a professional virtue of researchers? Reflections on promoting the responsible conduct of research. *Ethics and Behavior, 14,* 383–395.

Dunn, C. M., & Chadwick, G. (2002). *Protecting study volunteers in research*. Boston: Centerwatch.

Eckenwiler, L. (2001). Moral reasoning and the review of research involving human subjects. *Kennedy Institute of Ethics Journal, 11,* 37–69.

Elms, A. C. (1982). Keeping deception honest: Justifying conditions for social scientific research stratagems. In T. L. Beauchamp, R. R. Faden, R. J. Wallace, Jr., & L. Walters (Eds.), *Ethical issues in social science research* (pp. 232–245). Baltimore, MD: John Hopkins University Press.

Epley, N., & Huff, C. (1998). Suspicion, affective response, and educational benefit as a result of deception in psychology research. *Personality and Social Psychology Bulletin, 24,* 759–768.

Fisher, C. B., & Fyrberg, D. (1994). Participant partners: College students weigh the costs and benefits of deceptive research. *American Psychologist, 49,* 417–427.

Geller, D. (1982). Alternatives to deception: Why, what, and how? In J. E. Sieber (Ed.), *The ethics of social research: Surveys and experiments* (pp. 38–55). New York: Springer–Verlag.

Gillespie, J. F. (1999). The why, what, how, and when of effective faculty use of institutional review boards. In G. Chastain & R. E. Landrum (Eds.) *Protecting human subjects* (pp. 157–177). Washington, DC: American Psychological Association.

Greenberg, J. (1993). Stealing in the name of justice: Informational and interpersonal moderators of theft reactions to underpayment equity. *Organizational Behavior and Human Decision Processes, 54,* 81–103.

Hertwig, R., & Ortmann, A. (2001). Experimental practices in economics: A methodological challenge for psychologists? *Behavioral and Brain Sciences, 24,* 383–451.

Hoeyer, K., Dahlager, L., & Lynöe, N. (2005). Conflicting notions of research ethics: The mutually challenging traditions of social scientists and medical researchers. *Social Science and Medicine, 61,* 1741–1749.

Hogg, M. A. (2003). Intergroup relations. In J. Delamater (Ed.), *Handbook of social psychology* (pp. 479–501). New York: Kluwer Academic.

Holmes, D. (1976a). Debriefing after psychological experiments I: Effectiveness of post deception de-hoaxing. *American Psychologist, 32,* 858–867.

Holmes, D. (1976b). Debriefing after psychological experiments: Effectiveness of post experimental desensitizing. *American Psychologist, 32,* 868–875.

Israel, M., & Hay, I. (2006). *Research ethics for social scientists: Between ethical conduct and regulatory compliance*. Newbury Park, CA: Sage.

Kelman, H. C. (1982). Ethical issues in different social science methods. In T. L. Beauchamp, R. R. Faden, R. J. Wallace, Jr., & L. Walters (Eds.), *Ethical issues in social science research* (pp. 40–97). Baltimore, MD: John Hopkins University Press.

Korn, J. H. (1997). *Illusions of reality: A history of deception in social psychology*. Albany, NY: State University of New York Press.

Landrum, R. E., & Chastain, G. (1999). Subject pool policies in undergraduate-only departments: Results from a nationwide survey. In G. Chastain & R. E. Landrum (Eds.), *Protecting human subjects* (pp. 24–42). Washington, DC: American Psychological Association.

Martin, M. W., & Sell, J. (1979). The role of the experiment in the social sciences. *The Sociological Quarterly, 20,* 581–590.

McBurney, D. H., & White, T. L. (2004). *Research methods*. Belmont, CA: Wadsworth/Thomson.

Mead, G. H. (1934). *Mind, self, and society*. Chicago: University of Chicago Press.

Milgram, S. (1974). *Obedience to authority: An experimental view*. New York: Harper & Row.

National Commission for the Protection of Human Subjects of Biomedical and Behavioral Research. (1979). *The Belmont report: Ethical principles and guidelines for the protection of human subjects*. (Federal Register Document No. 79-12065). Washington, DC: U.S. Government Printing Office.

Neuman, W. L. (2007). *Basics of social research: Qualitative and quantitative approaches*. Boston: Pearson.

Oakes, J. M. (2002). Risks and wrongs in social science research: An evaluator's guide to the IRB. *Evaluation Review, 36*, 443–479.

Ortmann, A., & Hertwig, R. (2002). The costs of deception: Evidence from psychology. *Experimental Economics, 5*, 111–131.

Rosenthal, R. (1994). Science and ethics in conducting, analyzing, and reporting psychological research. *Psychological Science, 5*, 127–134.

Rosnow, R. L. (1997). Hedgehogs, foxes, and the evolving social contract in psychological science: Ethical challenges and methodological opportunities. *Psychological Methods, 2*, 345–356.

Sieber, J. (1992). *Planning ethically responsible research: A guide for students and internal review boards*. Newbury Park, CA: Sage Publications.

Sieber, J. (1999). What makes a subject pool (un)ethical? In G. Chastain & R. E. Landrum (Eds.), *Protecting human subjects* (pp. 43–64). Washington, DC: American Psychological Association.

Sieber, J. (2003). *Risk, benefit, and safety in human research*. Unpublished manuscript. California State University, Hayward.

Sieber, J. (2004). Empirical research on research ethics. *Ethics and Behavior, 14*, 397–412.

Sieber, J., Iannuzzo, R., & Rodriguez, B. (1995). Deception methods in psychology: Have they changed in 23 years? *Ethics and Behavior, 5*, 67–85.

Sieber, J., & Saks, M. J. (1989). A census of subject pool characteristics and policies. *American Psychologist, 44*, 1053–1061.

Singer, E., & Levine, F. J. (2003). Protection of human subjects of research: Recent developments and future prospects for the social sciences. *Public Opinion Quarterly, 67*, 148–164.

Tyler, T. R., Boeckmann, R. J., Smith, H. J., & Huo, Y. J. (1997). *Social justice in a diverse society*. Boulder, CO: Westview Press.

Tyler, T. R., & Lind, E. A. (1992). A relational model of authority in groups. *Advances in Experimental Social Psychology, 25*, 115–191.

Veatch, R. M. (1987). *The patient as partner*. Bloomington: Indiana University Press.

Waite, B. M., & Bowman, L. L. (1999). Research participation among general psychology students at a metropolitan comprehensive public university. In G. Chastain & R. E. Landrum (Eds.), *Protecting human subjects* (pp. 69–85). Washington, DC: American Psychological Association.

Walker, H. A., & Willer, D. (2007). Experiments and the science of sociology. In M. Webster & J. Sell (Eds.), *Laboratory experiments in the social sciences* (pp. 25–55). Burlington, MA: Elsevier.

Zelditch, M., Jr. (1969). Can you really study an army in the laboratory? In A. Etzioni (Ed.), *Complex organizations* (pp. 528–539). New York: Holt, Rinehart, & Winston.

Zimbardo, P. G. (1973). On the ethics of intervention in human psychological research: With special reference to the Stanford Prison Experiment. *Cognition, 2*, 243–256.

Zimbardo, P. G., Banks, W. C., Haney, C., & Jaffe, D. (1973). The mind is a formidable jailer: A Piradellian prison. *The New York Times Magazine*, April 8, Section 6, 38–60.

CHAPTER 7

Technological Issues Related to Experiments

LISA TROYER
University of Iowa

ABSTRACT

Technology is a fundamental element of social experiments, but not one that traditionally receives much explicit, systematic attention in discussions of experimental methods and processes. In this chapter, using examples from past and current research, I discuss how technology is used to create experimental settings, manipulate and represent independent variables, and operationalize dependent variables. Additionally, I discuss some emerging technologies that may change how experiments are conducted, the kinds of research questions our experiments can address, and how we train future social science researchers. I also discuss the potential liabilities of introducing new technologies in our experimental research and offer suggestions regarding how we might guard against these liabilities.

The study of the technological issues related to experiments involves an examination of the procedures and tools that we use to investigate social and behavioral phenomena experimentally, with particular attention to the advantages and potential disadvantages (such as the introduction of bias), as well as challenges (such as the need for researchers to learn new skills), that such procedures and tools (whether new or emergent) pose.

I. INTRODUCTION

As this chapter will illustrate, social psychologists have a long history of creatively using technologies to solve research problems and to open new avenues of investigation. Recent rapid advances, particularly in computational and digital technologies, have opened new opportunities for social and behavioral scientists to incorporate new and different technologies in their experimental research. Thus, it is timely for us to consider the issues that the technologies pose for experimental research in the social sciences; perhaps through such consideration, we can spark further gains in research by encouraging social psychologists to think consciously about how technologies may lend insight and efficiency to their work.

With this in mind, the purpose of this chapter is to examine the role that technology plays in laboratory experiments, with a focus on the advantages (and pitfalls) that it entails. While I will cite many examples of different technologies to illustrate these issues, the chapter does not exhaust all instances in which technology has played a role in laboratory experiments. To be sure, such a comprehensive overview would require summarizing every laboratory experiment conducted in the social sciences, because (as I will soon discuss) experiments themselves are a technology, protocols administered in experiments are technologies, and the knowledge produced by experiments is a technology. Thus, it would be difficult to find an experiment that does not introduce a unique technology.

In this chapter, I will begin by exploring what is meant by "technology" and the long-standing role that technology has played in laboratory experiments since the earliest social psychology experiments. The objective, however, is not to provide an exhaustive inventory of experimental technologies, but rather to generally discuss the role of technology in experimental studies of social psychological processes. This discussion sets the stage for examining issues related to how technology is used to manipulate and represent independent variables in experimental settings. Interestingly, technology has not been the outcome variable of interest in social science experiments. Yet, technology does play a vital role in operationalizing dependent variables, which I will discuss.

As this overview will demonstrate, technology affords researchers many advantages. With the increasing availability of new technologies to experimenters, however, it is also important to consider the pitfalls that they may pose carefully, particularly at the onset of the introduction of a new technology to an existing research program. The chapter will close with some words not only of caution related to the risks of technology, but also of optimism related to the enhanced and new social psychological insights on the horizon as technology is systematically introduced to laboratory experiments.

II. DEFINING TECHNOLOGY

Just what do we mean by technology? While dictionary and encyclopedia definitions vary somewhat, diverse sources converge on the notion that the term technology refers to the development or application of tools and processes for solving problems, primarily problems related to controlling elements in the environment or enhancing understanding of them (e.g., *Microsoft Encarta Online Encyclopedia*, 2006; *Webster's Dictionary of English Usage*, 1989; Wikipedia contributors, 2006). Such a general definition suggests that experiments, in and of themselves, are a technology; experiments are a methodological tool that social and behavioral scientists use to better understand social and psychological processes. Yet, researchers also employ technologies in the design and execution of experiments. While technologies can involve knowledge, processes, or procedures, in addition to material tools and mechanisms, the focus of this chapter will be on the material technologies (e.g., physical tools, instruments, hardware, and software) that researchers use in experiments.

III. THE ROLE OF TECHNOLOGY IN CREATING EXPERIMENTAL SETTINGS

Some of the earliest and most profound experiments in social psychology and group dynamics have revolved around the introduction of technologies to create a setting conducive to measuring particular social processes. Norman Triplett's 1898 study of the effects of competition on human performance, often cited as the first experimental test of social facilitation (e.g., Forsyth, 2006), is an excellent example of this. He constructed a "competition machine," which allowed him to study human physiological performance under different conditions of self-pacing versus competition with others.

The machine consisted of two fishing wheel cranks attached to boards arranged in a Y-shape, allowing subjects to participate alone or in competitive pairs. As subjects (children in Triplett's study) turned the cranks to reel in 16 m of line, a kymograph (which records the timing of changes in motion by etching a rotating drum) attached to the fishing line was used to record the movement of the line and operationalize the rate at which the line was being reeled in by the subjects (the dependent variable of interest). The objective presented to subjects was to achieve the highest possible rate of pulling in the line. Triplett's incorporation of the competition machine illustrates how technology becomes a central element of an experimental setting, which allows the experimenter to set into motion the process under study (in this case, performance).

As experimentalists are well aware, realism is an important concern (and often cited critique) of experiments. Realism refers to the extent to which subjects are exposed to a situation in an experiment that elicits perceptions, cognitions, feelings, and behavioral responses corresponding to those that the subject would experience if the situation occurred naturally outside the laboratory. There are two types of realism commonly considered in experiments, each of which may be facilitated (or impeded) by the introduction of technology in laboratory experiments: experimental realism and mundane realism. Experimental realism refers to the extent to which subjects are engaged in the study, and thus are responding to the situation in a manner representative of the types and magnitudes of responses they would generate in naturalistic settings. In contrast, mundane realism refers to the extent to which the experimental setting mimics the real-world environments.

In the case of the Triplett (1898) experiment, while mundane realism may have been relatively low (given that the competition machine was in a laboratory and deviated from devices that individuals commonly use in everyday life), the experimental realism was relatively high. For instance, subjects responded to competition—a social variable—much as Triplett (a cycling enthusiast) had observed that bicycle racers did.

Stanley Milgram's 1963 "shock generator," constructed for his path-breaking studies on obedience to authority, represents a more contemporary and perhaps a more striking example of the use of technology to enhance the realism of a setting. This machine, a console, featured a series of 30 rocker switches and lights with labels ostensibly indicating increasingly higher voltage, and hence increasing levels of electrical shocks, from a switch labeled, "15 volts" and "slight shock" to one labeled, "danger: severe shock"; the last two increments were labeled simply "XXX." The experimenter instructed subjects to administer successively higher shocks to a "learner" upon the learner's mistakes, ostensibly to increase learning.

Although the learner appeared to be attached to the shock generator by wires, the wires did not carry an electrical charge and the shock generator was not capable of delivering electrical shocks to learners. While the shock generator was also a means of operationalizing the dependent variable (i.e., by indicating how much shock subjects were willing to administer at the instruction of an authority [the experimenter]), its formidable appearance is likely to have contributed to the scientific salience of the learning context that Milgram sought to create for subjects, enhancing the experimental realism.

As in the Triplett (1898) experiment, mundane realism in the Milgram (1963) experiment was likely lower—after all, how often do we encounter shock generators in everyday educational settings? Yet, the explicit labels documenting the increasingly injurious nature of the shocks at each level likely contributed to the experimental realism in very important ways. For instance, subjects may have been aware that electrical shocks vary with voltage and that very large voltages can incur substantial injuries and even death. The labels on the console, however, further ensured that subjects would have this perception, effectively "stacking the deck" against Milgram. Thus, for instance, the obvious critique—that subjects were naïve about shocks and would not actually injure others at the instruction of a scientist if they knew what the shocks could do to the others—can be dismissed.

While the preceding two examples illustrate the role of technology in facilitating experimental realism, perhaps one of the most renowned examples of technology as a facilitator of mundane realism is found in Philip Zimbardo's well-known study of authoritarian role taking and aggression, the Stanford Prison Studies (e.g., Haney, Banks, & Zimbardo, 1973). The technology in this case might be considered "low tech"—steel bars and sparse furnishings, smocks, nylon stocking caps (to simulate shaved heads), and chains (locked around subjects' feet). Zimbardo used such materials in the basement of the building housing the Department of Psychology at Stanford University to create a physical space with characteristics akin to a jailhouse. As many readers are aware, the experiment provided exceptional insight on how people who are randomly assigned to social roles (in this experiment, guard or prisoner roles) adapt to those roles in an immersive experiment (i.e., subjects were confined to the basement "jailhouse" 24 hours a day during the experiment).

Indeed, it is likely that the simplicity of the technology Zimbardo used contributed to the effectiveness of the experimental setting; even Zimbardo and his research team fell victim to the degree of reality and began behaving as prison wardens at points in the experiment. Ultimately, the realism created by the use of this technology generated such intense psychological and emotional effects that the experiment, originally scheduled to run for up to 2 weeks, was terminated in 6 days. Despite the seemingly low-tech experimental setting, this study (and the one by Milgram in 1963 described previously) raised serious ethical concerns regarding the effects of realism on subjects.

More recently, researchers have begun to use computing technologies to render virtual environments—three-dimensional spaces, objects, and characters with which subjects may interact that give rise to the sensory experiences of subjects. Using these technologies, James Blascovich and colleagues (2002) launched a program of experimental research that offers the potential to study processes and variables that may be difficult or extremely costly to study in face-to-face settings. Furthermore, concurrent with advances in computing power, virtual reality technologies and digitizing processes are rapidly making it possible to increase both mundane and experimental realism. Virtual worlds can be photorealistic to the point that it is difficult to distinguish the virtual from the natural reality (as demonstrated in the popular "Matrix" movies). At the same time, as recent generations have become accustomed to video and computer games, the computer-rendered virtual environment may be increasingly more engaging for subjects today and into the future, thus positively affecting the experimental realism of research incorporating these technologies.

Despite the promise of emerging virtual technologies, Blascovich *et al.* (2002) are careful to note that much research is still needed to understand the relationships between "real" and virtual realities. Toward these ends, these researchers propose a model that suggests that in addition to realism, researchers need to attend to the extent to which subjects perceive social presence of other actors in virtual environments. Social presence refers to an actor's sense of awareness that another person exists and is sharing the interaction context with the actor (e.g., Short, Williams, & Christie, 1976). As Blascovich *et al.* caution, it is critical to consider systematically how introducing new or alternative technologies to our experiments may affect the socially relevant dimensions of the processes that we study, such as social presence. I take up this issue in a later section of this chapter.

While these examples vary with respect to the nature of the technology used, they all demonstrate how experimenters employ technology to contribute to the construction of settings that help them realize the factors contributing to the social processes they are interested in studying. Additionally, these examples illustrate that technologies need not necessarily be complex, modern, or sophisticated to facilitate the realization of social processes. Irrespective of the level of sophistication or contemporary nature of a technology, it plays a key role in the creating the settings and affecting the experiences of subjects. In the next two sections, I examine the ways that technology affects the realization of independent and dependent variables.

IV. THE ROLE OF TECHNOLOGY IN OPERATIONALIZING INDEPENDENT VARIABLES

Technology goes beyond contributing to the construction of the settings in which social processes are launched. In addition, technology can facilitate the representation and manipulation of independent variables. For example, in their classic study of communication networks, Alex Bavelas (1953) and Harold Leavitt (1951) operationalized communication networks by placing subjects in cubicles with slots on the cubicle walls representing the other group members with whom they could communicate. By opening or closing the slots, the experimenters could generate different communication networks and investigate how different types of network configurations and network positions affected the efficiency in group problem solving and group members' satisfaction (as well as other outcomes). Although a relatively low-tech method of manipulating the key independent variable (the configuration of a communication network), this nonetheless demonstrates the role that technology plays in representing states of independent variables.

Also, as computing technologies have become more accessible to researchers, low-tech operationalizations, such as the network operationalization that Bavelas (1953) and Leavitt (1951) employed can be recreated using computers. For example, communication networks can be relatively easily controlled using computer networks by simply determining which computers are networked to which other computers (either through programming or hard wiring). What advantages arise from such uses of computing technologies? One key advantage is the ability to control all other perceptual content (e.g., sound) to which subjects might be exposed that might affect perceptions and responses (including distractions in problem-solving exercises). While such content is theoretically likely to be distributed randomly across types of networks and positions (assuming experimenters randomly distribute participants), it nonetheless introduces random error, which may make it more difficult to detect effects and/or ascertain their magnitude in experiments. Thus, one major advantage of computing technologies as vehicles for operationalizing independent variables may be the level of control they potentially can afford to researchers.

A second advantage involves scalability. Using computers to operationalize communication networks, for instance, might enhance a researcher's ability to study larger and more complex networks and thereby generate theoretical refinements. Along these lines, the network exchange theory research program (e.g., Markovsky, Willer, & Patton, 1988) has benefited from the use of computing technologies to simulate power and social exchange in complex networks that are difficult and costly to study experimentally with human subjects in on-site laboratories (see, for example, Markovsky, 1995; Lovaglia & Willer, 2002). For instance, Michael Lovaglia and colleagues (Lovaglia, Skvoretz, Markovsky, & Willer, 1996) describe how network exchange theorists are using computer simulations to identify strategic networks (i.e., complex networks for which different theories of power and social exchange offer competing predictions). While the simulations are not themselves experiments involving human actors, they do facilitate the systematic design of experiments to adjudicate among competing theories and advance the field of study, by identifying appropriate test cases.

With the rise and proliferation of further computing advances, including the Internet, World Wide Web, and distributed computing technologies, the potential for experiments involving exceptionally large networks is now beginning to be recognized (e.g., Troyer, 2003; Willer, 1999). Along these lines, in 1997 the National Science Foundation convened a group of scientists from the social, behavioral, and economic sciences in a "NetLab Workshop." The NetLab report (National Science Foundation, 1997) proposed that the rise of broadband network computing would allow for experiments involving up to thousands of subjects.

Interestingly, while the technological capabilities for networked, worldwide collaboration appear to be well established, infrastructural, procedural, and normative barriers seem to have impeded the rapid proliferation of such research. For example, experimenters have experienced some difficulty obtaining resources to establish and administer networked laboratories. New procedures for institutional review boards must be developed to ensure that appropriate human-subject consent and protection procedures are in place, since this technology can lead individuals to enter experimental studies unknowingly. Furthermore, while Web-based experiments may facilitate broad participation of more diverse subjects in research, it may be difficult to verify that subjects are the individuals whom researchers intend to enroll in their experiments. It may also be difficult for researchers to control extraneous factors in Web-based experiments (e.g., speed and robustness of subjects' Web access, distractions in subjects' local environments, influence of others who may be present while subjects are participating in experiments).

Accordingly, researchers must think carefully about the importance of such controls, how they might be implemented when using this new technology, and what the consequences are of losing such control. Finally, the norms promoting independent research on the part of investigators are deeply entrenched in the social and behavioral sciences. Networked laboratories such as those described in the National Science Foundation report will be "collaboratories," requiring widespread collaboration among numerous researchers who may work in different states or even different countries.

Again, however, it is important to emphasize that sophistication and extensive computational power are not prerequisites to employing technologies effectively to facilitate the manipulation of independent variables in experimental research. By way of example, research in status characteristics theory (Berger, Fisek, Norman, & Zelditch, 1977) has used relatively low-tech methods of delivering status information to subjects. Status characteristics theory, widely recognized for its reliance on a standardized experimental setting (e.g., Cook, Cronkite, & Wagner, 1974; Troyer, 2002), posits that when attributes (such as race, sex, mathematical acumen, beauty) are known to differentiate individuals in a group, those individuals with preferred states of the attributes become more valuable to the group, are believed to possess greater competence at the group's specific task, and are more influential in groups.

In this line of research, differentiating attributes—status characteristics—are often the key independent variable that experimenters seek to study. To manipulate status characteristics, subjects in experiments testing the theory's hypotheses are given information about other group members, which is often delivered through relatively low-tech media, such as text information displayed on a computer screen, and sometimes accompanied with a Polaroid photograph of an individual depicted as the subject's partner, digital still image of the ostensible partner, or a prerecorded audio/video recording of an ostensibly "live" partner. The technology serves well insofar as it allows the experimenter to manipulate the independent variable (higher versus lower status) in a very controlled manner (e.g., avoiding introducing additional information, as might occur through use of a confederate).

Thus, the choice of technologies to facilitate the representation and manipulation of independent variables depends on many factors, including the accessibility of different types of technologies to researchers, the maturity of the technologies, the familiarity that subjects are likely to have with them, and, perhaps most importantly, the nature of the research issue being investigated.

V. TECHNOLOGIES AS INDEPENDENT VARIABLES

While technology is an effective tool that experimenters can use to represent and manipulate independent variables, it can also be the independent variable in experimental research. As electronic modes of communication have proliferated, researchers have begun to explore the effects of these different media on social processes. Yet, outcomes of the research are mixed with respect to whether alternative communication technologies facilitate or undermine group and individual performance. For instance, while it appears that the anonymity and/or lack of social cues that computer-mediated communication may afford can suppress some status dynamics, such as dominance (e.g., Dubrovsky, Kiesler, & Sethna, 1991), it may also undermine synergistic processes in groups that produce innovation (e.g., Silver, Cohen, & Crutchfield, 1994) and facilitate group performance (e.g., Griffith & Northcraft, 1994).

Theoretical difficulties can arise, however, when technological innovations related to experimental methods are introduced to a theoretical research program at the same time that new independent variables are being studied. An example of this arose in status characteristics research in the 1980s and 1990s, setting off a continuing debate over the robustness of a particular status characteristic: sex. As computing technologies became more accessible to researchers and the efficiencies they offered to experimenters became increasingly clear (for a summary, see Cohen, 1988), researchers studying status characteristics theory began to turn to them as a medium of choice for administering studies of status characteristics.

Martha Foschi (1996) developed a particularly provocative line of research documenting the higher performance standards to which females are held compared to males, attributing the difference to the differential status that women and men hold. Concurrent with this breakthrough research, she also introduced a breakthrough in methodology by administering her experiments through computers. Researchers rapidly adopted the software because of its efficiency and ease of use, but shortly thereafter investigators began to recognize that the studies employing the new medium yielded influence rates that were consistently lower for both females and males compared to earlier studies (influence is a key variable of interest in status characteristics theory).

Since shifts in gender attitudes also occurred during this period and other methodological changes were introduced (for a summary, see Troyer, 2002), it was unclear whether the lower influence rates reflected a change in how sex differences operate with respect to status-organizing processes in task groups. Through a series of systematic studies examining medium and other method-ological differences, which treat the technology as an independent variable, my colleagues and I have ascertained that the technology does have both direct effects on status-organizing processes and indirect effects on the extent to which individuals pay attention to one another and are engaged in the group's task (e.g., Troyer, 2001; Kalkhoff, Younts, & Troyer, 2006). That is, technology (in this case, the medium through which interaction occurs) has important effects on key theoretical processes. Although the debate over whether sex still operates as a status characteristic continues (e.g., Foschi & Lapointe, 2002; Hopcroft, 2002; Rashotte & Webster, 2005), it is clear that technologies can affect social psychological processes and are justifiably treated systematically as independent variables in laboratory experiments.

This suggests the importance of recognizing that while the technologies researchers use may impart efficiencies to their experiments, they may affect or interact with the theoretical processes, such as subjects' perceptions of an interaction partner's social presence, the extent to which the subject is engaged in an experimental task, or the subject's cognitive processing (e.g., perception, encoding, memory, retrieval). These are just a few examples of how technolo-gies may affect or interact with theoretical processes. To avoid confounding technology with other theoretical variables of interest and more rapidly advance the knowledge gains from experimental research, social psychologists must systematically assess whether and how the new technologies they introduce to their research generate such theoretically important effects.

VI. THE ROLE OF TECHNOLOGY IN OPERATIONALIZING DEPENDENT VARIABLES

Technology may also advance understanding of social and behavioral sciences through its role in operationalizing dependent variables. In particular, technologies that automate data collection often afford greater reliability and decreased risk of experimenter effects (including intrusiveness). Experiments that involve dependent variables of attitudinal and other self-report outcomes operationalized via pencil-and-paper questionnaires may reduce the experimenter and demand effects that arise in face-to-face interviews. Yet, computerized surveys may afford even greater benefits, since the data entry step that requires human input is removed (reducing the potential for human error). At the same time, experimenters must recognize that human error may still be introduced; subjects who click on the wrong button, press the wrong key, or overlook items on the computing screen may introduce inaccuracies that might be avoided on a paper-and-pencil instrument in which a subject may visually spot errors with more ease before submitting the questionnaire.

Increasingly, however, computerized survey technologies are becoming more accessible to researchers with built-in tools to reduce the likelihood of error (on the part of experimenter and subject). For instance, commercial products (e.g., WebSurveyor, Survey Monkey) provide useful templates to guide researchers in developing and administering computerized surveys using different question formats. The templates allow researchers to guard against (although not necessarily prevent) inadvertent errors by respondents (e.g., by authenticating respondents, requiring answers to one or all questions before the survey can be submitted). Once a survey is completed, researchers receive a machine-readable data file, which, as noted, vastly reduces the likelihood of data entry errors.

In many experiments, surveys are a fundamental means by which data related to dependent variables are collected and new computerized survey technologies offer tremendous advantages in the form of potentially more efficient and more reliable data collection. Yet, there can be pitfalls to these technologies. Computerized survey technologies (like software for administering experiments) can make conducting research easier; in some respects, the software behind these technologies has the potential to think on behalf of the researcher. This introduces a danger that individuals without adequate training in social science theory and research (including experimental and survey methods) may more frequently launch experiments, collect data, and publish or otherwise disseminate results that may be flawed as a result of a lack of training. Peer review, however, provides a safety net that researchers will be wise to promote and support as these technologies continue to proliferate.

Of course, self-reports are not the only dependent variables in which experimenters are interested. Furthermore, experimenters are well aware of the bias that may be introduced through self-reports. Are the attitudes, feelings, and reported behaviors truly the ones a subject is experiencing or would experience? Consequently, many research programs rely on technologies to capture behavioral outcomes also; these may be either direct measures of key dependent variables or indirect measures (i.e., manifestations) of dependent variables such as beliefs, expectations, attitudes, or emotions. Status characteristics theory, described earlier, is one such research program that is an interesting example because of the evolution of the technologies through which a key dependent variable, social influence, has been measured over time. According to the theory, higher status actors will exercise influence over lower status actors (even when the basis for the status differences bears no direct relation to the task at which the actors are working).

The standardized experimental setting employed in status characteristics theory research has long relied on a particular type of task (a binary choice task) to operationalize social influence. In one variant of the task developed by James Moore (1968), subjects view two rectangles, each composed of about the same number of smaller black and white squares. They are advised that one of the rectangles has more white squares than the other and are asked to indicate which of the two has more white area. Next, they receive feedback regarding a contrived partner's initial judgment.

In 14 to 22 of 25 trials of this task (depending on the experimenter's instantiation), subjects are advised that a task partner's initial judgment is different from their own. In reality, there is no real partner and this information is conveyed through technology. Early studies used an interaction control machine (ICOM), which consisted of a console with buttons through which a subject could indicate either one or the other rectangles as having more white area, alongside four lights (two corresponding to the subject's initial choice of either the top or bottom rectangle; and two ostensibly representing the subject's partner's initial choice). The buttons were connected to a panel in another room, which displayed the subject's initial choice to the experimenter (who recorded it on paper). The experimenter would then use a switch on the panel to trigger which light on the "partner's" section of the subject's console would light up. This allowed experimenters to generate "critical" trials in which the subject was led to believe that the partner provided a different initial judgment.

Subjects were then asked to generate a final judgment by again pushing one of two more buttons on their console. The final decision of the subject would be displayed to the experimenter on the panel in the experimenter's room through a light and recorded on paper. The experimenter could then compare the initial and final subject choices recorded on the paper to determine how often subjects were influenced by the partner's choice. Changes in the initial answers of subjects on critical trials operationalized the subject's deference to the partner, while the consistency of the subject's response across the initial and final choices (i.e., a "stay" response) operationalized the subject's rejection of influence.

As this description suggests, the ICOM is a highly specialized research device, and experimenters usually constructed their own ICOMs. As a result, experiments related to status characteristics theory were conducted only in a few locations, by relatively small groups of researchers. Furthermore, the dependent variable was subject to greater risk of human error, since subject responses were hand recorded by experimenters observing the ICOM.

The increased accessibility of personal computers in the 1980s and 1990s resulted in a dramatic shift. As described earlier, Foschi (1996) developed software to automate the delivery of the binary choice stimulus (e.g., the rectangles comprising contrasting squares) and to record subjects' initial and final choices. As also noted before, the software introduced a new variable to status-organizing processes, but the important point here is that it provided a much more expedient and reliable way of operationalizing the dependent variable, social influence. Furthermore, the combination of the software and proliferation of personal computers has allowed more researchers to conduct experiments related to status characteristics theory.

Once more, however, there is a risk that individuals without adequate theoretical and methodological training will foray into conducting experiments because the technology has made it easy to do so. Aside from peer review, another solution to mitigate this risk is to disseminate software manuals that go beyond merely providing step-by-step instructions for installing and launching the software. Such risks will also be reduced if the manuals explain the link between standardized features of the experimental methodology embodied in the software and elements of the theory that the software has been designed to test (e.g., Troyer, 2004).

When it comes to capturing behavioral instances of dependent variables in experiments, advances in audio/video technologies have also made important contributions. Researchers have had a long-standing interest in tracking communication exchanges (primarily, spoken words) between actors in small groups (e.g., Bales, 1951; Fisek & Ofshe, 1970). From such behavioral data, researchers can study such key social processes as dominance and deference, role playing in groups, and the emergence of social networks. Early studies examining these processes relied on paper-and-pencil recording of exchanges to generate raw counts of who said what to whom. Soon, however, researchers crafted machines that would scroll paper at a set rate of speed, and as counts of interaction content were entered on the scrolling paper, a coarse-grained operationalization of time was made possible. This, for instance, permitted researchers to assess not only what was occurring in the experimental groups, but also when events occurred. Although still relatively low tech, this methodology generated important early insights on group dynamics—for example, on the crystallization of leadership roles in groups.

During these earlier stages of research on group dynamics, audio/video equipment was prohibitively expensive and the relatively massive footprint of the technology made it highly intrusive. Over time, however, video cameras became smaller and less expensive, and researchers began to generate analog recordings of experimental groups (e.g., Smith-Lovin, Skvoretz, & Hudson, 1986; Skvoretz, Webster, & Whitmeyer, 1997). The advantages that resulted were numerous: observations could be replayed to improve reliability of the interpretations researchers were making. Not only could raw counts be studied, but also finer grained analysis of the temporal aspects of group interaction also became possible. A serious disadvantage of the analog recording technology, however, is its limited shelf life. Analog recordings decay, even if not played. Each time an analog tape is played, it decays more quickly. Thus, one of the key advantages—the ability to review and repeatedly reanalyze video—in some senses has been a liability, since it leads to more rapid decay of the data.

More recently, the evolution of digital audio/video technologies has begun to remedy the problem of decay (although digital data can also decay). The longer shelf life of digital recordings may facilitate data sharing and archiving of research in the future, provided an important hurdle—formatting standards—is addressed. Some vendors of recording equipment use proprietary formats for the data, which tends to limit the accessibility of the data for researchers who do not use the equipment or who do not have licenses for software that can decode the data into a format they can access. In some cases, the formats fail to gain a wide audience or are abandoned by manufacturers (although not a digital technology, Sony's Betamax format for video recording is an example of this).

The movement toward standards for the digitization of data (whether audio, video, or text) is critical to facilitate data sharing, which, in turn, is likely to promote meta-analysis and may allow researchers to investigate more efficiently future research questions that have not yet emerged, using archived recordings from prior research. This movement may be gaining momentum, as federal grant agencies, such as the National Institutes of Health and the National Science Foundation are placing increasing pressure (and in some cases, requirements) on scientists to share both data and publications resulting from federally funded research.

Digital audio/video technologies not only impart a longer shelf life for data, but also allow for smaller footprints, and thus less intrusive equipment during experiments. In addition, software advances are starting to permit automated analyses (e.g., through voice-recognition and motion-recognition software). While these technologies are still somewhat crude and require careful calibration by experimenters to avoid automating the introduction of error in data, the future looks promising as they are rapidly being refined.

To be sure, advances in computational and digital technologies are making a host of other instrumentation available to researchers. Dawn Robinson, Lynn Smith-Lovin, and Christabel Rogalin (2004) describe a number of instruments/methodologies allowing experimenters to assess physiological responses, which are believed to be linked to affective states (see also Guglielmi, 1999). For instance, stress is assessable through hormone (i.e., cortisol) levels in human saliva; electromyographical activity (EMG) in the facial regions can distinguish positive and negative affect as well as affective intensity; respiratory activity (rate, depth, and cadence) appears to provide fine-grained indications of emotion type (e.g., disgust; see Boiten, 1998); autonomic nervous system (ANS) activity (e.g., electrothermal activity, cardiovascular reactivity) can detect motivational states, such as challenge reactivity (e.g., Blascovich, Mendes, Hunter, Lickel, & Kowai-Bell, 2001); and functional magnetic resonance imaging (fMRI) of the brain offers tremendous potential for pinpointing emotional responses (e.g., Damasio, Adolphs, & Damasio, 2003).

In a striking experimental study of the potential of these types of physiological measures of affect, Michaël Dambrun, Gérard Després, and Serge Guimond (2003) demonstrated the utility of triangulating facial EMG, ANS, and respiratory activity to pinpoint prejudice attitudes and the affective states they generate more precisely. At the same time, they demonstrated that refinements to both instruments and their interpretations are still needed.

Aside from facilitating the more accurate measurement of affect and attitudes, physiological measures hold great promise for also refining our ability to measure interaction dynamics. For example, Stanford Gregory and colleagues (e.g., Gregory, 1994; Gregory & Gallagher, 2002; Gregory & Webster, 1997) have introduced voice frequency analysis to the study of social interaction. They show that, below the 0.5-kHz frequency of vocal pitch in the voice spectrum, a number of social phenomena can be very precisely measured, including social accommodation (adjusting one's voice frequency to interaction partners) and dominance. Jean Decety et al. (Decety, Jackson, Sommerville, Chaminade, & Meltzoff, 2004) found distinctive patterns in fMRI analysis of brain activity, depending on whether experiments were engaged in competitive or cooperative activities.

This discussion of physiological instrumentation suggests that these technologies will likely foster increasing interdisciplinary research between social and natural scientists. The development and accessibility of new forms of instrumentation are already forging new interdisciplinary fields of inquiry that bring social, physical, and natural scientists together to develop new strategies for investigating long-standing topics such as how economic decisions are made, or relatively new topics, such as how social organization affects and is affected by information technologies. The former is represented by the emerging field of neuroeconomics (e.g., Glimcher, 2003), and the latter corresponds to the emerging field of social informatics (e.g., Kling, Rosenbaum, & Sawyer, 2005).

Thus, as the increasing accessibility of new technologies motivates new research to address existing theoretical questions, it may bring together researchers from disparate disciplines like engineering, physical sciences, computer science, biological and life sciences, and social and behavioral sciences, leading to new theoretical questions. For this reason, it may be timely to consider how we are training the next generation of social and behavioral scientists to conduct experiments in our laboratories. Skills in interdisciplinary collaboration as well as technical skills in developing and using instruments are likely to be increasingly important in the very near future.

VII. SUMMARY AND CONCLUSION: PREPARING FOR THE FUTURE OF TECHNOLOGY IN LABORATORY EXPERIMENTS

This chapter has examined the many different and important roles that technology assumes in laboratory experiments. This examination has not exhausted all instances in which experimenters use technology in their laboratories. In fact, such a comprehensive inventory would most likely require summarizing all experimental research because, as noted at the onset of this chapter, experiments are themselves a form of technology. Yet, we have covered many different ways that technologies contribute to experimental research.

As this overview has suggested, technologies are most often used to bring efficiency, control, and precision to experimental research. As researchers decide whether to use a technology, they must carefully weigh whether and how the technology will contribute to our understanding of social and behavioral processes against how they may alter those processes. As described previously, I have attempted to explore this question systematically as I consider introducing new technologies in my experimental research (e.g., Troyer, 2001, 2002). In the absence of such systematic assessment of how technologies affect social and behavioral processes, we may risk muddying social and psychological understanding of those processes. Consequently, care must be taken as new technologies emerge and become accessible to experimenters.

In addition, because new technologies can make it easier and less costly to conduct experiments, there is some risk that individuals without adequate training and knowledge (including ethical training, methodological training, and theoretical knowledge) may undertake experimental research. We can guard against the introduction of flawed data and analyses through existing and improved peer-review processes and by disseminating material that explicitly discusses the theoretical rationale behind the methodology embodied in our technologies (such as user manuals or technical reports). Finally, it is timely to consider how we are preparing the next generation of experimental social scientists and to look for ways to foster their (and our own) interdisciplinary and technical knowledge and skills to ready the field for what is sure to be an exciting and insightful future for experimental work in the social sciences.

These cautions related to current and future uses of technology in experimental research should not be taken as a call to shy away from the many benefits that technologies have the potential to impart to our research. Rather, they are a reminder that, as new technologies emerge and are incorporated into social science research, experimenters should not lose sight of important virtues of our methodology: controlled, systematic, theory-driven study of social and behavioral processes.

ACKNOWLEDGMENTS

I thank Michael Lovaglia for sharing insights on the role of technology in facilitating advances in network exchange theory. Jane Sell and Murray Webster, the editors of this volume, provided outstanding guidance on the preparation of this chapter and constructive critique, for which I am grateful. I also appreciate the technical and intellectual resources made available by the Center for the Study of Group Processes at the University of Iowa, which have allowed me to pursue my interests in investigating the many valuable roles that technology can play in social science research and in generating social scientific knowledge.

REFERENCES

Bales, R. F. (1951). *Interaction process analysis: A method for the study of small groups*. Cambridge, MA: Addison–Wesley.

Bavelas, A. (1953). Communication patterns in task-oriented groups. In D. Cartwright & A. Zander (Eds.), *Group dynamics* (pp. 493–506). Evanston, IL: Row, Peterson, and Company.

Berger, J., Fisek, M. H., Norman, R. Z., & Zelditch, Jr., M. (1977). *Status characteristics and social interaction: An expectation states approach*. New York: Elsevier.

Blascovich, J., Loomis, J., Beall, A. C., Swith, K. R., Hoyt, C. L., & Bailenson, J. N. (2002). Immersive virtual environment technology as a methodological tool for social psychology. *Psychological Inquiry, 13,* 103–124.

Blascovich, J., Mendes, W. B., Hunter, S. B., Lickel, B., & Kowai-Bell, N. (2001). Perceiver threat in social interactions with stigmatized others. *Journal of Personality and Social Psychology, 80,* 253–267.

Boiten, F. A. (1988). The effects of emotional behavior on components of the respiratory cycle. *Advances in Experimental Social Psychology, 28,* 1–51.

Cohen, B. P. (1988). A new experimental situation using microcomputers. In M. Webster, Jr., and M. Foschi (Eds.), *Status generalization: New theory and research* (pp. 383–398). Stanford, CA: Stanford University Press.

Cook, K., Cronkite, R., & Wagner, D. (1974). *Laboratory for Social Research manual for experimenters in expectation states theory.* Unpublished manuscript, Stanford University, Laboratory for Social Research, Stanford, CA.

Damasio, A. R., Adolphs, R., & Damasio, H. (2003). The contribution of the lesion method to functional neuroanatomy of emotion. In R. J. Davidson, K. R. Scherer, & H. H. Godsmith (Eds.), *Handbook of affective neuroscience* (pp. 66–92). New York: Oxford University Press.

Dambrun, M., Després, G., & Guimond, S. (2003). On the multifaceted nature of prejudice: Psychophysiological responses to ingroup and outgroup ethnic stimuli. *Current Research in Social Psychology, 8,* 187–205, <http://www.uiowa.edu/~grpproc>.

Decety, J., Jackson, P. L., Sommerville, J. A., Chaminade, T., & Meltzoff, A. N. (2004). The neural bases of cooperation and competition: An fMRI investigation. *NeuroImage, 23,* 755–751.

Dubrovsky, V. J., Kiesler, S., & Sethna, B. N. (1991). The equalization phenomenon: Status effects in computer-mediated and face-to-face decision-making groups, *Human–Computer Interaction, 62,* 119–146.

Fisek, M. H., & Ofshe, R. (1970). The process of status evolution. *Sociometry, 33,* 327–346.

Forsyth, D. R. (2006). *Group dynamics* (4th ed.). Pacific Grove, CA: Brooks/Cole.

Foschi, M. (1996). Double standards in the evaluation of men and women. *Social Psychology Quarterly, 59,* 237–254.

Foschi, M., & Lapointe, V. (2002). On conditional hypotheses and gender as a status characteristic. *Social Psychology Quarterly, 6,* 146–162.

Glimcher, P. (2003). *Decisions, uncertainty, and the brain: The science of neuroeconomics.* Cambridge, MA: MIT Press.

Gregory, S. W., Jr. (1994). Sounds of power and deference: Acoustic analysis of macro social constraints on micro interaction. *Sociological Perspectives, 37,* 497–526.

Gregory, S. W., Jr., & Gallagher, T. J. (2002). Spectral analysis of candidates' nonverbal vocal communication: Predicting U.S. presidential election outcomes. *Social Psychology Quarterly, 65,* 298–308.

Gregory, S. W., Jr., & Webster, S. (1996). A nonverbal signal in voices of interview partners effectively predicts communication accommodation and social status perceptions. *Journal of Personality and Social Psychology, 70,* 1231–1240.

Griffith, T. L., and Northcraft, G. B. (1994). Distinguishing the forest and the trees: Media, features, and methodology in electronic communication research. *Organization Science, 5,* 272–285.

Guglielmi, R. S. (1999). Psychophysiological assessment of prejudice: Past research, current status, and future directions. *Personality and Social Psychology Review, 3,* 123–157.

Haney, C., Banks, C., & Zimbardo, P. (1973). Interpersonal dynamics in a simulated prison. *International Journal of Criminology and Penology, 1,* 69–97.

Kalkhoff, W., Younts, C.W., & Troyer, L. (2006). Effects of communication medium and gender on status-organizing processes. Manuscript under review.

Kling, R., Rosenbaum, H., & Sawyer, S. (2005). *Understanding and communicating social informatics: A framework for studying and teaching the human contexts of information and communications technologies.* Medford, NJ: Information Today, Inc.

Leavitt, H. (1951). Some effects of certain communication patterns on group performance. *Journal of Abnormal and Social Psychology, 47,* 38–50.

Lovaglia, M.J., Skvoretz, J., Markovsky, B., & Willer, D. (1999). Part 1: An automated approach to the theoretical analysis of difficult problems. In D. Willer (Ed.), *Network exchange theory* (pp. 259–269). Westport, CT: Praeger.

Lovaglia, M. J., & Willer, R. (2002). Theory, simulation, and research: The new synthesis. In J. Szmatka, M. Lovaglia, & K. Wysineska (Eds.), *The growth of social knowledge: Theory, simulation, and empirical research in group processes* (pp. 247–262). Westport, CT: Praeger.

Markovsky, B. (1995). Developing an exchange network simulator. *Sociological Perspectives, 38,* 519–545.

Markovsky, B., Willer, D., & Patton, T. (1988). Power relations in exchange networks. *American Sociological Review, 53,* 220–236.

Microsoft Encarta Online Encyclopedia. (2006). Technology. <http://encarta.msn.com>, Microsoft Corporation.

Milgram, S. (1963). Behavioral study of obedience. *Journal of Abnormal and Social Psychology, 67,* 371–378.

National Science Foundation Workshop Participants. (1997). Report from the knowledge networking NetLab Workshop, National Science Foundation, Washington, DC.

Rashotte, L.S., & Webster, M. (2005). Gender status beliefs. *Social Science Research, 34,* 618–633.

Robinson, D. T., Smith-Lovin, L., & Rogalin, C. (2004). Physiological measures for theoretical concepts: Some ideas for linking deflection and emotion to physical responses during interaction. In S. Thye & E. J. Lawler (Eds.), *Advances in group processes* (Vol. 21, pp. 77–115). San Diego: Elsevier.

Short, J. A., Williams, E., & Christie, B. (1976). *The social psychology of telecommunications.* New York: John Wiley & Sons.

Silver, S. D., Cohen, B. P., & Crutchfield, J. H. (1994). Status differentiation and information exchange in face-to-face and computer-mediated idea generation. *Social Psychology Quarterly, 57,* 108–123.

Skvoretz, J. V., Webster, M., Jr., & Whitmeyer, J. M. (1999). Status orders in task discussion groups. In S. R. Thye, E. J. Lawler, M. M. Macy, & H. A. Walker (Eds.), *Advances in group processes* (Vol. 16, pp. 199–218). Stamford, CT: JAI Press.

Smith-Lovin, L., Skvoretz, J. V., & Hudson, C. G. (1986). Status and participation in six-person groups: A test of Skvoretz's comparative status model. *Social Forces, 64,* 992–1005.

Triplett, N. (1898). The dynamogenic factors in pacemaking and competition. *The American Journal of Psychology, 9,* 507–533.

Troyer, L. (2001). Effects of protocol differences on the study of status and social influence. *Current Research in Social Psychology, 6,* 182–204, <http://www.uiowa.edu/~grpproc/>.

Troyer, L. (2002). The relation between experimental standardization and theoretical development in group processes research. In J. Szmatka, M. Lovaglia, & K. Wysineska (Eds.), *The growth of social knowledge: Theory, simulation, and empirical research in group processes* (pp. 131–147). Westport, CT: Praeger.

Troyer, L. (2003). Incorporating theories of group dynamics in group decision support system (GDSS) design. *IEEE Proceedings of the International Parallel and Distributed Processing Symposium,* Nice, France, pp. 182–189.

Troyer, L. (2004). MacSES/WinSES: *Software for the standardized experimental setting of status characteristics theory.* Unpublished software manual. University of Iowa, Center for the Study of Group Processes, Iowa City.

Webster's Dictionary of English Usage. (1989). Springfield, MA: Merriam–Webster.

Wikipedia contributors. (2006). Technology. *Wikipedia, The Free Encyclopedia,* http://en.wikipedia.org/w/index.php?title=Technology&oldid=55926197 (accessed April 25, 2006).

Willer, D. (1999). Developing network exchange theory. In D. Willer (Ed.), *Network exchange theory* (pp. 285–307). Westport, CT: Praeger.

CHAPTER 8

Funding Experiments, Writing Proposals

MURRAY WEBSTER, JR.
University of North Carolina–Charlotte

ABSTRACT

Proposal writing is integral to virtually every mid-level and high-level job in the social sciences. Chapter 8 describes the importance of proposals in the research process, with special reference to situations facing experimental social scientists. Although there are many funding sources and a great deal of money available, the process is competitive and success is much more likely with adequate preparation, skillful presentation, and persistence. I consider steps in the proposal writing process, offer suggestions for successfully integrating proposal writing into a research career, and provide some tips on successful proposal preparation.

I. WHY WRITE A PROPOSAL?

Times have changed. Until around the 1950s, most aspects of contemporary research in social science had not yet appeared. Many social scientists spent their lives teaching; they never even collected data. Some of the few who did research relied on government documents found in libraries when they needed information. Others reported on settings they observed themselves. Most often, publication meant producing interpretive theoretical discussions. Occasional small-scale investigations that relied on unpaid student assistance, even unpaid participation, occurred in early experimental research. However, research in those days bore little resemblance to the social science research enterprise today.

Today, research is not optional; it is expected as part of the responsibilities of virtually all social scientists, whether employed in higher education, government, industry, nonprofit institutions, or the military. (Colleges and universities expect teaching and service as well as research, but in other settings, research is likely to be nearly 100% of the job description for a social scientist holding an advanced degree.) Furthermore, contemporary empirical research requires extensive planning, difficult and expensive data collection, a team of researchers and research assistants, sophisticated analytic tools, computers and other technology, and data archiving and retrieval. In short, research today differs both quantitatively and qualitatively from that in earlier days. Its character is very different and there is much more of it. The changes make research much more expensive.

Where does the money come from? There are actually many more sources of it than most people realize. The largest source, not surprisingly, is the United States government. However, the federal government is far from the only source. Most national governments fund research, including governments of very poor countries that have little or no surplus money. Every branch of the United States Government—the Departments of State and Agriculture, the FBI and CIA, Defense, Homeland Security, Treasury, etc.—has programs to support research as part of its mission. The so-called "independent agencies," including the National Science Foundation, the National Institutes of Health (NIH), and the National Institute of Justice, exist explicitly to support research. In the United States, every government level below the federal—state, county, city—also funds research. Private foundations large and small fund research. Most people have heard of the large Bill and Melinda Gates, Ford, and Carnegie Foundations, but they might be surprised at how many funding agencies there are. It is a rare community and an even more unusual topic that does not have a funding foundation dedicated to supporting it.

Those sources provide what is called "external funding"—that is, a source external to the institution that receives the funding, as when the National Science Foundation awards a grant to an investigator at a university. Internal funding is another source of funding. While few universities are rich enough to support much of their faculty research, most offer small grants, often referred to as "seed money," in hopes that recipients will grow larger external grants from the initial results. For all of the other institutions mentioned, internal funding constitutes the majority of their expenses. Drug companies and other for-profit corporations fund their own research internally; so do nonprofit organizations. However, both for-profit and nonprofit research firms are heavily engaged in writing proposals to the funding agencies that provide most of their budgets. Federal government agencies such as NIH fund internal research along with funding much research done by others; county welfare agencies may pay for studies by their own employees or contract employees.

Research funding is so widely available and so differentiated that we may say with only slight exaggeration: whatever topic you might wish to investigate, there is someone who wants to support you. The first task is to find out who that might be, and the second is to convince the funding agency that you are a better candidate to invest in than someone else might be. Finding the funders is accomplished with a little investigation online, at a library, or, most effectively, with the help of someone from a university's or a corporation's research office, sometimes called an Office of Sponsored Programs and Research.

Research administrators use a variety of tools to stay abreast of current funding opportunities. The most common are subscription-based electronic databases such as The Community of Scholars (COS) or The Sponsored Programs Information Network (SPIN). If your organization subscribes to one of those services, you can post your research profile and select keywords that describe your research interests, and once you have done this, you will receive weekly e-mail updates of new funding opportunities in your area. Most federal agencies also allow you to sign up to receive e-mail notifications when new funding opportunities are released. A good research administrator knows all the published sources, as well as having access to much more information received in the mail, found in professional journals and other publications, and gleaned at meetings of professional research administrator societies. This person can become an invaluable partner in helping with all phases of your proposal preparation and submission.

The second task is to develop a research proposal. A proposal is an offer to do a particular research project in return for monetary support. As any economist could tell you, if an institution gives away money, people are likely to ask for it. Your task is to make your proposal more appealing than anyone else's. While it may be obvious that externally funded research requires proposals, internally funded projects usually require proposals also.

Research support falls into two broad categories, depending on how much discretion a researcher will have over its topic, methods, personnel, and outcomes. Sometimes an investigator thinks a topic is worth investigating and the proposal must convince his or her supervisors to spend the money on it; this is called an "investigator-initiated project." At other times, a corporation or a supervisor wants something investigated and asks teams within the agency and/or outsiders to design competitive proposals that will determine who gets to do the work. Investigator-initiated projects usually are supported by "grants"; the term means that the funding agency solicits proposals to do research, perhaps within a broad area (e.g., something related to employment opportunities of women), but otherwise does not specify details. The other kind of award is a "contract." This term applies when a funding agency specifies the outcome (or "product") expected, and may also specify personnel, research methods, and other details of the work (e.g., to design ways for new female hires at XYZ corporation to progress as fast as or faster than men in their careers).

Experimental research is supported by both grants and contracts, though the greater definition of contract research means that much of this chapter will contain information on preparing grant proposals. While much of this information should be helpful for writing proposals to do any kind of research, I will focus special attention on the needs of experimental proposals.

Besides the obvious benefit (finding money to support research), proposal writing confers a number of indirect benefits. It helps focus and clarify ideas for oneself and others. Anyone who has tried to explain something—in a presentation to colleagues or to students in a class, for instance—knows that explaining requires and fosters deep understanding. Similarly, there is a big difference between having a good idea and developing that idea into a proposal so that others can appreciate its value. For that reason, ideas in proposals tend to be much better than ideas that live only in someone's head.

Proposals receive feedback from other competent individuals. While anyone can ask others to give feedback on an idea, or even to read a paper and comment on it, most of us find out quickly that that is asking quite a bit. Most of the time, most competent people are busy with their own work and they hesitate to take time to think deeply about someone else's ideas. Reading a paper for someone is quite an imposition on one's time and work habits, and thinking about it and offering suggestions for improvement multiplies the time and effort required. Yet proposal reviewers have agreed to do just that. Often the review process generates the most thoughtful feedback available.

Along with supporting research and thereby advancing knowledge, successful proposals generate support for research assistants. Not only do research assistants benefit from a salary, but also funded projects often provide an important way for them to learn techniques and strategies of knowledge production. For graduate students, research involvement is invaluable; for research assistants in nonuniversity settings, it expands their skill sets, eventually enhancing earning capacity and job mobility.

There are many other indirect benefits of proposal writing, including making a researcher more visible in the discipline, reflecting favorably on one's university or firm, permitting an experimenter to pursue his or her interests rather than tasks assigned by someone else, and even leading to advancement of knowledge. Proposal writing is destined to be a very important part of the professional activity of most social scientists, and certainly so for every experimenter. Everyone needs to know how to write a proposal, and successful researchers need to know how to do it well. Yet proposals are competitive. Most of the federal funding agencies can support fewer than 25% of the proposals they receive. The situation is a little better with some private foundations, though certainly not with all of them. I do not know of any agency that funds over 50% of the proposals it receives, though there might be some that do.

Proposals will be read by reviewers who are other professionals from the discipline of the proposal writer. Sometimes, as with the larger federal agencies, reviewers come from the national pool of scholars. At other times, as with smaller agencies and many private foundations, reviewers are members of the funding agency—in particular, the program officers who receive proposals. Foundation proposals are also often read by members of the foundation's board of directors to assure concordance with the foundation's purposes. Successful proposals must appeal to reviewers, in other words. Success is not magic. It is competitive, and it is surprisingly fair. Researchers who succeed at getting proposals funded write proposals that reviewers regard highly. That is really the sum of the matter.

Because proposal writing is competitive, it makes sense to invest some time developing winning strategies and techniques. Short of winning a lottery or marrying extremely fortunately, researchers really have no other choice than to enter the competition. As with other parts of life, success is never guaranteed and is not always achieved. However, preparation, effort, perseverance, learning from one's own and others' experiences, and viewing any setbacks as temporary can enormously improve the odds of success.

II. SPECIAL PROBLEMS OF EXPERIMENTAL RESEARCH IN THE SOCIAL SCIENCES

Experimental researchers may face particular hurdles in explaining this kind of research to funding agencies. The most common misunderstandings include concerns that experiments have low external validity, or that results gained from experiments on college sophomores do not apply to individuals in other settings; however, there are others. It is important for experimentalists to anticipate them where possible and try to keep them from harming a proposal's chances for success.

Misunderstandings and concerns about experiments often are misplaced, and they may come from a reader's ignorance about research methods and the relation of theory and research. However, a research proposal is not really the place to try to educate one's colleagues (even assuming they want to be further educated). The best approach usually is to construct a scientifically valid proposal and emphasize what knowledge it can produce, being careful not to inflate claims. It is important to avoid appearing arrogant, as if an experimenter believes that this method produces better knowledge than other methods. A modest approach, including, when appropriate, acknowledgment that there are limits to the knowledge that the proposed experiments will produce, often defuses readers' concerns about supposed experimentalist arrogance.

An experimentalist can anticipate and sometimes avoid unfounded concerns about generality of findings by having a good grasp of the place experimental research plays in developing social science knowledge. Other chapters of this book and several sources at the end of this chapter should be useful here. Anyone who regularly conducts experimental research will have had numerous opportunities to discuss its philosophical foundations with other social scientists. The way an experimentalist writes a proposal can benefit from what he or she learns in those discussions.

Some misplaced criticisms appear regularly enough that it is wise for experimenters to anticipate them and, to the extent possible, to reduce their harmful impact on their proposals. Following are some common concerns and some suggestions that may be especially useful for experimental proposals.

- Experiments are artificial, in the sense that they are created by humans rather than by nature. This is an advantage of the method, for it makes strong tests of theoretical predictions possible. However, that fact is not widely appreciated outside experimental circles. In a proposal, there is no need to apologize for artificiality and it is useless to pretend it is not a fact. Any attempt to claim that an experimental design seems "natural" or "realistic" misses both audiences. Experimentalists see naturalism as a design weakness; those hostile to experiments see any attempts to create naturalism as inevitably falling short of the mark. Explaining an experimental design in terms of the conditions the experimenter intends to create, and why those conditions will permit tests of the project's hypotheses, is a much better strategy than entering into artificiality debates in the proposal.
- Experiments should not be contrasted to "the real world." An experiment is as real as any other setting, and participants engage in experiments as fully as they do in any other interaction. A proposal to do this sort of research deserves a serious description of the design. The research should not be treated like a game or some other situation where participants know their actions do not really count.
- Experiments are one research method—not the only method. They are well suited for certain kinds of questions, especially for testing hypotheses derived from well-developed theories. Other methods are better suited for other kinds of problems. An experimental proposal should describe why the experiment constitutes a good way to get knowledge in the particular case described, but there is little merit in triggering a discussion of what kind of research method is the best one in general.
- Experimental research is hard to do, if done well. People who do not conduct experiments sometimes do not appreciate how much effort and planning go into a well-designed experimental study. They may not understand why experimental research takes as long as it does, or costs what it does, unless the proposal carefully describes all parts of the

empirical work. Simply to write about studying the effect of some independent variable trivializes the design. An experiment reveals the effects of an independent variable under certain initial conditions and with certain operational measures, and many aspects of the design need pretesting to be sure they work as intended. The proposal should contain enough information that uninitiated readers can appreciate the many steps in the research. Every step in the research process incurs costs for personnel and supplies, so the proposal should show why those costs are justified. If incentives are necessary for subjects' payments, for instance, provide information about why they are important.

- A few social scientists think that experimenters are likely to mistreat human subjects. Perhaps this suspicion comes from the widespread showing of videotapes of shock experiments or prison research in introductory social science classes. While suspecting the motives and ethics of experimenters is unfair to and misplaced for every experimenter I have known, it is wise to think about the suspicion when writing a proposal. This is especially important when the design involves any deception, as many social science experiments do. The author should mention steps taken for protection of human subjects, including institutional review board approval (discussed in Chapter 6). The design should go beyond the minimum required by law, however, and the proposal should describe full explanations to participants at the end of the work, offering to answer any questions a participant may have about the experiment, any other design features to reduce anxiety, the fact that participants will be volunteers and whether they receive pay or other remuneration, and whatever else shows that the investigator recognizes the responsibility to deal ethically with people.

While those are some of the most common misunderstandings and prejudices about experimental work, there are many others. The best way to deal with all of them in a proposal, I have found, is to take great care in explaining all aspects of the research design and showing how each element is related to creating important conditions or gaining a significant piece of empirical knowledge. Of course that is good advice for any research proposal. However, part of the special challenge of doing experiments is that not everyone has a good understanding of this kind of work, and not everyone automatically presumes experimenters are concerned with human welfare. Nevertheless, the job of a proposal writer is to make the value of the work and the motives of the researcher clear, even to readers who may not feel disposed towards this sort of work, or who misunderstand crucial features of it. An investigator can do a better job of that if she or he thinks a bit about what kinds of presumptions and background knowledge others in the discipline are likely to bring to the table when they read a proposal.

III. THE STRUCTURE OF RESEARCH FUNDING; ROLES AND ROLE BEHAVIORS AND SOME TERMS

For understanding the research enterprise, it is useful to begin by making explicit the larger picture of institutional structures and roles that I will allude to throughout this chapter. Researchers can be more successful in writing proposals and getting funding if they are clear about the structures and role behaviors of everyone involved in the process.

At the largest level, two kinds of institutional structures are important. One is the university or research organization at which the researcher works. The other is the governmental or private agency that pays for the research.

While there are several differences between working at a research organization or a university, one thing applies to both institutions. A research grant or contract is a legal agreement between parties that are organizations, not individuals. In other words, if the National Science Foundation makes a grant after you submit a proposal from the University of North Carolina, that grant is part of a legal agreement between the U.S. Government and UNC. This means, for instance, that if you change jobs and leave UNC, UNC has the option of deciding to send the grant funds with you to your new job or to retain them and appoint another competent person to fulfill the research. You may hope they will do the former, but it is not automatic and you should investigate if this circumstance might apply to you.

However, for all of us, the status of the agreement has a significant benefit, and that is that the institution, not the investigator, must fulfill many legal requirements that most researchers are not equipped to fulfill. Among them, Congress requires that contractors guarantee a drug-free workplace. While a researcher probably will supervise his or her research staff to be sure drugs are not part of the picture, no faculty member is able to supervise the entire campus to make sure it is drug free. The nature of the legal agreement means that it is the university's problem, not yours, to certify.

There are other legal requirements that employers must comply with also. Some of those include employment access for members of under-represented groups and individuals with physical limitations, review by an institutional board for the protection of human participants or animal subjects, and compliance with various payroll laws such as minimum wages and medical benefits. Researchers do not need to keep up with all of these requirements, but they do need to understand that before they can submit a proposal from their employer, someone is going to review it to be sure nothing in it will violate a federal, state, or municipal law.

Institutions that provide money for research are called funding agencies. As noted before, they include both government and private agencies. They provide money as part of their mission. Government agencies all have missions, as noted earlier. Private agencies often want to affect society or some of society's institutions such as schools or families. Thus, funding decisions are made first and foremost in terms of whether a particular project seems to have a reasonable chance of furthering the funding agency's mission, and that justification must appear in every proposal, either implicitly or explicitly. Researchers sometimes add irrelevant justifications, such as they or someone they propose to pay "deserves" the payment, or the researcher is tired and needs some relief from his or her normal work duties. Those things just waste space in a proposal because the agency is prevented (by law, in the case of government agencies; by internal regulation for private agencies) from taking them into account.

Both government and private funding agencies provide money in the two categories described in Section I. Excellent experimental work has been supported both by grants and contracts in recent years.

Within a funding agency, two roles are important: program officer and reviewer. The program officer is less well understood by many researchers, many of whom have reviewed manuscripts and even proposals, but program officers are very important and it is worthwhile understanding something about their roles and role relationships to researchers.

Most program officers are highly trained; the large majority of them have doctoral degrees in the field they oversee, or in related fields. They read the scholarly journals and keep up with research; they are likely to know the work of many of the scholars in various fields. Program officers know what good research looks like, and they have an excellent understanding of the relations between theory and research design. Some program officers serve temporarily at funding agencies and then return to their universities or research organizations, while others make their careers within funding agencies. All, however, are as well qualified in their disciplines as are university faculty members.

The program officer is the "face" of a program, its public representative. Reviewers, by comparison, are less likely to be visible to outsiders, and they may wish to be anonymous to avoid inappropriate contact from researchers seeking privileged access to funding. (I will write a little more about reviewers in a later section of the chapter.)

However, while a program officer represents a program to a researcher, the really important relationship flows the opposite way. The best program officers conceive their job as representing members of the discipline to the funding agency—that is, advocating for the needs and interests of the researchers. It is a common mistake to think of a program officer as a gatekeeper, keeping a researcher away from the money that he or she needs. Far more often, a program officer is more than sympathetic, and looks for (and often finds) ways to help researchers achieve their goals.

I once heard a program officer summarize her relationship to researchers in this way:

> My job is to place six million dollars with researchers in my discipline. I will do that; no question. Your job is to make it easy for me to place some of that money with you.

Of course much of this chapter is devoted to offering information and suggestions to help experimental social scientists do that job.

A researcher who submits a proposal is an "applicant." It is not important to remember that term, except to remember that "supplicant" would be inappropriate. In other words, the role relationship of program officer and applicant is an exchange in which both parties hope negotiations will succeed. Do not think of a program officer as hoarding money; the program officer is going to allocate whatever budget she or he has. Young applicants sometimes appear to regress and become adolescents asking a parent for an allowance. Whining, excessive politeness, flattery—all techniques that may have worked within the family—are unprofessional and embarrassing for all. Avoid them.

Perhaps more commonly, applicants sometimes act as if all they have to do is convince a program officer to give them money. That is almost never true. For many reasons, program officers want to fund projects that others in the discipline—the reviewers—recommend for funding. While program officers make the final decisions, and for any of several good reasons those may not coincide perfectly with reviewers' recommendations for program officers do not act alone. Nagging a program officer and other pressure tactics do not work because, in addition to making a researcher seem unpleasant to deal with, the program officer's views are only part of the equation.

Earlier I wrote that the program officer represents researchers (you and me) to the funding agency. Part of our responsibility, thus, is to avoid making the program officer's job harder than it needs to be. Social scientists, especially the young, inadvertently cause trouble for program officers in government agencies when the titles of their proposals raise flags at Congress. While social science has some great friends in Congress, there also are a few who view us with suspicion, and those few often ask federal agencies to give them lists of proposals from our disciplines. They are looking for reasons to cut funding for the social sciences, and inflammatory titles—no matter how cute the researcher thinks they may be—are dangerous to everyone. It is in everyone's interest for applicants to think about possible consequences of what we write in proposals that go to the government—in fact, in everything we write for all audiences.

In Section I, I suggested that a researcher begin the quest with the research office at her or his institution. Research officers can help clarify topics and identify possible funding agencies. Once a researcher has settled on a few agencies, based on their mission statements and other information from the research office, she or he should contact those agencies. As noted previously, it would be a mistake to contact a funding agency without first learning its mission and other aspects of its functioning from the research office—at best, a waste of time, and at worst an embarrassing ignorance that can come back to haunt the researcher.

The program officer is the first point of contact with a funding agency for a prospective applicant. Phone calls and e-mails constitute the initial contacts in which a researcher can confirm understandings of the regulations of the agency, its mission, and other important details of proposal submission. This is a good time for a prospective applicant to outline (briefly!) the topic of the contemplated proposal and hear whether the program officer has any thoughts. With large agencies such as the federal agencies, program officers often can suggest exactly the right program to target a proposal to, and can clear up any number of potential misunderstandings right at the outset. It is well worth a researcher's time to initiate this contact.

Is the program officer too busy to bother with someone as insignificant as I am? Program officers certainly are busy. However, because of their roles, and by personal inclination, they really like to talk with us. Most program officers are passionate about research and their agency's mission. When you understand that a large part of a program officer's day, like almost everyone else's, is spent processing paper and attending meetings, you can appreciate how welcome an interlude to discuss research is for her or him. Yes, she is busy and yes it is an interruption when a researcher calls, but a good intellectual discussion (a brief one, remember) can restore a program officer's energy for the remainder of the day. Call.

There is one caveat here. Your call or e-mail to a program officer deserves to be well thought out, just as any other conference should be. It would be rude to call up anyone at work and free associate for a while. The sort of call a program officer dreads begins with "I would like to bounce some ideas off you." Program officers do not like to bounce.

Now that we have some of the basic facts about relations between universities and research organizations and funding agencies, about roles and role relationships within and among the institutions, we interpret how these play out in the life of experimental social scientists seeking funding for their work.

IV. RESEARCH PROGRAMS AND PROPOSALS

A research project, like a proposal, should be seen as an element in a larger plan or pattern. A proposal is not an end in itself; rather, it facilitates doing research and publication, and all the indirect benefits alluded to before. A research career, ideally, will include many proposals at different times for different purposes. Some of them will succeed, leading to funding; others may not. Wherever possible, writing a multiyear plan for research topics, considering what to do first, what follows, what is next, and the like, makes for a more coherent career with less wasted time.

If you are the sort of person who likes planning, it may make sense to draw a timeline for, say, the next 5 years. What research topics would you like to address in that time? What experiments will help with them? How long will each experiment take to complete? When does a proposal have to be finished in order to support an experiment? Questions like these can help someone to organize more efficiently, and, in most cases, they also provide reality checks on vague or unrealistic plans. For instance, if it will take 2 years to write a proposal and conduct one experiment, it is not likely someone is going to do more than two such experiments in the next 5 years.

Another use of a timeline is to help thinking about programs of sequential experiments. Often, results of one experiment suggest new experiments, or new experiments can build on established findings. A timeline can show the structure of such an experimental program; in doing that, it facilitates thinking in terms of programmatic development of knowledge about related topics.

Proposals should appear throughout a career in research. Just as it would be unfortunate to abandon proposal writing prematurely, it is a serious mistake to put off beginning it. A researcher should begin thinking about proposals at the very outset of his or her career. One reason for this is to develop proposal and research habits, which will be surprisingly difficult to pick up later. Another reason is that proposal reviewers tend to feel positively towards investigators at the start of their careers. If someone waits several years—for instance, until achieving tenure at a university—before writing the first proposal, he or she handicaps himself or herself. You can bet that person's proposal will be considered in competition with a proposal from someone else who is just starting her career. Reviewers then will ask, "How do we know this older person really is going to do the work?" After all, the record for the past 6 years may show no evidence or promise of research. Or reviewers may just want to help a beginning investigator: "Let's go with the young person and help her get her career started right." That kind of age discrimination may well be unfair, but we are not going to change that now. It is better to start early and thus let reviewers see you as the new hotshot.

If you are one of the people who come to experimental methods late, or if you are writing your first proposal after some years developing other aspects of your career, the preceding suggestion should not discourage you. A better way to use the information is to acknowledge the concern some readers might have, and explain what you were doing up to this point. Often some of that experience is relevant to the proposed work, and pointing that out can strengthen the proposal at the same time that it explains why it did not appear earlier.

By the way, discouragement can appear at any time. Later, I will address discouragement following proposal rejection. However, discouragement ahead of time is extremely self-defeating because it prevents even entering a competition. There are many good excuses and bad advice to discourage proposal writing. Funding agencies suffer budget cuts; someone may think he works in an unpopular area; there are many other demands competing for scarce time; older colleagues may not appreciate how crucial proposal writing has become since they began their careers. None is a good enough reason to stay away. If you look hard enough, it is possible to find good reasons not to do anything at all! But except for the few who spend their lives in semidarkened, fortified rooms, most of the rest of us take chances and engage with life.

Presuming someone has an idea for a worthwhile experimental study, how should she or he proceed with the proposal process? Generally, cast as wide a net as possible to enlist others' help. Talk the ideas over with colleagues as much as they are willing to do, to find weak points, refine good points, and find ways to explain things clearly. Invite a colleague to become a collaborator, especially if, as is often the case, that person brings talents complementary to your own. For instance, you might need someone with mathematical modeling skills, or if you are a new investigator, you might ask a more experienced experimenter to be part of the team.

After enlisting help with the project itself, take advantage of practical help from people whose professional skills are in the area of proposal development and management. Go to your organization's research office and find out about potential funding agencies for your proposal. When you actually begin writing, ask the research office to help with editing and preparing parts of the proposal, such as the budget, that may be mysterious. Research officers are usually eager to help, and when they are good at their jobs, they can be invaluable.

V. PREPARATION FOR WRITING A PROPOSAL

Once you have settled on one or a few potential funding sources, study their proposal requirements. Some issue RFPs (requests for proposals). Others have regulations on proposals, and those may include forms to complete as requested. Know for whom you are writing and how they want it done. Two things are crucial. First, learn and comply with their regulations. Those often include submission dates, page limits and formatting, and other material such as an abbreviated list of publications. (In a lecture, I once mentioned to a graduate class that one of the federal agencies specified margin widths [at least $1\frac{1}{4}$ in. all around], paper color [white], and font [Times New Roman or Arial], and wanted the proposal fastened with a single staple no more than $1\frac{1}{2}$ in. from the upper left corner. A student objected that seemed very picky. I responded that if you are asking someone for a quarter of a million dollars, it makes sense to ask in the way he or she wants.)

Second, learn and write to the funding agency's purpose or mission. Every institution has a mission statement; ask for it if they have not already provided a copy. For instance, the National Science Foundation (NSF) supports (mostly) basic research to advance scientific knowledge. The National Institutes of Health (NIH) support basic and applied research to improve the health of Americans. Do not submit a proposal for an experimental study of health behavior to NSF; do not propose an experiment with no connection to health to NIH. Private foundations have missions and goals that may change from time to time; they seek proposals to advance their goals. Show in your writing that you understand the funding agency's mission and how your proposed research can help achieve it.

Having completed the preliminary work, a researcher can begin to think seriously about planning the proposal. Here it definitely helps to list items on paper, for there are a lot of them and most are interdependent. Who will be responsible for writing the proposal, or the different parts of it? What resources will the writer(s) need; for instance, access to publications or consultation with others? What kinds of resources and permissions have to be in place before submitting the proposal? For instance, experiments require approval from institutional review boards (IRBs) (discussed in Chapter 6), and someone probably has to agree to provide space for and time of the researchers and other workers on the project. How long will every part of the writing tasks take? Working backwards from the agency's submission date, identify how long each part of the process will take, allow a little extra for unanticipated delays such as someone being on vacation or problems with the university computing system, and that tells the start date for the writing.

Proposal writing is hard work, and doing it well requires setting aside considerable time for the task. Successful proposals require planning and work, for a very good reason: a research proposal has at least as much intellectual content as a paper submitted to a scholarly journal. It contains a brief summary of relevant existing work, describes the new proposed research, shows evidence that it meets current standards of evidence, tells what it will mean in terms of changing the state of knowledge, and tells what comes next. Furthermore, a proposal is more demanding for several other reasons. Most importantly, it must stimulate enthusiasm. A reader must not only understand the quality of thought in it, but also must come to share the writer's belief that this work really ought to be done. A successful proposal conveys the writer's enthusiasm and brings some of it to the readers. Without enthusiasm, reviewers are not going to recommend putting an agency's scarce money into the work.

Previously I noted that funding agencies provide guidance about what proposals to them should contain. Every proposal, however, has certain broad sections to it. Each section leads to the one following it. More importantly, the justification for each part of a proposal depends on what has been established in the preceding topic. A well crafted proposal "flows" from an orienting statement, through the various parts of doing the work, to a conclusion in a way that seems, if not inevitable, at least reasonable and thoughtful.

VI. SECTIONS OF PROPOSALS

Proposals contain five large groupings of topics, or five parts. The five parts can be seen as analytic components. The actual sections of a proposal may be specified in the agency's RFP or its guidelines. The "parts," as that term is used here, are things that should be included in every proposal, whether or not they get labeled as sections of the proposal. Be sure the proposal attends to the issues of each part.

The first part is the proposal's overall *topic,* and it is probably the most important part of a proposal. A proposal's topic should appear early and clearly enough so that readers have a context for all that follows. For example, a proposal's topic might be to extend theories of exchange in networks—that is, to develop theoretical understanding of those processes. The topic statements of a proposal should help answer (or avoid) questions such as, "Why are they telling me this?" at later stages of the proposal. That is, every sentence in a proposal should obviously help to deal with the topic as initially stated.

Second, every proposal contains one or more *research questions*. These deal with a part of the general topic just described. A research question is more specific than a general topic because no research could fully explain every question in a topic. For example, the proposal might contain a research question such as, "Do participants who engage in negotiated exchange come to trust their partners as much as those engaging in reciprocal exchange?" (In this discussion, do not worry about what those terms mean; trust that the proposal defines them somewhere.) Research questions may be stated as hypotheses from explicit theory or other sources, or they may be stated with less background. Their crucial property is that they must be answerable from research. That point is not always appreciated; I will mention some instances of unanswerable research questions below. At the same time, a proposal must justify its research questions as significant, questions whose answers will help explain the area of the topic.

Third, every proposal contains a *research plan*. For an experimenter, this includes the experimental design and operations, as described in Chapter 1. The research plan should be tied to the research question so that a reader can clearly see why doing this would produce answers to the questions. Contrasting conditions of the experiment ought to have a reasonable chance of showing whether the research question is to be answered "yes" or "no," or whether a research hypothesis is confirmed or disconfirmed. For example,

Condition 1 will have people participate in negotiated exchanges, Condition 2 will have them participate in reciprocal exchanges, and afterwards participants in both conditions will complete some measure of the degree of trust they feel towards their partners. The research plan is the heart of the proposal and receives the most space.

Fourth, a proposal identifies *personnel* who will conduct the research. This includes the principal investigator (PI) and any co-PIs, research assistants, consultants, and others. Of course every person's contributions to the project must be spelled out and his or her inclusion briefly justified. The ultimate justification for every person identified must be that she or he contributes to the research plan described earlier.

Fifth, a proposal contains the related topics of *timetable* and *costs*. Some agencies want timelines; others accept descriptive text with tasks and estimated times for completion. Time is money because most of the cost of research is in hiring personnel. Proposal budgets often follow strict format restrictions. Most researchers are not familiar enough with all the restrictions to complete their own budgets, turning that task over to the skills of professionals in the research office.

One way to think about the five parts of a proposal is to recognize that readers will be evaluating each part. Does the topic deserve further research? That is, is there a foundation to the topic so that others will work on it, or has its value played out? Is the research question a good one? That is, has the writer found a question whose answer will advance understanding of the topic? Readers ask questions like these of each part of a proposal. When a proposal shows that the PI has thought through the reasons at every stage, it tends to generate positive impressions and even enthusiasm among readers. Such a proposal is likely to attain the enviable state of being considered "competitive." Now let us expand each of the proposal's parts and examine them in greater detail.

Topic. The topic of the proposal is the general theoretical context into which the research falls. Perhaps a PI plans to investigate power use in certain network structures, or identity processes in mixed-gender groups. Whatever the area, the research must be placed into a research tradition. This is the place to show that the PI understands how others have thought about the topic, and what related research has shown. Then the PI will go beyond what is known to propose new questions and answers. The background material is sometimes called a "focused literature review." The purpose here is not, as often appears in journal articles, to display the encyclopedic knowledge of the writer. Rather, the purpose of reviewing literature is to tell a story of increasing understanding by describing work that leads up to what the PI will propose doing.

The topic of research affects everything that follows in a proposal. For experimenters, that topic is often something in group processes, broadly conceived.

Whatever the topic is—network structures, status processes, organizational legitimacy, power relations in informal groups, identity processes, or anything else—the proposal ought to identify the topic early. Early statement sets the stage for the work to be described, setting context and suggesting possible specific questions. If readers understand the topic and know it is important, the research questions have a better chance of looking worthwhile also, and so on.

The topic also suggests the skill sets needed by qualified reviewers—namely, people who have demonstrated their interest in and knowledge of this topic. Clearly stating the research topic increases the chances of getting qualified reviewers. If a program officer cannot tell what a proposal's topic is, besides considering the proposal weak on that ground alone, he or she finds it difficult to know who would be an appropriate reviewer. It is in the PI's interest to see that the program officer can clearly tell who understands the proposal's topic.

Do not overlook research topics that might have interest for more than one discipline. Interdisciplinary topics often are very appealing to reviewers because they can contribute to knowledge growth in multiple fields and may help create links among them. Writers of an interdisciplinary proposal often benefit from getting chances to appeal to, and receive funding recommendations from, more than one set of reviewers. At the same time, good interdisciplinary work is difficult because it requires good knowledge of a discipline beyond one's own. Further, the interdisciplinary character must be real. A proposal that simply says "both psychologists and political scientists will be interested in this research" can look naïve or worse if the proposal does not adequately review what has already been done in one of the disciplines.

As this point suggests, it would be foolish to choose a research topic for which the PI has little or no training or competence. Just because a topic is in the news or looks important on its face does not make it a good choice for a proposal. For instance, the publicity given second-hand smoking, responsiveness to sexual predators, appeal of murderous ideologies, and dozens of others topics does not by itself make them good topics for everyone. Unless someone has demonstrated competence related to answering research questions related to the topic, the proposal is not going to fare well with reviewers or funding agencies.

Research question. This is what the proposed research will, we hope, answer. It is a piece of a jigsaw puzzle whose overall picture is the research topic. In fact, the research question in a proposal is an important part of the topic, and the proposal must show why that is so. One of the things reviewers will ask is whether the proposed research question will contribute meaningfully to the topic of the research.

When it is possible, a good way to present previous work is to use the image of a funnel. Begin with broad ideas appearing in general theories or even in topics of the theoretical social scientific classics. Then trace an idea through increasingly specific theoretical views, and into empirical findings. Show what has been investigated and what has not yet been explored. Then you can situate your proposed work in that body of theory and research. A variant idea is to trace two literatures through funneling, and to show how your work can illuminate two traditions. Not every research tradition is orderly enough to permit funneled reviews, but where possible, it provides a strong heritage for the proposed work. Of course any organization of a literature review depends on the reviewer's intellectual work in discovering and imposing orderliness on a messy historical record.

For much experimental research, there is a more or less well developed, explicit theoretical foundation. If the research topic includes presentation of general principles, then the research question section may identify one or more of the principles that can be modified to deal with the subject of this proposal. Alternatively, the research questions may flow from a new model that is presented in this section. When the research questions come from explicit theoretical foundations, it is important to outline the reasons why these particular questions will be informative for assessing the theory presented.

There are several pitfalls in choosing research questions. First, of course, the questions must be answerable by empirical (or, sometimes, by theoretical) research. While that might seem obvious, it is surprising how often inexperienced researchers do not consider it. I once knew a student who wanted to answer the question "Why is there war?" in her M.A. thesis. It's a wonderful question, but despite centuries of thought, nobody has come up with a convincing answer. Had that student actually begun that project, she would have died of old age before coming to a scientifically valid answer.

Questions that might look more focused can also be unanswerable. For instance, questions such as "How should an organization set up a gender-equality program?" do not have one empirical answer because organizations are complex and any program is likely to have both intended and unintended consequences. In other words, any findings would be conditional on particular, probably unmeasured, characteristics of the setting in which they were found. The "should" suggests moral questions that are beyond social science research. A more focused question such as "How much effect did a particular state law have on promotion rates of women and men in large and small corporations?" might be answerable with empirical research.

Another common mistake is to propose research for which an investigator already believes he or she knows the answer. Examples (nonexperimental) I have seen recently include: "Did the administration create a 'moral panic' to justify the invasion of Iraq?" "Are Mexican immigrant women in New York hard working and resourceful?" "Is community support helpful as released convicts adjust to living outside the institution?" "Would it be a good thing if more communities developed recycling programs?" If you already know the answer, it is hard to justify spending money to find the result. Put differently, if an investigator is already convinced of an answer, is he or she the right person to design research that has a fair chance of finding disconfirmation? Most reviewers would conclude that the answer is no. Someone might wish to spend money to convince others of what the proposer believes, but such a proposal would go to a different kind of agency. Propaganda and dissemination are activities different from research.

A third kind of problematic question choice comes from the tenets of a philosophical position that had some currency in the social sciences recently: social constructionism. Sometimes the term denotes study of ways in which meanings of interaction situations and social identities come to be shared by actors, and ways some may try to control the meanings others assign. Research within that belief system is certainly possible. But some extreme versions of social constructionism hold that what we call scientific knowledge is no more or less than what we can get scientists to agree to, and that any evidence is shaped by investigators' theories. If someone truly believes that, there is no reason to do empirical research because the outcomes will do nothing more than confirm the investigator's pre-existing ideas. Again, if there is no possibility of disconfirming or modifying the state of knowledge by research, funding the research is not a good use of scarce dollars.

Research questions always need justification. Again, there are good and bad justifications. Here are some very bad justifications:

- "Nobody has yet studied ..." There are plenty of trivial topics that remain unstudied for the best of reasons: nobody cares to take the time.
- "This is related to my life (ethnic background, family, life experience)." While this justification is seldom stated this baldly, readers can spot it easily. However it may be disguised, this justification appears surprisingly often in social science and it is always a mistake. Reviewers' eyes glaze over.[1]
- "I found a hitherto overlooked setting (or group, or practice)." This one combines both of the preceding justifications. Does it persuade you? Probably not, and it certainly will not do well with reviewers, either.
- "This is really interesting to me." While understandable, this is really an incomplete justification for doing the research. The real question is whether a topic is important to other researchers. Things I have recently discovered are often interesting to me, but often that fact just reflects my prior ignorance.

[1]On the other hand, if your topic is theoretically justified (as it must be to have a chance of success), and if some aspect of your life makes you especially well suited to investigate it, then by all means highlight that fact. For instance, if your experimental design requires some conditions of participants who speak Spanish as well as English, and you are bilingual, be sure to point that fact out. That advantage makes you better qualified than a monolingual person to conduct this research.

What makes a good justification? The best, probably the only, good justification is that the social science community is interested in a topic. In other words, published research has dealt with that topic or closely related topics, and what you propose to study—your research question—looks like it provides a missing piece to others' understanding of the topic. A research question is justified by being placed within the intellectual matrix of its discipline. The relevant community has seen this question and wondered about it. To say the same thing differently: reviewers of your proposal can see that you are dealing with a topic they are interested in. They are on board intellectually. They are primed to want to know what your research will find, and they may even feel that hoped-for enthusiasm.

Picking research questions in this way requires seeing oneself as part of a community of scholars, people who have some broad interests in common and whose work sometimes helps the community at large understand things better. This view is very different from the hubris that leads some to think they are entirely unique thinkers, dealing with questions and answers nobody else has ever thought of. Science is most often a cumulative, shared enterprise. The lone wolves are far more likely to be crackpots than overlooked geniuses. If that view is dismaying, at least wait until you have proven you can contribute to understanding things other social scientists want to know, and then go on to prove your unique brilliance with unique research questions. (But remember that Isaac Newton—one of the greatest intellects ever and justly praised for his theoretical and experimental studies of gravity—thought he had just such unique theological understandings in his old age.)

How can you be sure that your reviewers have already dealt with your topic? That is the way reviewers are selected, in part. Funding agencies ask people whose own work has something in common with the proposed work to judge its merits. With any luck, one or more of your reviewers will note approvingly that you have reviewed their published research and are proposing to take the next step, intellectually speaking, to build on it. Work is significant precisely because it contributes to cumulative growth of knowledge, and the better a PI can demonstrate how work will do that, the more likely that work is to receive support.

Another useful question to keep in mind is what some have called "the so-what factor." In describing a research question, it is helpful occasionally to imagine someone asking, "So what?" While most readers, like most listeners, are too polite to formulate that question quite so directly, it is always likely to be in the backs of their minds. A writer has the responsibility of making the case for a project, and if few readers or listeners are convinced, that is probably due to a writer's not doing a good job of it. Keep "so what?" in mind, and try to show in the writing that there are very good answers to that question.

Research plan. This is the heart of a proposal: what the PI will do if awarded the money. The most important consideration here is whether the research plan has a good chance of answering the research questions identified earlier. If you follow this research plan, will you get reasonably clear answers to the research questions?

The research plan is probably the section of a proposal that most social scientists are best prepared to write. Their training is precisely in how to design and conduct research. However, there are sometimes gaps between what a scientist knows and what that person writes in a research plan. The guiding directive for this part of a proposal is to explain everything; *never* assume that "a reader will know what I mean." Even if a reader does know how something ought to be done, the proposal has to show that the PI also knows how to do it. The amount of detail should be sufficient to allow a competent reader to assess the quality of design and operations.

In Chapter 1 we distinguished research design such as experimental conditions, independent and dependent variables, and characteristics of participants from research operations telling how design elements will actually be realized in an empirical setting. Design and operations must both appear in the research plan of a proposal.

Besides describing independent and dependent variables, an experimental research plan makes clear how the proposed situation meets the scope conditions of a theory. The proposal describes experimental conditions and clarifies the ways they differ from each other. The operations tell who will be participants, whether they will be volunteers, whether paid, and how they will be recruited. No federal agency will release funds until an IRB approves the design, and most private foundations impose similar requirements. It is important to show awareness of the importance of that review by including a sentence similar to this one: "This project has been submitted for review by our IRB, and if this proposal receives funding, no data will be collected until it has approved the design."

Wherever possible, offer brief justifications for details. For instance, if an experimental proposal includes four conditions with $N = 25$ in each, that number should be justified by a power analysis showing that it is sufficient to find predicted differences. It is quite helpful to sketch what the data will look like and how you will analyze them. It is important to show that the PI has thought about such questions and has reasons for whatever decisions she or he has made in regard to them.

In experimental research, it is often possible to adapt a previously used basic design for the new research questions. When possible, that approach is highly desirable for scientific reasons described in other chapters. It also is desirable for a proposal because any previous uses of the design constitute pretests for the proposed modifications. Reviewers will know that most aspects of the design have already been used and they work successfully without unanticipated problems appearing. Whenever possible, building on existing research methods strengthens a proposal.

Experimental research usually produces patterns of results. Beyond stating predictions, a proposal needs to review how the predictions were derived, and what results the PI will consider confirmation and disconfirmation. It is very helpful in a proposal to include discussions of what both predicted and unpredicted outcomes of an experiment would mean. For instance, a particular experiment might have predicted outcomes of conditions ordered $1 > 2 > 3 > 4$ for the specified dependent variable. The proposal should make clear that that is what is predicted, and why that ordering is expected from the theoretical foundations. However, the proposal should go beyond explaining its predicted outcome; it also should mention what the data will look like if the new ideas are wrong, and what other outcomes might mean. For instance, disconfirmation might produce $1 = 2 = 3 = 4$. The proposal ought to include a discussion of how disconfirmation might come about. It might mean, for instance, that the new theoretical ideas simply are wrong. It also might mean something is wrong with the empirical design or operations. A strong proposal will discuss possible meanings of disconfirmation and outline follow-up studies in case disconfirmation appears.

Many proposals neglect other possibilities, and often that is a mistake. Stronger proposals note some additional reasonable outcomes (e.g., $1 = 2 > 3 = 4$ or $1 < 2 < 3 < 4$) and tell how they might be interpreted. While not all possible outcomes mean something (other than poor design or operational error), some patterns do have meaning. Strong proposals show that the PI is aware of possible outcomes other than those predicted and those showing that the predictions are clearly wrong, and the PI is prepared to follow up if they appear.

Finally, the research plan concludes with what will be known as the result of the work, and what some next steps might be. This means returning from the operational level (the actual data) to the design and theoretical levels (what they mean and how they affect the state of knowledge about a topic). Next steps usually are further theoretical questions that may be pursued once results of the proposed experiments have come in and been evaluated.

Personnel. A proposal describes who will be associated with the project and in what capacity, usually in a separate section. Personnel includes the PI and any co-PIs, of course. In addition, there will probably be research assistants who help run experiments and interview participants, and, possibly, consultants. The proposal should briefly indicate why individuals other than PIs have been chosen, and what they will contribute to the project. Reviewers assess both roles (e.g., Why does this work contribute to the project?) and individuals (Is this the right person to do this work?). A strong proposal makes both questions easy to answer.

Reviewers assess whether the PI's training or experience qualifies him or her to do the proposed research with a reasonable probability of doing it well.

They also assess whether research assistants are well qualified, and whether consultants are needed. As with other parts of the proposal, personnel should be justified. The fact that reviewers will assess the qualifications of the PI and others to do the work, which they sometimes call their track records, helps to explain why picking a research topic based on popular consciousness usually is not a good idea. Just because a topic is in the news, or because a PI thinks it is important, does not make that PI the right person to investigate it. If someone has no relevant training or experience studying, say, global warming or sexual predators, reviewers are not likely to recommend funding a proposal on those topics by that person. The odds are that a newcomer does not know what others have already done and will overlook some well-known research pitfalls; thus, the chances of such a person making a useful contribution look small.

Budget and timeline. Earlier I suggested getting help with budget preparation from a professional in the institution's office of research. Budgets are complicated, and most researchers do not understand some parts of them. For instance, most university faculty do not know their university's indirect cost recovery rates and fringe benefits, or how those may vary between the academic year and the summer. Nor do they need to know those things. That is what research professionals know, along with other things. The main things a researcher does need to think about are all the tasks that need to be done to complete a project, and how long each of them will take (called person hours, or person months, and, of course, directly reflected in costs of salaries and wages).

In the meeting with the research officer, the PI can describe who will do the work and how long it will take, and for research assistants and consultants, what the PI would like to pay them. The research officer knows the institution's policies regarding such esoteric topics as indirect cost recovery (sometimes called overhead), fringe benefits, tuition remissions, and other costs that a PI probably does not understand well. A research officer also can give the PI some reality checks about how much money a funding agency typically awards—for those times when a PI thinks it will take a few million dollars really to do the job right, or where he or she wants the funding agency to pay salary for 2 years so that full time can be devoted to the research. (It does not happen.)

In thinking about a timeline for doing the work, the starting point is a careful estimate of how long each part actually will take. To that, a PI should judiciously add some percentage because, if anything unexpected occurs—and the unexpected almost always appears in research—the one thing we know is it will take time to adjust. Time, in most cases, is money. That is, research assistants and others will have to be paid during the adjustment period as well as when they are doing the actual work. Remember also to include time for pretesting of experimental designs and for data analysis.

Now that we are considering money, I suggest looking back over the entire proposal as what it really is: a request for money. Read it again and focus completely on the money. This is not to be crass. Everyone knows PIs are motivated primarily by theoretical questions and the search for truth. But think of it from the funding agency's point of view. *The only thing they can do for the research is to write a check.* They cannot provide the statistical consultant that a project might need, or arrange laboratory space to conduct the experiments. All they can do is send money. This means that the proposal must convey that the PI is ready to go as soon as he or she receives the money. Other than money, anything not already in place weakens a proposal, sometimes fatally. A funding agency official might wish to help with design and operations, but he or she cannot do that. All the official can do is approve sending money, so the proposal must convey that lack of money is the only thing holding back the important work it describes.

VII. SOME TIPS ON PROPOSAL PREPARATION AND WRITING STYLES

Proposal writing, like research itself, is an uncertain activity. Proposals and PIs are in competition with other proposals and PIs, and it is wise to presume that much of the competition will be very good indeed. Not every excellent proposal receives support, and there is no magic formula for writing a successful proposal. Still, here are some suggestions from someone who has seen a lot of proposals; I hope some of them are helpful.

Many proposals contain so much justification, especially of topic, that they begin to look defensive. Do not waste space with excessive justifications. In fact, begin the proposal with what you want to do, and let justifications take care of themselves as you describe the intellectual background of the topic. A beginning sentence like this can focus a reader's attention on everything to follow: "I propose to conduct a six-condition experimental study to assess a new model of power at a distance in network exchanges." That is clear, and you can explain any possibly unknown terms such as "power at a distance" later in the proposal. At least the readers will know what you are proposing to do.

What I called a defensive beginning that unnecessarily handicaps a proposal is some version of "This general area is really, really important in people's lives." Many proposals begin with a paragraph or even a page or two that seem to have no purpose other than convincing a reader that the topic matters. That always looks desperate. Most of the time, reviewers are willing to grant that a topic is worthwhile, or at least to suspend judgment until they have read your justification for the research questions (which you now know has little to do with saying you think it is important). Seeming desperate to be taken seriously is not a strong beginning for a proposal.

Develop ideas clearly and logically. Put the essence of the work at the beginning, and fill in gaps later. Reviewers want to know *what* you are going to do, and once you tell them that, they will be interested to know *why*. Doing things the other way around (why first; then what) invites confusion. Worse, it may invite irrelevant (from your point of view) ideas about other research questions as reviewers fantasize how they might study the topic.

Organize the writing to permit skimming. Include headings so readers can find things quickly if they want to review. When a reviewer gets to the end of a proposal, perhaps while reading the PI's views of what the research means and what might be done next, the reviewer might wish to review exactly what the different experimental conditions are. Clear organization and boldface headings make that easier to find. Reviewers meet to discuss proposals at Study Sections of NIH and Advisory Panels of NSF, and also for some private foundations. During discussion, if someone asks a reviewer for a fact about your proposed research—why you propose to include only female participants in certain conditions, for instance—the reviewer must be able to find the answer quickly. Make it easy.

Get someone else to read and comment on your proposal before you send it in. This will help you find parts that are left out, justifications that need to be added, and details that are obscure. The best reader is someone whose research is in a field different from yours because that person is unlikely to mentally fill in the gaps when you have omitted something. Your schedule for proposal preparation has to include time for getting these opinions and revising the draft once you get them.

Aim for the style called "technical writing" in English departments. This style is far removed from the evocative style of creative writing or atmospheric essays. Technical writing is literal. To someone trained in creative writing, technical writing may be boring; that really is irrelevant. The only important criterion is clarity; a reader should have no doubt at all about what you mean to say. Convey excitement through the quality of ideas rather than through the language used to express them.

One idea per sentence; one thought per paragraph. Each idea gets its own sentence, and no sentence tries to convey two ideas (as this one just did).

Avoid synonyms. Use one word for each idea, and use that word every time you intend that idea. Is that not boring? Maybe. Remember clarity is the criterion, not "fun to read." A proposal that sometimes says "stratification" and sometimes "social inequality" invites confusion. At the very least, a reader will wonder whether the two terms refer to the same thing and then look back and try to figure it out. After doing that a few times, this reader will be annoyed at the writer for making him or her put in that unnecessary effort.

Write simple declarative sentences. Readers should not have to wade through subordinate clauses and qualifications to get your meaning. Use active voice. Passives make it hard to tell what you propose to do. Consider the difference between "It is hoped that respondents will see ..." and "I will instruct respondents that ..." The latter is a proposal; the former is a vague wish. Identify all acronyms and abbreviations the first time they appear. Avoid jargon as much as possible except where a term is so well known and widely used that every competent social scientist would understand it in exactly the same way.

Proofread and ask someone else to proofread. *Do not* just use a spell checker. Spell checkers, helpful as they are, cannot identify misplaced words, misused words, or syntactical errors. They often cannot tell whether "its" or "it's" should be used or if you used "affect" where "effect" would have been correct. Any errors in syntax or grammar make a proposal look sloppy— definitely not the impression you wish to convey to readers.

Number the pages so that readers can refer to them easily. Do not try to evade page restrictions by using tiny fonts or tiny margins. That really irritates reviewers, who often have many other proposals to read! Why would you want to irritate them when they are deciding whether to recommend your proposal for funding? Except in rare circumstances (such as when a funding agency requests them), avoid appendices of any sort. They may look like an attempt to evade page restrictions, and even if they do not, you cannot be sure all reviewers will read them. Make the proposal self-contained. Finally, follow all the funding agency's regulations regarding treatment of human and animal subjects, data archiving, etc.

VIII. SOME STYLISTIC SUGGESTIONS

Technical writing style restricts some of the imprecision and distractions that can plague proposals. Here are some additional thoughts on how a proposal communicates ideas and expresses its writer. Consider this section as suggestions on avoiding problems that can make a writer appear differently from how he or she might wish to appear.

Title. The purpose of a proposal title is to tell what the proposal is about. Simple, right? Yet some writers treat a title as more of a teaser than a description. Advertisers and local news anchors may use teasers, but nobody really enjoys hearing them. Reviewers are going to read the proposal, no matter what its title. There is nothing to be gained, and some potential losses, from writing a title that does not really tell the subject of the proposal. If the title accurately conveys what a proposal is about, readers' thoughts are primed to think about it. Readers are unlikely to imagine some other topic and have to return to your topic later, something that at least some will find irritating.

Closely related to how a title reads is the unfortunate tendency to believe that every title must contain a colon. Colon writers often try to be clever with one half of the title, while the other part tells what the project is really about. For examples, look at titles in most social science journals. On one side of the colon, usually on the left side, is a phrase the writer thought was cute. "Cute" has no place in a research proposal if the writer wishes to be taken for a serious scholar. If one part of a title tells what the proposal is about while the other part attempts to be cute, discard the cute part and use only the descriptive part. You never know when a cute part will antagonize a reader unnecessarily, and, as I said, it gives an impression of the writer that does not help get funding. Eschew titular colonicity, or as a friend puts it, perform regular colonectomies.

Avoid distracting points. When we are emotionally involved with a topic, it is easy to throw in phrases and even whole paragraphs that tell how we feel but are irrelevant to proposed research. Try hard to stick to the point, and when you ask colleagues to read your proposal, ask them if they spot irrelevancies. For instance, I recently read a proposal to study gender bias in American Sign Language (ASL), which was overall a worthwhile topic. However, the author of that proposal could not resist dropping in sentences such as "However, through the years, deaf people have become an informed and empowered community." I'm sure that is true; it just has nothing to do with how ASL or people using ASL handle gender.

Besides wasting space, irrelevancies always carry the potential danger of antagonizing a reader. Sociologists may drop in a phrase telling that they disapprove of some government or corporate policy that has nothing to do with the research, and it is foolish to do so. Ask yourself how your comments will look to a reader whose political or other philosophies differ from your own. The answer should be obvious.

Use bullets and summarize. Often it is useful to enumerate a number of steps in research, or a number of variables or predictions, etc. Bulleted lists provide an economical way to do that. They also tell readers that they do not have to remember every item (e.g., the wording of every prediction) as long as they understand the category to which it belongs. It is quite likely a reader will want to return to your bulleted list to find specifics about a category, and this makes them easy to find.

Summaries are one of the main places to use bullets. At the end of a description of experimental conditions, bullet points might summarize the conditions so that, for instance, a reader carries them into a later section on data interpretation. The end of a proposal is a good place to use bullets listing what the project will accomplish. Frequent summaries, whether or not they use bullets, are very helpful to readers.

Please do not … Following are some writing habits that can be off-putting in a proposal, and probably in other writing as well. In a proposal, as frequently noted earlier, bad habits can cost.

Do not write "see" if a reader cannot easily find the object it refers to. A particularly egregious instance occurs when a writer refers to his unpublished dissertation for important details, such as an experimental manipulation or measurement technique. In fact, it is probably wise to avoid the command to "see" anything. Even if a reference is to an article published in the largest journal in the discipline, readers simply do not have time to search out additional sources when they are reviewing. Every proposal must stand on its own.

On references, be sure everything cited in the proposal actually appears in the citation list. You might be surprised at how often the section called "References Cited" omits one or more of the papers a proposal text cites.

While I have sometimes begun book chapters with a series of questions that I implicitly promise to answer in the chapter, questions usually are not a good way to begin a proposal. Questions are wordy, and if there is one fact true of all proposals, it is that their page limits always constrain PIs. Get right into the answers and let someone else imagine how to ask the questions that you are answering.

To make your proposal stand out and make the writing vivid, try hard to avoid clichés, of both expression and content. There is no point in writing the sentence "More research is needed," unless you can imagine sometimes writing its opposite. I recently sat on a review panel in which someone offered to buy lunch for any reviewer who had a proposal to do interviews or conduct focus groups *without* claiming they would be "in depth." He did not have to buy. Calling a book, a research project, or a publisher "major" adds nothing. (When was the last time someone asked you for "a minor credit card"?) Finally, watch out for unnecessary sentences like the following, from a proposal to study hospice usage, which are vacuous, redundant, and not even true: "Dying cuts across all categories of humanity. It is the one activity we all participate in, regardless of our gender, race, SES, or ethnicity." (How about using language or socializing the young?)

Watch for and remove weasel words—those that negate part of a sentence. Common examples are "may," "might," and "necessarily." To write that something "may" happen is to write that it might not happen. If both outcomes would make the sentence true, it just wastes space in the proposal. Even worse, putting "may" in a hypothesis or prediction means no pattern of data could disconfirm it. To say that something is not "necessarily" true usually does not matter. What is important is whether it is true or not; the "necessity" part is not empirical.

If weasel words do not say anything, wildly inflated claims say too much. The problem with inflated claims is that readers can spot them; rather than being impressed, as the writer probably hopes, they do not take that part of the proposal seriously. A proposal that claims results will apply to all individuals of all ages, genders, psychological makeup, and ethnic backgrounds reveals philosophical and methodological naiveté. A proposal to employ both experimental and survey methods and that claims the results of mixed methods will be so powerful that all subsequent research will have to use them, invites ridicule. (I have seen both claims.) Mixed methods and studying a range of diverse individuals are both good ideas. We all would like to think that our research will help others' understanding of the world and of how to do research. However, to believe—and even worse, to write—that we expect to revolutionize a field with our work either reflects extremely youthful naiveté or access to some very good drugs.

Do not neglect first and second impressions. In other words, try to make your proposal inviting to read. Use a large font. If the agency specifies 10 or 12 point (as most of them do), use 12. That is easier to read. Leave open spaces. Many agencies suggest single-spacing text; if you do that, leave double spaces between paragraphs. Keep normal margins. Insert boldface occasionally—for instance, in headings—to break up the "gray" look. If the agency includes points that proposals must include (as most do), make those points easy to find in your proposal because reviewers may be asked to check them off. For instance, write, "The PI will devote half-time during the summer to this project," and put that sentence at the beginning of a paragraph instead of burying it somewhere in a paragraph that primarily deals with a different topic.

IX. WHAT HAPPENS NEXT?

After you have produced the best proposal you can, using all the help and advice you can persuade colleagues and others to provide, you turn over the file to a research officer who submits it electronically or prints it and sends copies to the agency through the U.S. Postal Service. Knowing what happens next reduces anomie during the ensuing waiting period. It may also affect the way you prepare and write the proposal in the first place.

Federal agencies and large foundations rely on external reviewers—the readers mentioned in many places in this chapter. Smaller foundations and smaller governments usually ask one of their in-house program officers to review proposals. Naturally, the more people who will review your proposal, the longer the process will take. For federal proposals, you should expect 6 months to elapse between submission and decision, though sometimes the time is somewhat shorter.

If a funding agency uses external reviewers, where do they come from? Most of them are academic social scientists, though scientists in research firms also review some proposals. These reviewers are picked by program officers, and there are many places program officers look for them. Here are a few:

- Scientists known to the program officer. This may be because of their own earlier proposals or because of the program officer's knowledge of the discipline and files maintained by the agency.
- Publications in professional journals and papers read at professional societies. The program of a society's most recent annual meetings is an excellent source of potential reviewers who understand a topic area.
- Citations in the proposal. A PI's citation of someone's work as part of the background to the proposed research generally endorses that person as a reviewer. Notice whom you cite in the first few pages of the proposal, for the program officer is not going to send it to everyone you cite, and you may wish to make certain that some particularly relevant candidates do get asked.
- Suggestions from the PI. Yes, most agencies allow PIs to suggest potential reviewers, and some even permit requesting some names that should not be used. It is good to use this option when it is available, but a PI who does it without good faith does himself or herself serious damage (suggesting your mother to review is a mistake). Program officers can spot unfair suggestions, and they have long memories of PIs who tried to trick them.

There are other sources of reviewers, but those four give a fair range of the places they come from. I hope they also help explain some of the earlier suggestions about proposal preparation. A focused literature review, for instance, not only places the proposed work in intellectual context. It also suggests some of the scholars who might be suitable reviewers for it, for they know better than anyone else whether a proposal makes appropriate use of their work.

When writing, it is helpful to try "taking the role of the other," as George Herbert Mead might say. That is, try to imagine a reviewer reading your proposal. We have already considered what makes a worthwhile topic—namely, one related to topics others have investigated. This means there is a good chance reviewers will already have an interest in finding out what the proposed research will show. The PI does not, as noted before, need to spend a great deal of effort convincing them it is worthwhile.

Now consider what this part of the life of a reviewer might be like. Members of NSF panels and NIH Study Sections may have a couple of dozen or more proposals to read, think about, and then write reviews of. These reviewers usually have about a month to do those chores, often less. I hope that helps explain the significance of clear, vivid writing.

We can go further. Imagine a reviewer with a stack of proposals. In the evening, after the dishwasher is loaded and running and the kids are put to bed, he will sit down with the stack and eventually read yours: *one time*. That is right: he or she will probably only read it once. When I have given talks on proposal writing, I always tell people this, and I always dread the strong negative expressions that inevitably follow. "You mean I could spend half a year of my life writing a proposal and some SOB is only going to read it once?" Yes, welcome to the world.

If a proposal stands out from the pack enough that a reviewer considers recommending funding, then he or she might go back to it later. But the first impression comes from the first quick reading, and that has to be a good impression. The ideas should leap off the page and excite the reader. Leave aside your own emotional reactions and "shoulds," knowing this may provide motivation to put a little more time on a proposal. Make it clear, write for good flow, make it easy to review key points, and do everything else so that your proposal gets the most out of the one-time pass from a reviewer.

While your proposal is under review you are free to communicate with the program officer to see where it is in the process, whether you should provide additional information, or for any other reason. Most of the time program officers welcome such contacts. If circumstances make him or her too busy when you call or e-mail, you will get a return message soon. Program officers are dedicated people, and many of them have been on the other side writing proposals. They are sympathetic. Most program officers see their task as representing researchers to the funding agency, not the other way around. They really consider themselves your representatives. They know writing a proposal is hard work and that it has high stakes for the PI. They are much more welcoming, helpful, and friendly than many PIs presume, so any contact you initiate is likely to lead to a happy surprise.

X. SUCCESSFUL AND UNSUCCESSFUL PROPOSALS

If your proposal succeeds, congratulations. That means you will have the opportunity to do the work, with intellectual satisfaction and possibly career enhancement. Tell people. Your immediate supervisor at work (your department chair or dean if you are a university faculty member) will want to know. You might point out that fewer than half the proposals considered receive funding, and you could further point out that fewer than half of social scientists even ever write a proposal. A successful proposal moves you to the top of your class.

What about the less happy situation of a declined proposal? Most scholars I know read quickly through decline notifications (from journals as well as from funding agencies), and then put them aside for a few days. It is not pleasant having one's work rejected, after all. When you feel ready to get back to it, you might begin by looking on the bright side. First of all, look at the benefits you have received, even though funding is not one of them just yet. Writing a proposal clarifies your ideas, and it is as much an intellectual product as a research paper is. Also, by submitting a proposal, you receive a great deal of feedback on your work from very competent people. Reviewers' comments and often also the thoughts of a program officer will come your way. In a world where most of one's immediate colleagues do not have time to think about each other's work or ideas, that feedback is invaluable.

Treat the decline as a "revise and resubmit" if you think you can do better next time. Take the reviews seriously, think about them, and try to respond to them in revising your proposal. Incorporate changes to address what reviewers saw as weaknesses, and play up the parts they saw as strengths. Try again. People who succeed persevere. The people who are most often funded are the same ones who are most often declined. They are the ones who do not give up. A sure route to failure is to submit a proposal and if it is declined to conclude that "I guess I don't know how to do this." Nonsense! Nobody knows for sure how to do it. As with most things in life, a wise person makes attempts, watches the results, tries to do better next time based on them, and does not give up.

Bob Lucas, a research professional who writes advice on research and proposal writing, wrote, "When all is said and done, the best way to get a grant is to write a proposal." My observation is: "They seldom send you money that you haven't asked for." Try! And good luck to you.

ACKNOWLEDGMENTS

Lesley A. Brown, Director of Proposal Development at UNC–Charlotte, provided much useful information and many helpful suggestions for this chapter. The chapter's weaknesses are mine.

FURTHER READING

Chapin, P. G. (2004). *Research Projects and Research Proposals: A Guide for Scientists Seeking Funding.* New York: Cambridge University Press.

Friedland, A. J., & Folt, C.L. (2000). *Writing Successful Science Proposals.* New Haven and London: Yale University Press.

Ogden, T. E., & Goldberg, I.A.(2002). *Research Proposals: A Guide to Success* 3/e. San Diego, CA: Academic Press.

CHAPTER 9

Developing Your Experiment

LISA SLATTERY RASHOTTE
University of North Carolina–Charlotte

ABSTRACT

This chapter addresses several important elements in conducting experiments: design, pretesting and pilot testing, and data interpretation. Good design requires attention to variables, conditions, and manipulations. Pretesting and pilot testing help to avoid costly errors. Issues in data interpretation include power statistics and experimenter effects.

I. INTRODUCTION

For decades, at least, social scientists have discussed *whether* we should do experiments, *why* we do experiments, and even *when* we do experiments. But rarely do we discuss—particularly in print—*how* we do experiments. Reports of experimental research do not regularly describe all the minor and major decisions that went into the work. Yet a social science experiment is made up of myriad details,

225

and most of the details have to be done well or the outcomes of the experiment can be misleading or useless for the purposes intended. With this chapter, I hope to take some of the mystery out of conducting a social scientific experiment by presenting some particulars about how one should design, conduct, and analyze results from an experiment in order to maximize the usefulness of its outcomes.

Elsewhere in this book, you can read about certain elements of how to conduct an experiment (e.g., technological issues, ethical concerns, training those who will be conducting your experiment, recruiting participants, maintenance of records), but translating abstract considerations of good design into an actual, workable experiment can seem a daunting task, especially for new investigators. I hope this chapter can help with the process of doing a real experiment. I hope also to make clear some elements of experimental design that are often overlooked in more philosophical or abstract discussions.

For convenience, the steps in creating and conducting an experiment can be divided into several stages: *designing the experiment, pretesting* the operations and *pilot testing* the experiment, and *analyzing and interpreting the data* it produces. Every stage in the execution presents challenges and requires decisions on the part of the experimenters. As an overview, it is helpful to keep in mind that the essence of experimental design is to create a situation or multiple situations that include *all* the factors described in a theory, and *only* those factors. In most cases, an experiment will contrast multiple conditions, and those will ideally be identical to each other except for differences required by contrasting hypotheses.

II. DESIGNING THE EXPERIMENT

Good experiments begin with an explicit theory, which has the structure to permit predictions of derived consequences. Theoretically derived consequences are sentences telling outcomes a theory predicts, given a specified kind of situation. However, derived consequences contain abstract theoretical terms, not concrete terms that are immediately observable.

For instance, a derived consequence of David Willer's network exchange theory (NET) (Willer *et al.*, 2002) might be, "A person occupying a central node in a network will have more negotiating power than someone occupying an isolated node." While the sentence's meaning may be clear, it does not tell us in terms of operations just what being "central" means, or how to observe "power." On the other hand, "A person with two potential exchange partners will gain more points in negotiation than someone with only one partner" translates the theoretical terms into observable facts in an experimental situation. The first sentence is a derived consequence of a theory; the second is a testable hypothesis. In designing an experimental test of NET, it is necessary to create such testable hypotheses.

No experiment can test all of the derivations of a theory; one must choose some of those derived consequences for hypotheses, preferably a set with some range of theoretical assumptions. For instance, if the theoretical foundation of the experiment is a theory having five general propositions, it is wise to examine which propositions are used in the derivation of each derived consequence. Usually, any two, three, or four propositions will yield many derivations. The experimenter must choose to test a few derivations from among a large set. While the choice is somewhat dependent on personal preference and empirical simplicity, it is usually wise to be sure that the experiment tests as many of the propositions as possible. Thus, testing two derivations that are both implied by propositions two and three is redundant, and if no derivation that uses propositions four and five is tested, the experiment will provide only a partial test of the theory.

The design task is then to translate the conceptual terms in which the theory is couched into a realistic, although not usually real-world, situation where the experimenter controls many of the elements. By "realistic" I mean that the situation must be understandable to the subjects, and it cannot be so bizarre that they feel they have entered the "Twilight Zone." On the other hand, an experiment is not a natural setting. If a natural setting existed that provided a good test of theoretical derivations, there would be no need to design the experiment. Thus, most experiments seem a little strange to subjects, but so long as they understand the important aspects and their behavioral options, realism is neither needed nor desirable.

The more "realistic" an experiment seems, the more likely that some (but probably not all) of the subjects will fall into familiar role behaviors in it. If an experiment reminds some subjects of a high school classroom, for instance, they may activate ways they typically behave in classes: some will be attentive, some defiant, some bored, etc. Those role behaviors and the variability across subjects, of course, are not what an experimenter wants. What she wants is for the situation to present all and only the previously set initial conditions of her design. The situation should seem real to the subjects in that it is understandable and having some consequence, although it may be unlike anything they have encountered before.

The important practical consideration is how to make the experimental situation understandable and relevant to the subjects, with thought to their culture and background, while staying true to the theory. A number of abstract design elements thus come into play, particularly variables and conditions, manipulations, and manipulation checks.

A. Variables and Conditions

I will discuss, in turn, a number of abstract design considerations with which researchers must deal when conducting a social scientific experiment.

Primary among these considerations are manipulations, where the researcher puts into motion the initial conditions and independent variables as specified by her or his theory. It is important that the researcher is clear from the outset just what are the independent and dependent variables in the hypotheses and which ideas are being tested in the experiment, in order to create the experimental conditions. In other words, it is important to be clear just which ideas from the theory are to be tested, and what sorts of situations are appropriate for that purpose.

Hypotheses (like derived consequences) can usually be stated in the form "If X, then Y." More completely, they say, "Given a situation of a specified sort, if X, then Y." The first part of that sentence, a situation of a specified sort, describes the *initial conditions* of the situation. This governs the kind of experimental situation the researcher will create. (Of course, the experimental situation must also instantiate the scope conditions of the theory, as discussed by Foschi in Chapter 5, this volume.) The "X" represents the *independent variables*. They are elements that will be introduced in some experimental conditions and not in others, or introduced at different levels in different conditions. The "Y" is the *dependent variables*. These are what the researcher will measure once she has devised a suitable measurement instrument.

Once the variables are clear, one can determine how many experimental conditions are required. This requires a good understanding of the variables and the relevant number of levels each has. An incomplete number of conditions can be the downfall of an otherwise well-conducted experiment if the important comparisons cannot be made with the data. This includes the problem of not having baseline conditions when needed. However, not every experiment requires a full crossing of all of the levels of the independent variables in order to have a set of conditions that is complete for the purposes of testing the theoretical hypotheses under consideration. Again, it is important to be clear about exactly what one is testing when designing the experiment.

Knowing the predicted relationships among the values on the dependent variables is also important. Yet, again, the theoretical concerns allow one to determine just what the relevant comparisons are across conditions. The design of the experiment should allow for, and lead inexorably to, the experiment making comparisons among conditions that will give a true and meaningful test of the hypotheses and therefore of the theoretical concepts.

B. MANIPULATIONS

Manipulations are the process by which an experimenter creates the independent variables operationally within the experimental setting. Manipulations in social science experiments frequently fall into the category of information that is given to the subjects about themselves, anyone with whom they might be interacting, the situation, the task, or the social world. Other types of manipulations include the behavior of others in the situation (often computer programmed or performed by a confederate) and an imposed social network or structure.

The process of manipulations often includes a cover story or process of "setting the stage" as the researcher creates the scope and initial conditions and independent variables in an experimental condition. This cover story may or may not include deception. Often, it is through this cover story that the experimental manipulations are made. It often comes in the guise of the instructions that subjects receive regarding their participation in the study: what they are to do, and when, how, and with whom they are to do it. Manipulations can come in the form of commission—what is said in the given condition—or omission—what is not said in one particular condition that is said in others.

It is best to make sure that subjects hear all of the relevant pieces of information at least three times during the cover story. As a rule, experimental subjects are not especially attentive, and they often miss crucial pieces of information if they are said only once or even twice, so three times are required. They are also usually not very suspicious. Thus, while they might find the repetition of hearing something three times slightly tedious (if they notice it at all), it rarely causes them to disbelieve the cover story. It is better to err on the side of saying things too much, even with a risk of irritating the subjects, than to err on the side of not saying things enough and failing to properly create the conditions needed to create useful data.

(Please notice that I wrote "three times" three times in the preceding paragraph. If you even noticed the repetition, did it bother you? Probably not—and you are attending to the topic right now. Experimental subjects certainly are not bothered by this kind of repetition, especially when it does not take place in just one paragraph!)

Here is an important rule about creating experimental manipulations: SUBTLETY IS OUT OF PLACE IN EXPERIMENTAL DESIGN. I trust I do not need to repeat that point. Sometimes investigators try to create subtle manipulations in a misguided attempt to preserve "naturalness" or, they think, to avoid drawing subjects' attention to hypotheses under test. The problem with subtlety is that it goes against the goal of creating a situation that instantiates conditions and variables of the hypotheses. Subtle elements of a situation are missed by some people, and can be interpreted differently by different people. This means that, if the manipulations are subtle, some subjects will fail to notice them (and so will not be in the situation the researcher thinks they are in), and some will interpret them differently. Both these effects will introduce variance in the data, for people will be responding to different sorts of situations.

For instance, I once heard about an experimental design in which the researcher was interested in whether white subjects would play a competitive game differently when they thought their opponent was white than when they thought the opponent was black. Since subjects would never see the opponent, who in fact existed only as computer program, the researchers intended to identify the opponent's skin color by giving him what they thought was a "typically black" or a "typically white" name. These researchers wrote that they did not want to identify the opponent's ethnicity explicitly for fear that it would activate either stereotypes or concerns about appearing egalitarian.

The problem is that the researchers do not know whether subjects in the experiment will code the names they chose as revealing ethnicity; probably some would and others would not. Worse, many subjects may not even attend to the name of the opponent; who cares, if we are never going to meet? If it is important that subjects classify their opponents in the game, then good experimental design makes that element unmistakable.

Generally, the "fuller" a picture of an element the researcher can paint, the better the design will be. For instance, in the preceding design, the researchers could identify partner's skin color with an instruction such as, "Your partner today is named ___. He is, like you, a white student here at State University." That, at least, is clear and unambiguous. However, to really activate any behavioral tendencies that subjects may have so that they may be seen in this situation, it would be even better to show a photograph, or a videotape with action and speaking cues, to instantiate this variable. The clearer and the more complete the instantiation of important design elements is, the better.

Knowing who the subjects are, in terms of their background and culture, is also useful in creating the manipulations of an experiment. Making the situation, as presented in the cover story, relevant to the subjects creates a more believable situation, and one that they are more likely to take seriously. Students at elite universities, for example, might be more motivated by studies presented as basic science, while those at less elite schools may be more focused when the study purports to help them learn something about themselves. Subjects who are not used to the laboratory setting may need more friendly and repetitious instructions.

C. MANIPULATION CHECKS

One of the greatest strengths of laboratory experiments is the control the researcher has over the independent variables. However, researchers often fail to fully realize the potential of their experiments because they do not create the situation they intended to create. Careful experimental manipulation is important, but not difficult. One tool all researchers should employ is the manipulation check.

Manipulation checks can take several forms. During pretesting (discussed in detail later), experimenters should be sure to discuss with subjects what they heard, how they interpreted it, and how it affected their behavior. Additionally, a part of all experiments should be a questionnaire or interview (or both) in which the subjects are asked about the experiment. The experimenter should verify that the information given to subjects during the cover story was heard correctly and believed. This check should include any subtle, embedded information about partners, the task, the situation, or any other manipulations.

III. "THE GENDER EXPERIMENT": A PRACTICAL EXAMPLE OF ABSTRACT CONSIDERATIONS

It may be easiest to understand how these abstract considerations look in an experiment by examining an actual experiment that has been conducted. I will discuss is detail an experiment I designed and conducted (Rashotte, 2006). I use an example of my own work not out of ego, but rather because it is only for an experiment of my own that I will know fully the considerations and decisions that went into its design, pretesting, and conduct.

I designed an experiment to examine how to control status beliefs associated with gender. This experiment was intended to test several related hypotheses from the status characteristics branch of the expectations states theoretical research program within sociology. I was not concerned with *whether* or not gender was associated with status for my subjects; rather, I wished to demonstrate that *when* status beliefs were present, they could be controlled through certain mechanisms described in the theory. Thus, I made sure that status beliefs—favoring men or favoring women—were present in every condition of the experiment.

The theory posits a number of mechanisms that might allow general status beliefs to be overcome. I tested two in this study (1) by presentation of status information about a task that contradicts the generally held status beliefs (i.e., saying that women are generally better at the task at hand), and (2) by providing specific evidence that the generally held status beliefs do not hold for these individuals (i.e., saying that, while men are generally better at the task, in this case the male is not very good at it and the woman is exceptionally good at it).

A. VARIABLES AND CONDITIONS

My independent variables were gender of participant, status information regarding gender and performance on the task, and performance feedback.

Subjects always interacted with purported partners of the opposite gender. In some conditions, participants were told that males would do better at the task to be completed; in others, they were told women would do better. In certain conditions, participants were given (fictional) feedback on a pretest.[1] I was interested in comparing the effect on my dependent variable of the general information that women did better versus the specific feedback that the female partner did better (but not how the two things combined). I thus needed six conditions:

- male subjects, told males generally do better, with no feedback
- male subjects, told females generally do better, with no feedback
- male subjects, told males generally do better, but with feedback that the female partner did better
- female subjects, told males generally do better, with no feedback
- female subjects, told females generally do better, with no feedback
- female subjects, told males generally do better, but with feedback that the female subject did better than her male partner

My dependent variable was how often the subjects deferred to their partners in making decisions on the task when the partner disagreed with the subject.

B. MANIPULATIONS

To design this experiment, I began with a design that has been used in dozens of previous experiments and thus had a number of known properties. (For more on properties see Berger, Chapter 14, this volume.) The task at hand, the delivery of experimental instructions, and the cover story have been well established over decades of research. Technological advances have allowed for recent improvements as well (as discussed by Troyer in Chapter 7, this volume).

Subjects were brought into an isolated room containing a computer monitor, a television, and a video camera. They were told that the study would begin when everyone was settled into the various rooms (leading them to believe that there were other participants nearby)[2] and that, when the time came, they would need to look into the camera to introduce themselves.

The subjects then saw a videotape of instructions (said to be live via closed circuit television). The instructions were presented by a "Dr. Gordon" who

---

[1] All subjects completed the short trial version of the task in order to maximize comparability among the conditions. Only in the "feedback present" conditions were participants given (fictional) scores for the trial version.

[2] In fact, I usually did conduct several participants at the same time in order to support the belief that they were interacting with a real other, even though in reality each was interacting with the same fictional partner.

claimed to be an expert in the task at hand and the ability underlying good performance at that task. The tape included a "live" introduction from their "partner" and a chance for the subject to introduce himself or herself, at which time the subject appeared on the television in the room. The introductions included information about the school attended (always our institution, to eliminate other status-related variables)[3] and hobbies, to make the partners seem more real to the subjects.

The instructions delivered all three of the independent variables. The gender of the partner was first introduced when Dr. Gordon said, "I see we have two people working together today, a man and a woman" and reinforced by seeing the partner on screen and by the partner reporting gender stereotypic hobbies. The status information ("previous studies have shown that men/women are generally better at this task") was repeated three different times, including once just before the data collection phase began. The feedback information was provided, in those two conditions, by Dr. Gordon's colleague, "Ms. Mason," who was an expert at scoring performance at this task. Ms. Mason repeated the scores, and their meanings (unusually high or unusually low), three times. Ms. Mason was also videotaped, but said to be just down the hall.

C. Manipulation Checks

I conducted several kinds of manipulation checks. The interviews mentioned here were conducted in order to verify that the subjects heard all of the relevant information regarding the independent variables and that they understood the task they completed. The interviews were also used to determine if the scope conditions of the theory were in place. During a pilot testing phase of the study (more on this later), these interviews were even longer and covered other topics, such as whether "Dr. Gordon" was pleasant yet scholarly, if the session was an appropriate length, and how much of the instructional detail the subject could recall (beyond the basics related to the independent variables).

Subjects also completed a questionnaire just prior to the interview. The questionnaire served as a double check to the interview and also provided some guidance for the experimenter in terms of where there might be some problems with the subject. The questionnaire covered factual information as well as impressions and affective responses. Extreme emotional responses could indicate a subject for whom the study was problematic and not properly prepared.

---

[3]Subjects were also similar to their partners on race and age in order to eliminate other status effects.

IV. PRETESTING AND PILOT TESTING

In addition to the abstract considerations described before, pretesting and pilot testing are important elements of good experimental design and conduct that are frequently overlooked by researchers. Pretesting involves examining certain elements of the experiment in isolation; pilot testing involves conducting complete experimental sessions with an eye to what is and is not working as expected.

Both pretests and pilot tests are different from actual experimental sessions in that the subjects are required to act as informants. They let the researchers know what works and what does not work in the cover story, the task and/or interaction, the data collection, and all other parts of the experiment. This information can be gathered through questionnaires, interviews, or free-response surveys. They allow the researcher to fix unanticipated problems in the design.

The main elements of the design that need to be examined during both pretesting and pilot testing are scope conditions of the theory, the initial conditions of the experiment, and the instantiation of the independent variables. Researchers need to find out from the subjects if the scope conditions are holding in order for the theory's predictions to have any validity. The initial conditions, the cover story, must be understandable and believable for the data to have any value. Thus, for example, some groups of subjects may require that information be repeated more than three times to be comprehended. The independent variables must be clear and reasonable to have an effect on the dependent variables.

Additionally, researchers must ascertain that the measures are working as expected. Subjects must be paying attention, so the task must be somewhat interesting. Usually, experimenters wish to have a task that challenges participants without being so difficult as to cause undue frustration. If subjects become distracted or emotional, often they will not perform the task in the manner expected by the experimenter. The measures—especially key measures of the dependent variables—must be both valid and reliable.

Subjects must buy into the cover story, the situation, and the task. They must also believe that the experiment has some importance—if not to them personally, then to the researchers and to science generally. Cultural factors come into play here. Experimenters must determine, through pretests and pilot tests, how to frame the situation in order to get their subjects to believe it and want to take it seriously. Different populations of subjects will require different frames for the cover story, but often it is not possible to know how subjects perceive the encounter until pretests and pilot tests are conducted.

It is also in the process of pretesting and pilot testing that experimenter effects (discussed more fully later) can be identified. By having multiple experimenters conduct pretests and pilot tests, with thorough double checking, experimenter expectancy and observer effects can be identified early, before they can contaminate the data collected. Technology can be introduced when needed to reduce experimenter–subject interaction. Double-check systems can be implemented and training can be increased if necessary. (For more on training, see Shelly, Chapter 11, this volume.) New measures, less subject to observer effects, can also be introduced if other steps are not effective at minimizing experimenter effects.

When pretests or pilot tests show things are not working as expected, experimenters must determine what to fix and how to fix it. It is much easier to determine *that* things are going wrong than it is to determine just *how* things are going wrong. Pretesting various elements of the design, before or after starting pilot tests, can allow experimenters to isolate where the problems are. Thorough interviews with subjects, with specific questions related to important elements of the experiment like scope and initial conditions, can also help pinpoint the issues.

Once identified through pretests or pilot tests, problems must be fixed by the researcher. Issues with scope and initial conditions can usually be corrected by altering the cover story to resonate more with the subjects. Issues of hearing, comprehending, or believing the independent variables can be corrected by rewording—and/or reiterating—the information. If the measures of the dependent variable are not working as expected, new measures can be added to or used to replace existing measures. If subjects are reacting in an emotional way that is distracting them from the experiment, additional information must be provided to the subjects to help them contextualize what they are experiencing.

A. PRETESTING

Pretesting most frequently is used for experimental instructions, but can also be used for tasks, confederates, and instruments. These elements are isolated from the rest of the experimental setting, and subjects are asked to evaluate them independently. The important considerations are often the means of conveying the situational definitions and all of the interaction elements. Pretesting is essential to ensure that the abstract and theoretical concerns are translated into a practical reality for the subjects.

Again, I will use an example from my own research to illustrate details about pretesting. In a different experiment from the same tradition described earlier, new actors were used to portray Dr. Gordon and Ms. Mason. In order to be sure that this new portrayal created the right situation, as well as the desire on the part of the subjects to be focused and serious, pretests were conducted just on the videotape of instructions (Rashotte, Webster, & Whitmeyer, 2005). In fact, only a short 10-minute segment of the tape was tested (this is long enough to get a sense of the situation but not so long as to require students to be brought into the lab to view it).

Dr. Gordon and Ms. Mason were intended to be authoritative and pleasant and to hold the attention of the subjects. The short segment of tape was shown to students in classes at the same university where the experiment was to be conducted. Students also saw a 10-minute segment from another experiment previously conducted where the "host" experimenter on camera had been shown to be effective at creating the right situation.

Students rated each person they viewed on 40 seven-point semantic differential items. An ideal answer for each item was determined by the researchers (but not shared with the students doing the rating). A final open-ended question was also presented to allow the students to raise any issues that might not have already been covered. The 40 items fell into four general categories: authority and competence, absence of distractions, clarity, and serious manner. Comparisons between the mean ratings for each individual to the ideal rating showed that the new Dr. Gordon did not perform as well as the previous experimental host, but the new Ms. Mason performed well.

The particular failings of the new Dr. Gordon indicated that the problem was with the demeanor of the actor, and not with the instruction script. Thus, a new tape was produced with a different actor (in fact, it was the same actor who was in the tape from the previous experiment). This tape was also pretested, and the ratings were then compared with those from the previous experiment and the original actor. Things were then satisfactory, and the experiment could proceed.

B. Pilot Testing

Once the various elements of the experiment have been pretested, pilot tests can begin. Pilot tests are complete experimental sessions, designated "test groups" or "test sessions," in which the researcher spends additional time questioning the subjects about their participation. Pilot tests help to identify problems that did not arise in pretests.

It is not until pilot tests that competing processes are usually discovered. Competing processes include things like fatigue, hostility, and withdrawal. Experiments that are too long—from the subjects' points of view—or take place at the wrong time of day can lead to fatigue, which can cause the subjects to be less focused and less serious about the session. When part of the cover story involves providing the subjects with information about themselves or others in terms of abilities or other such characteristics, emotional responses can occur. Sometimes this leads to anger and hostility; sometimes it leads to sadness or other negative affect and withdrawal on the part of the subjects.

If these emotional reactions are not a relevant part of the experiment, they can be distracting and lead to corrupted data. Thorough questioning of pilot test subjects can lead researchers to detect these competing processes.

Once pilot tests are completed and all identified problems are addressed, the conducting of the experiment can begin. Once problems are fixed, sessions can be called "experiment" rather than "test groups." If no problems arise and no changes are made to the procedures of the experiment, the "test groups" can be turned into "experiment groups" retroactively.

V. ANALYZING AND INTERPRETING DATA

The final stage of an experiment is the analysis and interpretation of the data it produces. I will not address here statistical methods for experimental data generally, as those have been well covered elsewhere and most social scientists are well versed in them. There are two elements of data interpretation that I do believe are frequently overlooked by researchers, however. First, power analyses are often skipped altogether, and that may lead to researchers missing the evidence that their hypotheses are supported, even in an otherwise excellently designed experiment. Second, experimenter effects must be considered during data interpretation in order to rule out competing explanations for one's findings.

A. POWER ANALYSES

Statistical power analyses are easy calculations that allow one to determine the number of subjects that will be required in an experiment in order to detect meaningful differences reliably in the dependent variable. Calculators to determine statistical power are readily available online.

Statistical power is best thought of as the likelihood of not making a Type II error (failing to reject a null hypothesis that is not true). As one reduces the chance of making a Type II error, one increases the statistical power and thus the test is more sensitive (Keppel, 1991). The likelihood of a Type II error can be lessened by having a sufficient number of subjects in the experiment and reducing the variability within conditions.

Statistical power depends on three factors: the significance level, α (representing the probability of making a Type I error, or rejecting a null hypothesis that is true), the magnitude of the differences across conditions on the dependent variables, and the sample size n (Keppel, 1991). Most often, researchers are only concerned with the sample size, since the effect sizes are predicted by the theoretical constructs and α is set by convention at 0.05. Thus, many of the statistical tools that have been developed, including those online, are geared toward determining needed sample sizes.

Researchers should conduct power analyses in order to ensure that the data will be useful. If one does not have enough subjects in each condition in order to detect the differences between the conditions on the dependent variable, then all will have been for naught. The calculation of "how many is enough" requires knowing the expected differences between conditions, the variability within conditions, and the desired level of significance.

For example, let us think about a simple experiment. This experiment only has two conditions, and the dependent variable is measured as a proportion. The value of the dependent variable for each condition will be compared to a fixed value, 0.60. The predicted mean of the dependent variable is 0.65 for Condition 1 and 0.54 for Condition 2, with a standard deviation in each condition of 0.15. We will use the traditional 0.05 alpha level and a beta level (statistical power) of 0.50, which is also conventional.

By entering all of this information into an online calculator,[4] we find that we need 24 subjects in Condition 1 to detect the difference between 0.60 and 0.65 with a standard deviation of 0.15. We also find that we require a sample size of 17 in Condition 2, where our predicted value is more different from our comparison value. The more different the values we are comparing are, all else being equal, the smaller the sample sizes required are. If we have decided a priori to collect data from 20 subjects in each condition, we would not be able to determine if our predictions about the dependent variable were correct.

Statistical tests are available to allow one to determine sample size for comparisons to a fixed value (as discussed earlier), for comparing two groups to each other, for comparing groups in ANOVAs, for regressions, for surveys, and for many other situations. It is also possible to compute statistical power and confidence intervals for a given sample size. Power calculations are even available for Poisson distributions, Latin square designs, and survival analyses.

B. EXPERIMENTER EFFECTS

Another concern in creating useful data is experimenter effects—meaning how the experimenter behaves and how subjects respond to the experimenter. Experimenter effects have been found in studies on a wide range of topics, from religious attitudes (Hunsberger & Ennis, 1982) to sex (Winer, Makowski, Alpert, & Collins, 1988). Experimenter effects may include differences in the ways different experimenters in a team handle the experiment, or attempts by

---

[4]Here I use one available from HyperStat Online, found at http://davidmlane.com/hyperstat/analysisf.html.

subjects either to please or to annoy the experimenter. Whenever experimenters are creating a situation through manipulations, it is necessary to anticipate, recognize, and deal with experimenter effects.

Most social scientific experiments are at least single blind, in that the subjects are not told the study's hypotheses and they do not know which condition they are in. They will know only the information they have been given—not how it systematically varies from the information subjects in other conditions are receiving or even that it is systematically varied from that other information.

However, very few social science experiments are double blind; usually, the experimenter knows the assigned condition for each subject. Often, this is unavoidable because the design of the experiment requires different actions on the part of the experimenter in various conditions. In this case, though, steps must be taken to avoid experimenter effects that might bias the results of the experiment. Experimenter effects can be one of two kinds: observer/interpreter effects and expectancy effects.

Observer or interpreter effects can occur when the dependent variable requires judgment of some kind on the part of the experimenter, but the experimenter is not directly interacting with the subject at the time the judgment is made.[5] For instance, experimenters might be interested in coding instances of anger or of interpersonal influence from videotapes of discussion groups. No matter how detailed the instructions for coding may be, and no matter how well trained the experimenters are, there is always some ambiguity in recognizing such concepts in an actual discussion. It is possible, in those cases, for the experimenter to see what he or she expects to see based on the hypotheses.

To reduce this effect, it is often helpful to have a double check on the data, done by another researcher blind to the condition. This double check can be done for all subjects or, when that is not feasible, for a randomly selected set of subjects. Then reliability of the two coders can be compared and, when a satisfactory level is attained, the researcher can be more confident that he or she has data that accurately record phenomena of interest. The double check is especially useful when conducted during pretesting, as discussed in more detail previously.

A variant kind of reliability can be calculated in experiments where experimenters are called upon to make decisions in fairly complex situations, such as deciding after an interview whether a particular subject met all the scope and initial conditions of the experimental design. Again, the presence of ambiguity and the requirement for judgment make it important to assess whether interviewers are applying criteria uniformly. One way to assess that is to

---

[5]This is the case when the experimenter is viewing the subjects on camera or via computer while the experiment is in progress, or when the data are recorded in some fashion for later coding.

see whether, for instance, they are classifying about the same proportions of their interviewees in the same ways. Natural variation among subjects will make the proportions vary somewhat, but, on average, classifications ought to be fairly uniform across interviewers. As always, the researcher will have to decide how much variability across interviewers is acceptable and what level suggests further inquiry, perhaps with additional training of interviewers.

Expectancy effects—those showing up in subjects' interpretations and behavior—can be much more problematic, and often more subtle, than observer effects. They may occur because of subjects' desire to please the experimenter, to behave in ways they think she wants them to. On the other hand, subjects might also try to act in ways they think the experimenter does not want. Neither is desirable.

Trying to please the experimenter is actually more common than trying to annoy her. Experimental subjects may be paid volunteers, and they usually want to help "science" or at least the authority figures conducting the experiment. While trying to confirm the experimenter's hypothesis might seem like a good way to please her, it is by no means an easy thing to do. As noted, subjects are not told what the hypotheses are or how experimenters think or hope they will behave. However, another way to please the experimenter—by presenting a positively evaluated self in the situation—is available and quite common. A subject wanting to make a good impression may treat an interaction partner more cooperatively than he or she otherwise would do, or might take a long time filling out a questionnaire and so "overinterpret" the questions.

Trying to annoy the experimenter is the other side. If subjects feel they have been mistreated—perhaps by being coerced into giving a certain number of hours of research as a class requirement—they may want to show their displeasure by taking things out on the experimenter. Here, interestingly, they may well try to disconfirm hypotheses of the study. While they cannot be sure what the hypotheses are, they can be reasonably assured that, if they act in a bizarre fashion, it will be disconfirmatory.

Thus, trying to please the experimenter is likely to generate unusually prosocial behaviors; trying to annoy the experimenter is likely to generate very odd behavior. If an experimenter suspects either of these is a significant factor in the experiment—which, I might hope, will be determined in pretesting—it will be important to take steps to deal with it. Generally, an experiment should not be generating negative emotions and hostile behaviors. If those appear, it is worth taking the time to interview subjects at length to learn the sources of the feelings, and then taking appropriate steps to reduce or eliminate them. Paid volunteer subjects are likely to feel better about their participation than subjects forced to take part because they enrolled in a course. Positive feelings and attempts to "help," on the other hand, may be harder to eliminate. If an experiment generates considerable concern with self-presentation, it may be desirable to add that factor to the theoretical foundation of the design.

For instance, a decision-making model developed by Camilleri and Berger (1967) included three sources of utilities in a situation: getting the right answer, pleasing the experimenter, and pleasing one's partner. It was probably not possible to eliminate the experimenter effect here—the experiment was about getting right answers to problems, after all—but if it can be conceptualized and measured in the situation it becomes part of the theoretical foundation of the work.

Expectancy effects may also occur because experimenters may, unconsciously, treat subjects differently based on experimental condition. Because they are unconscious, often the experimenter is unaware of the changes in his or her behavior, and thus it is difficult to control them. However, there are ways to reduce these effects. One way is to increase the number of experimenters. Some people will alter their behavior more than others; since it is hard to detect when it is happening, one cannot simply just eliminate those who do it more. By adding experimenters, the likelihood of this type of experimenter effect decreases. However, good training and observation of experimenters is also helpful in reducing this effect.

When possible, it is also helpful to reduce experimenter–subject contact. Technologies in place today can aid in that goal; several of them are described by Troyer in Chapter 7. Presenting the cover story via videotape or digital recording still provides the subjects with something compelling (more so than reading on paper or a computer screen does), but eliminates any condition-to-condition variability, such as that from experimenter fatigue, behaviors, and responses to subjects. Videotaped instructions allow for the experimenter to replicate exactly the information subjects receive that is not varied across conditions, while also allowing her or him to edit in segments that do vary by condition. Data collection and measurement of the dependent variable by computer also help to eliminate experimenter effects.

Let us return to the extended example regarding the "gender experiment." In order to reduce experimenter effects as much as possible, I did a number of things. The dependent variable data were collected electronically by computer and were not subject to observer effects. I limited experimenter–subject interaction by presenting the instructions, which were consistent across conditions except for short segments that related to the independent variables, via videotape.

Additionally, I used a total of nine graduate and undergraduate student experimenters, all of whom underwent extensive training, to conduct the experimental sessions. The sessions included extensive interviews after the dependent variable data were collected; these interviews included debriefing about the deceptions presented in the cover story. These interviews were audio taped. When an experimenter first began working on the study, I listened to a number of these audiotapes for each experimenter and provided feedback on reducing experimenter effects. Later, I randomly selected tapes for review to ensure continued control.

VI. SUMMARY

With this chapter, I hope I have provided both the novice and the experienced experimenter with some useful advice. By paying careful attention to the many details involved in the design of an experiment, we can be surer that the data that result will be useful and meaningful. Whether you are planning your first or your twentieth experiment, the devil truly is in the details.

By paying careful attention from theory to hypotheses to variables to manipulations, you can increase the strength of an experimental design. You can strengthen the design further by considering the subjects in your population (Chapter 10), through thoughtful use of technology (Chapter 7), and with thorough training of the experimental staff (Chapter 11). Of course, no design is ever perfect, but the use of manipulation checks, pretests, pilot tests, and power analyses can help improve upon even the most well thought out experiments. Careful attention must not cease when the design is in place or even after the elements have been pretested; good experiments are an exercise in vigilance.

ACKNOWLEDGMENTS

The author wishes to thank the editors of this volume for their improvements to this chapter. This work was supported in part by National Science Foundation Grant SES 0317985.

REFERENCES

Berger, J. (2007). The standardized experimental situation in expectations states research: Notes on history, uses, and special features. In M. Webster & J. Sell (Eds.), *Laboratory experiments in the social sciences* (pp. 355–378). Burlington, MA: Elsevier.

Camilleri, S. F., & Berger, J. (1967). Decision making and social influence: A model and an experimental test. *Sociometry, 30,* 365–378.

Foschi, M. (2007). Hypotheses, operationalizations, and manipulation checks. In M. Webster & J. Sell (Eds.), *Laboratory experiments in the social sciences* (pp. 113–135). Burlington, MA: Elsevier.

Hunsberger, B., & Ennis, J. (1982). Experimenter effects in studies of religious attitudes. *Journal for the Scientific Study of Religion, 21,* 131–137.

Keppel, G. (1991). *Design and analysis: A researcher's handbook.* Englewood Cliffs, NJ: Prentice Hall.

Rashotte, L. S. (2006). Controlling and transferring the status effects of gender. Presented at the meetings of the International Society of Political Psychology, Barcelona.

Rashotte, L. S., Webster, M., Jr., & Whitmeyer, J. (2005). Pretesting experimental instructions. *Sociological Methodology, 35,* 151–175.

Troyer, L. (2007). Technological issues related to experiments. In M. Webster & J. Sell (Eds.), *Laboratory experiments in the social sciences* (pp. 173–193). Burlington, MA: Elsevier.

Willer, D. *et al.* (2002). Network exchange theory. In J. Berger & M. Zelditch, Jr. (Eds.), *New directions in contemporary sociological theory* (pp.109–144). Lanham, MD: Rowman & Littlefield.

Winer, G. A., Makowski, D., Alpert, H., & Collins, J. (1982). An analysis of experimenter effects on responses to a sex questionnaire. *Archives of Sexual Behavior, 17,* 257–263.

CHAPTER 10

Human Participants in Laboratory Experiments in the Social Sciences

WILL KALKHOFF
Kent State University

REEF YOUNGREEN
University of Massachusetts–Boston

LEDA NATH
University of Wisconsin–Whitewater

MICHAEL J. LOVAGLIA
University of Iowa

ABSTRACT

Recent technological advances require an expanded definition of laboratory experiments to include theory-driven fundamental research that occurs in a variety of physical settings and uses a variety of participant interface and data collection techniques. In this chapter we focus on strategic, theoretical, and methodological issues related to the recruitment and management of laboratory experiment participants in psychology, sociology, and political science. We provide interested researchers with information and resources to begin or enhance research with human participants in their own laboratories.

I. INTRODUCTION

Laboratory experiments in the social sciences are concentrated in four disciplines: psychology, sociology, political science, and economics. We focus on the recruitment of participants in the first three. (Kagel and Roth, 1995, provide extensive details of experimental methods in economics.) Experiments using human participants represent a relatively small proportion of the research in each of these disciplines. For example, much of psychology is not social, focusing instead on physiological aspects of human behavior unrelated to social interaction or on the social behavior of nonhuman animals. The subdiscipline of social psychology is primarily experimental in psychology but spills across the disciplinary boundary with sociology. In sociological social psychology, experiments have always had their place, but in uneasy coexistence with less intrusive observational methods. In sociology, experiments are concentrated in a subdiscipline known as "group processes." Laboratory experiments in political science have only recently become influential in the direction of research in the field. In all four disciplines, the influence of experimental research on the field has been disproportionately large relative to its volume.

Szmatka and Lovaglia (1996) explain the peculiar "convincingness" of the results of laboratory experiments compared to results from other research methods as a combination of reproducibility, incremental adjustment of research design to counter criticism, and transparency. A critic can be invited into the laboratory to see for himself or herself with relatively little investment of time and resources. (See also Lovaglia, 2003, on the power of experiments.)

Methodological issues are often couched in theoretical and even moral terms as practitioners of a particular technique seek to carve out a resource-rich niche for their work (Szmatka & Lovaglia, 1996). The treatment of participants in experiments is no exception. Whether participants are volunteers, paid for their efforts, or required to participate to complete a class has been a hotly contested issue that sociological experimenters have used to distinguish their research from that of psychologists. Sociological researchers have argued for greater care in avoiding coercion to participate, while psychological researchers have maintained that participation in experiments is a vital part of undergraduate education and thus should be required.

Within sociology, the debate over the ethics of deceiving participants has polarized experimental and nonexperimental social psychologists. Should "informed consent" require that all the relevant issues involved in the research be explained to participants before they agree to continue with it? Experimenters who use deception point out that few important topics could be effectively researched in the lab without deception. For example, explaining in advance that the study investigates racist tendencies of participants would certainly alter their behavior during the study, thereby masking any relevant effect.

Those eschewing deception maintain that the damage to the reputation of the discipline caused by deceiving participants far outweighs any contribution to knowledge that such techniques produce. Experimenters counter that participants should expect the reality they encounter in an experiment to match the reality outside the lab no more than they expect the same congruence from a theatrical production. (Ironically, opponents of constructing alternate realities in the lab often come from a social constructivist ideological position that also denigrates experimenters as positivist.) In practice, the issue has been resolved by institutional review boards (IRBs) that approve experimental designs. IRBs have consistently upheld the value and ethical soundness of experiments that use deception when approved procedures are followed and participants thoroughly debriefed after the experiment.

While theoretical and moral debates about participants are interesting, the main purpose of this chapter is to describe the ways that participants are recruited for laboratory experiments in enough detail to allow a researcher setting up a laboratory to recruit participants efficiently.

II. HUMAN PARTICIPANTS IN PSYCHOLOGY

Psychology has been concerned with the problem of "recruiting participants" for decades. The uncertainty (how do we get participants?) faced by other disciplines that have more recently come to utilize laboratory experimentation has encouraged a "mimetic process" (DiMaggio & Powell, 1983) whereby the new players have tended to model psychology's tried and true procedures. In this section we begin by describing these procedures in terms of what is most used around the country. We then turn to a discussion of important ethical and methodological issues that present themselves in relation to methods for recruiting participants. Next we discuss some recent Web-based technological innovations for managing participants once they have been recruited. We close with a summary of recommendations for developing procedures for recruiting participants and maintaining subject pools based on lessons from psychology.

A. How Are Participants Solicited?

Two major surveys of subject pool practices in psychology include specific questions concerning how research participants are "solicited" (i.e., recruited). Over 25 years ago, Miller (1981) surveyed the top 100 most cited universities in the United States, Canada, and the United Kingdom based on total citations in the Science Citation Index for 1978. Of the 76 universities that responded (though no responses were received from the United Kingdom), 70 included some form of introductory psychology subject pool. Of these, 8.6% relied on *volunteerism* as the basis of participation, 12.9% gave *extra credit* for experimental participation, 14.3% offered *extra credit with other options* for earning the same credit, 2.9% *required participation*, 58.6% *required participation but offered options* for fulfillment of the requirement, and 2.9% used some *other means* of recruiting participants.

More recently, Landrum and Chastain (1999) reused Miller's (1981) instrument in an exhaustive survey of psychology departments without graduate programs ($N = 1238$).[1] Of the 570 universities responding (47.6%), 478 provided "usable responses" (Landrum & Chastain, p. 31). Their results indicated that 34.7% of the departments in their sample currently relied on *volunteerism* as the basis of participation, 18.9% gave *extra credit* for experimental participation, 24.8% offered *extra credit with other options* for earning the same credit, 3.3% *required participation,* 14.1% *required participation but offered options* for fulfillment of the requirement, and 4.2% employed some *other means* of recruiting participants.

Psychology departments today may also be less likely to require participation (with options) than in previous years. The apparent increasing reliance on volunteerism makes sense given the proliferation of federal and professional association rules that govern the involvement of human subjects in research along with university IRBs that operate to ensure researcher compliance with these rules (see later discussion). By that argument, though, it is surprising that so many departments (even more than in 1981) are in discord with federal and professional association rules by requiring participation without offering options to fulfill the requirement. Admittedly, however, the differences between Miller's 1981 sample of elite institutions and Landrum and Chastain's 1999 sample of undergraduate-only psychology departments complicates interpretation of the observed trends. That being said, we turn now from a description of what departments have been doing and what they are currently doing to a discussion of what perhaps they *ought* to be doing.

B. Ethical Considerations

In developing ways to recruit participants, set up subject pools, and carry out experiments, researchers have confronted two main ethical concerns: (1) how to avoid coercive practices, and (2) how to ensure that research participation has educational value.

Coerciveness. As indicated before, roughly one-third of psychology departments rely on the use of "totally volunteer" subjects for research participation (Landrum & Chastain, 1999). On the other end of the spectrum, just over 3%

---

[1]Landrum and Chastain offer the following explanation as to why they surveyed departments with only an undergraduate program in psychology: "[T]here seems to be some indication of a general increase in research productivity at undergraduate institutions (based on informal conversations with colleagues at conferences) and because our own university has a psychology department without a graduate program" (1999, p. 30).

of psychology departments require experimental participation. Requiring participation (without options) is problematic because federal rules, as well as American Psychological Association and American Sociological Association guidelines, state that individuals should not in any way be coerced to participate in research. Accordingly, some departments (roughly 14%) offer other options to fulfill research requirements (Landrum & Chastain). Most often these involve having students who "opt out" to write a paper, complete additional coursework, or take a quiz (Sieber & Saks, 1989).

However, while the availability of such alternatives may seem to minimize coerciveness (i.e., insofar as any alternative is presented), a student faced with choosing between participating in an experiment and completing a dull and potentially more time-consuming alternative is actually faced with a "Hobson's choice" (i.e., an apparently free choice that is not really a choice). In other words, "Limited and unattractive alternatives [still] constitute an element of coercion" (Sieber & Saks, 1989, p. 1057). To limit coercion, researchers must think more conscientiously and creatively about developing meaningful alternatives to required participation in experiments. (For a list of suggestions, see McCord, 1991.)

As another kind of work-around to the problem of coerciveness, around 44% of psychology departments do not require student participation in experiments, but instead offer some kind of extra credit in exchange for participation (Landrum & Chastain, 1999). Of departments that make use of extra credit, less than half do so without offering other options to earn equivalent credit. Again, students faced with this situation may feel that they have no choice but to participate in experiments (i.e., if they wish to improve their grades). Leak (1981) showed that students in one program found this kind of extra credit "a temptation somewhat hard to refuse" (p. 148). The bottom line is that it is a mistake for researchers to assume that coerciveness is avoided by offering extra credit for participation instead of making participation a requirement. The situation is somewhat improved when departments offer extra credit *with options to fulfill the requirement*. Here again, though, researchers must think carefully about developing meaningful extra credit options instead of going no further than just presenting some kind of option that, while perhaps easy to evaluate and tally (e.g., a paper), is unattractive to students.

Educational value. In the early 1980s, Miller pointed out that "little has been done to provide evidence that experimental participation is a valuable pedagogic device" (1981, p. 213). An exception noted by Miller was Britton, who found that students required to participate in psychology experiments at a large southern state university rated the educational value of the requirement "considerably below the maximum" (1979, p. 197). Miller's observation that scant evidence exists regarding the educational aspects of experimental participation, along with Britton's discouraging finding, constituted something of an embarrassment because psychology departments had long before then justified such participation mainly in terms of its ostensive educational value (Jung, 1969). Consequently, Miller's work inspired a number of studies dealing with the issue of the potential educational value of experimental participation. With a few exceptions (e.g., Coulter, 1986; but see Daniel, 1987, for a critique of Coulter's research), the weight of the evidence that has accrued over the last 25 years suggests that research participation does have educational value, but clearly some methods of soliciting subjects prove better than others at achieving this goal.

As mentioned previously, Britton (1979) found that students in the department he investigated did not rate their experimental participation favorably in terms of its educational value, but these students were *required* to participate in experimental research. Indeed, coercive policies and student perceptions of educational value appear to be negatively related. Nimmer and Handelsman (1992) present results from a quasi-experiment showing that students reported more positive attitudes toward the learning value of research participation when they were exposed to a "semivoluntary" research requirement as opposed to a mandatory research requirement (they could participate in research in exchange for extra credit toward their course grade).

Yet while offering extra credit as an incentive for research participation may produce more favorable student perceptions of educational value vis-à-vis mandatory participation, it does not seem to go a long way on its own toward meeting learning objectives. In a recent study, Padilla-Walker, Zamboanga, Thompson, and Schmersal (2005) found that among 193 students enrolled in an introductory psychology class at midwestern state university, those who opted to participate in research for extra credit were already high academic achievers. Those who ostensibly had the most to gain from the experience were the least likely to participate. Specifically, of those who took advantage of the extra credit research opportunity, 70% were students with excellent or good grades, while only 3% were students with below average grades.

The most promising results seem to come from studies examining psychology programs that require participation but also offer options to fulfilling the requirement. Landrum and Chastain (1995) examined the educational value of research participation among students enrolled in a general psychology course at Boise State University where each student had to "complete some sort of outside-of-class *activity* exposing him or her to psychological research" (p. 5; emphasis added). Most students opted to be research participants, but in the context of being able to choose between experimental participation and a thoughtful alternative, they tended to agree that their participation (1) helped them to learn about psychology, (2) helped them to understand research better, and (3) added variety to the course. Furthermore, these students tended to disagree that their research experience was a waste of time.

However, when faced with less meaningful alternatives to participating in research (e.g., writing a paper), the educational benefits of a research requirement may not be as robust. In a study of 774 general psychology students enrolled at Central Connecticut State University over a 4-year span, Bowman and Waite (2003) found that students who chose to participate in an actual research study were more satisfied with their experiences than those who wrote an optional paper. This finding underscores the point we made earlier that researchers should think carefully about devising meaningful alternatives to required participation in research. Doing so appears both to reduce coerciveness and enhance student learning.

However, the best way to know whether one's own departmental policies effectively address these ethical concerns is, of course, to check and see. Toward doing so, Landrum and Chastain (1995) describe a short, easy to administer "spot-check" form developed to assess the educational value of undergraduate participation in research. Leak (1981) describes a brief 10-item post-study questionnaire that can be used to gather information on students' perceptions of both coercion and educational value in relation to their participation in research activities.

C. METHODOLOGICAL CONSIDERATIONS

The most common criticism of laboratory experiments is that their findings lack external validity. Not surprisingly, then, methodological issues in relation to subject recruitment practices and subject pool policies tend to revolve around how the unique characteristics of college-student research volunteers threaten the generalizability of such data (see, e.g., Rosenthal & Rosnow, 1969). But if the goal of laboratory experiments is *not* to generalize findings statistically, then such criticisms constitute a "red herring" insofar as they draw critical attention away from other truly relevant (and potentially serious) methodological problems.

We share the view that the purpose of laboratory experiments is to *test theories*. In the sense that we are using the term, a theory is a set of logically related propositions specifying expected cause-and-effect relationships among variables under specific scope conditions. Scope conditions, the linchpin of scientific theory testing, are statements about a theory's domain of applicability. For example, status characteristics theory (e.g., Berger, Fisek, Norman, & Zelditch, 1977) predicts that a group's higher status members will be more influential than its lower status members, but only when the group is *task oriented* and *collectively oriented*. We expect the claims of status characteristics theory to hold true whenever and wherever its scope conditions are met. Therefore, as Lucas (2003) puts it, "Criticizing an experimental test of status characteristics theory that employs undergraduate students as having low external validity because results cannot be generalized to a larger population misses the point. The theory makes propositions on human behavior unbounded by the particulars of population parameters" (p. 241).

With such an understanding, one can begin to appreciate the *advantages* of employing homogeneous nonprobability samples (e.g., college-student volunteers) in the service of theory testing. The main advantage is summarized by Calder, Phillips, and Tybout (1981):

> Homogeneous respondents ... are preferred because they decrease the chance of making a false conclusion about whether there is covariation between the variables under study. When respondents are heterogeneous with respect to characteristics that affect their responses, the error variance is increased and the sensitivity of statistical tests in identifying the significant relationships declines. Thus, heterogeneous respondents ... increase the chance of making a Type II error and concluding that a theory was disconfirmed when, in fact, the theoretical relationship existed but was obscured by variability in the data attributable to nontheoretical constructs (p. 200).

From this point of view, we are encouraged to think about subject recruitment practices as a potential source of *measurement error* (see Thye, 2000). Thus, rather than attempt to develop subject recruitment practices that increase heterogeneity and maximize the generalizability of findings (as conventional wisdom might advise), the researcher seeking to conduct a laboratory test of a well-specified theory should strive to do just the opposite. The key methodological issue concerns how subject recruitment practices can be tailored to *increase* (rather than decrease) the homogeneity of subject pools.

To illustrate the problem, let us consider a fable describing the usual exigencies faced by experimental psychologists in relation to recruiting participants:

> The interior decoration of psychology departments is done on an ad hoc basis, but the departments share a remarkable commonality. Toward the beginning of each year bulletin boards contain a few notes about joining clubs, renting spare rooms, and cheap prices for yesterday's computers. After the first few weeks of the term, new notices go up asking for volunteer subjects saying it only takes 10 minutes. You remember these notices; they attract the "good of science" volunteers. A few weeks later, because these good hearted volunteers are a scarce resource, the notices start offering either credits for your courses or monetary payment (Wright, 1998, p. 99).

If we regard subject recruitment practices as a possible source of measurement error, the problem as illustrated by the preceding fable is that such practices are often treated in an arbitrary way. The experimentalist who makes it a practice of doing whatever she or he has to do at a given point in the semester to amass research volunteers will likely increase the heterogeneity of the sample. She or he will probably recruit different kinds of volunteers at different times, depending on the particular recruitment method that is used to build up the subject pool. The result at the study level is increased variability in measurements and consequently, as Calder *et al.* (1981) point out, an increase in the likelihood of Type II statistical error (i.e., false acceptance of the null hypothesis). In sum, while not widely recognized, subject recruitment methods have important implications for theory testing. By employing recruitment procedures uniformly, the researcher ensures that the subject pool will be as homogenous as possible. Consequently, "the researcher can be more confident that any negative results reflect failure of the theoretical explanation" (p. 200).

D. Managing a Subject Pool

Regardless of how research participants are solicited—whether by volunteerism, requirement, offering extra credit, or some other method—they of course still have to be scheduled for experimental sessions once they decide to participate. Traditionally, psychology departments have used sign-up sheets posted on public bulletin boards to schedule student appointments for participation in research. This method is becoming increasingly problematic, however, as federal regulations concerning the privacy rights of research participants become more stringent. Emerging protections for participants mandate that students should not have access to a list of who else has signed up for studies. As such, the days of public sign-up sheets are numbered.

As an alternative to paper-and-pencil sign-up sheets, researchers are increasingly turning to automated systems administered via the Web to manage human subject pools. We identify four currently popular systems, but in general they all work as follows. After being informed in their classes about existing research requirements, volunteer opportunities, or the chance to earn extra credit in exchange for research participation, students are directed to a secure Web site where they can read more about their department's research policies and procedures, create a personal account, fill out any prescreening surveys, view a listing of studies currently available to them, and schedule an appointment online by selecting among vacant time slots posted by researchers. Students receive appointment confirmations and reminders sent to an e-mail account of their choosing, and they can cancel their appointments up to a certain point if allowed to do so by the researcher.

After an experimental session, the researcher uses the Web interface to document a student's participation and, if appropriate, award credit. Most importantly, this information can be retained in a database to prevent students from signing up for a given experiment more than once. Finally, in cases where credit is awarded, instructors can access the Web system through a separate password-protected entry point to retrieve information regarding their students' participation in research (e.g., which studies they completed, how much credit they earned, etc.).

Overall, Web-based experiment management systems are quite an improvement over sign-up sheets. First, by allowing students to create their own private accounts on secure Web servers, the new systems are more compliant with stricter government protections concerning individuals' privacy rights. Second, the newer automated systems save time and money and reduce errors associated with manual processes. Researchers no longer have to spend their time preparing sign-up sheets and worrying about paperwork errors and do not have to set aside funds to pay phone schedulers and cover the costs of photocopying sign-up sheets and reminder slips.

Third, Web-based experiment management systems are convenient for participants, instructors, and researchers alike. Participants can sign up for experiments (or cancel appointments if given the option) from any place where they have Web access. Instructors can log on to the system from the convenience of their offices or homes to generate reports indicating which of their students have been awarded credit for participation. Site administrators and research personnel can access the system through their own password-protected entry points in order to edit studies and time slots and to check to see which sessions in the schedule have been filled, canceled, etc. The latter feature is particularly useful because most systems allow administrators to prevent students from signing up too late for experimental sessions. That is, most systems can be set up to "close" an experimental session if it is not filled by a certain point (e.g., 12 hours in advance). Experimenters can check the system remotely and be certain of their appointments, or lack thereof, well in advance.

A fourth important general benefit of Web-based experiment management systems is that they work to enhance experimental/methodological integrity in a variety of ways. As mentioned before, most systems include automated checks that prevent students from signing up for an experiment more than once. Many systems also allow researchers to post "prescreening surveys" that determine each student's eligibility for experiments and automatically provide her or him with a customized listing of available studies. Furthermore, with the conveniences they afford, and by sending appointment confirmations and reminders via e-mail, the newer Web-based experiment management systems may increase overall participation rates by as much as 50% and reduce the rate of "no-shows" to 5% or even less (Sona Systems, Ltd., 2006). At the time of this writing, there are three major developers of Web-based experiment management software: (1) Experimetrix (http://www.experimetrix.com), (2) Sona Systems, Ltd. (http://www.sona-systems.com), and (3) Human Participation in Research (HPR) (http://hpr.msu.edu).

E. Lessons from Psychology

The task of developing procedures for recruiting participants and managing subject pools, even though it may seem easy on the surface, can be quite daunting when a researcher comes face to face with all of the ethical, methodological, and technical issues at stake. Based on our review of psychology's contributions to addressing these important issues, we close by offering a summary of six recommendations to serve as a primer for those new to the process of building an ethically and functionally sound infrastructure to support laboratory research. Researchers seeking to improve their existing procedures for recruiting participants and managing subject pools may also find the following suggestions useful. Our recommendations are listed in Table 1.

TABLE 1 Summary of Recommendations for Developing Procedures for Recruiting Participants and Managing Subject Pools

Recommendation	Details
1. Incorporate what works.	Sieber and Saks (1989) provide copious details on a set of "exemplary subject pool documents." Researchers will find helpful the guidelines that they propose for subject pool documents concerning (1) instructions for student participants, (2) instructions for subject pool users, and (3) the administration of subject pools (see pp. 1058–1061).
2. Develop ethically responsible policies and procedures for recruiting participants.	In line with federal and professional association rules, researchers must ensure participant privacy and avoid coercive subject recruitment practices. Privacy issues can be addressed by using Web-based experiment management systems as opposed to public sign-up sheets. Coerciveness is minimized when researchers provide meaningful alternatives to research participation such as attending research colloquia or serving as observers in studies (see McCord, 1991, for more suggestions). Offering these kinds of meaningful alternatives should also serve to enhance the educational value of such research experiences.
3. Get your policies and procedures approved.	Once departmental policies and procedures for recruiting participants and subject-pool maintenance have been developed, they must be approved by your institution's IRB prior to being implemented. Also, researchers can get answers to compliance-related questions from their respective IRBs.
4. Implement quality control to evaluate the ethicality of recruitment policies and procedures.	Landrum and Chastain (1995) describe a short "spot-check" form developed to assess the educational value of undergraduate participation in research. Leak (1981) describes a 10-item poststudy questionnaire for gathering information on students' perceptions of both coercion and educational value in relation to their research participation.

(Continues)

TABLE 1 (*Continued*)

Recommendation	Details
5. Implement a Web-based experiment management system.	Automated experiment management systems like Experimetrix, Sona Systems, and HPR (1) save time and money, (2) reduce scheduling and record-keeping errors, (3) are convenient for students, instructors, and researchers to use, and (4) enhance experimental/methodological integrity . Visit each company's Web site to choose the system that works best for your department's research needs.
6. Use uniform recruitment practices as much as possible.	Recruitment practices are a potential source of measurement error. Approached arbitrarily, recruitment practices may increase subject pool heterogeneity resulting in increased variability in measurements and consequently an increase in the likelihood of Type II statistical error. Participant recruitment practices are thus an important, though often overlooked, part of the development of theoretical knowledge.

The practical issues of recruiting participants for sociology experiments are similar to those in psychology. Before beginning a new experiment, researchers must make several decisions: (1) how to secure a pool of participants, (2) what restrictions might be needed on the type of people invited to participate, and (3) how to secure and maintain a pool of participants from which to draw.

III. PARTICIPANTS IN SOCIOLOGY

A. VOLUNTEER OR REQUIRED PARTICIPATION?

In practical terms, the central issue of incentive to participate concerns participant motivation. Monetary payment in exchange for voluntary participation in an experiment is an efficient way to ensure that participants focus on the experimental task. This is particularly true if money is incorporated *as part of the experimental manipulation*. For instance, in an experiment in which participants are told that they are working with another participant and that their task is to see how many problems they can jointly solve, researchers can motivate participants to try their best to solve the problems by telling them that their pay in the study depends on how many problems they and their partner solve. In such an experiment, participants will likely pay close attention to the contributions of others (typically desirable in sociological experiments) and focus on problem solving.

In contrast, participants who receive course credit for their appearance may pay little attention to the task at hand. Requiring experimental participation for course credit virtually guarantees participant availability, but it may decrease the quality of participation. The important point here is that using different incentives in the same experiment or across replications may unwittingly affect participant behavior, thereby complicating the interpretation and comparison of study results. As such, the issue of participant motivation as it relates to recruitment incentives should not be approached in an arbitrary way.

B. SELECTION CRITERIA: WHO WILL PARTICIPATE?

It is helpful to gather basic information about future participants before contacting them for a specific study. Some experiments require participants that possess certain characteristics. When these characteristics are necessary for the successful completion of an experiment, researchers target people with those characteristics for participation. For example, consider the hypothesis that educational attainment (as a status characteristic) affects influence. An experimental test might include only first-year undergraduates who are matched with a (fictitious) more educated partner in a "lower status" condition, and a less educated partner in a "higher status" condition.

C. METHODS OF RECRUITING PARTICIPANTS

There a various ways in which participants might be recruited to participate in sociology experiments. The appropriate method may be determined by available resources or the aims of the experiment. The two detailed next include in-person recruiting and semester-commitment recruiting.

In-person recruiting has historically been the primary method of making contact with potential research participants in sociology. This method provides participants a brief synopsis of the sort of phenomena being examined in the laboratory, an introduction to one or a number of researchers and research assistants conducting experiments in the laboratory, and the opportunity to provide researchers a way to make contact in the future to schedule a date and time to serve as a participant in an experiment.

In-person recruiting requires a coordinated and well-organized team effort to visit high-enrollment classes. The first step is to create a list of high-enrollment courses. With the advent of electronic course enrollment, this process became relatively simple. Most institutions make enrollment information available before classes begin. The next step is to establish contact with the instructors of the high-enrollment courses on the list and request permission to visit their classes. E-mailing each of the instructors has proven efficient. There are several important points to convey in a message requesting a class visit. While specifying that a class visit would be most effective during the first week or two of classes, it is important to allow instructors the freedom to select the date for a recruitment visit. It is also important to mention to instructors that no class time is required for the visit, but that having 5 minutes at the beginning of class would be most effective. Finally, it is important to mention that students will be told that their choice to participate in an experiment is voluntary and will not affect their course grade.

After receiving approval to make a class visit, a team of research assistants is assembled to assist with the visit. Generally, classes with 100 students require two or three research assistants. Larger classes (300–700) might require as many as five assistants. Prior to making the visit, a contact flyer is created. The flyers provide space for a prospective participant to fill in her or his name, gender, year in school, phone numbers, best time to be contacted, and a description of any experiments the student may have participated in previously. Information on past participation can be used to avoid redundant participation in the same or similar studies.

With enough contact flyers for each student enrolled in the class, the team of research assistants arrives at the classroom location approximately 10–15 minutes prior to the start of class. As students enter the classroom, research assistants hand out contact flyers. Two or three minutes prior to the beginning of class (or the minute class begins if the instructor permits), one assistant stands at the front of the class, requests students' attention, briefly introduces the recruitment team, and explains the nature of research participation. The content of the description of research participation includes information concerning (1) the sorts of topics being investigated in the laboratory, (2) the typical duration of studies, (3) whether monetary or other incentives to participate are being offered, and (4) the assurance that students are not obligated to participate in an experiment. Furthermore, it is critical to emphasize that participation or nonparticipation in an experiment in no way affects students' grades in the class (unless the instructor offers course credit for participation).

After completing the presentation, research assistants quickly collect the contact flyers and exit the classroom. To increase the likelihood that instructors will continue to work with experimenters in the future, research assistants must take care not to disrupt class time. Efficient administration of the recruitment process is of paramount importance to maintaining an enduring subject pool. Another concern is that extra credit can be problematic if only some instructors consent to offering an incentive to participate while others do not. In that case, subjects in the same study from different classes have different incentives to participate, which may, as we discussed previously, unwittingly affect participant behavior in ways that might confound study results. Having discrepancies in participation incentives across classrooms is also ethically questionable.

Notwithstanding these concerns, an important step in maintaining a pool of potential participants occurs *after* classroom visits are conducted. All of the contact information obtained via in-class recruitment efforts should be entered into an electronic database. The ability to access contact information electronically allows researchers conducting different experiments who use the same participant pool to avoid contacting potential participants more than once and to keep track of those who have already participated in an experiment. Furthermore, establishing a password to protect potential participants' contact information adds a layer of security to this sensitive information. After the contact information for each potential participant has been entered manually into a database, the completed contact flyers should be destroyed.

The next step in securing participants for sociology experiments using the traditional method requires a dedicated and tenacious scheduler. The individual filling this role references the database of potential participants and contacts them via telephone or e-mail. The scheduler's goal is to fill all available experimental sessions with participants from the list. Prior to calling or e-mailing potential participants, the principal investigators for each experiment receive approval of a "script" used by the scheduler to make appointments. Approval of a script is granted by the university's IRB. The scheduler then uses the approved script to contact potential participants. In our experience, contacting participants during the weekend before the week they would be scheduled to participate is most effective. Phoning in the late afternoon is also effective.

After the scheduler fills all possible experimental sessions for the upcoming week, it is good practice to make reminder calls to participants the day before they are scheduled to participate. We find that this strategy decreases the rate of "no-shows" for experimental sessions. Paid schedulers can also be offered a bonus for each scheduled participant that successfully completes participation in the study. Another useful strategy, albeit a more costly one, involves overbooking experimental sessions and arranging payments for those who come but do not take part in the experiment.

After an experimental session concludes and the participant has been fully debriefed, participants in experiments offering monetary compensation must be paid (or the steps required that result in payment must be taken). In some cases, researchers may have direct access to cash payments that may be distributed before the participant leaves the laboratory. Increasingly, university accounting procedures require a voucher that is submitted for payment. In such cases, mailing information is collected from the participant after the debriefing session. The researcher explains that the mailing information will be forwarded to the office responsible for paying participants and that the participant should expect a check from the university to arrive in the mail. Alternatives to cash payments that participants seem to value include gift certificates to local restaurants and music stores. Distributing these alternative forms of compensation allows the researcher to reward participants immediately, which can have a positive effect on participant satisfaction and future recruitment.

A second method of recruiting a pool of research participants in sociology is semester-commitment recruiting. As its name suggests, researchers using this technique secure a commitment from participants to participate actively in experiments for an entire semester. This method is only valuable in situations where participants in the pool may take part in multiple experiments or in the same experiment multiple times. As such, this technique is not recommended for experiments in which deception is used early in the term. David Willer has used the semester-commitment approach in his network exchange theory research program at the University of South Carolina.

Making initial contacts with potential participants for semester-commitment recruiting can be achieved by way of e-mails or the in-class technique. The primary way in which this method differs from others is that researchers obtain a commitment from participants to participate in as many experiments (or sessions of the same experiment) as they are able during a semester (i.e., as is reasonable given their schedules). If participants are paid for their participation, the pool of semester-commitment participants can be thought of as employees hired to contribute to the research being conducted. If participants receive course credit for their participation, these participants may be thought of as students enrolled in a semester-long course whose grade depends on the frequency of their participation. One way to manage course credit in lieu of payment is to create a research practicum course in which participants may register. Though participants in this framework may not receive instruction in the traditional format (i.e., with lecture and discussion), their experiences and observations of the research process justify college credits in the same way that internship credits often count towards obtaining an undergraduate degree.[2]

We should also note that David Willer has pioneered the implementation of *Web-based experiments* in sociology in a way that involves a new means of recruiting participants. While participants have so far been recruited in traditional ways to participate in most Web-based experiments, there is no reason that a participant in a "Web lab" experiment be located at or have any connection to a specific university.

In addition, popular role-playing games on the Internet represent a new frontier in social science laboratory research (Lovaglia & Willer, 2002).

---

[2]Though, as we discussed in the previous section, the required nature of such participation raises ethical concerns.

For example, the game EverQuest has more than a million subscribers, many of whom participate daily to create a place for themselves in a virtual society where power and status are manipulated in the context of shifting network coalitions. The principal advantage of the Internet laboratory is that complex social situations created by investigators are ongoing as subscribers play out their roles over the course of months or even years. Another benefit is that every behavior is accurately recorded in real time. Control over the characteristics of participants, however, is virtually absent. The drawback of participant heterogeneity is compensated by the huge number of participants and the amount of demographic information that can be collected to establish statistical control.

University undergraduates are not ideal for studying some research questions, as dictated by theory and sometimes practicality. Web-based experiments and online games and simulated communities are a promising alternative for studying questions not amenable to analysis with recruits from university student populations.

IV. PARTICIPANTS IN POLITICAL SCIENCE

Political science studies research questions for which undergraduates often make suboptimal research participants. Participation is reserved for adults and undergraduates who have little or no previous experience as research subjects. A variety of techniques are used in political science to recruit participants for experimental research that could be classified as laboratory based, whether those experiments occur in a university laboratory, in the field, are survey based, or online. Typically in political science, potential participants are told they will receive some form of reward, usually monetary pay, for participation.

A. LABORATORY LOCATIONS

Recruitment varies greatly even in laboratory experiments. As is often the case in psychology, one method is to use a Web site where individuals can register and sign up for participation. An example may be seen at http://pless.princeton.edu/ (the Princeton Laboratory for Experimental Social Science [PLESS]). Princeton University graduate and undergraduate students are directed to the PLESS Web site several times a year. The lab posts flyers around campus, sends out an annual e-mail with information about how to sign up, and occasionally will run advertisements in the campus newspaper directing students to the Web page. The lab has even passed out pencils with the Web site's address stamped on them. All of this generates a fair amount of word-of-mouth references. There are no limits on how many experiments students may participate in, as long as they do not participate in the same experiment more than once. Specifically, following the usual procedures, students first visit the Web site and register an account. They then sign up for an experiment by reviewing a dynamically updated calendar.

Another common means of recruiting participants in political science involves visiting college dormitories to collect personal information from students who want to participate over the course of the year. This information can be entered into a database, and a selection of prospective participants may be generated from that list. The next step is to send an e-mail to each student directing him or her to a Web site to sign up for the experiment (see http://brl.rice.edu for a sample Web site and experiments). Wilson and Eckel (2006) have used this method to recruit participants to explore beauty and expectations in trust games.

Non-Internet methods to recruit participants are common as well. In an interdisciplinary project (sociology and political science), Sell and Wilson (1999) recruited participants from introductory social science and humanities classes. Students were told they would be paid in cash for volunteering in "decision-making" experiments. Those that volunteered were scheduled at their convenience and randomly assigned to experimental conditions.

Another example of non-Internet recruiting is seen in Bottom, Eavey, Miller, and Victor's (2000) work. They recruited 240 participants from undergraduate and graduate classes in the school of business, the school of engineering, and the college of arts and sciences. They advertised an experiment in "collective decision making" in classrooms, through an electronic bulletin board, and through sign-up sheets posted in the student union. All methods mentioned a minimum payment of $3 plus an opportunity to earn more based on group decisions.

B. Laboratory Locations Using Nonstudent Participants

In laboratory experiments when the student population is not desired, researchers may also draw from the general public. For example, Berinsky and Kinder (2006) enlisted participants through posting advertisements and also recruited from local businesses and voluntary organizations. Participants reflected great diversity (compared to the college-student sample), though as discussed earlier, for theory-testing purposes this is not desired. In addition, Ansolabehere, Iyengar, Simon, and Valentino (1994) examined the effects of negative campaign advertising on voter turnout. During an ongoing political campaign (therefore featuring actual candidates and voters), they recruited participants by putting advertisements in local papers, handing out flyers in shopping malls and other public venues, posting announcements in employer newsletters, and telephoning people from voter registration lists. All participants were promised payment of $15 for an hour-long study. While the sample was not random, descriptive statistics suggested that it reflected the population from which it was drawn.

Another study by Iyengar, Peters, and Kinder (1982) recruited participants from a specific city using classified advertisements that offered $20 to those who participated in "research on television." Interested citizens responded by phone and were randomly assigned to experimental conditions and scheduled at their convenience. Descriptive statistics suggest this method also produced a roughly representative sample of the city population. Redlawsk (2002) recruited participants in a large city by contacting different organizations (including the YMCA and a senior citizen center) and requesting that they invite their members to volunteer in experiments in return for a $20 donation to the organization per member who participated.

C. Laboratory Experiments in the Field

In some field experiments, a community becomes the laboratory. For example, Eldersveld's often cited early work (1956) examined the effects of personalized versus impersonalized propaganda techniques on voting behavior. Eldersveld mailed out different forms of propaganda and followed up with postexperiment interviews. Local participants came out of a sampling frame of city clerk records. He selected all people living in four precincts of a central area and who had voted regularly in both state and national elections (but not in local elections, for reasons related to his research question). While not perfectly representative, the sample size of 187 in two conditions allowed much statistical power for the use of statistical control variables.

In a more recent study, Gerber and Green (2000) randomly selected households and exposed them to direct mailings, telephone calls, or personal appeals before a general election to determine which method had the most impact on voter turnout. From a complete list of registered voters, they created a sampling frame of households. This technique generated a sample of 22,077 households. The effectiveness of randomization was checked using voter turnout data from an earlier election, a technique based on statistics and that showed there would be no significant difference between current and past voting behavior. The benefit of this technique is the large sample size that allows statistical control to overcome the loss of experimental control that occurs with a heterogeneous sample.

Bahry and Wilson (2006) recruited participants for their field experiment using a sampling frame of individuals who had participated in an earlier interview pool in Russia. A total of 646 participants were included, with 252 from

Tatarstan and 394 from Sakha. Experiments were conducted in small villages, medium-sized cities, and large urban areas within these Russian republics. Experiments were limited to villages and medium-sized cities where at least 20 individuals had been interviewed earlier. Some medium-sized cities where travel was difficult or impossible were skipped. Payment for approximately 2 hours of participation reflected a week's wage or more for 62% of their participants.

Finally, Wantchekon (2003) conducted an experiment in the Republic of Benin in West Africa. Working with a team of consultants who helped him contact the leadership of selected parties, he communicated directly with them and campaign managers who then agreed to run an "experimental political campaign" in select districts. From his list of 84 districts, Wantchekon chose eight and divided each into three subgroups. Each subgroup was exposed to either one of two experimental conditions or served as a control.

D. SURVEY AND ONLINE LABORATORY EXPERIMENTS

In survey-based experiments, investigators use secondary data while adding a manipulation. Gilens (1996) did this to examine whether white Americans' opposition to welfare is rooted in prejudice against African Americans or in nonprejudice reasons. Using the National Race and Politics Study dataset— a national telephone survey—he applied a manipulation in the survey where half of the respondents were asked a specific attitudinal question about whites, and the other half were asked the same question about African Americans. Nelson and Kinder (1996) also used a secondary data source to recruit participants and create an experiment. In their work, participants were recruited from the sampling frame of respondents who completed the 1989 National Election Study (NES) and who also had provided their telephone numbers. The researchers created a representative sample of the American adult population randomly drawn from this frame. Advantages of survey experiments are large sample size, and the ability to assign the respondents to questions randomly and to generalize results to a larger population if desired.

Online experiments in political science are also performed using a variety of recruiting techniques. Investigators may rely on "drop-ins," where participants come across the experiment while surfing the Internet. Another method uses banner ads that offer some kind of incentive for participation. Iyengar (2002) has used a market research firm, Knowledge Networks, to reach a nationwide representative sample. Through standard telephone methods, Knowledge Networks recruits a continuous sample of individuals between the ages of 16 and 85 who are provided free access to WebTV. In exchange, these individuals agree to participate on rotation in different studies.

Iyengar (2002) examined online self-selection and found that drop-in Internet experiment participants reflect reasonably well the online user population, but participants still differ from the general population since non-Internet users are not reflected in the experiment sample. Iyengar also noted that among subjects in online experiments, Republicans outnumbered Democrats and Independents compared to the broader online population. This is an important issue for political scientists and others who may prefer a "party-representative" sample for their research. Using the Internet as a platform for experiments offers many advantages (e.g., a worldwide geographic domain, the ability to reach diverse populations, and low cost). As with any format, there are drawbacks as well (e.g., sample selection bias, excluding the population with no Internet access, or lack of participant homogeneity for theory testing).

V. CONCLUSION

In describing the methods used by laboratory researchers in several social science disciplines to recruit and work with human participants, we hope to have gone into enough detail to allow interested researchers to begin research with human participants in their own laboratories. As we have noted, recent technological advances require an expanded definition of laboratory experiments to include theory-driven fundamental research carried out in a variety of physical settings and using a variety of participant interface and data collection techniques.

REFERENCES

Ansolabehere, S., Iyengar, S., Simo, A., and Valentino, N. (1994). Does attack advertising demobilize the electorate? *American Political Science Review, 88*(4), 829–838.

Bahry, D. L., & Wilson, R. K. (2006). Confusion or fairness in the field? Rejections in the ultimatum game under the strategy method. *Journal of Economic Behavior and Organization, 60*(1), 37–54.

Berger, J., Fisek, M. H., Norman, R. Z., & Zelditch, M., Jr. (1977). *Status characteristics and social interaction: An expectation states approach.* New York: Elsevier.

Berinsky, A. J., & Kinder, D. R. (2006). Making sense of issues through media frames: Understanding the Kosovo crisis. *Journal of Politics, 68*(3), 640–656.

Bottom, W. P., Eavey, C. L., Miller, G. J., & Victor, J. N. (2000). The institutional effect on majority rule instability: Bicameralism in spatial policy decisions. *American Journal of Political Science, 44*(3), 523–540.

Bowman, L. L., & Waite, B. M. (2003). Volunteering in research: Student satisfaction and educational benefits. *Teaching of Psychology, 30*, 102–106.

Britton, B. K. (1979). Ethical and educational aspects of participating as a subject in psychology experiments. *Teaching of Psychology, 6*, 195–198.

Calder, B. J., Phillips, L. W., & Tybout, A. M. (1981). Designing research for application. *Journal of Consumer Research, 8*, 197–207.

264

Coulter, X. (1986). Academic value of research participation by undergraduates. *American Psychologist, 41,* 317.

Daniel, R. S. (1987). Academic value of research participation by undergraduates: Comment on Coulter. *American Psychologist, 42,* 268.

DiMaggio, P. J., & Powell, W. W. (1983). The iron cage revisited: Institutional isomorphism and collective rationality in organizational fields. *American Sociological Review, 48,* 147–160.

Eldersveld, S. J. (1956). Experimental propaganda techniques and voting behavior. *American Political Science Review, 50*(1), 154–165.

Gerber, A. S., & Green, D. P. (2000). The effects of canvassing, telephone calls, and direct mail on voter turnout: A field experiment. *American Political Science Review, 94*(3), 653–663

Gilens, M. (1996). "Race coding" and white opposition to welfare. *American Political Science Review, 90*(3), 593–604.

Iyengar, S. (2002). Experimental designs for political communication research: From shopping malls to the Internet. Work presented at the *Workshop in Mass Media Economics, Department of Political Science, London School of Economics,* June 25–26.

Iyengar, S., Peters, M. D., & Kinder, D. R. (1982). Experimental demonstrations of the "not-so-minimal" consequences of television news programs. *American Political Science Review, 76*(4), 848–858.

Jung, J. (1969). Current practices and problems in the use of college students for psychological research. *The Canadian Psychologist, 10,* 280–290.

Kagel, J. H., & Roth, A. E. (Eds). (1995). *The handbook of experimental economics.* Princeton, NJ: Princeton University Press.

Landrum, R. E., & Chastain, G. (1995). Experiment spot-checks: A method for assessing the educational value of undergraduate participation in research. *IRB: A Review of Human Subjects Research, 17,* 4–6.

Landrum, R. E., & Chastain, G. (1999). Subject pool policies in undergraduate-only departments: Results from a nationwide survey. In G. Chastain & R.E. Landrum (Eds.), *Protecting human subjects: Departmental subject pools and institutional review boards* (pp. 25–42). Washington, DC: American Psychological Association.

Leak, G. K. (1981). Student perception of coercion and value from participation in psychological research. *Teaching of Psychology, 8,* 147–149.

Lovaglia, M. J. (2003). From summer camps to glass ceilings: The power of experiments. *Contexts, 2*(4), 42–49.

Lovaglia, M. J., & Willer, R. (2002). Theory, simulation, and research: The new synthesis. In J. Szmatka & K. Wysienska (Eds.), *The growth of social knowledge* (pp. 247–263). Westport, CT: Praeger.

Lucas, J. W. (2003). Theory-testing, generalization, and the problem of external validity. *Sociological Theory, 21,* 236–253.

McCord, D. M. (1991). Ethics-sensitive management of the university subject pool. *American Psychologist, 46,* 151.

Miller, A. (1981). A survey of introductory psychology subject pool practices among leading universities. *Teaching of Psychology, 8,* 211–213.

Nelson, T. E., & Kinder, D. R. (1996). Issue frames and group-centrism in American public opinion. *The Journal of Politics, 58*(4), 1055–1078.

Nimmer, J. G., & Handelsman, M. M. (1992). Effects of subject pool policy on student attitudes toward psychology and psychological research. *Teaching of Psychology, 19,* 141–144.

Padilla-Walker, L. M., Zamboanga, B. L., Thompson, R. A., & Schmersal, L. A. (2005). Extra credit as incentive for voluntary research participation. *Teaching of Psychology, 32,* 150–154.

Redlawsk, D. P. (2002). Hot cognition or cool consideration? Testing the effects of motivated reasoning on political decision making. *Journal of Politics, 64*(4), 1021–1044.

Rosenthal, R., & Rosnow, R. L. (1969). The volunteer subject. In R. Rosenthal & R. L. Rosnow (Eds.), *Artifact in behavioral research* (pp. 61–112). New York: Academic Press.

Sell, J., & Wilson, R. K. (1999). The maintenance of cooperation: Expectations of future interaction and the trigger of group punishment. *Social Forces, 77*(4), 1551–1570.

Sieber, J. E., & Saks, M. J. (1989). A census of subject pool characteristics and policies. *American Psychologist, 44,* 1053–1061.

Sona Systems. (2006). Product: Web-based subject pool management. Estonia: Sona Systems Ltd. Retrieved September 19, 2006 (*http://www.sona-systems.com/product.asp*).

Szmatka, J., & Lovaglia, M. J. (1996). The significance of method. *Sociological Perspectives, 39,* 393–415.

Thye, S. R. (2000). Reliability in experimental sociology. *Social Forces, 74,* 1277–1309.

Wantchekon, L. (2003). Clientelism and voting behavior: Evidence from a field experiment in Benin, *World Politics, 55,* 399–422.

Wilson, R., & Eckel, C. C. (2006). Judging a book by its cover: Beauty and expectations in the trust game. *Political Research Quarterly, 59*(2), 189–202.

Wright, D. B. (1998). People, materials, and situations. In J. Nunn (Ed.), *Laboratory psychology: A beginner's guide* (pp. 97–116). Hove, East Sussex: Psychology Press.

CHAPTER 11

Training Interviewers and Experimenters

ROBERT K. SHELLY
Ohio University

ABSTRACT

This chapter describes training of experimenters and interviewers for roles in social science experiments. Topics covered include recruiting assistants, developing scripts and roles for the experimenters, and fostering good habits for the collection and analysis of experimental data. Discussion of how to encourage assistants to develop work habits that contribute to the investigator's experimental program concludes the chapter. Experiments are treated as theatrical productions in the discussion.

I. INTRODUCTION

Preparing research assistants to carry out an experiment that will lead to an informative test of research hypotheses involves several important elements. These require substantial thought on the part of the investigator to design the study, time to develop materials and train experimenters, and the recognition that supervision is important. Working with the students we employ for these roles often requires that we take on the role of a socializing agent similar to that of a parent when we first involve them in our work. Recognizing and executing this complex pattern of identity management sometimes requires great patience and the willingness to exercise restraint in the face of monumental errors.

As I develop the plans for a program to develop skills and abilities in new experimenters and interviewers in the following pages, it is important to remember that this situation involves a relationship in which both parties are learning how the work of the experiment will be carried out. I have worked with a large number of students as research assistants over my career. They have each had different levels of ability and prior preparation when we started our work. Many have had a course in research methods and nothing more; others have had extensive experience working with other investigators. Some have developed new skill sets that have led them to be able to contribute creatively to my research program. Others have been effective research assistants but for various reasons decided that their contributions will be limited to carrying out the tasks I set for them. As an investigator, I value both types of experiences, even though I would like to think that every student I train will become a "junior me" and "take over the business" of my research program when I retire.

II. PRELIMINARIES

The first step in any training program is to determine what the research assistants are to do—in other words, the job description. This is often set by the type of data one wants to collect. Is the study designed to collect open interaction data with video tape of groups discussing a problem for an hour? Will the data then be transcribed or will they be coded directly from the video record in the twenty-first century equivalent of a Bales (1950) real-time coding sessions? Good presence in presentation and the ability to "perform" to a script may be sufficient for collecting data in this situation. If assistants are to code these types of data, then a more thorough training program that addresses how to recognize behavioral categories, their start and stop events, and the passing of turns to another group member may be appropriate. Part of the training program should include details on how to recognize and code nonverbal and paraverbal behavior if this is part of your research interest.

Skills needed to conduct an open interaction study are different from those needed to conduct a study employing computer-mediated (pseudo) interaction in which the data are stored directly to a file and minimal coding is needed to pursue an analysis. In the former, the research assistant may need to start and stop the session and make sure the tape is properly stored. In the latter, he or she may need computer skills that will allow him or her to troubleshoot problems that develop in the course of the study. The possibility that errors may be introduced by equipment failures and the ability to redress these faux pas quickly may save experimental sessions and reduce the need to schedule additional data collection.

We often collect questionnaire and interview data in the course of an experiment. This is done frequently to determine if participants meet the scope conditions of the theory. The research assistant needs to understand the purposes behind this effort and appreciate that it is an integral part of the experiment. Skills required here are highly variable. In some cases, it may be sufficient if the assistant is able to collect, sort, and store paper-and-pencil instrumentation. In others, it may be desirable for the assistant to conduct a complex open-ended interview that is tape recorded and then review the recording and paper-and-pencil records to make a decision about whether or not to include the subject in the pool of "good" data. Decisions about these issues affect the training program one implements.

Recruiting assistants depends on available pools of personnel. I frequently find that courses I teach serve as a fertile ground for interesting students in the chance to participate in data collection. This is particularly true for courses in research methods and in my substantive area of interest in group processes. Both environments have proved to be fertile grounds for cultivating and recruiting future research assistants. Since I teach these courses on the undergraduate and graduate levels, I recruit potential research assistants with varied levels of background from my classes. This can present training problems if I am working with students who have diverse backgrounds when I start an experiment.

The other option is to take potluck with assigned graduates from the department pool. This has proved to be less satisfactory since students are often assigned based on availability and not on interest. Assistants assigned in this way may require more motivation and supervision. From time to time, I have had to alter the nature of the assignment if the student proves to be unsatisfactory as a research assistant. In general, working with the department chair or director of graduate studies so that he or she understands your interests and needs may be very beneficial. Requests for particular students or sets of skills are more likely to be honored if the responsible person understands the requirements of your projects.

It is also important to train assistants in research ethics. An understanding of federal and institutional policy regarding the treatment of human subjects is critical for the entire research staff. There are a variety of means available to accomplish this end. My own institution has an online training program that qualifies an individual for the lifetime certification as having completed the appropriate training. I usually try to provide some preparation for research assistants in this area, but it is highly variable and depends on the roles I expect them to play. If the person is only doing data entry, I emphasize the importance of keeping data confidential. If the person is carrying out experiments that involve deception, I emphasize the importance of careful and thorough preparation for the study, the reasons for the deception, and the importance of careful debriefing. For research assistants whom I expect to be involved heavily in my projects, such as those that involve longer term relationships or the expectation that the assistant will be involved in developing the protocols for the study, I ask that they become familiar with institutional policy, federal policy, and the appropriate professional code of ethics.

Another area of interest that affects the training program is the nature of the raw material one starts with. In many ways this is critical. Beyond the obvious point that we want someone who is reliable, conscientious, and at least a bit compulsive, good preparation is essential. I prefer to work with students who have completed a general research methods course that covers the major techniques of data collection as well as some discussion of scope and initial conditions, and data analysis. This preparation has the effect of creating a common language of discourse so that I do not have to teach them how to think about the role they play. To borrow an analogy from the theater, I do not expect Paul Newman as my lead, but I hope the aspiring experimenter has at least been to acting school.

III. EXPERIMENTS AS THEATER

I wish to continue the analogy with the theater for a time by pointing out that the experiment is a dramatic event. As investigators, we often plan our studies so as to restrict the interaction to a relatively narrow domain of activity. We are interested in the verbal and nonverbal activity of our participants as they try to solve problems in open interaction or make decisions about offers and counteroffers in a negotiation. Our objective in creating experimental situations is very much like the effort in the theater to create a suspension of disbelief. We wish to convince our participants that the situation they confront is real, that the other participants are genuine in their actions, and that consequences of action are real. All of these elements of a good experiment require careful planning and a good cast to carry them out.[1]

The mode of data collection is particularly important in designing the training program. If the project is a vignette study that requires collection of questionnaire data in classrooms or laboratories with minimal interaction, it may be sufficient

---

[1]Provocative discussions of issues relating to casts and production in theater are contained in Willis (1994).

to work with a researcher to read a prepared script. If the study requires that the research assistants serve as experimenters or confederates who play the role of a "real subject," you may need to train them to enact these roles. This training should emphasize that assistants must follow the script and that the experience must be real for the subject. The experiment is designed to create a believable experience for the subject. The research staff must accomplish this goal. Simply reading the script is not sufficient; managing emotional tone and impressions given off as material is presented to subjects are critical elements in creating a successful data collection effort.

The theatrical analogy includes the creation of space for the experiment, the selection and proper placement of props to carry off the cover story, a script that details actions and identifies what to do if unexpected contingencies develop, knowledge of roles that must be filled, and a cast that has been well rehearsed. Space for the experiment can be almost any environment that meets the needs of the study. I have run discussion group studies in a physical anthropology laboratory with a cased skeleton in the wings,[2] in an archeology lab with tools and artifacts lying about, and in classrooms with all of the leftovers at the end of the day scattered around the room. Other investigators I have shared these stories with tell of similarly Spartan or unusual accommodations for their work. The key is to recognize how to use the space, arrange props, and create a believable social experience for the participants.

It is also advisable to create a sense of "onstage" and "offstage" for the research staff by designating areas where assistants can be away from the experiment, and yet remain available for their parts in the study. It is important that distinctions about where "backstage" and "frontstage" behavior are appropriate be well understood and enforced. Inappropriate activity in either onstage or offstage areas can have negative consequences for the experimental session where the event occurs or jeopardize the entire study if the breach is egregious.

The risks of inappropriate "frontstage" behavior can often be managed with rehearsals and practice sessions, but "backstage" problems may not appear until sessions are actually carried out with live subjects. For instance, inadequate soundproofing in a laboratory space may require that the staff remain silent while the subjects engage in a discussion or work on a computer for the experimental sessions. It is also advisable to train your staff in appropriate language to be employed when discussing the study with subjects and one another in public. For instance, references to subjects should be as "participants." Emphasizing the importance of a professional demeanor is also important as experimenters learn how to conduct sessions.

Props include a variety of objects. They may include writing instruments, tape recorders, video cameras, supplies such as audiotape or videotape,

---

[2]The skeleton was behind a moveable partition, out of the view of the participants in the study.

research questionnaires for recording responses in interviews, and the mundane such as tables and chairs. Management of these items and training in their use should be a part of any preparation to carry out an experiment. The training program should include discussions of what to do if equipment fails, and the possibility that data may be lost due to failure to record properly information about sessions, conditions, and days and dates a session was completed. Finally, some words about costume: most experiments involve the presentation of information from a legitimate researcher. Because legitimacy is often tied to position and appearance, it is likely that the experimenters will have to dress as "professionals." I have found that some simple direction about presenting a neat or professional appearance is often sufficient to encourage experimenters to leave the ragged T-shirt and cut-off jeans at home and appear in clothes that would get them seated in most restaurants in the United States.

Developing a script for the experiment, and its rehearsal, is a significant part of any preparation for a study. The exact content of the script will depend on the experiment and data being collected. Rehearsals can take a variety of different forms. One could videotape practice sessions of the study, review audiotapes of the exit interview and debriefing, or observe rehearsals in real time. Whichever of these techniques one chooses to employ, feedback early and often is important. Theatrical directors employ notes to their actors to encourage character development and adherence to the script. You may wish to develop a similar strategy about rehearsals. Formal notes have the advantage of requiring one to identify specific behaviors that must be corrected, and may form the basis of action if it is necessary to terminate a relationship with an experimenter.

The degree of prior experience and preparation of your experimenters will affect the need to rehearse and refine the performance for participants. This may actually be a two-edged sword, however. Experienced assistants may feel they have little to learn, but may have bad habits to unlearn from your point of view. Training and rehearsals are the places to solve these problems; losing data because of bad behavior by an experimenter is not acceptable.

IV. PREPARING THE ASSISTANT FOR THE ROLE

Determining how to train an assistant depends upon the nature of the data collection task. The development of skills needed for supervising an entire data collection enterprise is more time consuming than training for only one or two tasks. In addition, how much to reveal to a new experimenter about the goals of an experiment is always a question with technical and substantive import. I discuss these issues in turn, beginning with theoretically important issues on which beginning researchers need to be prepared.

The experimenter should understand the role of scope conditions in the research so that he or she can effectively make decisions about the quality of data from a particular experimental session. For instance, expectation states experiments require that participants are collectively focused on arriving at the best solution to the group task. If they are not, the data they produce are suspect and generally not included in the set for analysis. Learning how to determine whether an individual is collectively focused is a key part of the preparation of the experimenter. Are decisions to be made on the basis of volunteered information or at the end of a detailed interview conducted at the conclusion of the experimental session? What decision criteria are to be employed to include or exclude a participant in the final data set?

All of these points require that experimental assistants have a grasp of the theoretical points under investigation. This frequently requires that they become familiar with research literature that many graduate and almost all undergraduate students will find unfamiliar and daunting. Careful selection of materials and a planned reading program are frequently valuable in preparing a potential assistant in this aspect of conducting studies. The extent of this preparation will depend on how you expect the person to work with you on your research. If the students are to become a part of an ongoing research team, extensive preparation is in order. If they are to serve only as experimenters for a single study, then brief, targeted materials may be sufficient.

The creation of a reading program for training assistants will depend on the role you wish to have them fulfill. Students who are to collect and enter questionnaire data in a vignette study may be exposed to very little research literature if that is their only role. If the student is to carry out coding of open interaction data, it may be desirable to expose him or her to some material on content analysis and coding of interaction. The particular work chosen will depend on your objectives. If students are expected to carry out experimental sessions, code and enter data, and make decisions about including and excluding subjects from the final data analysis, more extensive reading lists may be in order. Decisions about how far to carry this education will depend on the possibility that experimenter effects are a concern.

Any experimental study has a set of scope conditions and a set of initial conditions that define the social situation within which the subject acts. Scope conditions define the situation of action such as collectively oriented group activity aimed at solving a task. Initial conditions specify the relationships between social actors in the situation such as initial advantages and disadvantages group members have with respect to resources or status in the group. Student experimenters should have a grasp of initial conditions so that they can determine if subjects believed they were sufficiently similar to or different from one another to satisfy the conditions of the experimental design.

In expectation states experiments, participants are informed that they occupy status that is the same as or different from that of their partner in the study. Learning how to assess whether this information was internalized by the participant is a key element in determining whether or not to include data from a particular participant for analysis. Did the participant understand the information that was presented? Did he or she accept and believe it? Did he or she act on it? These are all questions the experimenter must answer to reach a decision about including data for a particular individual in the final data set. Explicit criteria for assessing a postsession interview help, but prior training will produce higher quality decisions.

My reading list for novice experimenter trainees includes Bernard Cohen's (1989) *Developing Sociological Knowledge*. The discussion of initial and scope conditions, the use of indicators as measures for concepts, and the linkage between these key elements of testing hypotheses is one of the best available. If students have difficulty understanding or appreciating the material, I often consider them for more limited roles in the research enterprise.

Providing a novice experimenter with explicit and detailed knowledge about hypotheses under investigation may be more problematic. Ideally, a well-prepared research assistant will have a detailed understanding of the measures, hypotheses, and expected values of dependent variables in the study he or she is conducting. Robert Rosenthal (1976; Rosnow & Rosenthal, 1997) and his colleagues raise questions about introducing artifactual biases into the data collection process when the experimenter has this detailed knowledge. Such biases may contaminate data and lead to false acceptance or rejection of research hypotheses.

An impressive array of studies over the past 40 years has detailed how human experimenters produce subtle, adverse effects on outcomes of their studies with subtle, often hard to detect communication to research participants. This work is variously accessible as many of the books are out of print, but you must be aware of the problems that inadvertent, nonverbal, and paraverbal communication may create as participants are instructed, supervised, and interviewed by an experimenter. This problem is by no means limited to beginners working on their first experiment. Investigators will be aware of these problems and prepared to train personnel so that they may be minimized or avoided.[3]

The possibility that experimenters may have unintended effects on results is a significant issue for social science experiments. Recognizing this should not paralyze us, however. The use of multiple experimenters and the replication of conditions across experiments will reduce the risks of making incorrect decisions about the quality of data from experiments. One solution often advocated

---

[3]*People Studying People: Artifacts and Ethics in Behavioral Research* (Rosnow & Rosenthal, 1997) is a remarkably readable treatment of the issues in this area and should be on the reading list for new experimenters as they begin their training.

to the possibility that experimenters may inadvertently affect outcomes is the use of double-blind experiments. These may be appropriate in some circumstances, but not others. When the experiment requires an interview or questionnaire specific to particular initial conditions, keeping the information from the experimenter is not possible. Under such circumstances, it may be appropriate to ensure variation in assignment of experimental sessions to experimenters so that an assessment of possible experimenter effects can be carried out.

As a research assistant develops skills and takes on responsibilities during the conduct of investigations, you must confront the issue of the extent to which the person is contributing intellectually to the project. As I mentioned earlier, I try to prepare research assistants as if they will take up my research program. This often involves training not only for the limited roles of conducting experiments, but also for the intellectual tasks of contributing to the development of the research program. Ideally, such preparation takes on the qualities of an apprenticeship program, with roles and responsibilities increasing as the assistant gains experience. The ultimate goal is for the student to be able to develop his or her own research project that answers important theoretical questions based in your program. Often this requires at least a year of work with you on a project before the student is ready to proceed on his or her own. Guiding the student through this process requires well thought out plans on your part. Students do not develop the skills to accomplish this end until they have "practiced" on several projects. The goal should be to identify and work with the various elements that go into translating a theoretical idea into a study, identifying how to develop the study, and implementing the logistics of the study.

V. TRAINING FOR SPECIFIC TASKS

A. CONDUCTING EXPERIMENTAL SESSIONS

When working with individuals who have no experience in collecting data or who have not previously conducted experiments, I design a ladder of opportunities to prepare them to supervise the experimental sessions. The first set of experiences involves simple tasks such as telephone scheduling of participants as the first step. This generally allows them to learn to follow a script and interact with participants; by asking for feedback about their experience, it offers an opportunity for me to assess their skills. Depending on the nature of the experimental protocol for the session, I then move the assistant into a more responsible role by having him or her participate in sessions I conduct. Usually, this involves asking the assistant to conduct postsession interviews, serve as a secondary experimenter if the role exists for the study, or act as a greeter if multiple locations are in use for participant arrival. This process usually does not require many sessions to provide some rudimentary training so that the new assistant can move on to more complex tasks.

In addition, I often ask the trainee to take part in the study as if he or she were a naïve participant. Not only does this increase the understanding of the trainee, but it also leads to important information about the effectiveness of the manipulations and the clarity of instructions. Questions and comments of the new experimenters are often more informative for the experimental protocol than comments gleaned from postsession interviews of naïve participants. This information can highlight problems of implementation since the trainee has developed some knowledge about the intent of the study and is often aware of the expected effects of the experimental manipulations. The trainee is also less likely to be intimidated by the experimental setting than the typical naïve participant would be. The earlier in the sequence of collecting data this can be accomplished, the better. Information gleaned from such an experience at the pretest stage of a study is much more valuable than information secured as data collection is concluding.

Once the assistant has completed all of these steps, I feel comfortable working him or her into the overall rotation to run sessions. I sometimes continue to run sessions myself and compare notes for a few days to ensure the data collection is going as intended. It is also valuable to plan some retraining from time to time to keep research assistants sharp and to ensure they are following the protocol as expected if the data collection run is to extend for a long period of time, such as across several academic terms. Some occasional monitoring of sessions may be in order as well. This should be done with the knowledge of the assistant to ensure he or she understands that it is done to ensure a quality execution of the protocol.

B. Coding Interaction Data

Learning to code open interaction data is in many ways much easier than it was when I began working in this area as a graduate student. We were still using Bales Interaction Recorders (1950) and began training by tapping pencils on the table to signify acts as we observed groups through a one-way mirror. This had the advantage of creating some reliability in our identification of events, but we still had to classify the activity. Interaction coders today have the advantage of being able to replay a video record to determine if the behavior they are observing is the initiation of a new act, if it is in response to an action of a specific other member of the group, and how it should be classified according to the coding scheme in use in your investigation. One key issue to confront is whether you wish to code the data from a video record or from a transcript prepared from the record. Both have advantages and disadvantages, and different training demands.

Coding directly from the video record has the advantage of employing codes for verbal, nonverbal, and paraverbal behavior. Techniques for implementation of this scheme are relatively easy to train with software that is available for today's computers. Two such routes are available. One approach is to employ software that is specially developed for coding continuous time data such as that produced in open interaction.[4] The other option is to employ video playing software on the computer in conjunction with a word processing program. Time stamping of the video record and repeated coding that is time sensitive are not as easily accomplished with this approach. Whichever route is chosen requires that you work closely with the person you are training to be sure he or she understands the importance of maintaining a secure file of the original data. I have had an assistant inadvertently erase files when trying to code data. As a precaution, I copy the video record to a CD/DVD and have the research assistant work with the copied disk, maintaining a copy of the original file on my office system as well as a videotape stored independently of any computer system.[5] As technology advances, these approaches may appear as archaic as tapping pencils to signify acts.

The alternative to coding videotaped discussions involves preparing a transcript from the video record and coding interaction from this. The preparation of a transcript can be a time-consuming and enervating process. It also presents an opportunity to engage in some "precoding" by the transcriber. For instance, the transcript may need to identify pauses in interaction of specific length, the interruption of one speaker by another, and the use of back channel comments as the transcript is developed. All of these are time consuming, often require repeated replays of specific short segments of interaction, and require that the transcriber have a good grasp of your intent. Even if you are not concerned with this level of detail in the transcript, you must still confront how to manage inaudible utterances, the display of overt body language that affects interaction, and environmental noise that may affect the quality of the recording. Time invested in training the transcriber may be as much as or more than that invested in the actual coders employed for the analysis. The advantage is that the record can be shared easily with other researchers.

---

[4]Two options are currently available for use in this approach. One involves the use of software specifically developed for this purpose by the Noldus Corporation. This solution is relatively expensive (ca. $9,500 plus a high-end personal computer). The other is MacEvent Coder, a program developed by Lisa Troyer of the University of Iowa. Both have the advantage that the record can be time stamped for each coded event.

[5]The use of DVD or CD storage also has the advantage that audio distortion is less likely, time stamping of the coding record is easier, and the data are less likely to be compromised by stray magnets or heat.

Once the transcript is produced, training a coder is a matter of identifying coding rules such as what the unit of observation is to be, how to classify the activity, and how to recognize regular turn passing as opposed to interruptions. This can often be accomplished by providing the trainee with a set of the coding rules and some instances of coded transcripts you have developed for this purpose. The ideal state of affairs is that you be able to provide this practice coding material from data that are not a part of the actual study material. If this is not possible, it is useful to develop checks to ensure that the work of the assistant is not simply a copy of what you provided. In this case, I try to stress the importance of reliability, but also make the point that my examples are intended to provide guidance and not to be used as exact models that are to be reproduced. Assessments of reliability with transcript data are often quite high, but there should be some discrepancy between your examples and the data from the assistant.

Assessment of reliability is particularly important in establishing the quality of the training for research assistants and ensuring the integrity of the analysis of the data. Training for coding data should not be done on data you wish to analyze to test hypotheses as it may lead to artificial outcomes. Transcripts from prior projects or transcripts of interaction from groups that do not fit the scope or initial conditions of the study are appropriate alternatives. The coding of the data that is employed to test the hypotheses should be assessed to ensure that it is appropriately recorded. Various alternatives exist for this and the selection of appropriate measures depends on the approach taken. The simplest method is to correlate the coding of two or more coders of the same data. More complex measurement models are available and depend on the nature of the coding employed by the investigator.

C. Coding Other Types of Data

Many computer-assisted experiments produce a data file that is ready for analysis or that requires minimal effort to reach this point. In this case, it is often desirable to be sure the new assistant has the necessary skills to translate the data records from one format to another. This may require that you ensure the person has the requisite training to move data from a text file to a spreadsheet and from a spreadsheet to a statistical analysis program. This level of skill is common among undergraduates on most campuses today. If this is not the case, these skills can be developed fairly quickly if the person has some level of computer skill. Here again, the raw material makes a difference. The more automation you can design, the better. The possibilities for mistakes in translation are reduced or removed if it is possible to automate the "recoding" process. If they do occur, they are likely to be much easier to detect as they are more likely to be systematic as opposed to random.

D. ANALYSIS

Once the data files are completed, analysis with statistical software is possible. My experience has been that this is where I am on my own. Most students I have worked with have only rudimentary understanding of statistics and any analysis more complex than a simple one-way analysis of variance is beyond their understanding. There is variation in this expertise and I have had some undergraduate students work with me who have the skills and experience to carry out complex analyses. That said, it is advisable to develop skills for students by asking them to take up more complex problems.

Careful supervision is in order, however. This is particularly the case if your work requires time series analysis, the testing of stochastic models based outside the scope of the standard linear statistical model, or other more advanced techniques such as structural equation models. Preparing the typical undergraduate to work in this area may be out of the question, but advanced undergraduates and graduate students should be able to work at these more sophisticated levels. At this point, course work in a good statistics program is desirable before the assistant starts working on such analyses. Variations in the levels of expertise students bring to the project lead me to conclude that enhancing statistical skills of an assistant within the context of a research project is best answered on an individual basis.

VI. COMPENSATION

Compensation for work in the project may take one of several forms. One of the easiest to grasp is the awarding of academic credit for work on a project. I usually equate hours worked and credit so that the student will be working about 3 hours a week for each unit of credit awarded. Financial compensation is governed by local labor market conditions and institutional policy. Navigating this issue is sometimes a daunting process. For instance, I usually pay about twice the minimum wage for coding and transcription work but would be willing to go higher if the student was exceptionally skilled. Students I have employed to conduct experiments are usually compensated at the same rate. When I have research assistantship available from institutional sources, the rate of pay is fixed by other compensation policies and may include substantially more than this amount. These figures are in a labor market with a scarcity of jobs for students. Intangible compensation is also available and may be more desirable for some students.

A careful plan that specifies how to reward professionally for various levels of participation in a study is always desirable before the study begins. Acknowledgment of contributions to a study in a footnote is common in the field. This level of recognition is appropriate for research assistants who have faithfully carried out experiments and contributed to the project by highlight-ing protocol problems during the collection of data. I usually reserve recogni-tion as a coauthor for individuals who have gone above and beyond the call of duty. Contributions of several types fit this category. For instance, data coding that is particularly onerous and requires the research assistant to participate in the development of a coding scheme and its implementation is the sort of work that may deserve recognition as a coauthor. In other instances, student assistants have suggested analyses that have led to full-blown papers in their own right. In this case, there is no question in my mind that the student deserves credit for this with a role as the coauthor. Finally, if a student were to develop his or her own research project that resulted in a solid study with a publishable paper, I would encourage him or her to take on the role of lead author in the project and I would assume the subservient role as secondary author. A key element in the rubric of how to reward student contributions professionally is that you are open with students about how recognition is tied to their contributions.

Some prior understanding of the appropriate rules in your field is in order. One of my physicist friends related to me that he is one of 206 authors on an article in his field. The laboratory in which this work took place has a specific set of rules about the inclusion of individuals who oversee its technology and hence 30 to 40 individuals appear on all of the publications when data origi-nate in this establishment. Others are listed as they contribute to the work. While sociologists are not likely to encounter this level of complexity in decid-ing who is listed on a publication, explicit rules are needed. Clarity on this issue will reduce conflict after the fact.

VII. EXPERIMENTAL STAFF AS A GROUP

Thus far, I have addressed training and preparation for research assistants as if there is only one person in the project at any one time. This is often not the case; some projects require a large staff. For instance, experiments may collect data from groups of participants who arrive and participate at the same time, such as a discussion group study or a computer-mediated interaction study in which the appearance of real-time interaction is critical in convincing participants that their interaction is taking place as they participate. In these circumstances a staff large enough to greet, seat, and conduct postsession interviews with each participant is necessary. The research staff thus could be as large as the groups one is studying. Depending on the availability of staff, you may employ even more individuals due to scheduling problems. Employing a large staff raises significant questions about how to train a number of individuals to work in the study, the roles they will fulfill, and an assurance scheme to be certain that all of them are executing the study in the way intended when you designed it.

There are several significant issues. For instance, managing the schedules of experimenters can be problematic. Frequently, times of the day, days of the week, and the academic calendar play a role in when participants are likely to be available if you are working on a college or university campus. These concerns typically determine your needs for personnel. I collect data with volunteer participants who receive a nominal payment for their time commitment. My colleagues in psychology employ a subject pool model in which students gain points for their course grades by participating in studies. This difference creates a situation in which they often have difficulty getting participants to sign up and appear early in the academic term; students are engaged in other activities at these times. On the other hand, the cash inducement I pay often ensures a pool early in the term.

At the end of the term, the situation is reversed. The psychology subject pool is full of participants scrambling to improve their course outcomes late in the term, but the interest in earning a few dollars disappears in the face of term papers and reading assignments as final examinations approach. These "facts" of data collection lead to time management concerns for research assistants and the creation of good work environments. Your staff may be as pressed for time at the end of the calendar as your participant pool.

Another issue present in employing a large staff is the dynamic as a social group. Some of your personnel may be "working" for the first time in a cooperative task setting. As a consequence, you may need to engage in socialization about how to work well with others. Lessons in punctuality, the courtesy of informing others if one is ill, how to correct mistakes one makes as well as those made by others, and the necessity to accomplish tasks that are uncomfortable may be necessary for some trainees.

You will also want to determine the extent to which you wish to employ a sink-or-swim strategy as opposed to a strategy that involves careful lessons about how groups of peers work together to accomplish a task. Whichever model you adopt, my experience is that "getting into the trenches" with the staff is a good idea, at least for a few cycles of data collection. One gets a chance to model good behavior in carrying out the study and observe the strengths and weaknesses of members of the staff. This may also give you a chance to communicate the importance of reporting problems to you, as the chief experimenter.[6] I think that is particularly good strategy to be present during these early phases of the data collection process. This is especially true if the

---

[6]Students often think of such reporting as "ratting" on their peers. You must communicate the importance of good information in these situations, even if there are unpleasant consequences for members of the staff. The collection of experimental data is costly; errors make it more costly.

experiment involves implementing a new protocol. Your presence allows you to catch problems as they occur, rather than after several sessions have been completed. It also affords the opportunity to determine if there are problems in the way the project is being carried out by the staff, even if they are experienced. The time spent now may be cheap insurance for the entire data collection enterprise.

Perhaps as important as the training that you put in place is the issue of who is in charge. Several different models may be employed, but it is imperative that you, as the senior investigator, communicate how decisions are made, who is to report to whom, the role of protocol in constraining behavior in the laboratory, and your position as the final arbiter of the project. I have worked with students who, once well trained, were provided with the appropriate materials and left in charge of a significant portion of the data collection enterprise, including supervision of a small staff. I have worked with others who required constant supervision on my part, even though they were well trained to begin with. Individual differences in responsible behavior and the ability to work with others frequently determine which model you employ.

Staff members who work together develop their group identity. This can be both helpful and harmful, depending on how it develops and how group members make use of this identity. For instance, some staffs I have worked with have focused on the task at hand and are thoroughly committed to the success of the project, but have no links outside the collection of data. This is certainly a sound example of task cohesion (Cota, Evans, Dion, Kilik, & Longman, 1995). Other staffs have developed strong social bonds with each other, using "down time" between sessions to play games, compose songs, and otherwise socialize with one another in the laboratory. This may carry over outside the laboratory as well. Staffs of this type are clear examples of groups based on social cohesion (Aiken, 1992; Cota et al., 1995). Both types have been very successful collecting data and, in my experience, produce high-quality material. The one concern for staffs with high levels of social cohesion is that monitoring may need to be a bit more intentional to ensure the group stays on task. Deliberate efforts to ensure one of these outcomes are probably not likely to be successful. My experience is that interaction dynamics within the staff lead to one outcome or the other independent of my actions.

A. COMMUNICATING WITH A STAFF

You may find that periodic meetings with staff are valuable for reinforcing training, communicating about problems encountered in collecting data, and learning about how effectively the staff is working together. The timing of such meetings and their content are variables under your control, except in emergency situations. A simple rule of thumb is that such meetings should be planned in advance, notice given to the staff so that they can attend, and a productive agenda created. Meeting simply for the sake of meeting is not likely to endear you to your experimenters. That said, it may be useful to schedule meetings frequently, early in a data collection sequence, and use them as training and reinforcement events and then taper off as the collection process hits its stride. Later meetings may be designed to catch up on progress in completing the data collection design, identification of issues of suspicion, lack of satisfying scope or initial conditions in the study protocol, or other problems in the collection of data.

The use of telephone calls, e-mails, and instant messages may be useful from time to time, but I am always concerned about the tendency for these "conversations" to become monologues on my part about the experiment. I prefer face-to-face meetings when issues of substantive significance arise in the data collection process. Learning that a particular day and time are experiencing large numbers of "no-shows" is best handled with these communication techniques, but brainstorming how to solve a general no-show problem may not be.

Encouraging staff to talk in these meetings is critical. Good experimenters will identify problems in the protocol. You want to know about these before they create significant artifacts in your data that imperil the validity of your study. Develop a strategy to encourage participation in these discussions as peers so that all of the members of the group contribute. Going around the table asking for contributions may work, but you may need to do more to break down status differences, especially if your student helpers see you as the "boss." An ice breaker can be helpful in initiating these conversations. Often a tale of a missed opportunity to ask a question in a project or a tale of the experiment that was a complete mess creates the climate to encourage candid and informative discussion.

VIII. TRAINING FOR POSTSESSION INTERVIEWS

In addition to translation of computer-generated records, you may ask your assistant to collect and code data from postsession interviews. This step is particularly critical when deciding which data to employ in testing your hypotheses, and it requires as much care in training as coding open interaction does. The objective for the research project is to identify those subjects for inclusion in the analysis who meet the scope conditions of the theory and who also satisfy the initial conditions for their assignment to a particular condition of the theory. For instance, in expectation states research, individuals must be task focused and collectively oriented as well as believe that they are better than, equal to, or worse than their partners in order to include their data in the analysis.

Listening to tape recordings of interviews with the trainee may provide an important starting point. Working with the assistant to identify key terms, verbal cues such as tone of voice, and other paraverbal markers such as long pauses may provide valuable information for both you and the trainee. Learning to identify body language in the interview may require that you video tape some interviews or group discussions to provide examples for your trainees. Generally, this is time well spent, as is the careful development of a protocol that identifies appropriate decision rules for excluding data from a particular subject. It may also be advisable to provide a set of guidelines about when to consult senior investigators about including or excluding data when an interview produces ambiguous information.

Supervised practice is always advisable until the trainee develops proficiency in conducting the interview. Periodic checking is advisable once the person is working by himself or herself. I have begun to review all tapes of postsession interviews to determine if the appropriate questions have been asked, and the appropriate decisions made by the interviewer. This can be an arduous and tedious task if the project is a large one and it may be desirable to develop a sampling regimen if this is the case. It may be advisable to carry out this decision making as part of group meetings with the staff. Employing multiple decision makers will help increase the reliability of these choices. If the project has a large number of cases, sampling may be the most expedient way to maintain a consistent set of rules among staff members.

At the very least, the assistant should be able to identify whether or not an individual participant satisfies the scope and initial conditions for inclusion in a data set for testing hypotheses. This is often a subtle and complex decision-making process. Subjects may meet scope conditions but not initial conditions, for instance. In this case, the individual should be excluded from the analysis. It is often valuable when training a new assistant to ask him or her to determine whether this outcome is unique to this subject or is part of a pattern in the data collection that requires systematic attention and perhaps intervention in the protocol. As I write this, I am pretesting a new experiment that is not successful in its current implementation. I have encouraged the assistant who is in training for this study to identify information from the interviews to assist us in revising the protocol. This has proved particularly valuable as he has begun to develop a deeper understanding of the project, and has also become much more sensitive to the success and failure of our attempts to manipulate feelings of our subjects toward one another and their orientation to the collective task.

This end-of-session data are also valuable when you assess the relative capabilities of various assistants as well as the success of the experimental protocol. If the project is large enough to warrant several assistants, comparison of rates of rejection of participants from the analysis pool may be in order. These comparisons allow you to identify experimental conditions that are not accomplishing the intended effects if rejection is consistently high across several experimenters. Alternatively, an experimenter whose rejection rate is inordinately high in comparison to rates of other experimenters may require special attention to determine if the person is applying rules too stringently. The complementary situation may apply if the assistant is not rejecting as many as his or her fellows. This monitoring is valuable as it may signal a variety of problems with decision making by the research staff. At the very least, it will allow you to monitor the possibility that experimenter effects may occur for some and not others of your research staff.

IX. ESTABLISHING A MENTOR ROLE

In the introduction, I mentioned wanting to produce research staff as passionate about my research program as I am. This does not happen often, but it is possible to create conditions that lead students to an appreciation of the value of social science experimental work. This often is possible if a mentor–mentee relationship can be established with individuals on your research staff. A mentoring relationship requires an intentional effort on your part to create a task ladder of increasing skill development. Associating it with increasing responsibility in the research program is necessary. I must confess that I have not had much success creating research assistants who go on to independent careers in experimental social science, due in part to the nature of the student population at my institution. Most of these students are focused on careers in the criminal justice system.

Simple ideas about how to create a task ladder to increase skills are best. Such a series of tasks may begin with the "easiest" script, the shortest time commitment, or the simplest interviews. Whatever the approach one takes, it should be apparent to the person who is learning in this situation that there is a sequence of skill development and responsibility to be mastered. As one is successful at one level, advancement is possible. One marker of this may be the written report.

I ask my assistants to keep a log of their experience with a particular subject. This log includes information about the conduct of the experimental session, the interview, notes about how the decision was reached to keep or reject data of a particular participant, and a summary of the session. This is very informal note taking, but serves as a record for the construction of other written reports. The mechanism of keeping this log is open. It is possible to store this information in computer files. I have used large catalog envelopes for each experimental session in the past, but these begin to present storage problems and their discrete character and ease of separation defeat the purpose of using them as the data for descriptive and analytic material. I have become a convert to hard copy in a three-ring binder.

The three-ring binder has[7] the virtue of having all of the information in one place, as well as allowing investigators to rearrange pages as needed to prepare reports. For instance, it may be desirable to review data from all of the rejected participants at the same time, of all accepted participants, from only one condition, from a range of dates, or collected by a particular assistant. The ease of manipulation of loose-leaf records makes this relatively easy.

The point of keeping detailed records of experimental sessions becomes apparent to the trainees when you ask them to write reports. These reports may be of several sorts. One possibility is a report for the other members of the research staff about what is going well and what is not going well as the experimental sessions proceed. The frequency of preparing such reports will depend on the rate at which data are being collected and your interest in being able to monitor progress. It may be desirable to have a report each week or at the conclusion of particular phases of data collection, such as after the first 50 acceptable cases, the first month of data collection, or the completion of a particular set of experiments.

A key issue in this report preparation is the audience that is being addressed. Initially, it is appropriate to have reports prepared for internal audiences only. My preference is to ask that the first report be prepared for me. This gives the new research assistant a chance to write for a known audience. It can also be a nonthreatening experience if you present the task as one of providing information about how the process of data collection is proceeding. The expectation is that the report will provide valuable information about progress, but it may be written in an informal prose style.

Subsequent reports may be prepared with different audiences in mind: colleagues in the same research area who are interested in the progress of your work; colleagues in other fields interested in the work, but relatively uninformed about the approach to data collection you are undertaking; and, finally, a section of a research report intended for professional presentation in a conference paper or research article. The key to this process is to stress the developmental nature of the report-writing process. I usually do this by relating the fact that one of my publications went through something on the order of 20 drafts before it saw print. This tends to reduce the anxiety of the new research assistant and also to make the point that even seasoned professionals do not "get it right" the first time.

A matter of pure judgment must be addressed at this point. It is always desirable to consider when to raise the bar for the research assistant as he or she acquires new skills and develops maturity. Part of the answer to this question is an easy one. Once the person has developed appropriate skills and is

---

[7]Employing computerized records for this purpose is becoming more common. My preference for paper is idiosyncratic. With increased availability, shifting to a word processor for this purpose may be advisable.

confident in their application, it may be time to ask him or her to learn new, more complex skills. On the other hand, issues of maturity and motivation play an important role in decisions about increasing responsibility and the appropriate level to identify for further advancement. Some of my assistants have mastered all of the skills I have asked them to but their interest in going further is not present. Others have wanted to take on more responsibility but have not demonstrated the maturity to do so.

X. CONCLUDING REMARKS

A good experiment is like a good play: sound script, excellent staging, good use of props, and a superb cast. There are some differences: the script in the experiment is open ended. No one, including the researcher, knows precisely how it will come out in the end. We may know how we want things to come out but the actual outcomes depend on our ability to stage a good production, have the cast put on a convincing performance, and create the suspension of disbelief critical to the production of high-quality data. An experimenter also encourages a strict, repetitious performance of the "play." Idiosyncratic actions, improvised lines, and variation in emphases may be valued in the theater, but they are cause for substantial concern in an experiment where one is trying to create consistent social conditions and the only variation is that created and authorized by the experimenter.

Every so often some of these elements fail. If the script is not doing what we want, we may revise it after the pretest. In many ways this is the easiest element to adjust. If the staging is wrong (lighting is bad, the computers fail too often), we can effect repairs. Sometimes it is necessary to remove a research assistant who is not performing satisfactorily. This is always unpleasant, but it may be necessary. If the person is routinely late or misses assigned work times regularly, this may be a message that he or she wants to leave the study but does not wish to quit. Dismissal is in order. Sometimes the product will be poor in that data from sessions executed by an individual turn up bad too often. Again, dismissal or reassignment may be in order. This is always a difficult task, but if you can have a conversation specifying the problem and how to solve it, the experience may prove valuable to both you and the assistant who is being released.

Having said all of that, I have not dismissed many assistants over 30+ years of experimental work. I suspect that requiring the classroom experience in a research methods course may account for some of this stability. Once the student has been through the methods course, he or she typically understands the data collection process on some level. The training and work you provide are practical experience that enhances and extends classroom learning. Good students will appreciate this. Those unlikely to appreciate the experience generally do not apply. Training those who do is an exciting and valuable enterprise. Success often leads to collaborators in your research program.

ACKNOWLEDGMENTS

I wish to acknowledge the contributions of individuals who have shared their experiences and laboratory resources with me. Hans Lee, my dissertation advisor, provided valuable experience in how to train personnel for a laboratory setting and emphasized the value of keeping carefully prepared laboratory manuals. David Wagner of the University at Albany provided his laboratory manual so that I was not working only from my own memories and materials. Alison Bianchi of Kent State University graciously read parts of this manuscript. My wife, Ann Converse Shelly, has provided valuable feedback as the manuscript was developed. Jane Sell and Murray Webster provided valuable comments on an early draft. I wish to thank them for their careful and thoughtful suggestions. Finally, I wish to thank Murray Hudson, professor emeritus of theater at Ashland University, for his generous sharing of ideas about theatrical production and the role of the director in developing a performance that depends for its success on the suspension of disbelief.

REFERENCES

Aiken, L. R. (1992). Some measures of interpersonal attraction and group cohesiveness. *Educational and Psychological Measurement, 52*, 63–67.

Bales, R. F. (1950). *Interaction process analysis.* New York: Addison–Wesley.

Cohen, B. P. (1989). *Developing sociological knowledge: Theory and method.* Chicago: Nelson Hall.

Cota, A. A., Evans, C. R., Dion, K. L., Kilik, L., & Longman, R.S. (1995). The structure of group cohesion. *Personality and Social Psychology Bulletin, 21*, 572–580.

Rosenthal, R. (1976). *Experimenter effects in behavioral research.* New York: Irvington Publishers.

Rosnow, R. L., & Rosenthal, R. (1997). *People studying people: Artifacts and ethics in behavioral research.* New York: W. H. Freeman.

Willis, J. R. (1994). *Directing in the theatre: A casebook* (2nd ed.). Metheun, NJ: The Scarecrow Press.

CHAPTER 12

Common Problems and Solutions

KATHY J. KUIPERS
University of Montana

STUART J. HYSOM
Texas A&M University

ABSTRACT

While the development of procedures for experimental research and the implementation of the design often appear to be simple and straight forward projects, the steps involved are far more time consuming and complicated than one may except. This chapter discusses some of the common problems encountered when conducting laboratory research, including bias in experimental manipulations, possible contamination, issues with using and training confederates, recruiting and scheduling subjects, videotaping, managing time, and maintaining records. We offer suggestions and a "hands-on" approach to setting up and running an experiment primarily in university settings using undergraduate subjects.

I. INTRODUCTION

Typically, the goal of experimental research conducted in a controlled setting is to test theory or construct theoretical explanations. Independent variables are manipulated or introduced and all other variables (extraneous) are carefully controlled in order for the experimenter to measure the dependent variable and make conclusions about how the variables are related. With attention to design details such as the introduction of variables, the measurement of dependent variables, and random assignment of subjects into control and experimental groups, what could go wrong? In this chapter, we suggest that there is much that experimenters must anticipate in making decisions about experimental design issues, maintaining the constancy of conditions across groups, and avoiding many of the common problems in conducting laboratory research. We discuss common problems and issues and offer some suggestions taken from experimental practice that may provide guidance for anyone designing and implementing an experiment.

To do this, we draw on our own experiences and training as researchers and experimentalists. While the problems on which we focus come from training and research experiences within a university setting and our examples therefore are drawn from the university setting as well, many of these problems also may be encountered in other research settings with other types of subject pools.

II. RELATIONS WITH THE LARGER DEPARTMENT OR PROGRAM

We begin our discussion with some important considerations for setting up an experimental lab facility, designing space, and deciding upon access issues. While initially a researcher may be grateful for any space that is quiet in which to conduct experiments, unless that space is clearly designated for experimental research, it is difficult to control extraneous variables across conditions. When others have access to the space, the room may change over time, so the first subjects in an experimental run will participate in a room that is very different from the room where the final subjects participate. An old poster or a stack of books may lead subjects to make assumptions about the hypotheses being tested or activate irrelevant emotional responses, as some political posters can do.

All such effects are undesirable. Chairs stacked to the ceiling in a corner may give the impression that the project is less legitimate than other social scientific research and that subjects do not need to worry about doing their best on the task. Because space is often at a premium in academic settings, departments are eager to "dump" supplies and equipment into laboratory rooms for temporary storage, students or other "squatters" may set up an office in a laboratory space that appears unused, and experiments may be interrupted by someone looking for an empty room. Laboratory space should be clearly dedicated for research and should be separate from the comings and goings of academic departments and programs. The space should be visually and spatially neutral and external noise should be eliminated.

Experimenters, schedulers, and other laboratory jobs must be done by people designated for those positions. It may be tempting to use university secretarial or administrative help to schedule or refer arriving participants to a laboratory room, but that is not a good idea. For one thing, it would be an additional burden on a receptionist. Also, someone who is not a member of the research team cannot be counted on to do the job as it should be done—that is, avoiding unnecessary cues and other information, and treating every person who arrives in almost exactly the same manner. Not having dedicated personnel handling all arrangements detracts from the seriousness, legitimacy, and importance of the research itself. Participants may view their participation as just an extension of their coursework and experimenters risk subjects' task orientation and focus when their research is not separated from their teaching.

The experimental treatment begins when subjects first are contacted and consequently everything should be held constant except for the experimental independent variables. That means that much of normal interaction, such as greeting, asking how someone feels, commenting on the weather or on sports, etc. is inappropriate because these things cannot be done uniformly for every experimental subject. While an experimenter who receives subjects can learn not to engage in small talk, most departmental receptionists will not avoid it. The problem is to treat everyone exactly the same except for the independent variables, and that treatment starts with the first time anyone interacts with a subject in recruiting, scheduling, and arriving at the laboratory.

Because pay may be a part of the manipulation in an experiment (such as in a status-construction experiment where pay level is a part of the status manipulation), another potential problem for experimenters is in how the payment will be distributed to subjects. Of course, careful bookkeeping records are required for the distribution of subject monies and many complications can be avoided by handling the pay within the postexperimental interview. Pay also may be an important part of the debriefing. Typically, pay is distributed during the postexperimental debriefing when subjects are told about the purpose of the experiment, given an explanation for any deceptions, and have an opportunity to ask questions. The manner in which the debriefing is conducted will have an effect on how subjects feel about their participation (which may involve being deceived) and receiving their pay at this time, rather than later, will make the experience more positive. But rules for payment are not universal. For example, in many social dilemma studies, subjects are told that they will be paid in private so that their decisions are not made public during the experiment. In such cases, subjects can be paid one at a time as they leave.

Equipment and facilities are often sources of problems encountered in a laboratory. While most research involves the use of computers for data storage, analysis, or reporting, laboratory equipment is unique in that it must be operating correctly while subjects are present. Conduct during the experiment, from the arrival of the subject until the completion of the debriefing, can be seen as a performance with the subjects as an audience. (Shelly, Chapter 11, this volume, develops the theatrical analogy.) Equipment failure creates an impression of incompetence and unscientific work. Also, because all features of the experiment must be controlled and constant (with the exception of treatment variables) in order not to contaminate the experimental results, extraneous variables such as broken or misbehaving equipment must be eliminated. It is therefore important to check equipment thoroughly before the experiment begins, monitor it in pretests, and make periodic checks during the experimental run for any malfunctions.

Some labs are able to pay for technical support for maintenance and service to equipment. While this option can be a great help, particularly when subjects are tightly scheduled and repairs need to be done quickly, it is important that the relationship with information technologies staff is explicitly discussed. If staff is involved in maintenance, it is important to discuss scheduling so that everyone knows when he or she may or may not enter the subject rooms. If someone stumbles in while a study is being conducted, his or her presence would have an effect on the measurement of variables in a way that is impossible to determine and the contaminated data would be unusable. In addition to information technologies support, other equipment in the lab may also malfunction. Experimenters have found it wise to buy tools and a toolbox in advance for quick fixes to furniture, tacking up cords, and making minor repairs to equipment.

In order to avoid most of the problems with unwanted visitors, especially while an experiment is under way, the experimental lab should be locked at all times. In our experimental labs, we find it helpful to post signs clearly identifying the area. (This also provides research legitimacy from the subject's point of view.) We also post a sign requesting that no one enter the subject rooms, including custodial personnel for cleaning and services. Unless an experimenter can be certain that custodians only work at times when data are not being collected, and that they will never move anything in the laboratory—which usually cannot be guaranteed—it is best to ask them not to enter and to do his or her own routine dusting and emptying of wastebaskets.

Locking the lab and any file cabinets and computers where data are stored is also important for protecting the confidentiality of participants. Institutional review boards will require that all records be kept in a secure location, as Hegvedt describes in Chapter 6 of this volume. Signed consent forms should be stored separately from other experimental data in locked file cabinets or cupboards. Locking the lab will also minimize the possibility of items or records being misplaced, moved, or taken by someone who is not a part of the research project. Potential subjects and payment money will be wasted if subjects show up for a study and experimenters are unable to run the experiment because any of the needed items are missing.

Because experimental manipulations are often complicated, requiring several forms and questionnaires, and because experimenters are less likely to bias subjects when they run more than one condition during a time period, it is easy for mix-ups in materials to occur. An experimenter may grab the wrong form for a condition if everything is not clearly labeled and stored. To minimize such mistakes, we post clear instructions and orders of procedures in the laboratory control room. We also label and separate forms and documents for the different conditions even if some of the same are used for more than one condition. Many mix-ups can be avoided by keeping materials for each condition together and forms and questionnaires clearly labeled.

III. EXPERIMENTAL MANIPULATIONS AND DECEPTION

In laboratory research, manipulation refers to the construction of events or information in a controlled environment. Most often, experimenters manipulate the independent variable by how, where, or when it is introduced into the experimental situation or by the level of its introduction. They also may manipulate information provided to the subjects about the task, their partners, and even the subjects themselves. This manipulation may involve the use of deception—deceiving the subject about the true nature of the study or its hypotheses—or deception may be used when other aspects of the experiment are manipulated in order to introduce scope conditions (that must be met in order for the theory to apply) or to control other extraneous variables that may have an unwanted effect on the dependent variable.

A. DECEPTION OR NONDECEPTION?

The first decision regarding deception is whether or not to use it. One way in which deception is used in experiments is to prevent subjects from learning the true hypotheses. Once subjects are aware of the hypotheses, they may not behave as they would without that awareness. They may consciously agree with the predictions and try to "help" experimenters or they may disagree with the predictions and try to show experimenters how they are wrong. In other cases, subjects simply may be unconscious of the influences that such knowledge has upon them, although its influence is well documented in experimental research (see, for example, Orne 1962, 1969; Roethlisberger & Dickson 1939; Rosenthal, 1967, 1969, 1976). Subjects may pick up clues about the hypotheses or goals of the experiment and this may influence a change in their behavior to go along with what they think is demanded of them. Subjects create demand characteristics with knowledge or guesses about the expectations that researchers have for their behavior. Experimenters should be keenly aware of how demand characteristics may form. One way to minimize the effects of demand characteristics is to make sure that subjects are unaware of hypotheses and of experimental manipulations.

Ethically, experimenters are obliged to limit deception unless unavoidable (see ethical codes for various professional associations such as the American Sociological Association's *Code of Ethics*, 1999.) If it is possible to conduct the experiment without deceiving the participants, that is always preferable. It is acceptable to use deception only if it has a specific purpose in the study, if it does no permanent harm, and if the benefits from participation outweigh any negative effects. Deception should be used only to the minimal degree necessary. So, while experimenters may not reveal their complete hypotheses, they will try to give information that is not completely untrue to subjects about what subjects will be doing. It is always best to reveal everyone's actual role in the experiment and his or her purpose, if possible. The main exception to this rule is designs requiring the use of confederates.

Researchers must, of course, obtain informed consent from participants and this is underscored when deception is used. While some deception is acceptable, researchers must never misrepresent any potential risks to subjects. If there are risks or discomforts that are likely to affect a participant's willingness to continue with the study, those should be revealed. An important part of the postexperimental interview must include a debriefing where the participant is informed of exactly how he or she was deceived, given an explanation for why the deception was necessary, and assured that his or her behavior was consistent with that of others in the study. Experimenters explain to their subjects that they were fooled by deception because the experimenters went to a great deal of trouble to construct the measures and tests so that nearly everyone believed what they were told. After deception has been explained to subjects, the debriefing session also should include opportunities for participants to express concerns and ask questions in a safe environment.

Before participation in any experiment, the subject should give his or her informed consent. (See Hegtvedt, Chapter 6, this volume, for a more complete discussion.) Information in the consent form will reveal or clarify how many subjects are involved, what the study generally entails, and how responses are being recorded. Subjects give their permission for their participation in the study by adding their signatures to the forms. One form is kept for the experimental records and the subject receives the other form.

If deception is a part of the study, there are a variety of acceptable ways in which to present it. Subjects will want to know a little bit about the study before they begin and, for that reason, a cover story is typically given. The cover story is an explanation of the purpose or nature of the research and it accomplishes several things. First of all, it assures the subject and makes him or her more comfortable in an unnatural situation. The cover story also should arouse some interest for the subject whose attention to the task and the various elements of the study is essential. If the subject is not paying attention, the outcomes will be irrelevant. Additionally, the cover story will lessen a subject's natural preoccupation with the hypothesis and the true purpose of a study by providing an explanation for the setting in which he or she will be asked to participate.

Subjects may be intentionally misled through written or verbal instructions. In many experiments, subjects are told that the task in which they are about to participate measures a particular ability. For example, there is a tradition in expectation states research to use tasks that are intentionally ambiguous in order to prevent subjects from having special information or skills that might make them experts at such tasks and influence the outcomes. At the same time, they are told that there really are right and wrong answers (when there are none) in order for them to be task oriented. Subjects also may be told something about themselves that is not true. In some trust experiments, subjects are told that they will work with a partner who is trusted by the project director—trusted to make the best final decisions and to distribute pay fairly, for example. Subjects also may be deceived through the actions of others, either the experimenter's behavior or that of a confederate. Later in this section we discuss the use of confederates in laboratory experiments but experimenters will also be acting to create a certain impression.

The design of an experiment may require false treatments or dependent variable measures. In status-construction experiments (Ridgeway, Boyle, Kuipers, & Robinson, 1998), for example, in order to create a nominal characteristic that carries no prior meaning for each subject, subjects complete a "personal response style" test (a task adapted from social identity studies; Tajfel, Billig, Bundy, & Flament, 1971) and are told that this information will tell whether they belong to a group called S2s or to another group called Q2s. In actuality, there is no such characteristic and tests simply ask subjects to distinguish between reproductions of paintings by Klee and Kandinsky and indicate their preferences. Once the tests are "graded," a subject is told his or her "personal response style" type—making him or her dissimilar from his or her partner. Deception also may take place through manipulation of features of the setting such as arranging furniture to structure interaction or separating outcome measures into what appears to be two separate studies to reduce suspicion about the true goals of the study.

So, how do experimenters make choices about deception? If an experiment can be conducted without deception, that is always preferable. When deception is necessary, as little deception as necessary to create the needed conditions should be used. Not only are ethical concerns at stake when using deception, but also if the deception involves many aspects of a study, there is a greater chance for the deception to be discovered by the subjects. It may be best just to give subjects just enough information to draw conclusions but not to give additional false information if it can be avoided. Rather than telling subjects something that is untrue, experimenters can construct situations that will allow subjects to draw their own conclusions. Those conclusions may not be accurate but, compared with falsehoods and lies, they are easier to correct in a debriefing.

For example, in the status-construction experiment mentioned earlier (Ridgeway *et al.*, 1998), subjects are asked to fill out a background information sheet. After the information is collected and evaluated and another test is taken, experimenters tell subjects how much pay they will receive "based on the information the laboratory has about [them] and the other participants." Subjects do not know what information is being used but, by implication, some of that information comes from the background information sheet. In the debriefing, great care is taken to assure subjects that the pay they were assigned had nothing to do with the background information that they supplied.

B. STRENGTH OF THE MANIPULATIONS

In designing experiments, decisions must be made about the strength of the treatment variables and how, where, and when they will be introduced; it is important to consider how strong the presence of a variable will be. In general, we desire the most conservative test possible for our hypotheses, so experimenters introduce the independent or treatment variable in such a way as to allow the subjects to have choices in how they will behave. Laboratory interactions are structured to make alternative behaviors possible—not just those that are predicted, but also behaviors that would disconfirm hypotheses. When participants are allowed to make choices in their behavior and in their responses, there is a clear measure of the relationship between treatment and outcome decisions under controlled conditions.

On the other hand, experimenters must make sure that a manipulation is understood or interpreted correctly—not necessarily in a conscious sense—but subjects must be aware of experiencing treatments that the experimenter predicts will influence their responses. The tendency is to load up on the indicators of the treatment in order to make sure that they are not missed. In the status-construction experiments, subjects were told about their pay levels at several points in the study, they read about their pay levels on the computer and on a payment form, they were asked to write their pay levels on several different forms, and they were later interviewed about pay levels on questionnaires and in person. Not only was the information repeated to make sure that subjects were aware of their pay and their partners' pay, but also several manipulation checks were conducted to make sure that subjects understood the differences and how they compared with others.

Manipulation checks are conducted to assess whether an independent variable is experienced or interpreted in the way that the experimenter intends it to be and whether the scope conditions (discussed later in this chapter) have been created. Checks may be conducted right after a variable is introduced, when subjects will be more likely to recall its presence. In some experiments, these manipulation checks indicate if aspects of the experiment are well enough understood by the subjects for the experiment to proceed. For example, in many social dilemma experiments, manipulation checks ensure that subjects understand the payoff structure of the experiment. In other experiments, however, manipulation checks anywhere but at the end of the study would arouse suspicion or otherwise distract subjects from the task at hand. Of course, this especially would be the case for experiments involving deception. If a check is given in the middle of the experiment, there is also the potential that subjects will be influenced by the checking. It is safer to place manipulation checks at the end of the experiment in the postexperiment interview. Some argue that subjects may have forgotten the manipulation by this time. If a variable is manipulated as strongly as it should be for good experimental design, however, subjects will recall it and the risk of influencing the dependent variable measures is reduced.

A common term for actors who work for the experimenter and pose as other subjects or as bystanders is "confederates." Confederates deliberately mislead subjects and typically participate in the introduction of an independent variable, modification of the decision process, or control of the setting within which predicted outcomes are expected to take place. The use of confederates presents ethical problems related to deception, as discussed earlier. In addition, depending on how confederates are used, they may introduce more extraneous variables into the experimental setting if their introduction is not carefully scripted and controlled. Because each participant in all experimental conditions must have exactly the same experience under the same controlled conditions, it is vital that confederates behave in almost exactly the same way with each subject.

In the original status-construction experiment, subjects worked on a task with a partner in a doubly dissimilar encounter. That is, the encounters took place between actors who differed in two ways: on a nominal characteristic and in terms of the level of resources each would receive. Subjects were scheduled in groups of four and had an opportunity to work with a series of partners about whom they received additional information. Researchers immediately discovered problems in pretests with this design because subjects were able to view and take into account a variety of characteristics about each other that were beyond the control of the experimenters. These characteristics, such as dress, facial and physical features, voice, congeniality (real or inferred), etc., combined with the doubly dissimilar characteristics to influence outcomes—status beliefs and observable power and prestige in the group interaction. It was impossible to ascertain the effects of the doubly dissimilar encounters by themselves.

To control information that subjects received about their partners, confederates were used instead in the final experimental design. Each subject worked with two confederates at two different times, interacting through an audio connection for verbal interaction. Not only did the use of confederates reduce much of the contamination or extra noise that interfered with the doubly dissimilar encounters, but also it allowed us to control the behavior of the partner, an important cue for the determination of status construction.

In many social psychological studies, interaction is a necessary component of the process investigated. In such situations, if the interaction must be face to face, the use of well-scripted confederates will provide many of the necessary controls for extraneous variables. Additionally, the easiest way to control confederate behavior is to restrict the type of interaction he or she has with the subject. For example, in the experiment discussed previously, subjects were allowed only audio communication with their partners when discussing their decisions in order to cut down on the other cues that are provided in a face-to-face setting.

Often subjects work with a partner on a task. If it is not necessary that the subject see or hear the confederate with whom he or she works, the confederate/partner may communicate with the subject electronically through computer interaction. In that case, the confederate/partner may be programmed into the computer and all interaction apparently generated by the confederate/partner actually will be generated by the computer. It is important for a subject to have some indication that his or her partner is, indeed, a real person since computer interaction is not rare. Fortunately, it is now possible to locate software that allows an experimenter to program in specific responses on the part of a confederate/partner to keystrokes from a subject.

If subjects must see and/or hear the other people with whom they work on a collective task, again, technology can provide some solutions. Subjects may view on a monitor partners who are presumably working in another room. These partners could be previously audiotaped or videotaped so that their behavior and characteristics are exactly replicated for each case. (Later in the chapter we discuss other techniques and considerations in setting up an experiment using confederates. Troyer, Chapter 7, this volume, discusses many technological issues related to laboratory experiments.)

In most of our laboratory experiments in sociology, we are interested in small-group interaction. By small group, we usually mean from 2 to 20 participants—only enough for the participants to have face-to-face interaction. One consideration in designing experiments must be the size of the group. Generally speaking, the larger the group, the more possibilities there are for extraneous variables to affect aspects of the experiment. If groups are mostly or entirely made up of naïve subjects, there may be scheduling problems. Problems with scheduling only one subject and getting him or her to show up in the right place at the right time are magnified by the number of subjects in the group. It will require always scheduling additional participants as substitutes and, of course, paying them even if they are not needed. This increases the cost of an experiment.

There is no formula for deciding the exact size of a group, unless size is a theoretical variable, as it is in some network experiments. Experimenters usually try to limit the group size to the minimum number of participants possible without endangering the manipulations. It is most important to consider the theoretical concepts under study. For studies focused on interaction, it may be possible to observe most features of groups, such as subordination, in a dyad. For example, we can observe status differentiation with just two individuals working with each other even though we may conceptualize that differentiation in much larger groups. At the same time, the dynamics of group interaction differs greatly when going from a two-person group to a three-person group, and those features may be theoretically important to capture. Some experiments may require examining organizational aspects of groups and, in this case, the size is dictated by the structure of the organization in question. Johnson (1994), for example, creates organizational structure with only three people. Other experiments may require large groups per se. Recent investigations of such large groups have used computer networks to facilitate group size.

Additionally, if more than one confederate must be used, experimenters will try to limit the interaction as suggested previously. If participants must interact with each other, again, the nature of that interaction should be limited as much as possible.

IV. EXPERIMENTAL DESIGN ISSUES

We often have referred to the design of an experiment in discussing manipulations. Manipulations are not the only place where problems can occur, however. Experimenters need to consider the length of the experiment as it influences subjects' and experimenters' fatigue, the emotional stress experienced by the subject due to an unfamiliar experimental environment, and the level of involvement for subjects in the task.

A. LENGTH OF THE EXPERIMENT

It is important that the experiment not be too long for a variety of reasons. Subjects may become fatigued or sleepy. They may become bored and lose interest in the task and in performing well. They may get hungry or have other needs that necessitate the interruption of the experiment. Of course, if an experiment is interrupted, that subject's experience is no longer the same as the other participants' and the controlled setting is contaminated.

It is also more difficult to maintain deception over longer periods of time. The more subjects interact with a fictitious partner or under ambiguous circumstances, the more opportunities they have to become suspicious of the manipulations. They may grow suspicious over time. Subjects also have more opportunities to spot deception, to wonder about features that have been glossed over, or to be offended by someone's behavior. Of course, if an experiment is boring, requires many repetitions of a task, or is otherwise not very engaging, the likelihood for subjects to lose interest and focus will increase.

Therefore, experienced researchers try to keep their experimental tasks short and to limit experimental trials. The entire experimental session should not last longer than an hour and a half, and limiting experiments to no more than an hour will produce the most focused responses on tasks. If experiments must run longer because they require several manipulations with follow-up tasks or retests, subjects should be allowed to get up, change positions, move from one room to another, or to do different tasks.

B. STRESS AND DISCOMFORT

Related to the concern over time limits in laboratory experiments are stresses and discomfort levels for participants. Stress may come about as a result of emotional strain (embarrassing activities, information that negatively affects self-esteem), physical strain (long or fatiguing tasks), and mental strain (repetitious tasks, instructions that are complicated or difficult to remember).

To avoid potential problems, researchers should anticipate that subjects may be made uncomfortable by offensive words or situations. Pretesting or pilot testing of all materials and manipulations should be conducted well before the experiment is scheduled to begin. Pilot subjects should be run through the experiment as if they were real subjects. Afterwards, the researcher should interview the pilot subjects, looking for any aspects of the experimental design that may cause subjects discomfort. As suggested earlier, the length of the experiment and the number of repetitions for tasks should be limited. Additionally, subjects should not be asked to perform tasks that will bring on excessive fatigue or embarrassment. Pilot tests can be conducted to screen for any of these features.

C. INVOLVING THE PARTICIPANTS

Often, experimenters ask their subjects to perform a task that may or may not be related to the actual measure of the dependent variable. The involvement of the participants in the task is essential for experiments in several ways. Frequently, a scope condition for an experiment is that participants are oriented towards a valued task—they are not distracted and they want to do it well. Additionally, subjects often must work with one or several partners on a task in order for interaction to be measured. To ensure meaningful results, the task itself must be somewhat engaging. At the same time, in order to control familiarity with the task, expertise of subjects, and perceptions of self with regard to task performance, the task should not be one that subjects have done before (unless that is a condition of the experimental hypotheses). For example, Ridgeway *et al.* (1998) asked subjects to match early language terms, and expectation states researchers have asked subjects to assess contrasting black and white squares for which pattern had the most white color (Berger, Fisek, Norman, & Zelditch, 1977). These tasks are challenging and they hold subjects' interests without becoming too tedious. Subjects are often intrigued and curious about their abilities on such tasks and are anxious to find out how well they performed.

Another way that experimenters are able to ensure task focus is to offer a reward for doing the task well. In many social dilemma studies, pay for subjects' particular contributions is given and, in fact, often is the dependent variable. In status-construction studies (Ridgeway *et al.*, 1998), we offered to pay the group with the best score a bonus of $50 to be divided among the participants. It is often common to offer each group a bonus if they do particularly well. This possibility for an additional monetary reward oriented the subjects to the task.

Of course, experimenters should check for involvement, along with other conditions essential for the hypotheses, in the postexperiment interview. Experimenters ask subjects what they thought of the task, if they tried to do their best, and if they have any questions about it.

V. RUNNING EXPERIMENTS USING CONFEDERATES

Once a decision has been made to use one or more confederates in an experimental design, an experimenter will need to consider who will play that role. Confederates posing as subjects should match characteristics of the subject pool as closely as possible. To accomplish that, it is best to select confederates from the original subject pool. Professional actors are not necessary; however, confederates who are able to act as if they are subjects without arousing suspicion are essential.

In fact, actors present their own problems, which you may not expect. While they can be very good at portraying a role, that is not the same skill needed in experimentation. Part of a good theatrical performance is individuality—portraying a role with enough spontaneity that it creates a memorable "person." Individuality and memorability, of course, are inappropriate in laboratory settings. Furthermore, there is a reason stage plays have directors. The language of the stage is not ordinary language, and an experimenter is unlikely to know how to communicate to actors just what he or she wants them to do. Unless an experimenter plans to hire a director, he or she may be well advised to try training nonprofessional students to serve as confederates.

Additionally, confederates should not be people who might be known to the subjects before they participate in the study. In Ridgeway's status-construction studies at Stanford (1998), a large number of confederates were used. They were hired from the student body (since subjects were undergraduate students) and, although subjects would not actually see them, confederates were chosen who resembled and sounded like typical students (male and female). Confederates were told to arrive early but their physical resemblance to the subjects reduced any chance for contamination created by a quick glimpse from an uneasy subject. Confederates were also told to dress like typical undergraduates, unassuming, and without any props that might affiliate him or her with specific student subcultures. The goal with confederates (as it is with the physical layout of the laboratory) is to avoid any unusual or memorable features. They should appear like the subjects, without anything that may trigger associations of other people a subject may know, or may have seen.

Experimenters will need to determine how many confederates to employ. Obviously, consistency across experimental groups is essential, so one might imagine that one well-trained confederate could ensure that each subject's encounter will be exactly the same. The opposing argument is that if confederates are well trained, more than one may be advantageous. Multiple confederates may be advantageous because of problems of scheduling, confederate fatigue, and illnesses. When a research team includes multiple confederates, it is important to assess confederate performance. Confederates can be compared in their actual performance, in assessments on session reports and postexperiment interviews, and through evaluations on post-task questionnaires. If confederates are well trained and if the script (discussed later) is functioning effectively, results should be similar and consistent. Of course, the demands of the confederate role, complicated scripts, and elaborate procedures will all influence the decision about how many confederates to use. If the role is demanding and requires considerable nonscripted interaction with subjects, then fewer well-trained confederates will make fewer errors.

A. TRAINING CONFEDERATES

Thorough training is essential. Experts suggest beginning with a manual for confederates with the following sections: the importance of consistency and control in experimental research, the role that the confederate will play, what the confederate should say and do, what the confederate should NOT say and do, and what may be said and done in response to the subjects' behaviors and comments. For example, in the status-construction experiment, confederates' instructions included the following:

> Please be careful about mistakes, note them on your session report (at the end of the experiment) if any occur, and make sure that you follow the script and order of procedures EXACTLY ... Remember that it's not OK to ad lib your responses. Everyone MUST BE SURE TO SAY EXACTLY WHAT'S ON THE SCRIPT, at least initially when you first announce your choices. This is how we make sure that all subjects are having essentially the same experience. Of course, once you make your initial proclamation, you will have to do some acting. And you are the best judge of how to do that—different subjects require different types of responses for you to communicate your deferentialness or nondeferentialness ... Use the comments posted on the sheets on the wall above your desk.

Confederates should always work from a very clear script with all dialogue included. Of course, the less a confederate is expected to say, the fewer the possibilities for error there are. The behavior and speech of a confederate also must be as realistic and normal sounding as possible. If some of the information that a subject will receive about a confederate can be controlled, such as having the confederate in another room where his or her behavior will not be observed, it is advisable. Then, spoken interaction is the only arena for errors or miscommunications. In the status-construction experiments, what the confederates initially said was very carefully scripted. Additionally, confederates were given lists of possible comments to make depending on how the subject responded.

Practice sessions are critical. The practice sessions can be recorded in order for the experimental research team and the confederates to go over the tapes to critique them. This is also an opportunity for the confederates to comment on any problems they see with procedures or with scripts.

B. MANAGING AND ASSESSING CONFEDERATES

At the end of each experimental session, the confederate should fill out a session report giving comments and impressions about how the session went, his or her performance as a confederate, and whether any suspicion on the part of the subject was detected. These reports are valuable, not only before the actual experiment begins, but to evaluate confederate performance in pretests, and throughout the study as confederates become more confident (and, perhaps, less vigilant) in performing their roles. In the status-construction experiment, the confederate session reports included questions about whether the subject was cooperative, task oriented, seemed to try to do his or her best, seemed to believe that the confederate was a real subject, and took the task seriously.

Confederates also were asked if they recognized the subject (since that would influence the acting), if they detected any characteristics of the subject other than what they had been told by the experimenter, and whether they had any technical problems.

Confederates benefit from monitoring—observation and listening in on their acting—and from frequent meetings about problems encountered by both confederates and experimenters. Audiotaping verbal interactions and reviewing those tapes after the experiment allows the experimenter to focus on the subject while the experiment is being run and attend to confederate performance later. In the status-construction study, we listened for specific types of encounters and for any language that was not allowed in the script. We also conducted periodic analyses of each confederate on the dependent variables to make sure that his or her influence was consistent (not necessarily the same) within a range and that there were no confederates who were influencing very unusual or unique responses from subjects on a consistent basis.

We also assessed confederate performance in a post-task questionnaire and in our postexperimental interviews with subjects, asking about their perceptions of the people with whom they worked. Subjects were asked similar questions about their own performance and about the confederates' behaviors: evaluating task orientation, team orientation, and believability of the manipulations. While our subjects and confederates never had face-to-face interaction, concerns about individual features of confederates such as physical attractiveness or charisma also may be checked by comparing subjects' evaluations of confederates and dependent variable results across confederates.

VI. DEVELOPING PROCEDURES

Experimental procedures must be developed such that each and every interaction between laboratory personnel and research participants is replicable. This requires careful development of word-by-word scripts, complete with blocking (i.e., stage directions) and intonation information for the entire experimental session and all scheduling and reminder phone calls. Developing these scripts is a time-consuming process. A common error when planning experiments is to schedule too little time for the development of these materials. The best way to save time and other resources here is, of course, to locate and use procedures that already exist and that have been used successfully by others. Using an existing design that can be modified for a new experimental purpose is far less likely to cause unanticipated problems than designing an entirely new experiment.

When constructing a new script, it is useful to break the task into smaller pieces. Distinguish between those parts that are standardized across conditions (e.g., phone calls, participant arrivals and greetings, securing informed consent, paying participants, task instructions, and measurement) and those parts that are related to the manipulation of the independent variable, which differ by condition. It is perhaps easiest to work first on those parts that are shared across conditions, and then move on to the manipulation.

Rashotte, Webster, and Whitmeyer (2005) suggest beginning with a list of *scope conditions* and *initial conditions* for the experiment. Scope conditions abstractly define the domain, or type of social situation, to which a theory is intended to apply (Walker & Cohen, 1985). Initial conditions describe concretely what will be done in the laboratory to instantiate the scope conditions and other features of the situation such as the type of interaction that will be used (Rashotte et al., 2005).

Expectation states theory's scope conditions state that actors are (1) collectively oriented (2) toward completing a valued task (Berger et al., 1977). In the standard expectation states setting (described by Berger, Chapter 14, this volume), initial conditions instantiate a collective task by asking participants to work with a partner (who may be simulated by a computer) on trials of one or more tasks, each of which is said to involve a different newly discovered ability (e.g., contrast sensitivity), and which are said to be unrelated to other well-known abilities. Instructions usually include a presentation of "scoring standards" about how previous teams of similar students have scored on the test (see Troyer, 2001, and Chapter 7, this volume), so that subjects have a point of reference from which to interpret their own scores relative to others', especially those with whom they will be interacting. Collective orientation is encouraged by telling participants that individuals working with others often make better decisions than they would when working alone, and that the study at hand is interested in learning about "just this type of situation."

Thus, the list of initial conditions would include the following information being delivered to the participant: (1) individuals working together often make better decisions than they would working alone; (2) several newly discovered abilities are being studied, and they are important and interesting because they are known to be unrelated to other more common abilities; (3) each task problem has a correct answer, and individuals with high levels of the ability typically make more correct final choices than do individuals with low levels of the ability; (4) instructions for using equipment (to indicate initial and final choices and receive feedback on partner's initial choice); and (5) scoring standards. In the expectation states protocol, standardized instructions have developed for delivering this information (see Berger, Chapter 14, this volume; Berger & Zelditch, 1977; Cook, Cronkite, & Wagner, 1974; Troyer, 2001; Chapter 7, this volume).

When developing new procedures, simple phrasing is best so that each important idea is clearly and deliberately communicated (though initial phrasing may require modifications based on information gained from analyses of pretest data; see later discussion). Other general guidelines include expressing one idea in each sentence and avoiding complex grammatical constructions. If the procedures do not instantiate scope conditions for the participant from the participant's point of view, then her behavior will be irrelevant to assessing derived hypotheses. Therefore, key aspects of the instructions are typically repeated at least three times, using the same or slightly different phrasing each time. Because laboratory experiments aim to create a simplified, and therefore an "artificial," social situation, it is not necessarily important that the instructions sound natural or conversational. Rather what is most important is that, from the perspective of participants, the procedures instantiate scope conditions. The extent to which a set of procedures do, in fact, instantiate initial conditions can only be assessed through pretesting the procedures, the task we consider next.

VII. PRETESTING

Once an initial script (and an initial videotape, if one is being used; see later discussion) is complete, it should be assessed during a pretesting phase. The overall goal in pretesting is to learn how well the procedures create the required initial conditions for study participants and whether the independent variable manipulation is adequate. Some specific questions to address during pretesting might include: Did the participants treat the situation seriously? Did they understand what they are being asked to do? Did they notice and believe manipulations?

Even when the experiment is only a slight modification of a set of standardized procedures, it is desirable to run initial "practice" sessions. For these, it is usually most straightforward to run the simplest condition (often the control condition) first and use graduate students or undergraduate students involved in conducting the research as stand-ins for participants. Such an initial run-through will allow quick identification of problems in stage directions (e.g., failure to note in the script that research assistants must deliver to the participant a writing implement along with any paper-and-pencil instrument) and timing (e.g., allocating 5 minutes for participants to complete a postsession questionnaire when the instrument actually takes 40 minutes to complete). Initial groups also help to assess the usability of data collection instrumentation in real time, and provide training opportunities for research assistants.

To assess other aspects of the instructions, pretest participants should be drawn from the same population as will be participants for the study itself. Manipulation checks (as discussed previously) provide important information for assessing design. For example, if information about the participant's and partner's scores is provided as part of a manipulation, a postsession questionnaire item might ask a respondent to recall and write down her and her partner's scores. To assess collective and task orientation, an item might ask, "When working on the contrast sensitivity task, how important was it for you to get the *correct* final answer?" along with a seven-point anchored scale running from "extremely important" to "not at all important." Collective orientation might be assessed using similar items asking how important were "getting the right answer," "sticking with initial decisions during disagreement trials," and "changing your initial choices just to agree with your partner when your initial choices were different." Each would have anchored scales running from "extremely important" to "extremely unimportant."

Multiple measures for each important concept are usually desirable. Questionnaire items are an economical way to gather information about how the participant understood and interpreted the situation, but to assess the effectiveness of procedures fully, pretest participants usually are interviewed at some length immediately following the experimental session. When an investigator suspects that the situation did not, in fact, instantiate the desired initial conditions for a given pretest participant, follow-up questions should ask carefully about the participant's experience to learn as much as possible about what happened and how the participant experienced the situation.

In Hysom's dissertation research (2003), for example, postsession interviews indicated that several participants in a low-status condition did not accept their low scores at a fictitious ability (meaning insight). These individuals did not appear to suspect that the ability, the task, or the scores were fictitious; rather, they appeared to discount the importance or meaning of having received a low score. One possible explanation for this was that these particular students, attending a select, private university, might have had little personal experience doing poorly in other (mostly academic) testing situations. It seemed possible that for these participants, the inconsistency of their low task score with previous scores on other tests might have led them to downplay its importance.

In a revised version of the script, the disassociation between meaning insight and academic performance was made explicit. Phrases were added, such as that the ability was also unrelated to scores on standardized exams such as the SAT, one's GPA, "or even past performance in a classroom situation." This information was reinforced multiple times. In subsequent pretesting of the revised script, no participants in the low-status condition said anything to indicate that they did not accept their scores. Procedural changes based on pretest feedback also usually are pretested, in order to verify that goals of changes are met.

Postsession interviews with pretest participants are typically much longer than are interviews with actual study participants. Orne (1969) recommends, for example, enlisting pretest participants as "co-investigators," informing them of the study's purpose; explaining in some detail what is being measured, and how; and asking, from their perspective, how they made their decisions, arguing that such information is useful in assessing demand characteristics. Another assessment that can be made during pretesting is reactivity of participants to the procedures: does anything in the instructions offend or anger participants? If yes, procedures should be adjusted to avoid such reactions, just as they should be changed if, for example, pretest participants find the task instructions confusing or difficult to follow.

Many participants are eager to discuss their experiences. Early nondirective questions can encourage this (e.g., "Well, what did you think?"), with later questions becoming more specific (e.g., "After you made your *initial* decision, and then saw your partner's answer, what did you do?"). Care should be taken that no early questions inadvertently provide information to participants that may alter their responses to later questions.

The focus of postsession interviews changes once pretesting is complete. During pretesting the function (beyond the essential debriefing and payment functions) is to assess the procedures so that they can be improved. Once a study commences, procedures must not be altered, so this function becomes the determination of whether, for each given participant, scope and initial conditions were *not* met. If for a given participant, the theory's scope conditions are not met, then predictions derived from the theory are not expected to apply. Therefore, the experimenter should exclude data from that individual, as his or her responses are irrelevant to evaluating the predictions under test. Of course, excluding participants introduces a possibility of biasing results, and such decisions seem especially likely to be vulnerable to experimenter effects (discussed in a later section). Therefore, the default decision is to include a participant unless a specific, pre-established reason for exclusion definitely exists.

Decision rules for exclusion are developed before any study sessions are run, are uniformly applied to participants, are conservative so that the subject is not easily excluded, and are explicit. A decision rule for "lacks collective task focus," for example, might state (1) that a participant must provide unambiguous information that she came to a conclusion that it was not necessary, important, and/or legitimate to take her partner's initial choices into account (e.g., "I didn't believe my partner was real, so just I ignored him"); *and* (2) that she therefore changed her behavior in some concrete way (e.g., "So I just chose answers by alternating between the left and right buttons"). Additionally, exclusion decisions are made at the end of each interview. Such decisions may be reviewed by the experimenter later—for instance, by listening to a recording of the interview—but such second guessing suffers from a lack of the many nonverbal cues that are present during an interview (e.g., smiling, avoiding eye contact, or rolling one's eyes).

Other important information can be gained during pretesting. One advantage of using a standardized protocol, for example, is that pretest results can be compared with past studies if comparable conditions are run. Troyer (2001; Chapter 7, this volume), for example, notes that some recent research employing the standard expectation states protocol has obtained lower P(s) scores than were found in previous research using the same protocol. In an experiment designed to explore possible reasons for these differences, she compared P(s) scores, collective orientation, and task focus between actors in theoretically comparable conditions, but where particular aspects of the procedures were subtly altered. Initial choice feedback to the participant on critical trials, for example, said either that the partner had "disagreed" with the participant's initial choice or that the partner had made an initial choice that was opposite to the participant's initial choice. The subtle difference (feedback presented as disagreement versus feedback presented as difference) in fact affected P(s) scores for high-status actors. Troyer's research alerts us to the importance of comparing different ways of phrasing and presenting information when developing effective procedures.

Laboratory personnel who are authoritative, friendly, believable, and competent help ensure that study participants accept the information that is presented and follow directions. It is therefore important to train assistants carefully, before commencing a study, and stress that every interaction with every participant is "part of the study," and that these interactions, therefore, must to be standardized. Experimental instructions are delivered at a relatively slow rate so that subjects understand each aspect of the study. Along these lines, it is important that all research assistants speak at the same rate (slower, more measured than everyday speech), emphasize the same words (bold typeface in a script can indicate words to be stressed), pause at the same places for the same duration (pauses are noted using boldface slashes) and maintain eye contact similarly. During training, modeling these behaviors multiple times is an effective way to develop standardization.

VIII. VIDEOTAPE

Videotapes, and before that, films, have been used by experimental researchers for decades, both to record interaction or, more generally, to collect behavioral data, and to present stimulus material and instructions to participants (see Dowrick & Biggs, 1983). No matter how videotape is used, it is advisable to have back-up videotape players/recorders, cables, blank tapes (if recording interaction), and instruction tapes (if using prerecorded instructions). Next, we first discuss in some detail the recording of group discussions, and then the production and testing of video instructions.

A. TAPING INTERACTION

Poor-quality recordings can lead to many difficulties. Audio and lighting problems, for example, can make analysis of recorded interaction much more difficult, for if a coder cannot understand what participants are saying, it is impossible to transcribe the audio record accurately. If a participant pushes her chair back from a table, and thus out of the camera's frame, then her behavior is not visible, and it will also be impossible to code. Bright fluorescent lighting can wash out contrast, so incandescent desk lamps should be considered to improve tape quality. Before running any groups, it is useful to test the quality of tape that results from different placement options for cameras, lights, and microphone. Verify also that any props provided to participants cannot easily be placed where they will block the camera. If facilities are not soundproof, outside noise may distract participants and may make the audio more difficult for coders to understand. Directional microphones are a good way to avoid picking up this type of noise. To ensure high-quality audio recording, it is generally advisable to use microphones that are external to the video camera. Lapel microphones are good for capturing speech, but problems arise if participants touch or play with the microphone, as the noise this produces in the audio record can make analysis of the tape more difficult. When purchasing microphones, there are different types with different pickup patterns. Although somewhat dated, Wallbot (1983) provides an introduction to microphone use in laboratory settings that still is useful.

When recording interaction such as a discussion group, it is usually necessary to have at least two cameras so that participants sitting across from one another at a table can be captured. Johnson, Fasula, Hysom, and Khanna (2006), for example, arranged three discussion group participants so that one sat on one side of a desk, while the other two sat on the other side. One camera was mounted high on the wall behind the desk pointed at the two participants on one side, and another camera was mounted high on a wall across the room, pointed at the single person on the inside of the desk. Cables were routed inside the walls. Another option for discussion groups is to use a dedicated camera and lapel microphone for each participant. When multiple cameras are used to record discussion groups, a screen splitter is a useful piece of hardware. It takes as input the video signals from two or more cameras, and sends the images from each camera, arranged one in each half (horizontally or vertically) or quarter of the monitor screen. This eliminates the need to synchronize multiple tapes of a single group, postrecording, which is necessary for coding reactions of participants toward one another during interaction.

Camera position is determined most by the types of data that will be extracted from the recording, taking into account the physical facilities available. If a task requires participants to move about, dedicated cameras may not be possible. Cameras and microphones should be unobtrusive. Tripods take up a lot of space and are awkward to work with. Loose cables can be potential disasters. Consequently, cameras mounted on the wall and cleanly routed cables can be important for subject interaction. When arranging chairs and surfaces for discussion groups, it is useful to limit their possible movement. It is also important that laboratory assistants verify, for each experimental session, that all participants are in frame before taping begins and that microphones are turned on.

The development of a procedure to uniquely identify and store tapes and other physical artifacts (completed paper-and-pencil instruments, for instance) produced during an experimental session will help ensure that no data are misplaced or damaged. One way to do this is to create a session information sheet that can be completed quickly by research assistants after each group session. Provide space on the sheet to record the study name, session date and time, condition, technician's name, comments, and, importantly, a unique code number for each experimental session. This unique code then can be placed on the session's videotape and all other saved session artifacts. The labeled items and the record sheet then can be placed in a large manila envelope or "zip-type" plastic storage bag, and these can then be secured in a locked cabinet or drawer. Such procedures ensure that data are clearly linked and that the confidentiality of the subjects is maintained. Procedures for quickly backing up recorded videotapes can avoid potentially costly disasters. If using digital video, the video can be transferred by cable to a computer hard drive and burned onto CD. A simple VHS tape duplicator can be created by connecting the output of one player to the input of a second player.

Videotape can also be used to record research assistants' interaction with participants. Johnson *et al.* (2006), for example, videotaped each of two experimenters as they ran discussion groups to check that key information in the script was correctly presented, and that their demeanor (coded as very confident, mostly confident, mostly hesitant, very hesitant), tone of voice (coded as firm versus soft), and use of verbal qualifiers were similar.

B. Using Videotape to Deliver Experimental Instructions

Videotape can also be used to deliver instructions to groups or to single individuals. The advantages of using videotape are: increased experimental control, increased standardization, and ease in running sessions. In a simple setup, participants might sit at study carrels set in rows with a large video monitor placed so that each participant can see it over the top of the carrel. A more elaborate setup might have several small testing rooms, each with a television monitor placed where the participant can easily view its screen and a camera positioned to capture a person watching the monitor. In the control room are a videotape player and, if needed, another camera for delivering "live" instructions to the participants (see later discussion). Cables connect testing room and control room cameras, through a signal switcher, to a videotape recorder and monitor. The signal switcher is a piece of hardware that allows signals from multiple sources to be independently routed, quickly, using a simple push-button interface, to any monitor or recorder.

Prerecorded videotape can be presented to participants either as prerecorded or as "live, from another room in the laboratory." One advantage to presenting prerecorded videotapes *as* prerecorded is that it allows for the use of subtitles and graphics, which can emphasize important information. Note that it is important here not to go overboard; use only a few simple effects specifically to stress important pieces of information.

In Hysom's (2003) research, part of the instructions were described as live because it was necessary for the participant to "introduce" himself to his (prerecorded) partner. In other parts of the session, video was presented explicitly as a prerecorded video "about the abilities we are studying today." Parts of the video that were presented as prerecorded had much higher production values than did sections presented as live. In acknowledging prerecorded sections, a narrator in a suit and tie sat at a desk in front of a neutral background and spoke directly into the camera.

During key parts of the instructions, subtitles were used to emphasize the material. For example, the words "contrast sensitivity" faded into view along the bottom portion of the screen as the ability was being described. Similarly, when scoring standards were presented, the scene cut from the narrator to a graphic of a scoring chart. As each score category was discussed, the part of the chart containing information about that category was highlighted in a contrasting color. Sections presented as live were recorded in an office setting, with bookcases filled with various equipment and books as the background, and the male researcher wore a lab coat over a white shirt and tie. During experimental sessions, the researcher wore the same clothes, so the transition from in-person interaction (as participants arrived and were taken to their testing rooms) to ostensibly live instructions was seamless.

Introductions between the participant and a prerecorded "partner" were accomplished using a signal switcher. The prerecorded researcher presented as live on the participant's monitor asked the partner to introduce himself to his partner, and then reached off screen and ostensibly pressed a button. The participant immediately saw his "partner" on screen. The partner was prerecorded, looking into his own camera, as he introduced himself by name. The researcher then asked from off screen, "And you are a student here at _____? Is that right?" to which the prerecorded partner replied, "Yes, I am a student here at _____." Using the signal switcher, the output from the camera in the participant's own testing room was immediately routed to the participant's monitor so that the participant saw himself on his own screen. The researcher's voice asked the participant what his name was, and then verified that he was also a student at the university. Once the participant said his name and identified his university, the screen returned to an image of the researcher in the control room. Because the actor playing the partner can be of any race, age, gender, etc., the use of video introductions allows for the control of status and other characteristics displayed by the partner.

If individuals appearing in a videotape are presented as live, they must wear the same clothing for each session. The effort associated with uniformity can be minimized if individuals in the videotape are shot only from the waist up. Also, it is important to avoid an outward facing window in any shot that is ostensibly live because seasons may change.

C. PRODUCTION OF VIDEOTAPED INSTRUCTIONS

When producing videotaped instructions, it is crucial to prepare thoroughly before beginning to tape. As with most aspects of experimentation, the more work that is completed up front, the fewer problems will be encountered in running the study. Previously we discussed the use and training of confederates. When confederates are videotaped, there are some additional considerations. Meet with and train confederates appearing in the videotape to verify, before the scheduled shoot date, that all are fully prepared and can recite or read through their lines without difficulty. Scheduling practice sessions before the scheduled shoot can help make shooting go more smoothly.

During recording sessions, at least one person generally keeps notes about deviations from the script, and multiple shots of each section of the script are made. Minor mistakes (disfluencies, minor deviations from the script) can later be edited out, as needed. (In video presented as live, some disfluencies actually can be expected and may make the tape more believably live.) Once all scenes are recorded, the videotape is carefully reviewed, and the best shots are identified for use in the final instruction tape. When reviewing tapes, the creation of a log, noting the time stamp at which each shot begins, is useful so that, later, the experimenter and/or the editor can return quickly to shots chosen for the final tape. Those selected are then edited together so that the final tape flows smoothly.

Videotapes can be produced in house, or all or part may be completed by professionals. The availability of affordable digital video cameras and several commercially available software programs makes possible the in-house production of high-quality videotapes. Most cameras and many of the currently available software programs are fairly easy to master, at least for basic editing operations. The main advantage of producing tapes in house is that a researcher has more control over the process and, if needed, can make changes to a tape more easily and quickly during pretesting. A potential disadvantage to in-house production is that it requires knowledge of the video software. When production and editing are done by hired consultants from outside the lab, verify how much time they will need in order to edit and return the tapes. It also will save time and avoid many problems to explain carefully in advance of beginning work exactly what is expected from the consultants and to ask what they require in return. Because experimental instructions are a unique type of video production, additional time should be spent explaining the goals of the taped communications.

As with live research assistants, the narrator on videotaped instructions should appear serious, authoritative, and competent. Distractions should be minimized and speech clear and easy to understand. Rashotte *et al.* (2005) have developed a technique to assess videotaped instructions, playing them for large classes and asking students for their evaluation of the narrator's presentation using a series of semantic differential items (e.g., too fast/too slow, ignorant/knowledgeable, competent/incompetent, etc.).

Because digital video technology and computer editing programs develop over time, this may affect purchase decisions for equipment and software. Some suggestions here are to standardize as much equipment as possible, including cables, cameras, and media. It may save costs in the long run to avoid proprietary media formats. Note that one useful feature of camcorders is that they can be used as video cameras and as videotape players. For cabling, S-video cables and RCA type audio cables are good choices. One final note about producing videotapes that are to be shown as live is to make sure to turn off the time and date display so that it does not destroy the illusion that a performance is actually live.

D. Using Taped Instructions

Using tapes simplifies the running of experimental sessions. It is most useful to have a separate well-labeled tape for each different condition. Even though most of the instructions will be the same for each condition, it is awkward to switch tapes during a session. For each session, after randomly assigning the participant to a condition, the researcher selects the correct tape, loads it into the player, checks to make sure it has been rewound, and cues it. Before pressing the play button, it is important to make sure that the signal switcher (if one is used) is correctly set so that the tape will appear on the participant's monitor.

As with any mechanical or electronic hardware, videotape technology is imperfect. Tapes break or get eaten by equipment; players break down and assistants play the wrong recording. The best way to deal with such problems is to avoid them by regularly testing equipment and by having multiple backup tapes for each condition. Some errors, if they occur, safely can be ignored. A participant is unlikely to know, for example, if the wrong tape was loaded, and if this happens, he or she can most likely simply be moved into the condition that corresponds to the tape that was actually presented. When the worst happens, however, and equipment fails, or some procedural error occurs that is noticeable to the participant, it may be best to stop the session. If a session must be prematurely ended, debriefing is especially important. In such cases, the participant is informed of what happened, what should have happened, and why it should have happened.

IX. MAINTAINING A SUBJECT POOL

Here we briefly discuss some of the common problems in maintaining a subject pool, including subject recruitment. For a more detailed discussion of subject recruitment, the reader should consult Kalkhoff, Youngreen, Nath, and Lovaglia, Chapter 10, this volume. For most experiments, research involves undergraduate students attending the academic institutions where the research is conducted. While experiments may also use older adults, members of the community, or people with institutional attachments such as employees in work organizations or students in schools, we confine our discussion here to subject pools of college students and the unique problems with such pools.

Students are typically recruited from large classes for freshmen. Freshmen and, to a lesser degree, sophomores make the best subjects because they are naïve in the ways that are important for experimental research; they are unlikely, for instance, to have already participated in other social science experiments. They also should be naïve to the hypotheses or goals of the research, and freshmen typically have not had a chance to be exposed to current theoretical work in the social sciences. Subjects also should be naïve to experimental design, the possibility of deception, and the manipulation of variables in an artificial setting. Beginning college students also have not had a chance to be exposed to such details of experimental research. Additionally, they usually are eager to participate and to do well and are more likely to believe what they are told about the study. On the other hand, using subjects with such a great desire to please the researchers should put experimenters on the lookout for any misinterpretations or possible demand characteristics (discussed in the previous section on deception).

Generally, faculty throughout the university are supportive of academic research and most are willing to give recruiters access to their students. A letter and an explanation about the study being conducted usually provide the legitimacy needed to make a brief presentation in a class and collect sign-up slips filled out by students expressing an interest in participating in a study. In order not to take up too much of a generous colleague's class time, the recruiting presentation should be brief. The goal is not to educate students about experimental research or about the project. In any recruiting session, researchers are well advised to give as little information about the project as possible, while establishing their legitimacy as a social science researcher and clarifying the minimal risk for participants in the study.

Experimenters also may want to collect some information on potential participants at the same time. Precious time and subject monies will be wasted if an experiment is run with subjects who previously have been made suspicious by their participation in an experiment or from learning in a class about deception in experiment research. Those subjects can be weeded out by collecting some background information in the recruiting session. Such information also can reveal if the potential subjects are similar in terms of demographic characteristics or if they have had experiences that may put them outside the scope of the theoretical predictions being tested. The trade-off, however, is that the longer the form students are asked to fill out, the fewer the students who are likely to sign up for the project. The researcher's desire for the most appropriate subject pool should balance with the need to get a large (and unselected) subject pool.

An incentive for participation is often required. Some researchers will make an arrangement with colleagues who teach large sections of an introductory course to give extra credit for experiment participation. This may be an effective incentive but experimenters should be careful not to fill their subject pools with participants from only one major. A better strategy is to recruit from an introductory course or courses that are required of all freshmen. It may be possible to arrange extra credit opportunities with these instructors. Others find that paying subjects will increase the likelihood that they will volunteer to participate and will show up once they have been scheduled.

Pay will need to reflect the going rate at the university or college where recruitment takes place. Not only should it be compared to the minimum wage for working on the campus, but it also should be compared to the value students place on other activities such as studying, working, or socializing. If the value of the other activity outweighs the amount being offered, students may be reluctant to participate and keep appointments. At more affluent universities, if students are more active in sports or organizations or if the pressure for top grades is great, pay will need to be sufficient to lure them away (even briefly) from such activities. Importantly, if rewards are a part of the study, (such as in many economics studies and in sociological studies of network power), incentives should not be mixed. For example, students who are receiving course credit for a study should not be mixed with students who are being paid because control in incentive structure as an extraneous variable is sacrificed.

A. SCHEDULING

As discussed in other chapters, some studies may involve a sign-up process for participants and no scheduler will be necessary. However, in studies that require routinized scheduling, a well-trained scheduler with a clear script and a list of what to tell and what not to tell participants can minimize no-shows and reduce suspicions. Schedulers will need access to long-distance telephoning when scheduling from a student pool because many students likely will provide cell phone numbers with nonlocal (hometown) area codes in their contact information. All subjects should receive two telephone calls: one to arrange the appointment and one, the day before the experiment, to remind of or confirm the appointment. But before the scheduler telephones anyone, he or she must check the information on the potential participant to make sure that the scope conditions regarding subject characteristics are met. For example, if status-equal groups are being run, females must be grouped with females and males with males.

Schedulers should be provided with a manual similar to the manual for confederates. The manual will tell the scheduler how many and when subjects need to be scheduled. It also will explain how essential it is that subjects know and understand that researchers are counting on them, once they have agreed to participate, at a certain time and that if subjects do not show up, they may jeopardize the entire experiment. This point should be repeated several times to the subjects in the scheduler's script in order to be communicated effectively. (It should not be repeated, however, to the point that subjects feel coerced.)

The scheduler usually includes a very clear introduction identifying himself or herself with the research lab and the particular manner in which subjects were recruited. Subjects then are given a very general description of the study for which they are being scheduled. According to the particular experiment, subjects might be told how long the study takes and the approximate rate of pay.

Subjects should not be scheduled more than a week ahead of the time that they will participate. Things change in subjects' lives in longer time periods, and cancellations and no-shows become more frequent. In the first telephone call, the scheduler should always follow the script. If the potential participant is unable to participate during a time slot on the schedule, the scheduler will have clear instructions on what to tell him or her. The scheduler may say that she will call back another week and try again to schedule. Or she may say that she will telephone again during the next semester or session. This alerts potential participants to the possibility of another telephone call and, if they do not object to the scheduler telephoning them again at a later date, she will have a much better chance of keeping them in the subject pool.

Schedulers also will need to give clear directions to the building and room where subjects are to meet experimenters. It is best to repeat this information several times during the initial telephone call, try to get the subject to repeat it back, and repeat it in the follow-up reminder call. Subjects generally will not write down the information, so repetition is the best way to get them to remember the directions. Another way to help subjects to remember appointments is to create an opportunity for them to write it down. One scheduler's script that we use says, "In a moment I'll give you some important information. While I place you on our schedule, can you get a pencil or pen and paper so that you can take down this information?" After the participant returns, the appointment time is confirmed and directions are given.

The directions should also include special instructions for entering the building (if it is in the evening or on a weekend and doors are locked) or where to meet the experimenter if the participant cannot find the room. Subjects also can be given a telephone number at which to reach an experimenter if they must cancel their appointment. Sometimes subjects do write this number down or remember the name of their contact person. With the callback feature on cell phones and most land-line telephones, however, schedulers also find themselves being contacted at the number from which they originally scheduled the experiment appointment.

Schedulers will need to keep careful lists of who is scheduled for which time. The day before a session is scheduled, the scheduler should make a second telephone call to remind subjects about their appointments. This reminder call is especially important because potential subjects often will forget that they have made an appointment to participate. With cell phones, appointments may be made while participants are unable to write them down. A reminder call helps reduce the proportion of no-shows. The follow-up reminder phone call may be left on an answering machine or voice mail, or with a roommate. The message need only include the time of the appointment and where subjects are to go, along with a telephone number for cancellations. It is essential that everyone who is scheduled be contacted.

It is also important to mention that the experiment may begin with the appointment call made by the scheduler. In the status-construction experiment, for example, potential subjects were told that they would be working with a partner, so it was essential that they show up on time in order not to keep anyone waiting.

B. No-Shows

"No-shows" is the term for a subject who fails to make the experiment appointment and does not call to cancel. The experimenter must decide if he or she will try to recontact the potential participant and reschedule the experiment appointment. In general, unless our subject pool is dangerously small and we may run out of potential subjects, we do not recontact no-shows, but this is a matter for the experimenter to consider. While there are legitimate reasons for not making a scheduled appointment, we reason that a no-show may be less responsible and/or too busy to participate, and that the likelihood that he or she will make the effort to keep the second appointment is not great.

Student culture in terms of socializing activities will have an effect on the likelihood of no-shows and will need to be considered when schedules are set up. In general, no-shows are most likely to occur on Friday afternoons (or even Thursday afternoons) when students are beginning their weekend relaxation and partying. They are also most likely to occur at the midpoint and end of the semester or term when students are preparing for finals and writing papers and easily may overestimate the amount of free time that they will have. Even weather can affect the no-show rate. Rainy days and days with beautiful weather are worst; cloudy days seem to be best. Besides using reminder telephone calls, no-shows also can be reduced by a skilled scheduler who keeps track of experiment times that are consistently difficult to fill or that potential subjects are reluctant to commit to. Some modifications of the experiment schedule may be needed to reduce no-shows and late arrivals.

What about overbooking, which airlines and restaurants do routinely? We have used this technique, though it can deplete a subject pool quickly. If an experimental design requires several subjects for a group, such as a six-person discussion group, then overbooking may be the only way to keep the project running. If exactly the right number keep their appointments, of course all is well. If an "extra" person arrives, we greet that person, explain that because of (unstated) circumstances, we need to reschedule him or her, and express apology for inconveniencing him or her. We also pay a small amount for showing up and reschedule the person right then (to be sure that he or she will get into the new group!). We have not found any resentment at coming once without participating, and these people are very likely to keep the second appointment and to be fine subjects. Thus, the only negative aspect of overbooking is that it uses a subject pool faster. It does not seem to cause harm to individuals who come more than once, or to affect their behavior in the experiment when they do participate.

C. Contamination of the Pool

The subject pool may become contaminated if potential subjects receive biasing information beforehand. One nonobvious source of that problem is faculty who teach the class from which subjects are recruited. It is important that faculty do not give subjects negative information that would make them reluctant to participate. It is also important that faculty do not accidentally slip and tell subjects what the experiment is really about or, if there is deception, what it might involve. For this reason, it is best that as few people as possible have such information about the study. Colleagues should know the subject of the investigation only in the most general terms. Colleagues will also have respect for experimental research when recruiters are respectful of faculty time in the classroom and make their presentations brief. It is particularly damaging to recruitment rates if the instructor introduces recruiters as people "who are looking for guinea pigs." Try to get the instructor to allow you to introduce yourself, or at least ask him or her simply to give your name and say you are looking for help with an interesting research project.

The subject pool also may become contaminated if subjects who have participated in the study tell potential subjects what the hypotheses are, that deception is involved, or that they will work with a confederate. The experiment itself should be a pleasant experience for subjects so that they will be sympathetic to the goals of the research once they have participated. They should be treated with respect, paid for their work, and not kept beyond the time indicated. If the experiment must be canceled for any reason (equipment failure is the most common), it is worth the money to pay subjects for some of their time in order to generate good will and maintain a professional reputation on campus.

Some experimenters have found that subjects are less likely to reveal manipulations or deception if they are made to feel a part of the study. In the postsession debriefing, not only are subjects protected by the experimenters revealing the purpose of the research, but subjects are included in the research. Experimenters may include them by asking if subjects understand the predictions and if subjects think that the predictions will work. If subjects are allowed to comment on and share personal experiences about the variables under investigation, they are more invested in the success of the study and are more likely to keep it secret. In order to minimize this problem, we include the following statement in the experimenter script for our postexperimental interviews:

> Also, we would appreciate it if you would not tell your friends about what we are looking for in this study. Once people know the hypotheses, they don't behave the way they normally would in an experiment. Since we are recruiting from freshman lecture classes, you will probably know others who will be participating in the study. If they ask, please only tell them that it's a study about perceptions in groups

and nothing more. It's fine to discuss the study in more detail once they've finished participating.

The pool should be repeatedly assessed by experimenters for possible contamination and potential problems should be anticipated. If the size of the pool reaches such a low level that random assignment to conditions becomes difficult, it may be possible to contact additional faculty members and recruit from additional classes, increasing pool size. However, experimenters should be careful that all students at a particular level are included in the pool initially and that all have an equal chance of being selected. This helps to fulfill the random assignment requirements as well as meet ethical concerns about volunteers being able to participate.

X. PAYMENT AND CREDIT ISSUES

As noted before, participants are typically paid a small amount of money to encourage their participation. The establishment of a "petty cash" fund, maintained in the lab, from which to pay participants is common practice. Depending on the amount paid to participants and the total number of participants in a study, the amount of cash needed for payments may be substantial. Financial Services personnel may not be familiar with social science experimentation and may not, therefore, understand why a petty cash fund of such a large size is necessary. In one case we know of, a researcher was asked to submit information for each subject, requesting for each that a separate check be issued and mailed to participants. Considering the modest sums typically paid to participants in this type of research, that procedure, in addition to being costly, would be likely to make the payment less of an incentive. With this in mind, it may be useful to get to know your Financial Services personnel, the department funds custodian, and any other individuals involved in the process well before requesting funds. Positive interpersonal relations can be crucial in quickly resolving problems or confusion about funds if it arises.

Different institutions and funding agencies may have different rules and procedures for establishing and maintaining a petty cash fund. The first time you establish such a fund, it may take longer and be more complicated than you anticipate. Typically, procedures require a named "custodian" and the use of prenumbered receipts. The custodian is most often the faculty member conducting the study. Funds are released to the custodian (with options for payment to be made in cash or check, or electronically), who has legal responsibility for the funds. After spending part of the money, the custodian then presents valid receipts and/or cash, usually by a date set at disbursement, equal to the amount initially released. Policies also usually stipulate that funds be kept in a secure location, where access is limited to laboratory personnel; a locked file drawer in a locked laboratory room meets that requirement.

Research assistants will need to be trained to issue numbered receipts in the correct order, verify that participants correctly complete their part of the receipt, return the cash box to its secure location after each session, and lock the lab. Research assistants who are running the study will need to keep track of the funds so that they can be replenished before funds run out. When petty cash needs to be replenished, receipts totaling the amount initially disbursed and any required paper work are returned to the institution. Turn-around time for replenishment may vary, but 2 weeks is common. If someone forgets to replenish petty cash in time, an experimenter may, therefore, need to be prepared to stop running subjects for a few days.

To avoid delays and other problems, it is important to verify that all expenses are entered into proper categories. With this in mind, determine what expenses are allowable, and note any rules that differ between the funding agency and the university. A researcher has to incur only expenses that are allowed by both sets of rules. A common rule is not to commingle petty cash with other funds. For instance, nobody associated with a project should regard the petty cash box as a place to borrow money, and multiple accounts, such as a research petty cash fund and a departmental petty cash fund, should not be mixed. For questions about allowable uses of funds, check with the funding agency directly—preferably with the officer in charge of the grant, if there are questions regarding a given expense. (Remember that such officials are usually busy people in charge of many grants, so the grant number should always be included in any communications. Also try not to refer every question to the program officer. Handle what you can on your own, reserving contacts over handling money for cases that really cannot be dealt with locally.)

In the case of petty cash or working funds, it is likely that a fund's allowable uses are stipulated when the fund is established, and other uses will not be honored. It will be helpful to know the project (account, fund) numbers associated with the experiment. In larger departments, this information is most likely available from the department accountant; in smaller departments, the office manager may have this information. Finally, request amounts that are multiples of the amount you will be paying each participant, and get the correct denominations initially so that research assistants do not have to scramble for change during experimental sessions.

XI. EXPERIMENTER EFFECTS

Experimenter effects are errors introduced during the collection or analysis of experimental data due to the behavior of the experimenter. They can affect the data collected in an experiment and thereby confound the analysis of results in at least four ways (Rosenthal, 1976): through (1) subtle differences in participant treatment, (2) errors in recording data, (3) errors in selecting cases, and (4) errors in the analysis of data. In the vast majority of cases where experimenter effects have been studied, these effects bias results in favor of the hypothesized results, with the experimenter unaware that he or she is even making an error or treating participants differently based on expectations.

Rosenthal (1976; 2005) suggests several strategies for addressing experimenter effects. The best known is the use of experimenters who are blind to the condition being run. We mentioned using blind interviewers previously with reference to assessing the extent to which scope conditions are not instantiated for a given subject. When an experimenter is blind to condition, her expectations regarding the hypothesized outcome for the condition are controlled, and so too is any cuing behavior caused by such expectations. The careful development and assessment of standardized videotaped instructions can have the same effect of controlling for differences in experimenter behavior, since the behavior of the experimenter, shy of the manipulation of the independent variable, is the same for every participant.

Another technique for the control of experimenter effects is to minimize contact between the experimenter and participant. The less exposure a subject has to experimenters, the less likely it is that cues from the experimenter will be "picked up" by the respondent. Videotaped instructions greatly reduce the amount of contact between participants and experimenter. Because any errors or problems in a videotape will affect all participants, however, careful pretesting and development of these instructions to identify and rectify problems is that much more essential.

Conducting experimental research to test theories or construct theoretical explanations requires careful attention to all details from setting up the lab to developing procedures to implementing the experiment. Published summary reports of experimental tests may lead to conclusions that the data collection process was relatively easy because experimental research is often described in a clear and simple presentation. Readers of this chapter, however, likely will conclude that setting up an experiment, training assistants, and recruiting subjects are far more complicated and time consuming than a research report indicates. Underlying experimental operations are a variety of hidden problems and issues. In this chapter we suggest solutions that we and others have found helpful for some of the common problems and issues. Through sharing the details of our experiments, we hope not only to facilitate replication of social scientific research, but also to encourage laboratory experimentation for building social scientific knowledge.

REFERENCES

American Sociological Association. (1999). *Code of ethics and policies and procedures of the ASA Committee on Professional Ethics.* Washington, DC: American Sociological Association.

Berger, J. (2007). The standardized experimental situation in expectation states research: Notes on history, uses, and special features. In M. Webster & J.Sell (Eds.), *Laboratory experiments in the social sciences* (pp. 353–378). Burlington, MA: Elsevier.

Berger, J., Fisek, M. H., Norman, R. Z., & Zelditch, M., Jr. (1977). *Status characteristics and social interaction: An expectation states approach.* New York: Elsevier.

Berger, J., & Zelditch, M., Jr. (1977). Status characteristics and social interaction: The status-organizing process. In J. M. Berger, M. H. Fisek, R. Z. Norman, & M. Zelditch, Jr. (Eds.), *Status characteristics and social interaction: An expectation states approach.* New York: Elsevier.

Cook, K., Cronkite, R., & Wagner, D. (1974). *Laboratory for social research manual for experimenters in expectation states theory.* Stanford, CA: Stanford University Laboratory for Social Research.

Dowrick, P. W., & Biggs, S. J. (1983). *Using video: Its psychological and social implications.* Chichester, England: John Wiley & Sons.

Hegtvedt, K. A. (2007). Ethics and experiments. In M. Webster & J.Sell (Eds.), *Laboratory experiments in the social sciences* (pp. 141–172). Burlington, MA: Elsevier.

Hysom, S. J. (2003). *Reward, status, and performance expectations: An empirical test.* Paper presented at the annual meetings of the American Sociological Association, Atlanta, GA.

Johnson, C. (1994). Gender, legitimate authority, and leader–subordinate conversations. *American Sociological Review, 59,* 122–135

Johnson, C., Fasula, A. M., Hysom, S. J., & Khanna, N. (2006). Legitimacy, organizational sex composition, and female leadership. *Advances in Group Processes, 23,* 117–147.

Orne, M. T. (1962). On the social psychology of the psychological experiment: With particular reference to demand characteristics and their implications. *American Psychologist, 17,* 776–783.

Orne, M. T. (1969). Demand characteristics and the concept of quasi-controls. In R. Rosenthal & R. L. Rosnow (Eds.), *Artifact in behavioral research* (pp. 147–179). New York: Academic Press.

Rashotte, L. S., Webster, M., Jr., & Whitmeyer, J. M. (2005). Pretesting experimental instructions. *Sociological Methods, 35,* 163–187.

Ridgeway, C. L., Boyle, E. H., Kuipers, K. J., & Robinson, D. T. (1998). How do status beliefs develop? The role of resources and interactional experience. *American Sociological Review, 63,* 331–350.

Roethlisberger, F. J., & Dickenson, W. J. (1939). *Management and the worker.* Cambridge, MA: Harvard University Press.

Rosenthal, R. (1967). Covert communication in the psychological experiment. *Psychological Bulletin, 67,* 356–367.

Rosenthal, R. (1969). Interpersonal expectations: Effects of the experimenter's hypothesis. In R. Rosenthal & R. L. Rosnow (Eds.), *Artifact in behavioral research* (pp. 187–227). New York: Academic Press.

Rosenthal, R. (1976). *Experimenter effects in behavioral research.* New York: Irvington.

Rosenthal, R. (2005). Experimenter effects. In K. Kempf-Leonard (Ed.), *Encyclopedia of social measurement* (Vol I, pp. 761–875). New York: Elsevier.

Tajfel, H., Billig, M. G., Bundy, R. P., & Flament, C. (1971). Social categorization and intergroup behavior. *European Journal of Sociology and Psychology, 1,* 149–177.

Troyer, L. (2001). Effects of protocol differences on the study of status and social influence. *Current Research in Social Psychology. 16,* 182–204. http://www.uiowa.edu/~gpproc.

Troyer, L. (2007). Technological issues related to experiments. In M. Webster & J.Sell (Eds.), *Laboratory experiments in the social sciences* (pp. 173–191). Burlington, MA: Elsevier.

Walker, H. A., & Cohen, B. P. (1985). Scope statements: Imperatives for evaluating theory. *American Sociological Review, 50,* 288–301.

Wallbott, H. G. (1983). The instrument. In P. W. Dowrick & S. J. Biggs (Eds.), *Using video: Psychological and social implications* (pp. 73–87). Chichester, England: John Wiley & Sons.

PART III

Experiments Across the Social Sciences

Part 3 of the book describes experimental situations developed for particular cases, such as differences between applied and basic research purposes; designs common in economics, political science, and sociology; development and uses of two widely used experimental designs; and other issues. Readers from the different social science disciplines may find new ideas in these chapters as they learn what their colleagues are doing, either the same as or differently from them.

Chapter 13, by James E. Driskell and Jennifer King, describes experimental research for applied purposes, focusing on differences from (and some similarities to) the more common experiments for basic research purposes. Everything from seeking funding through design of the experiments to presenting the results may differ in applied experiments, and this chapter provides details on all the ways they differ. As with other chapters, these authors seek to provide useful guides for researchers doing, or who might want to do, applied experimental research. At the same time, this chapter can help even life-long basic researchers to understand the value of applied experimental research methods and what they contribute to social science knowledge.

Chapter 14, by Joseph Berger, describes his and his colleagues of over half a century creating and developing an experimental situation used by large numbers of social scientists worldwide to study status and expectation processes. Besides providing a fascinating narrative in the history of social science, this chapter identifies design considerations and methods that are relevant to any experimental design.

Chapter 15, by Linda D. Molm, describes research using social exchange theories from their outset through the present, including her own research designs and approach. Early experiments focused on two-person exchanges, but quickly moved to studies of alliance formation in social networks of differing sizes and connectivity. Social exchange has been the basis for our understanding of the effects of punishment and reward in negotiation, as well as how negotiated and reciprocal exchanges differ in their structure and outcomes. Status experiments (Chapter 14) and exchange network experiments constitute the two largest bodies of experimental research in sociology today, and Molm's chapter provides a detailed description of these experiments and the thinking behind them.

Chapter 16, by Andreas Ortmann and Giovanna Devetag, describes a particular subset of experimental economics experiments: experiments on coordination. Coordination experiments are of differing types, but all address the general problem fundamental to social action: how can organization be accomplished? Devetag and Ortmann describe the four types of coordination games or settings that have been the focus of experimental economists: pure coordination, Pareto ranked, mixed motive, and critical mass. They analyze these four settings and summarize how the experimental research has advanced the theoretical field. They also suggest future arenas for investigation.

Chapter 17, by Rick K. Wilson, describes how voting and agenda setting have been investigated experimentally within political science. The "canonical experiment" on voting in political science is one in which a group of individuals with widely disparate interests vote repeatedly in an effort to reach agreement. Wilson notes that experimental research in voting behavior shows the importance of sociological and psychological factors along with political; in this way, experimental political science may have similarities to experimental economics in showing how theories may benefit from incorporating processes previously studied in other disciplines.

Chapter 18, by Jane Sell, treats social dilemma experiments as studied from four social science perspectives: political science, economics, sociology, and psychology. These experiments study one of the oldest problems in social science—namely, how to provide benefits available to everyone (such as public radio) supported by voluntary contributions, while at the same time minimizing the problem of "free riders" who use the benefit without contributing to it. The topic has fascinated social scientists and theorists, and it also has practical and moral implications. While it is not particularly difficult to design an experiment in which actors are tempted to free ride, Sell shows that designing experiments whose results will be theoretically meaningful (and thus will add to knowledge) is much less straightforward. She describes a number of experiments investigating different aspects of the free-rider problem and provides contemporary understandings developed from those experiments.

CHAPTER 13

Conducting Applied Experimental Research

James E. Driskell
Florida Maxima Corporation

Jennifer King
Naval Research Laboratory

ABSTRACT

Applied experimental research is experimental research that applies or extends theory to an identified real-world problem with a practical outcome in mind. In this chapter, we distinguish applied experimental research from more basic theoretical research, discuss the linkages between basic and applied research, and describe practical considerations in conducting applied experimental research.

I. CONDUCTING APPLIED EXPERIMENTAL RESEARCH

The first task that we face in this chapter is to describe what we mean by the term "applied experimental research." Previous chapters have discussed "experimental research" in considerable detail. Our goal, then, is to define what the

329

descriptor "applied" implies as it describes a particular type of experimental research. This task is not as easy as one would hope. One way to approach this task is to refer to the journals in this field. In other words, how do the journals that publish applied experimental research describe this type of research? According to the *Journal of Applied Psychology*, applied research should "take ideas, findings, and models from basic research and use them to help solve problems that both psychologists and nonpsychologists care about" (Murphy, 2002, p. 1019). Applied experimental research should "enhance our understanding of behavior that has practical implications within particular contexts" (Zedeck, 2003, p. 3). According to the *Journal of Experimental Psychology: Applied*, applied experimental research constitutes "research with an applied orientation" that should bridge practically oriented problems and theory (Ackerman, 2002, p. 4). According to the *Journal of Applied Social Psychology*, applied research is research that has "applications to current problems of society" and that can "bridge the theoretical and applied areas of social research" (Baum, 2005, paragraph 1). Therefore, we have accomplished our first goal with at least some success: at a broad level, applied experimental research is research that applies or extends theory to an identified real-world problem with a practical outcome in mind.[1]

However, this picture becomes a bit more clouded when we consider the types of content areas in which applied experimental research is conducted. Again, scanning the relevant applied research journals, we find that applied experimental research may include research on perception, attention, decision making, learning, performance, health, race relations, group processes, leadership, work motivation, assessment, work stress, violence, poverty, legal issues, aging, gender, population, behavioral medicine, consumer behavior, sports, traffic and transportation, eyewitness memory, and other topics.

We feel that it would not be a particularly pleasant experience for the reader, or for the authors, to attempt to survey this broad domain of applications. Therefore, the objective of this chapter will be somewhat less ambitious. We intend to discuss applied experimental research in the context in which we are most familiar: applied social psychology. Our coverage of this topic is selective rather than comprehensive. Furthermore, in keeping with the title of this chapter, we intend our discussion to be practical rather than abstract. Thus, our focus is on practical issues involved in conducting applied experimental

[1]Cohen (1989, pp. 53–58) distinguishes basic research, applied research, and engineering. In this categorizing, basic research is oriented to producing and evaluating knowledge, applied research demonstrates the value of accepted knowledge for a practical purpose, and engineering is oriented to solving recognized problems using all available means, including well-established knowledge from basic research as well as other kinds of knowledge, such as intuitions and experience-based hunches. Webster and Whitmeyer (2001) describe somewhat different distinctions, with examples of each type. In this chapter we describe what Cohen would term "applied research and engineering." As we show, these kinds of research also develop and assess new knowledge in ways similar to those of basic research.

research in real-world contexts, such as the military. These issues include generating proposals and obtaining funding, conducting applied experimental research, and presenting research results to the user.

A. A BRIEF HISTORY OF APPLIED EXPERIMENTAL RESEARCH

We can easily trace early pioneering work in applied social science back over a century, including Triplett's (1898) field experiments to study the "dynamogenic factors" that affected work, Taylor's (1903) research for the steel industry to enhance worker productivity, Walter Dill Scott's (1903) research on advertising, Thorndike's (1903) work in education, and many others. We can further distinguish three historical periods of applied social science research. The first period is represented by these early attempts at application, in which researchers ventured out of academia to address problems of everyday life. The hope was that scientific knowledge could be applied "by enlarging its scope and making its experiments where people work and play as well as the laboratory" (Freyd, 1926, p. 314). Moreover, applied researchers hoped that their work could replace "the lore of the folk by an array of knowledge equally concrete and practical, but immeasurably wider, more accurate, more systematic, and free from personal bias (Bryan & Harter, 1899, p. 347; cited in van Strien, 1998).

Perhaps this pioneering spirit is most interestingly illustrated by the work of Hugo Münsterberg, who wrote one of the first texts referencing applied social science research in *Psychology General and Applied* (1914). Münsterberg was not always a proponent of applied research and in 1898 wrote a scathing attack on then current attempts to apply psychology to the educational system, claiming, "Our laboratory work cannot teach you anything which is of direct use to you in your work as teachers" (Münsterberg, 1898, p. 166). This statement is characteristic of Münsterberg, as he seemed to be a polarizing individual in general. Benjamin (2006) has noted that although Münsterberg was one of the leading applied psychologists of the time, he was also "one of the most despised individuals in America" (p. 414), due partly to his personal egotism and pomposity but primarily to his prominence as a leading proponent of German involvement in World War I. In any case, Münsterberg later reversed his position on the value of applied research, and in fact believed so strongly in the importance of applied research that he suggested that the government establish experimental research stations across the country, similar to the agricultural extension stations that were being introduced at that time. Although agricultural extension stations were established in every county or district in every state, the case for applied social science research stations was probably a bit harder to sell.

A second historical period of applied social science research is represented by a measured backlash or at least professional ambivalence regarding applied research. The early attempts to apply principles of social science to the workplace were admittedly "seat-of-the-pants" efforts and, as van Strien (1998) has noted, "'Bona fide' psychologists, as academics saw themselves, looked on the crude methods of nonacademic practitioners with great reservation" (p. 220). Again, reference to Munsterberg is illustrative of the ambivalence regarding applied research during this period, as he both conducted applied research and also railed against applied research that lacked strong scientific foundation. Yet as Benjamin (2006) has noted, Münsterberg himself rarely abstained from commenting (or publishing) on almost any topic whatsoever, and was broadly criticized by other scientists for forsaking scientific standards for the sensationalism and faddism of popular pseudoscience.

Moreover, during this period, academic scientists were particularly concerned with distinguishing themselves from the phrenologists, psychics, mediums, palm readers, and spiritualists that had captured the public imagination. Therefore, although applied scientists were to attack the problems of the day with some success, there was a stigma attached to applied research and an ambivalence regarding those who would forsake the academic corridors for the outside world that continues to this day. Although we do not wish to overstate the case, there exists in many instances a bias towards a "pure" academic culture in the social sciences that views applied research as suspect.

A third historical period paralleled the applied research activities that were initiated during World War I and conducted intensively during World War II. It would not be unfair to suggest that the "person in the street" in the early part of the twentieth century viewed the social science enterprise with at least some degree of indifference. In fact, when social scientists first approached the U.S. War Department in 1917 to propose research to support the war effort, the response was initially lukewarm (Driskell & Olmstead, 1989). However, the applied research that was conducted during WWI in areas such as selection, training, and motivation constituted the first large-scale effort in the social sciences to apply principles derived from the academic laboratory to address society's needs. In fact, the military provided a particularly fertile ground for the examination and application of scientific theories.

Thus, psychologists were able to study learning, human performance, and morale; sociologists could study conflict, race relations, and the military organization; and political scientists could study international relations and policy. One research project called for the assessment and testing of over 4,000 recruits arriving at a military camp. The statistical team for this project included such luminaries as E. L. Thorndike, L. L. Thurstone, and A. S. Otis, and the cost to the War Department for this work was less than $2,500 (Driskell & Olmstead, 1989). As the nation faced World War II, military program managers were able to draw on the relations formed with the scientific community in World War I, and the subsequent applied research conducted during World War II had a tremendous impact in both (1) contributing to the war effort, and (2) legitimating and demonstrating the value of applied social science research.

This success is best illustrated by reference to a classic wartime research program that became known as *The American Soldier* research, summarized in a four-volume series entitled *Studies in Social Psychology in World War II* (see Stouffer *et al.,* 1949). Contributors to this research included Samuel Stouffer, Robin Williams, Robert K. Merton, Edward Shils, Paul Lazarsfeld, Brewster Smith, Rensis Likert, Arthur Lumsdaine, Irving Janis, Carl Hovland, Quinn McNemar, Louis Guttman, and many others. Stouffer *et al.* summarized this work as a "mine of data, perhaps unparalleled in magnitude in the history of any single research enterprise in social psychology" (p. 29). Certainly, the broader contributions to social science made by these leading scholars was as impressive, including the work of Stouffer and Williams on relative deprivation; McNemar, Guttman, and Likert on scaling; Lazarsfeld on quantitative methods; and Merton (with Alice Rossi) on reference groups. In brief, the social and behavioral sciences stepped up to the plate in World War II and hit a home run. The successful application of social science knowledge in addressing practical wartime needs during World War II not only established this field in the public eye as having considerable practical value, but also contributed significantly to the subsequent postwar growth and development of the field as a whole.

II. BASIC AND APPLIED RESEARCH

Although we attempted to neatly define applied experimental research in the opening paragraphs of this chapter (and declared some measure of success in doing so), in practice we find that this type of work is not so easily categorized. In the following, we discuss the distinction between basic and applied research, and then briefly address the role of theory in applied research.

To address the topic of basic versus applied research, we first turn to a seminal figure in the development of science policy and the creation of a scientific infrastructure in the United States, Vannevar Bush. To briefly summarize an illustrious career, Bush was an MIT engineer who became president of the Carnegie Institution (now Carnegie Mellon University), and then director of the federal Office of Scientific Research and Development during World War II. After considerable success leading the wartime research efforts, Bush was approached by President Roosevelt at the conclusion of the war with the request to provide a recommendation on how the lessons learned and successes achieved could be continued in peacetime. Bush's response, summarized in *Science: The Endless*

Frontier (Bush, 1945), became the blueprint for the establishment of the National Science Foundation (NSF). In this document, Bush wrote:

> Basic research is performed without thought of practical needs. It results in general knowledge and understanding of nature and its laws. The general knowledge provides the means of answering a large number of important practical problems, though it may not give a complete specific answer to any one of them. The function of applied research is to provide such complete answers.... Basic research leads to new knowledge. It provides scientific capital. It creates the fund from which the practical applications of knowledge must be drawn (Bush, 1945, "The Importance of Basic Research," paragraph 1).

This is an exceptionally eloquent statement, but as is often the case, its elegance obscures some further qualifications regarding the distinction between basic and applied research. First, the terms *basic research* and *applied research* refer to ideal types. In its purest form, basic research involves the testing of theory for the purpose of understanding fundamental processes. The benefits of basic research are long term, and in many cases, the societal payoff may be in areas that were not even envisioned by the original basic researchers. In its purest form, applied research involves applying theory to identified real-world problems. The benefits of applied research are short term, and the results have an immediate and identified use or application. However, there is a considerable amount of research that is carried out between these two poles. Rarely are basic researchers not at least cognizant of the broader practical implications of their work, and rarely do applied researchers conduct their work without concern for how their results may extend or elaborate theory.

Moreover, there is no one criterion that separates basic from applied research. Some have noted that the motivation of the researcher is a primary factor. Basic researchers will see the primary value of their work as building theory or expanding a body of knowledge, whereas applied researchers will see the primary value of their work as solving a real-world problem. Others have argued that a primary factor separating basic from applied research is the temporal nature of the contribution of the research. The practical payoff from basic research may be years away or simply may not be seen as a significant or relevant question, whereas the practical payoff from applied research is its *raison d'être*, and the results are intended to be put into use in the short term in a specific context.

It is further useful to note that even the most basic research should have identifiable societal implications, especially from a political standpoint. For example, NSF funds basic research, so it is the primary funding source for basic research in the social sciences, though other agencies sometimes fund basic research also. Although NSF focuses on basic research at the frontiers of knowledge, some NSF-funded results are immediately useful, and NSF often goes to great length to tout these as success studies. Emphasizing these "discoveries" is one way to demonstrate to legislators and other interested parties that the national investment in research and development (R&D) is being put to good use. The fact that research, both basic and applied, must meet some foreseeable national good recalls a famous anecdote: when the British physicist Faraday was asked by the Finance Minister Gladstone in the 1850s whether electricity had any practical value, Faraday replied, "One day Sir you may tax it." It is quite likely that this reply resulted in a well-funded program of research as well as a delighted politician.

In one of our favorite quotes, Melton (1952) facetiously labeled basic research as "what I want to do" and applied research as "what someone else wants me to do" (p. 134). Although this may not be as elegant a proclamation as that provided by Vannevar Bush, there is certainly an element of truth in this statement. In practical terms, there is a clear and broad distinction between research designed to test theory and research designed to apply theory. Moreover, as we note in a subsequent section of this chapter, research designed to test theory and research designed to apply theory are conducted in a different manner. Nevertheless, basic and applied research go hand in hand, with applied research serving as a bridge between basic research and real-world applications.

III. THE ROLE OF THEORY IN APPLIED EXPERIMENTAL RESEARCH

Popular culture draws a sharp distinction between theory and application, depicting theorists as slightly befuddled intellectuals and applied researchers as unprincipled mercenaries. To counter this perspective, we offer psychologist Kurt Lewin's famous statement regarding theory and application:

> Many psychologists working in an applied field are keenly aware of the need for close cooperation between theoretical and applied psychology. This can be accomplished in psychology, as it has been accomplished in physics, if the theorist does not look toward applied problems with highbrow aversion or with a fear of social problems and if the applied psychologist realizes that there is nothing so practical as a good theory (1944/1951, p. 169).

This statement is most informative in the present context in that it evokes the close interplay between theory and application. As Vannevar Bush so pleasingly stated, basic research and theory provide the fund or capital from which practical applications can be drawn. Joseph Berger (1988) has noted that a theory is grounded in its applications and that these applications further shape the theory in its development. Thus, the relationship between theory and application is reciprocal.

On the one hand, applied research takes the concepts and principles from basic theory and uses them to solve real-world problems. Thus, applied research is guided by theory, and applied research that is devoid of theory is itself indistinguishable from the work of the charlatan or seer. On the other hand, theory can grow and develop because the results of applied research often present "problems" that the current theory cannot explain. Moreover, one criterion that determines a theory's value is the extent to which it can be applied usefully in various settings. Thus, in this sense, theory both guides and follows applied research, and applied research both guides and follows theory.

IV. DEVELOPING A PROPOSAL AND GENERATING FUNDING FOR APPLIED EXPERIMENTAL RESEARCH

Given our interests in addressing practical aspects of conducting applied experimental research, it does not get any more practical than asking: "Who funds applied research in the social sciences?" The overall level of federal support for applied research in the behavioral and social sciences in FY 2005 was approximately $1.3 billion, according to summary data gathered by the National Science Foundation (2006). Federal funding categories are separate for psychology and the social sciences, so we will summarize these data separately.

Within the field of psychology, funding for applied research comes primarily from the Department of Health and Human Services (providing roughly 77% of applied research funding, approximately $446 million) and the Department of Defense (DOD; 13%, $77 million). Other major contributors include the Department of Transportation (3%, $18 million), Department of Veterans Affairs (2%, $14 million), and the National Aeronautics and Space Administration (NASA; 4%, $20 million).

Within the social sciences, the largest funding sources of applied research include the Department of Health and Human Services (32%, $195 million), the Department of Education (25%, $154 million), and the Department of Agriculture (18%, $114 million). Other major contributors include the Agency for International Development (12%, $73 million), the Department of Labor (3%, $20 million), and the DOD (2%, $11 million).

However, those numbers indicate general funding levels for applied research, and do not necessarily reflect the extent or sources of funding for applied *experimental* research in psychology and the social sciences. There are no existing data of which we are aware that would provide this more specific information, so we again adopt the strategy of perusing relevant applied experimental research journals to scan authors' acknowledgments of funding support. It would seem that a considerable amount of funding for applied experimental research comes from federal agencies such as the Department of Health and Human Services, DOD, and NSF, along with a smattering of private funding sources. It may be surprising to see NSF as a significant funding source for *applied* experimental research, given NSF's primary basic research mission.

However, Gerstein, Luce, Smelser, and Sperlich (1988) have noted that whether research is termed basic or applied may have more to do with the funding agency's stated mission rather than some intrinsic aspect of the research being supported. In other words, even with a basic research mission, it is not surprising that NSF in practice supports a good bit of theoretically focused empirical research that has an applied orientation.

To a considerable extent, seeking funding for "theoretically focused empirical research with an applied orientation" from NSF does not differ from seeking funding for "theoretically focused basic research" from NSF. (Webster, in Chapter 8 of this volume, discusses seeking funding for basic research.) Therefore, we will focus the following discussion on funding for applied experimental research from DOD, for two reasons. First, DOD provides approximately $87 million in funding for applied research in the social and behavioral sciences annually (National Science Foundation, 2006), a substantial amount by any standard. Second, the authors have had considerable experience with applied DOD research, and one author (Driskell) has served on both sides of the fence—as an applied researcher and recipient of multiple research contracts as well as a government researcher/program manager responsible for funding such research.

The first thing that must be considered is that DOD is an applied organization. Although research performed for the military is often indistinguishable from that performed in the university experimental laboratory, military research has a special characteristic: it is always subject to "audit"—that audit being "how does this contribute to improved military operations?" Research is evaluated by military sponsors in terms of how well it will contribute to solving operational needs. In other words, most military research starts with a military requirement, and the end product of most research is an application (near or long term) to an identified military problem.

In seeking DOD research funding, you may deal with two types of organizations: *funding offices* (such as the Office of Naval Research) and *performing organizations* (various DOD research laboratories). The funding offices, most of which are located in the Washington, D.C., area, fund research directly. Typically, a broad agency announcement (BAA) will be issued periodically that describes the types of research that the organization is interested in funding, and research proposals are submitted according to the instructions therein. This approach represents a "formal" proposal mechanism, and the procedure of proposal preparation and submission is similar to that of other federal agencies, including NSF.

The performing organizations, which include various government laboratories and research centers scattered across the country, receive their funding from the funding offices to support both in-house research as well as external research. These organizations may also issue announcements describing their research interests, but often the procedure of proposal preparation and submission is more informal. For example, it is likely that an astute researcher seeking funding may scan research reports in an online database such as the Defense Technical Information Center's Scientific and Technical Information Network (STINET) service (http://stinet.dtic.mil/), find a government laboratory researcher doing related research, make contact, find out that researcher's interests and needs, and develop funded research out of that contact.

It may be worthwhile, in seeking funding, to know thy target. Typically, the government laboratory researcher/program manager is a Ph.D-level scientist who has gone directly from graduate school to employment in the research organization, and who has extensive in-depth knowledge of his or her relevant applied content area. The government researcher usually serves in a dual role, serving as an in-house researcher as well as a program manager with budget and funding authority. For an applied scientist seeking research funding, the government program manager can serve as your person "on the inside," familiar with the requirements of their organization and current research needs. From the perspective of the government program manager, the applied scientist represents someone who can help solve a problem.

As noted, there is considerable money invested in research each year, and the money draws the attention of some who are more interested in getting the dollars than in using them to conduct high-quality research. Government researchers are usually alert to such individuals. From experience, we note that this government researcher often has to endure two such types of persons seeking research funding. The first is the *businessperson/generalist*. This person may represent a private company (often staffed largely with master's level personnel) that can perform a variety of studies or analyses for the government organization, although it would typically not get involved in experimental research. The businessperson/generalist is a jack of all trades and master of none, but can provide resources and personnel to perform a number of lower level research tasks.

The second type is the *academic purist*. This represents an ivory tower scientist with the mindset of "how can you fund my work?" rather than "how can I apply my expertise to address your needs or interests?" In this case, the academician is interested in doing his or her own work without oversight, but wants the applied government scientist to fund it. This often does not go over very well. The third type of person is the one that the government researcher hopes to find, and often does: the *applied scientist*. The applied scientist offers expertise in a specific content area, is interested in finding out the interests and needs of the government researcher and the agency's mission, and is flexible and adaptive in adapting his or her expertise to address these requirements.

Certainly these types are caricatures, but they represent three different ways of interacting with the potential research sponsor: (1) "I can do anything"; (2) I'll do what I want to do, but I want you to pay for it"; or (3) "I have an ongoing program of research that I believe will be of interest to you, but first tell me about your research needs." It may be worthwhile for the applied researcher to consider the manner in which he or she chooses to approach a potential applied funding source. It is our experience that government sponsors welcome well-intentioned applied researchers who are willing to listen to the sponsor's needs and apply their expertise to address these problems.

In discussing the informal interaction that often occurs between the research sponsor and the applied researcher in discussing potential research, we further note that the applied researcher must be flexible in adapting his or her research interests to the requirements and interests of the sponsor. For illustration, let us consider the field of group dynamics, broadly defined. Group researchers have often drawn on an input-process-output model, as shown in Figure 1, to organize the variables that are relevant to group interaction (see Hackman & Morris, 1975). A basic researcher interested in group status processes, for example, may conduct research with a goal of developing a better fundamental understanding of status processes. However, an applied research sponsor will, almost by definition, be interested in performance outcomes also, as represented in the rightmost column of Figure 1.

Let us say that an applied researcher is interested in, for example, the role of status in determining challenging and monitoring behaviors in an aviation aircrew (referring to the difficulties first officers [copilots] often experience in

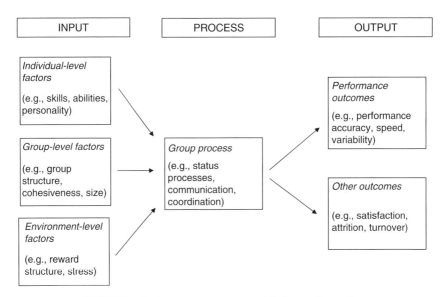

FIGURE 1   An input-process-output model of group interaction.

issuing challenges or warnings to captains; see Milanovich, Driskell, Stout, & Salas, 1998). However, the research sponsor who may be interested in this topic will also be interested in two further questions. The first question is how this phenomenon affects performance outcomes. An applied research sponsor interested in studying groups is ultimately going to be concerned with group outcomes, such as how accurately groups perform or how quickly groups perform. A second question is how this phenomenon is affected by other contextual factors that are important in the real-world setting of interest, such as the type of group examined, the type of task undertaken, or how effects are sustained over time. It is the nature of applied research that any single phenomenon must be examined in the broader context in which it occurs. Our intent in this discussion is not to discourage anyone from entering the challenging world of applied experimental research, but instead to note that the singular focus that serves the basic researcher so well may have to be widened as the applied researcher examines a specific phenomenon in context in the real world.

You are likely to hear the potential research sponsor describe research programs in numerical terms, as being 6.1 (six one), 6.2 (six two), or 6.3 (six three) research. Federal research and development is divided into separate activities denoted by account numbers. The types most likely to be encountered by the applied social scientist include basic research (6.1), applied research (6.2), and advanced technology development (6.3). The basic research (6.1) category consists of scientific studies that develop the knowledge base for subsequent research or application. The applied research (6.2) category consists of R&D oriented toward a specific military problem. It includes the exploration of new technologies or concepts that hold promise for application to specific military needs. The advanced technology development (6.3) category consists of advanced development and feasibility demonstrations in operational settings.

It is important to note that the term *basic research* has a somewhat different interpretation in an applied organization such as DOD than the term has in an academic environment. Even basic research is undertaken with an applied orientation and with a goal of transition to 6.2 and 6.3 R&D, and toward later application. In other words, the military does not fund research to enhance the state of the art; it funds research to enhance the state of the military.

The proposal. It is not clear or explicit what criteria government researchers use on an informal basis when evaluating potential new research efforts. Certainly, the proposed research must be scientifically sound, it must be relevant to some identified need and capable of being defended to others in the organization and in higher offices, and it must have a payoff or product that is seen as valuable. If the proposer is successful in making this case during preliminary discussions with the government sponsor, the proposer may be invited to submit a white paper. A white paper is a short, three- to four-page synopsis of a research idea. A white paper also represents a decision point—if a white paper captures the attention of the government program manager and other decision makers in the organization, the proposer may be asked to submit a more detailed or full proposal. If not, the white paper ends up in one of the government's many file drawers. Therefore, it is incumbent on the applied scientist to be able to capture and present complex ideas in a nutshell, which is a difficult task for many of us who are more accustomed to presenting complex ideas in an elaborate and comprehensive fashion.

Another document that is often requested with the submission of a white paper is a "quad chart," which is a single PowerPoint slide or page partitioned into four quarters. Each quadrant has a heading, and the quadrant headings may include (1) title and proposer, (2) technical approach, (3) operational requirement or need addressed, and (4) research cost and schedule. Each quadrant may have room for three or four short sentences. If you think it is difficult to put a complex research proposal into four pages, try putting it into four sentences! At any rate, a quad chart may be the only thing a potential decision maker scans, so it also represents a decision point.

Based on our experience, we would hazard a guess that a midlevel program manager evaluating multiple potential new research starts may consider the operational need first—does the quad chart or white paper identify an operational need that is of current importance?—and then consider the technical approach or research proposed to meet that need. Typically, when developing a white paper, the research proposer can rely upon the government sponsor's knowledge of current requirements to provide input to ensure that the white paper addresses these issues. Again, it is entirely appropriate, and certainly in the researcher's interests, to contact the government program manager to discuss these and other programmatic issues.

It is also the case that some research topics are at any point in time more glamorous than others, or are hot topics in the military R&D community, and a proposal that at least references these topics may be more successful than one that does not. It might be a stretch to tie, for example, proposed research on group status processes into current interests in "network-centric operations," but it might be advantageous to do so if possible. At the least, the appropriate use of relevant buzz words may indicate to the reader or evaluator that the proposer is cognizant of current interests and requirements. We do not intend to be cynical in making this suggestion—the applied researcher is not expected to be a military subject-matter expert, but, with a little bit of background research, should be knowledgeable enough to "know the lingo" and be familiar with the general research agenda of the funding organization.

A further consideration in developing a white paper or proposal is the identification of a potential user. Applied research is research that is intended to be put to use. Ideally, each research project will have a near-term user identified, even to the extent of having someone in authority in the field available who can pick up the phone and say, "Yes, I can put this research into use." Regrettably, this is not often the case in the social sciences; compared to a scientist who is developing a new computer chip, our research rarely gets the end user leaping out of his or her seat. Although the ideal is to have some identified "transition" for research results in place, this is often not the case for experimental research, and may be more the norm for advanced development efforts that are ready to be fielded. For applied experimental research, it is usually sufficient that the researcher and sponsor simply identify potential users or distinguish clear implications of one's research to establish a foundation for later transition of research results.

If your proposal is selected for award, you will be issued a contract or grant. In terms of how the research will be carried out, there is little distinction between whether research is conducted via a contract or a grant. There are, however, federal regulations that distinguish between the two. In general, a grant is issued by a federal agency to carry out a public purpose or stimulate a specific desired activity. A contract is issued by a federal agency to acquire goods and services for its direct benefit and use. Therefore, basic research is most often carried out via a grant, and applied research is often procured via a contract.

In practice, there are some distinctions between the contract and grant mechanisms. Contracts are essentially procurement instruments for obtaining goods or services, so they include a specific statement of work describing the services required, and they include a set of milestones and deliverables for accomplishing this work. Grants are financial support for a public purpose; the statement of work is more loosely defined and the milestones and deliverables are less explicit. In most cases, an applied researcher would be delighted to receive either a grant or contract to support proposed research. At any rate, whether a grant or contract is awarded is determined by the regulations of the funding agency and is largely non-negotiable.

V. CONDUCTING APPLIED EXPERIMENTAL RESEARCH

Unlike basic research, which is essentially a hands-off endeavor from the point of view of the research sponsor, applied research implies some type of relationship with the sponsor. At the least, applied research implies a reciprocal relationship between the applied researcher, who offers expertise, and the research sponsor, who offers an opportunity or requirement for application. In practice, this relationship can range from benign neglect to a desire on the part of the sponsor to be involved actively in the research. However, the research sponsor's involvement often takes the middle ground. It is often the case that the government program manager who is sponsoring and funding a particular project has shared interests with the applied researcher, but is too busy managing multiple projects to become heavily involved in any particular experimental study. Nevertheless, in applied research, the government sponsor is often a valuable resource in securing entrée to research sites, obtaining required permissions, and ensuring research participation.

In most cases, applied experimental research consists of research conducted in an experimental laboratory setting for applied purposes. Many have argued that the primary purpose of the experimental laboratory is to test theory (see Driskell & Salas, 1992; Mook, 1983; Webster & Kervin, 1971; Webster & Sell, Chapter 1, this volume). Moreover, one defining characteristic of the experimental laboratory, its artificiality, is advantageous in that it allows the researcher to create controlled conditions most conducive to testing theory. That is, the greater the artificiality of the experimental setting in terms of isolating only those variables relevant to the hypothesis being tested, the greater confidence we have that we have provided a clear test of the theory.

Experimental research to test theory, as we have noted earlier, is a common type of basic research. However, applied experimental research often has as its goal *applying theory* to a specific setting. To the extent that basic research designed to test theory seeks control over actors, tasks, and context (see Ilgen, 1986), applied research seeks to apply theory in situations in which we are concerned with specific actors, tasks, or contexts. Berger (1988) has noted that this type of applied research can serve to instantiate theory, establishing the extent of its application in various real-world contexts. Therefore, applied experimental research is often conducted to apply theory to address real-world problems by incorporating actors, tasks, or contexts specific to that setting. This type of applied research can be conducted (1) in academic laboratories, often incorporating realistic simulations or task abstractions; (2) in an experimental setting in the target or user community; or (3) in the field, utilizing the actual tasks that are performed on a daily basis.

These three types of applied research, in order, represent an increasing level of realism in terms of actors, tasks, and context reflecting the real-word environment of interest, as well as a decreasing level of control in experimentation. Howell (1998) has noted that the general research strategy of conducting applied experimental research using realistic simulations or abstractions of real-world tasks can serve as a bridge between laboratory experimentation on the "basic" side and field research on the "applied" side.

The following describes one example of this type of applied experimental research conducted in the user community. Driskell and Radtke (2003) conducted empirical research to examine the effect of gesture on speech production and comprehension. Previous researchers had hypothesized that conversational hand gestures improve communication by enhancing speech, conveying information that augments the information provided by the speech channel (Beattie & Shovelton, 1999; Kendon, 1983). However, others had hypothesized that the primary function of gesture is to assist the speaker in formulating speech by aiding the retrieval of elusive words from lexical memory (Krauss, 1998; Krauss, Chen, & Chawla, 1996). Moreover, these researchers argued that speech production plays a mediating role in the observed relationship between gesture and comprehension. That is, those studies finding that gestures enhance communication may have observed this result simply because they did not control for the possibility that speakers who are allowed to gesture produce more effective speech. From a practical standpoint, the research sponsor was interested in computer-mediated communication and the question of whether one should be concerned with designing a video interface to display gestural information in computer-mediated communication when some argue that gesture provides little communicative information in the first place.

We designed an experiment as a basic analog of technical communications to represent a setting in which the speaker knows something and is trying to convey it to the listener. We conducted the experiment at the sponsor's site using on-site personnel as research participants, varied the extent to which participants were allowed to gesture on the laboratory task, and developed measures of speech production and listener comprehension. The results indicated that gestures enhanced both listener comprehension and speech production, and that gestures had a direct effect on listener comprehension independent of the effects gestures had on speech production. Further, the research implied that the ideal visual field for a computer-mediated communication system should include information communicated via gestures. Although this research addressed a very specific question, it has implications for a variety of instances in which the scope conditions are the same. For example, in both the *Columbia* and the *Challenger* disasters, communication problems were central, and these problems concerned the lack of video connections that could have ensured more effective communication (see Vaughan, 1996; Langewiesche, 2003).

The purpose of this research was to apply hypotheses derived from communications theory to a military computer-mediated communications environment. We attempted to incorporate greater realism in this application in terms of the actors (using military personnel as research participants), the task (developing an approximation of the real-world task in the laboratory), and the context (defining the experimental situation in terms of an operational setting the participants would care about). We use this example as an illustration of applied experimental research conducted in the user community, and to set the stage for addressing some of the unique practical problems that can occur in this setting. Applied experimental research that is conducted at a user or real-world site differs in several respects from typical experimental research.

A. Recruiting Research Participants

The first required task is to find a user community that is willing to participate in the research. Sometimes this is already established and a user community is directly involved at the initiation of the research program.
However, as we noted earlier, often an applied research program does not
have an established user community at its beck and call, but is conducted
because it has clear implications for application. In this case, the government
sponsor may assist in locating a community that will cooperate by providing
participants and a setting for the research. For DOD, this is typically a military community or field organization convenient to the researchers that will
participate out of a sense of organizational support, although it may not
typically receive any direct or immediate payoff from the research itself.
Research is conducted at the host organization on a not-to-interfere basis.
This means that the researcher will often receive an unused classroom or
offices for laboratory space and can request that research participants be
scheduled at desired times, although this schedule must not interfere with
the day-to-day operations of the organization (which may mean scheduling
on evenings and weekends).

Typically, the researcher is assigned a point-of-contact person whose job is
to ensure research participants are available according to the research schedule.
In some cases, participants may not show up because, on any given day, organizational requirements or unexpected "fire drills" may supersede the research
schedule. This must simply be accepted. In some cases, such "no-shows" can
occur because the point of contact is too busy to track participants. This must
be dealt with diplomatically. Often, scheduling can go *too* precisely—the point
of contact may send enough participants for an entire day's schedule, all showing up at 8:00 a.m. and instructed to wait their turn. In this case, the researcher
must do the best he or she can to ensure that the participants are treated
with appropriate care and respect.

B. Laboratory Procedures

In contrast to the relative homogeneity of an undergraduate student research
pool, research populations in real-world organizations can vary considerably.
In the military, the participant population may range from new recruits with
a minimum of education or experience to high-level officers with advanced education and training. Care must be taken to develop laboratory procedures that
are appropriate for the participants. Furthermore, because of the heterogeneity of the population, it is especially important to ensure random assignment
of participants to experimental conditions. Unless precautions are taken,

it may be likely that one week, participants will come from a mechanical division, the next week from a military police division, and so on.

In applied research, it is often easier to garner research participants' interest and motivation because you are using a research task and an environment that is important to them. Therefore, it is important to establish clearly in the experimental instructions not only what you are doing and why you are doing it, but also why this research is important for the military. In general, we have found that military research participants are genuinely interested in research that is being done in "their" world, and are quite motivated to take part. On the other hand, one problem with using a research task that often only approximates the real-world task of interest is that the participants know from experience that this is not the exact task that they do in the real world, and they will let you know this in no uncertain terms.

Finally, in debriefing participants, the same theme of why the research is valuable to them and the organization should be emphasized. The research participants are part of a larger organization and, especially in the case of the military, it is an organization to which there is a high level of individual commitment. Thus, participants are particularly interested in the value of the research for the military. This emphasis on relevance can help avoid *contamination* of the research population, which refers to individual subjects telling others about critical features of the experimental situation, rendering these others unfit for participation in the study. Contamination of future research participants is less likely to be an issue if debriefing procedures are comprehensive and emphasize why the organization would be harmed if future participants are told of experimental details in advance.

C. ETHICAL CONCERNS

Ethical principles require that researchers obtain the informed consent of research participants. Informed consent includes the right to decline to participate in a research study. In the case of applied research conducted for the military, the host organization is a federal agency with its own regulations, which include the authority to have military personnel participate in research activities. In other words, research participants are generally "volunteered" by superiors to take part in a given research study. Nevertheless, the applied researcher is not relieved of ethical responsibility to ensure the welfare of research participants. As Haverkamp (2005) has noted, the researcher has an obligation to promote the best interests of the participant, and this obligation cannot be set aside by the participant's consent, or in this case, by the organization's authority.

Therefore, it is good practice to obtain informed consent from research participants regardless of whether their participation is mandated by the organization. Given that the individual has been directed to participate in this activity, the researcher does not have the authority to release him or her from this obligation. What we have done in practice is to provide an alternative for those who wish to decline to participate in a study, such as allowing them to sit in the waiting room during the scheduled period and then returning to duty as previously instructed. This helps to achieve two goals. First, those who do not wish to participate can decline and still meet the requirements of their organization. Second, it ensures that those who do participate are more likely to meet basic scope conditions of the research study, such as caring about the outcome of the task.

VI. PRESENTING RESEARCH RESULTS

When doing basic research, publishing the results of one's work is part of the scientific enterprise. When doing applied research, the motivation to publish can be a bit more prosaic. We noted earlier that applied research is research meant to be put into use. It has a strong product orientation, and in fact one of the most commonly asked questions from sponsors may be, "What is the product of this research?" One product of applied research is its findings, which may be disseminated in presentations or publications. Sponsor-requested presentations may occur at several points over the course of a research program. In fact, some research contracts or grants are funded incrementally, and continuation is dependent on demonstrating satisfactory results at yearly program reviews.

Research presentations provide one way for the government sponsor to garner or maintain project support from various users or stakeholders, who may include not only higher level program managers in their own organization, but also higher level program managers in the funding office or funding source, and representatives from the actual end-user organizations. The applied researcher presenting his or her research results to this audience may quickly learn that there may be various "agendas" in the room, only one small part of which may involve the results of his or her research. Again, guidance and support from the government researcher/sponsor is useful in this setting.

Research projects may also result in publications in various outlets. Research contracts or grants typically call for a final technical report to be submitted to the funding organization. The sponsor may also request that this report be published as an in-house technical report. In practice, this means that the final report that is submitted at the end of a contract will go through several more revisions and approvals at various levels within the organization, and eventually become published as an official technical report. The technical report will be disseminated through a distribution list (including the various stakeholders mentioned previously) and through the online repository, the Defense Technical Information Center's STINET. A technical report is a comprehensive write-up of research results, but also should include guidelines or recommendations for applying the research results to the research problem or application. Remember, applied research starts with a problem and ends with recommendations for addressing that problem.

Research results may also be published in scientific journals, including those noted in the introduction of this chapter that encourage applied experimental submissions. However, note that publication or other dissemination of results outside the government may require prior government approval. Any publication restrictions that may apply will be documented in the research contract or grant. This approval, if required, typically means a review of a draft manuscript by someone in the government organization prior to submission for publication. Approval is typically a formality, given that the research is unclassified rather than secret (applied experimental research in the social sciences is rarely classified as secret) and given that secrecy regarding the subject matter of the research is not deemed to be in the nation's interests. Therefore, although publication restrictions may be written into a contract or grant requiring prior government approval before publication, it would be extremely rare for publication to be disallowed. Finally, as we noted earlier, applied research implies a partnership between the applied researcher and project sponsor. In cases in which there may be an active and substantial contribution to a research project by the government program manager/researcher, it may be useful to establish an understanding for shared publication.

VII. SUMMARY

We have attempted to define applied experimental research and describe some of the common practical issues involved in conducting applied experimental research. We now consider if we were in Faraday's position and were approached by Gladstone or his modern equivalent with the question, "What is the value of applied experimental research?" Our answer, in keeping with the tone of this chapter, would address both the practical and theoretical value of applied experimental research. At a practical level, applied experimental research can solve problems that people care about. Howell (1998) notes that "when properly planned, theoretically grounded, carefully managed, competently executed, and adequately funded, behavioral science is as capable of yielding solutions to significant real-world problems as are the physical or biological or any other sciences" (p. 424).

One example of a successful practical application of applied experimental research involved a commercial aviation accident that occurred on July 19,

1989. United Airlines Flight 232 experienced the failure of an engine and complete loss of hydraulic pressure, leaving the airplane with no flight controls. The airplane crashed during an attempted landing at Sioux City, Iowa, and 111 of the 296 passengers and crew members were fatally injured. In this case, the fact that the flight crew was able to bring the airplane down under some measure of control was seen by most experts as just short of a miracle. Moreover, the National Transportation Safety Board (1990) noted in its investigation report that the flight crew of United 232 had recently received crew resource management (CRM) training, and this may have contributed to the outcome in which 185 lives were saved. Crew resource management training was developed from research applying principles drawn from small-group research to the aircrew environment (see Salas, Bowers, & Edens, 2001). Thus, applied experimental research can result in practical applications that make a difference in people's lives.

At a theoretical level, applied experimental research serves as a bridge between theory and application. It serves as a major component of Berger's (1988) concept of a *theoretical research program*, consisting of theory, theoretical research to extend and elaborate theory, and applied research that grounds and further drives theory. It has been implicated in cyclic models that link theory, testing, and application in a recurring cycle or loop. As Howell (1998) has noted, science is cumulative, and applied experimental research can lead to further refinements, further applications, and a deeper understanding of social phenomena.

REFERENCES

Ackerman, P. L. (2002). Editorial. *Journal of Experimental Psychology: Applied, 8,* 3–5.

Baum, A. (2005). Author guidelines. *Journal of Applied Social Psychology.* Available: http://www. blackwellpublishing.com/submit.asp?ref=0021-9029.

Beattie, G., & Shovelton, H. (1999). Mapping the range of information contained in the iconic hand gestures that accompany spontaneous speech. *Journal of Language & Social Psychology, 18,* 438–462.

Benjamin, L. T. (2006). Hugo Münsterberg's attack on the application of scientific psychology. *Journal of Applied Psychology, 91,* 414–425.

Berger, J. (1988). Directions in expectation states research. In M. Webster, Jr., & M. Foschi (Eds.), *Status generalization: New theory and research* (pp. 450–474). Stanford, CA: Stanford University Press.

Bryan, W. L., & Harter, N. (1899). Studies on the telegraphic language: The acquisition of a hierarchy of habits. *Psychological Review, 6,* 346–375.

Bush, V. (1945). Science: The endless frontier. A report to the president by Vannevar Bush, director of the Office of Scientific Research and Development. Washington DC: U.S. Government Printing Office. Available: http://www.nsf.gov/about/history/vbush1945.htm.

Cohen, Bernard P. (1989). *Developing sociological knowledge: Theory and method* (2nd ed.). Chicago: Nelson–Hall.

Driskell, J. E., & Olmstead, B. (1989). Psychology and the military: Research applications and trends. *American Psychologist, 44,* 43–54.

Driskell, J. E., & Radtke, P. H. (2003). The effect of gesture on speech production and comprehension. *Human Factors, 45,* 445–454.

Driskell, J. E., & Salas, E. (1992). Can you study real teams in contrived settings? The value of small group research to understanding teams. In R. Swezey & E. Salas (Eds.), *Teams: Their training and performance* (pp. 101–124). Norwood, NJ: Ablex.

Freyd, M. (1926). What is applied psychology? *Psychological Bulletin, 33,* 308–314.

Gerstein, D. R., Luce, R. D., Smelser, N. J., & Sperlich, S. (Eds.) (1988). *The behavioral and social sciences: Achievements and opportunities.* Washington, DC: National Academy Press.

Hackman, J. R., & Morris, C. G. (1975). Group tasks, group interaction process, and group performance effectiveness: A review and proposed integration. *Advances in Experimental Social Psychology, 8,* 45–99.

Haverkamp, B. E. (2005). Ethical perspectives on qualitative research in applied psychology. *Journal of Counseling Psychology, 52,* 146–155.

Howell, W. C. (1998). When applied research works. In J. A. Cannon-Bowers & E. Salas (Eds.), *Making decisions under stress: Implications for individual and team training* (pp. 415–425). Washington, DC: American Psychological Association.

Ilgen, D. R. (1986). Laboratory research: A question of when, not if. In E. W. Locke (Ed.), *Generalizing from laboratory to field settings* (pp. 257–267). Lexington, MA: Lexington Books.

Kendon, A. (1983). Gesture and speech: How they interact. In J. M. Weimann & R. P. Harrison (Eds.), *Nonverbal interaction* (pp. 13–45). Beverly Hills, CA: Sage.

Krauss, R. M. (1998). Why do we gesture when we speak? *Current Directions in Psychological Science, 7,* 54–60.

Krauss, R. M., Chen, Y., & Chawla, P. (1996). Nonverbal behavior and nonverbal communication: What do conversational hand gestures tell us? *Advances in Experimental Social Psychology, 28,* 389–450.

Langewiesche, W. (2003). *Columbia*'s last flight: The inside story of the investigation—and the catastrophe it laid bare. *The Atlantic Monthly Online.* Available at http://www.theatlantic.com/issues/2003/11/langewiesche.htm.

Lewin, K. (1951). *Field theory in social science.* New York: Harper. (Original work published in 1944.)

Melton, A. W. (1952). Military requirements for the systematic study of psychological variables. In J. C. Flanagan (Ed.), *Psychology in the world emergency* (pp. 117–136). Pittsburgh, PA: University of Pittsburgh Press.

Milanovich, D., Driskell, J. E., Stout, R. J., & Salas, E. (1998). Status and cockpit dynamics: A review and empirical study. *Group Dynamics, 2,* 155–167.

Mook, D. G. (1983). In defense of external invalidity. *American Psychologist, 38,* 379–387.

Münsterberg, H. (1898). The danger from experimental psychology. *Atlantic Monthly, 81,* 159–167.

Münsterberg, H. (1914). *Psychology general and applied.* New York: Appleton.

Murphy, K. R. (2002). Editorial. *Journal of Applied Psychology, 87,* 1019.

National Science Foundation. (2006). *Federal funds for research and development: Fiscal years 2003–05.* (NSF 06-313). Arlington, VA: Division of Science Resources Statistics, National Science Foundation.

National Transportation Safety Board. (1990). Aircraft accident report: United Airlines DC-10-10 engine explosion and landing at Sioux City, Iowa (NTSB/AAR-90/06). Washington, DC: Author.

Salas, E. Bowers, C. A., & Edens, E. (2001). *Improving teamwork in organizations: Applications of resource management training.* Mahwah, NJ: Erlbaum.

Scott, W. D. (1903). *The theory of advertising.* Boston: Small & Maynard.

Stouffer, S. A., Lumsdaine, A. A., Lumsdaine, M. H., Williams, R. M., Smith, M. B., Janis, I. L., *et al.* (1949). *The American soldier: Combat and its aftermath.* Princeton, NJ: Princeton University Press.

Taylor, F. W. (1903). Group management. *Transactions of the American Society of Mechanical Engineers, 24,* 1337–1480.

Thorndike, E. K. (1903). *Educational psychology.* New York: Lemcke & Buechmen.

Triplett, N. (1898). The dynamogenic factors in pacemaking and competition. *American Journal of Psychology, 9,* 507–533.

van Strien, P. J. (1998). Early applied psychology between essentialism and pragmatism: The dynamics of theory, tools, and clients. *History of Psychology, 3,* 205–234.

Vaughan, D. (1996). *The Challenger launch decision: Risky technology, culture, and deviance at NASA.* Chicago: University of Chicago Press.

Webster, M. (2007). Funding experiments, writing proposals. In M. Webster & J. Sell (Eds.), *Laboratory experiments in the social sciences* (pp. 193–225). Burlington, MA: Elsevier.

Webster, M., & Kervin, J. B. (1971). Artificiality in experimental psychology. *Canadian Review of Sociology and Anthropology, 8,* 263–272.

Webster, M., & Sell, J. (2007). Why do experiments? In M. Webster & J. Sell (Eds.), *Laboratory experiments in the social sciences* (pp. 5–23). Burlington, MA: Elsevier.

Webster, M., & Whitmeyer, J. (2001). Applications of theories of group processes. *Sociological Theory, 19,* 250–270.

Zedeck, S. (2003). Editorial. *Journal of Applied Psychology, 88,* 3–5.

The Standardized Experimental Situation in Expectation States Research: Notes on History, Uses, and Special Features

JOSEPH BERGER

Stanford University

ABSTRACT

The standardized experimental situation of expectation states research has played an important role in the growth of expectation states theory. In this chapter we briefly describe the construction of this experimental situation, its use in the testing of theories, its role in developing a body of comparable empirical data, and its role in the application of expectation states theories. We also consider some of the many substantive problems that have been studied in this situation and suggest that there is now the opportunity to develop more elaborated standardized situations given our current knowledge of status processes.

I. INTRODUCTION

Expectation states theory is a *theoretical research program*. As such it consists of a set of interrelated theories, bodies of relevant research concerned with testing these theories, and bodies of research that use these theories in social applications and interventions (Berger, 1974).

A major substantive concern of these theories has been with understanding how status processes operate to organize interaction in groups: the different conditions under which these processes are activated, the various forms these processes can assume, the stable states that emerge as status processes evolve over time; and the different types of behaviors determined by these stable status states.

In undertaking the study of status and other social processes within the expectation states program we have been guided by a specific theory building strategy. The first principle of our strategy has been to analytically isolate status processes from other processes with which they may be interrelated in some concrete setting, such as social control, affect, or power processes. The idea here is that the interrelation of different interpersonal processes is to be treated as a task for subsequent theoretical research. A second key principle in our strategy has been to describe the operation of status and other processes in terms of abstract theories that are formally rigorous and empirically testable. Finally, a third key principle of our strategy has been to develop theory-based empirical models in order to be able to apply our abstract theories to the realities of specific concrete social settings (see the appendix at the end of this chapter).

Testing the type of abstract theories we have been interested in constructing often involves creating special status conditions—for example, conditions in which status processes actually are separated from affect or control processes or creating social conditions that are rarely found in everyday social interaction. As a consequence, our theoretical strategy led us to the conviction that a major (although not exclusive) source of data for this program would have to be from experimental research. It is within this context that the standardized experimental situation (SES) has evolved as an important source of data for testing and developing theories in the expectation states program.

The major elements of this standardized experimental setting were devised in the late 1950s and early 1960s, although research on the nature and properties of this setting continues to the present (see Troyer, Chapter 7, this volume). In this chapter, I shall briefly describe the early history in the construction of this experimental situation as well as some of its special features. In addition, I shall consider some of the ways the experimental situation has been used both to develop and apply the theoretical knowledge of the different branches of the expectation states program.

Throughout, I shall be describing events that go back 40 to 45 years and have not been previously recorded as history. So, as is true of all such historical narratives, there are cautions that the reader should bear in mind in reviewing this history. Two are particularly important. I undoubtedly shall be presenting a much more simplified picture of events than in all likelihood was true, and I shall be presenting a more rational picture of our activities than probably was the case. Actually, there was a considerable amount of trial and error in our activities, which is an important fact to bear in mind in understanding the evolution of the standardized experimental situation.

II. ON THE CONSTRUCTION OF SES

The history of SES begins with my work following the completion of my Ph.D. thesis (Berger, 1958). The general concern of the thesis was with accounting for the properties of interaction hierarchies as they emerged in small problem-solving groups. These properties were most evident in Bales's observations of small, informal task groups whose members were presumably similar in status. (See Bales, 1953; Bales *et al.*, 1951; Bales & Slater, 1955; and Heinecke & Bales, 1953). Bales found that inequalities in initiation of activity, in receipt of activities, on ratings of best ideas, and in group guidance regularly emerged in such groups. Once they emerged, these inequalities tended to be stable, and with the possible exception of sociometric rankings they tended to be highly intercorrelated. Research by others (e.g., Harvey, 1953; Sherif, White, & Harvey, 1955; Whyte, 1943) had shown that the evaluations of specific performances of individuals were correlated with the individual's position in the established hierarchy of the group. Independent of actual performance level, high-status members are seen as performing better than low-status members.

I approached the problem of explaining the nature of such interaction hierarchies by conceptualizing an idealized interaction process as it occurs in status-homogeneous groups. This process was seen as involving sequences of behaviors such as action opportunities (chances to perform), performance outputs (problem-solving attempts), performance evaluations, and communicated evaluations (positive and negative reward reactions). Basically, I argued that expectations (or expectation sets as they were then called) *emerged* out of the evaluations of performances and created a status ordering that determined the distribution of subsequent behaviors. Furthermore, given my arguments on the way in which this idealized interaction process operates, I sought to account for the *interrelation* of distributions of action opportunities, performance outputs, and rewards to overall performance evaluations, as well as account for the *stability* of this structure across tasks, given this interrelation. In addition, I argued that all of these processes would be affected by the primary normative orientation under which the group operated. Specifically, in "instrumental" (or task) focused groups, the behavioral components of the status hierarchy would be more highly interrelated and the structure more stable than in "integrative" (or process) oriented groups.

At this stage, I felt that I had only a general conception of what appeared to be fundamental processes. I wanted to expand the scope of these ideas and at the same time get a much more precise understanding of what was involved in their operation. For example, in the initial formulation, expectations emerged out of the evaluation of performances. Were there other processes by which expectations emerged or by which they became significant in a situation of action? Also, in my thinking I had developed a conception of an underlying process (an expectation process) that was related to an observable process that, with the addition of influence behaviors, became known later as the *observable power and prestige order* (OPPO). An important unanswered question was, "How can we conceptualize the different ways in which these two kinds of processes are related?"

Eventually, following the dissertation, I came to believe that in order to study these processes with the precision that I was interested in I had to be able to create performance expectations. Specifically, I wanted to study the behavioral consequences of high or low performance expectations possessed by an actor and his interacting partner. Since I wanted to be able to *randomly assign* such performance expectations to self and other, it became clear that the techniques required to create these conditions would have to involve the manipulation of information given to subjects. Subsequently, I also realized that I had to devise a more highly controlled experimental situation in which to study the behavior of expectation state processes.

A. The Manipulation of Expectations

A solution to the first problem turned out not to be too difficult. I decided that I had to create an ability or set of abilities that is instrumental to solving a group task. This would have to be one that subjects would believe was an actual (real) ability, and one on which they had no idea as to their actual capacities. Inasmuch as the group problems we most frequently used in those days involved human relations problems, the answer was clear: social insight and prediction ability. This was defined to subjects as the ability to "get right into a social situation," and "understand what's going on," and "predict quite accurately what will happen next."

In 1957 and 1958, while I was at Dartmouth College, a colleague[1] and I devised and ran a two-phase pilot study that involved two-person problem-solving groups. In the first phase, subjects were informed about the discovery of social insight and prediction ability. They were then "tested" on this ability and were randomly assigned high and low scores. Thus, we created a situation where in each group one individual was in a high-self and low-other state and the second was in a low-self and high-other state. In the second phase, subjects were asked to solve in an open-interaction setting a complex human relations case that involved using their social insight and prediction ability. In addition, subjects were presented with individual and group standards, purportedly based on studies of college students like themselves, to enable them to evaluate their individual as well as group performance. Who-to-whom interactions were scored using the behavior categories formulated in the thesis, and we predicted that the distribution of these behaviors would be a function of the performance expectations we had created.

[1] This test study was done in collaboration with the social psychologist Harry A. Burdick.

We found that subjects differed on the distribution of action opportunities, with the low-status subject distributing more to the high than the high to the low; that they differed on performance outputs, with the high initiating more than the low; and that they differed on positive rewards, with the low distributing more to the high than the high to the low. Viewing these profiles from the standpoint of *performer–reactor differences* (problem-solving attempts initiated to distribution of action opportunities and rewards to the other), this meant that performer-reactor rates were higher for the high-status subject as compared to his low-status partner.

We were impressed with how effective these techniques were, and I considered that this problem was solved: that we could, using such methods, manipulate performance expectations and study their consequences.

B. THE BEHAVIORAL SETTING

At this point, I still did not have the kind of experimental situation that I believed would enable us to study expectation processes with the precision that I desired. Actually, I was getting ready to do a full-scale open interaction experiment based on what had been discovered in the Dartmouth pilot study. This experiment was intended to investigate the behavioral consequences of different types of expectation structures (Berger, 1960).

For a time we considered developing a modified Bavelas Box.[2] This would enable us not only to study the behavioral consequences of expectation states but also to control the rates at which action opportunities, performance outputs, rewards, and exercised influence occurred between interactants. The assumption was that differences in the rates of each of these behaviors would lead to the formation of different types of expectation states. In fact, I remember exploring this idea in the late 1950s with my colleague and future collaborator, Morris Zelditch.

However, just as I was leaving Dartmouth and preparing to join the Department of Sociology at Stanford University, I started to work on a highly precise formulation of how expectations determine behavior and how behavior determines expectations. This was a finite Markov Chain model developed in collaboration with J. Laurie Snell, a mathematician at Dartmouth College (Berger & Snell, 1961). The key behavior in this model was *influence behavior* that, from my standpoint, involved *changes in the performance outputs of an actor as a consequence of the behavior of another*. I realized that in order to study these influence behaviors, which by this time were conceptualized as part of the observable power and prestige order, I needed an interaction sequence that would involve a large number of disagreements between interactants, and so the decision-making structure of the standardized experimental situation was born.

[2]In 1950, Alex Bavelas formulated a set of concepts to determine the relative centrality of positions in communication networks. The Bavelas Box was a device used to create and study group performances in the networks of interest to Bavelas and his students (see Leavitt, 1950).

I immediately went to work on more fully conceptualizing this decision-making situation and in particular the social context within which it should take place—such as task and collective orientation. Fortunately, I had a graduate student working with me, Robert Z. Muzzy, who informed me that he could build a machine that would do what I wanted it to do—control the behavior of interactants in the ways that I wanted it controlled—and once I described the decision-making process to him, he built our first interaction control machine (ICOM). The ICOM allowed interaction between participants but eliminated the face-to-face component so that the only information known to subjects was that provided by the experimenter.

When finalized, the basic decision-making sequence had the following features: Each individual, say in a two-person group, is given on every trial from 5 to 10 seconds to study a problem and decide which one of two alternatives is the correct solution to the problem. His or her choices are then communicated to the partner. The communicated information on choices is controlled by ICOM so that subjects can be told on any trial whether they agreed or disagreed with their partner independent of what is actually the case. With this feedback information, each individual then has the opportunity to restudy the problem and to make a final decision. These final decisions are not communicated to partners. Initial choices are defined as only preliminary decisions to provide information to partners, while subjects are told that final decisions are the only decisions that count on their team's performance record.

While there has been experimentation on the number of critical (disagreement) trials used in experiments, typically at least 20 critical trials have been used with either three or five neutral (agreement) trials to allay suspicion. The basic behavioral quantity obtained in this setting is P(s), the proportion of times individuals stay with an initial decision, given disagreement with their partner.[3]

Two big problems still remained to be dealt with before we could actually do experiments. We had to develop tasks that were appropriate to the new experimental setting, and we had to construct standardized scenarios that would define the experimental situation to subjects in a meaningful manner.

C. EXPERIMENTAL TASKS

Developing an appropriate task for the new experimental situation presented problems. Initially we considered using the social insight and prediction ability task that had been so successful in the Dartmouth research. We actually con-

---

[3]It was assumed that it would be possible to modify this decision-making sequence to capture some of the other types of behaviors in the OPPO. For an example of just such a modification in decision-making sequence, see the study by Conner (1977), where the dependent variable is the likelihood of making performance outputs, given action opportunities.

structed such a task involving 20 to 30 short decision-making situations. But it proved to be unsuitable. There was simply too much information in each of these cases for subjects to process fully within the short time span of a 5- to 10-second experimental trial. So we kept looking for a new task—a new ability.

At about this time I learned of the research that members of the anthropology department were doing on the structure of a particular primitive language. This gave me the idea for constructing the "meaning insight" ability task. This task involved matching English words with phonetically presented words from a primitive language that presumably had the same meanings as the English words. On each problem trial, subjects were told to study the English words and associate whatever meanings they called to mind. They were then told to study the non-English words, sound them to themselves, and associate whatever meanings they called to mind. Their task was to decide which primitive words had the same meaning as the English word.

My students and I constructed a large number of meaning insight items in two forms. In the first, two primitive words and one English word were used and in the second two English words and one primitive word were used. The primitive words, of course, were fictional. We administered these items to students in classroom settings and selected as usable those items in which the choice alternatives were equally likely to be chosen or as close to that criterion as we could get.

In spite of some problems connected with it, we decided that meaning insight was a usable task.[4] Subjects accepted the idea that such an ability actually exists. They were interested in being tested for the ability, and they felt that they "could work on the task"—in the sense that there was cognitive activity involved in getting the right solution. Thomas L. Conner (1964) did some of the most important work on this task and other tasks that were being developed in our laboratory at that time. Still later, in collaboration with Bernard P. Cohen, Morris Zelditch, and graduate students, we worked on other tasks in particular, the different forms of the spatial insight task. James C. Moore (1965) played a major role in the development of this task.[5]

---

[4]As a condition on the construction of tasks, it was felt that the choice of alternatives from one trial to a second should be an independent trials process. For example, the choice of an alternative, A or B, on trial n should be independent of the particular choice, A or B, that had been made on the n–1 trial. This condition did not hold for the first three tasks that we developed, including that of meaning insight (Conner, 1964).

[5]"Spatial insight" tasks were constructed over a number of years and took different forms. Initially, each task involved a slide of a rectangular figure where the rectangle consisted of 100 subrectangles. By a random process, 50 of the subrectangles were selected to be colored black and 50 white. The subject's task was to determine whether there was more white or black in any given rectangle. To counter a bias toward a white response that was discovered in this first task, a second version was developed in which each task consisted of two rectangles, each with different

(Footnote Continues)

D. Standardized Scenarios

One big task remained—namely, to develop a standardized scenario for experiments in this new setting. This involved constructing appropriate experimental procedures.

We wanted to develop a set of standardized procedures that could be used in a large number of experiments. These experiments would, of course, differ in terms of the theoretically relevant conditions and variables involved. But the idea was to have a *core scenario* that would be used from experiment to experiment.

Basically, we wanted to define the experimental situation as socially meaningful and as embodying the information and features we regarded as important in doing expectation states research. Among other things this required:

- Subjects should accept the reality of the instrumental characteristic, meaning insight ability. This meant accepting the idea that it was a newly discovered and very important ability that task success or failure depended on their level of this ability and that it was not simply a result of guesswork or chance.
- Subjects should accept the idea that meaning insight ability was unrelated to other well-known abilities such as mathematical and verbal ability. The objective was to dissociate the instrumental ability from those that subjects would have knowledge about. This would facilitate the random assignment of the different levels of meaning insight as required for experimental study.

At this stage the notion of scope conditions, as abstractly defined conditions under which a social process is predicated to hold, was still being developed.[6] But already it was believed that the next two conditions were essential for the study of expectation states processes:

- Subjects should be task oriented. This meant that their primary focus was on the task and that they would be motivated to work for the correct solutions of the task.
- Subjects should be team and collectively oriented. This meant that taking into account the judgments of the other person on the team was both legitimate and useful to the team goal. Subjects' initial decisions were defined as preliminary choices and they were informed that only final choices mattered. To create task and collective orientation, we made extensive use of

---

arrangements of black and white subareas. In this case, the subject's task was to determine which rectangle had a greater amount of white area. Near-veridical versions of this task were also constructed where a response bias of a given magnitude favoring either white or black was built into the task (Conner, 1965). See Moore (1965) for a report of the development of this task.

[6]For discussions of the notion of theoretically defined scope conditions, see Berger (1974) and Walker and Cohen (1985).

performance standards, both individual and group, which were presumably based on studies of subjects like themselves, here and elsewhere.

The number of disagreements (or critical trials) was an important issue:

- It was necessary to provide subjects with enough critical trials so as to get stable measures of the subject's behavior, while at the same time reducing the subject's suspicion, due to the number of critical trials, as much as was possible.[7]

In 1962 and 1963, we ran our first status experiment in the standardized experimental situation. In this first experiment we manipulated expectations on C^*, the characteristic instrumental to the task, assigning individuals to high–low, low–high, high–high, and low–low self–other states. We predicted and found that the individual's likelihood of resisting influence was directly related to his expectation advantage over his partner (Berger & Conner, 1966, 1969).

As we conceptualized our task back in the early 1960s, all four components—ways to manipulate or identify status and expectation states, a core set of standardized scenarios and experimental procedures, a set of novel experimental tasks, and an appropriate decision-making process—were required to construct a standardized experimental situation. Developing this situation required a considerable investment of time and effort. We accepted these costs since we anticipated that there would be a long-term payoff for our research.

III. THE GRAPH FORMULATION OF STATUS CHARACTERISTIC THEORY

The graph formulation of the status characteristic theory was introduced in expectations state research in 1977 (Berger, Fisek, Norman, & Zelditch). The overall three-part structure of the theory, its metatheoretical component, its theoretical component, and its theory-based models component are briefly described in the appendix at the end of this chapter. With the introduction of the graph formulation, the relation between theory and experimental situation was still further developed. There are at least three ways in which this occurred that are particularly worth noting.

---

[7]In the earliest stages of our work, we anticipated finding trend effects in subjects' response data. As a consequence, we were interested in employing enough trials so as to detect these trends and to allow the process for each subject to reach a stable state. However, after the Air Force study (Berger, Cohen, & Zelditch, 1972), which involved an extended trials sequence, we became aware of the problems, such as boredom, weakening of task, and collective orientation, associated with such extended sequences; almost all subsequent studies at Stanford involved 23 to 25 trials, including three to five neutral trials.

To begin with, the graph theory allowed us to represent in a formal manner the different types of actor-situational structures that we were interested in. From the early 1960s, we had been using "heuristic graphs" to map status situations that we had determined from experience could be created and realized in the standardized experimental situation. The new theory allowed us to transform these heuristic graphs into mathematical structures. This enabled us to deal with an extremely wide range of status situations from the simplest to the most complex. We could now formally represent these situations, and make general as well as specific predictions as to what could be expected to occur in these status situations.

Second, the new theory facilitated the task of devising experiments to test theoretical predictions. The status graphs have been used to conceptualize in terms of their graphic structures experimental conditions required in such tests. In addition to this, the theory has been used to assess the feasibility of such experimental tests. The graph theory has four parameters, *two general and two situational parameters*, whose values have to be determined to make specific numerical predictions. The general ones are assumed to be applicable across a broad range of situations and are either empirically determined as in Berger *et al.* (1977) and Balkwell (1991) or determined on a priori grounds as in Fisek, Norman, and Nelson-Kilger (1992). The two situational parameters are applicable to and have to be determined for the specific situations that are of immediate concern.

Given our experience in working with the standardized situation, we have been able in general to assess the range of empirical values that we might expect to find for each of our two situational parameters. Using this information, we then can determine the range of empirical results that might obtain for given experimental conditions. On the basis of such analysis, we have been able in general to decide whether the likely results of a proposed experiment will enable us to make a decision or not on the theoretical arguments that were being tested (see Section B, following).

Third, on a more general level, the graph theory enables us to specify more fully what is involved in applying such a theory to a specific empirical situation. The graph theory, as is true of the original version and the second version of the status characteristic theory (Berger, Cohen, & Zelditch, 1966; Berger & Fisek, 1974), is an abstract theoretical structure. Such an abstract theory must be empirically interpreted. This is the role of what we have called *theory-based empirical models*. These models consist of a combination of factual information and assumptions—for example, factual information (or assumptions) that race or gender can be treated as instances of a diffuse status characteristic for a given population at a given time, or model assumptions on the translation of theoretical concepts to behavioral observations (see the appendix). Understanding the nature of and the specific elements in such models is essential inasmuch as such elements are involved in testing and applying an abstract theoretical structure to some concrete reality. Furthermore, it is the status of these elements that often is initially called into question with the failure of theoretical tests and applications.

IV. USES AND SPECIAL FEATURES OF SES

A. COMPARABILITY AND CUMULATIVENESS

One of the primary objectives in developing a standardized situation was to facilitate the acquisition of a cumulative body of comparable data to be used in theoretical research. Instead of being confronted with the results of a set of experiments—all presumably dealing with the same problem—in which tasks differed, settings differed, and dependent variables differed, the objective was to have such factors similar with variations being restricted to theoretically relevant factors and variables. Presumably, the results of such experiments would be particularly useful in developing, assessing, and testing theoretical knowledge. In fact, there are important situations where this has turned out to be the case.

When my colleagues, Hamit Fisek and Robert Norman, and I constructed the graph formulation of the status characteristics theory, we were able to assess the consistency of that formulation against the results of 12 experiments (involving 57 experimental conditions) that had already been conducted in the SES (Berger et al., 1977). The fact that the formulation was consistent with this extensive database was of enormous importance in allowing us to evaluate our work at that stage, in informing us that we were on the right track, and in enabling us to build further on our theory (see also Fisek et al., 1992).

B. STRONG TESTS OF THEORETICAL ARGUMENTS

It was assumed that the standardized experimental situation would be used to test and develop theoretical arguments across a wide range of substantive problems, and this in fact has been the case (see Section D). Among such experiments, we were particularly interested in constructing strong tests of theoretical arguments.

Strong tests of theoretical arguments are empirical studies that involve (1) direct tests of specific theoretical arguments in the context of alternative arguments where (2) the original arguments and their alternatives are formulated within a common theoretical language. Such tests, when they are experiments, may consist of a complex set of experimental manipulations and conditions and they may also involve a subtle pattern of predicted experimental results.[8]

---

[8]For related ideas, see Platt (1964) and Cole (2001).

Strong tests have been conducted successfully in the standardized situation. Three such experiments have been the Wagner, Ford, and Ford (1986) study on the confirmation and disconfirmation of gender-based status expectations; the Norman, Smith, and Berger (1988) and Berger, Norman, Balkwell, and Smith (1992) studies on status inconsistency and status organizing principles; and the Webster, Whitmeyer, and Rashotte (2004) study on second-order expectations. In each of these cases, the experiments were conducted to test for derived theoretical predictions while at the same time discriminating among alternative theoretical arguments.

In the case of the status inconsistency studies (Berger *et al.*, 1992), for example, the experiments involved a direct test of the *principle of organized subsets*. This principle argues that *all valued status information* that has become salient is combined by an actor so as to take into account its sign (whether it is positive or negative status information) and its degree of task relevance (its weight) in the situation. There are other theoretically reasonable principles that have been proposed, such as *status canceling* principles, where actors are seen to "cancel" oppositely signed and equally weighted items of status information, and *status balancing* principles, where actors are seen as eliminating inconsistent status information in forming expectations for each other. Our objective was to design an experiment where outcomes supportive of one of these principles are at the same time inconsistent with those that would be supportive of the other principles.

To do this, it first was necessary to translate each of the nonstatus characteristic principles—balancing and canceling—into the concepts of the status characteristics theory and then to determine the conditions under which these principles made alternative predictions. In translating all status-processing alternatives into the same theoretical language, we were creating multiple versions of the status characteristics theory (or *theoretical variants* in the language of Wagner & Berger, 1985, and Berger & Zelditch, 1993). As a consequence, we were assured that the *only differences* involved in deriving alternative predictions were those due to the alternative status organizing principles in these theoretical variants.

On the basis of extended experience in working with the SES, we had information on the range of empirical values that we might expect to find for each of the situational parameters of the graph theory. With this information we could then determine the possible range of empirical results that might be obtained under the operation of each principle. Based on this analysis, we determined that there was a good chance of getting results that would enable us to discriminate between the different status principles. Eventually, the experiment was conducted and the results provided support for the principle of organized subsets while at the same time they failed to provide support for the competing status-processing principles (Berger *et al.*, 1992; Norman *et al.*, 1988).

C. A HOLY TRIANGLE: TESTS, THEORIES, AND APPLICATIONS

Very early in my work with colleagues B. P. Cohen and M. Zelditch on the status characteristics theory, the issue arose of how one relates research in a highly controlled setting such as the SES to research in less controlled settings. For example, how does one generalize the results of experiments carried out in the SES to nonexperimental settings? In part, in response to such questions we started to talk about what we called within our group a "holy triangle": tests, theories, and applications. The basic idea is straightforward. In situations where it is required, highly controlled experimental settings are used to test theories, and it is theories and theoretical arguments that are applied to less controlled settings or used in social applications and interventions. This, of course, is an ideal conception of the development and use of theoretical ideas, and it is not clear to me how often in fact it has been realized. But one very important case where this ideal conception appears to have been realized is worth describing in this context.

When Hamit Fisek and I published the second version of the status characteristics theory (Berger & Fisek, 1974), we presented theoretical arguments that claimed that all of the status information that has become salient in a situation of action is combined. At that time we had the results of experiments that we had done in SES, starting with experimental findings published in 1970 (Berger & Fisek, 1970; Berger, Fisek, & Crosbie, 1970) that supported this argument. In addition, there was the as yet unpublished research of Zelditch, Lauderdale, and Stublarec that we knew about, which also supported the combining argument (Zelditch, 1980; Zelditch et al., 1975).

In 1977, Susan Rosenholtz, a student of Elizabeth Cohen, used the combining argument that appeared in the second version of the status characteristics theory to predict that groups exposed to multi-ability tasks would have less differentiated hierarchies than those working on standard unidimensional tasks. This research was done in an open-interaction laboratory setting and, indeed, Rosenholtz found what she had predicted: less differentiated hierarchies in groups exposed to multi-ability tasks (Rosenholtz, 1977).[9] Eventually, Elizabeth Cohen and her colleagues devised multi-ability status interventions that are suitable for classroom use, and have been able to study their effects as

---

[9]It is to be noted that in Rosenholtz's intervention (as has been true of other such interventions) both the definition of the task situation and the structure of the task involve multi-abilities. In a recent and important study, Goar and Sell (2005) have shown that differentiation in the group's interaction hierarchy can be reduced by multi-ability definition in the task situation while holding the structure of the task constant. See also Fisek (1991) for a theoretical formulation on complex task situations.

part of their "complex instruction curriculum" in actual classroom settings (see E. G. Cohen, 1982; E. G. Cohen & Lotan, 1995, 1997).

This, then, is a case where researchers have gone from experiments in SES on combining status information, to general theoretical arguments on combining, to predictions in open interaction laboratory settings based on these arguments, to field studies, and to interventions in actual ongoing classroom situations. A case similar to the preceding one is to be found in the research of Murray Webster on source theory. Webster first formulated source theory (1969), then tested the theory in SES, and subsequently he and Doris Entwisle devised interventions based on the theory that they used to raise children's expectations in actual classroom settings (Webster & Entwisle, 1974). In both cases, research in SES was either the basis of theoretical arguments or was used to test such arguments. Theoretical arguments, in turn, were then used as the grounds for applications and interventions.[10]

D. FLEXIBILITY OF THE STANDARDIZED EXPERIMENTAL SITUATION

The standardized experimental situation has proven to be a highly flexible setting in the sense that it has proven to be possible to adapt and employ the situation in the study of different theoretical problems. Here are some of the research problems, in addition to those concerned with the core status characteristics theory, that have been studied in the standardized situation:

- Research on decision-making authority. This research investigates how individuals who possess different levels of decision-making authority (decision maker versus advisor) and different expectation states are influenced by the behavior of the other (Camilleri & Berger, 1967; Camilleri, Berger, & Conner, 1972).
- Research on the effects of sources of evaluations. Sources of evaluation are individuals with the right to evaluate the performances of an actor and are individuals whose evaluations also matter to the actor. This research has investigated how consistent and conflicting sources of evaluation can create different types of expectations, which in turn lead to different types of interactive behaviors (Webster & Sobieszek, 1974).
- Research on reward expectations. Reward expectations are expectations for the possession of rewards allocated in a status situation. Of particular interest are studies on what has been defined as the "reverse effect" in the

---

[10]For further discussions of these and related issues on the external validity of experiments, see Lucas (2003b) and Webster (2005).

status value theory of distributive justice (Berger *et al.*, 1972). This is a theoretical argument that under appropriate conditions the allocation of rewards, in and of itself, creates task expectations (see Cook, 1975; Harrod, 1980; Stewart & Moore, 1992).

- Research on the transfer of status interventions. Status interventions are techniques that enable an individual to overcome invidious effects of status categorizations. This research has investigated the transfer of such interventions from an original status occupant to a second, from an original task to a new task, and from an original status category to a second and different category (e.g., race to educational attainment) (Berger, Fisek, & Norman, 1989; Lockheed & Hall, 1976; Markovsky, Smith, & Berger, 1984; Prescott, 1986; Pugh & Wahrman, 1983; Rashotte, 2006).

- Research on referent actors. This is research on how information about referent actors—individuals who are not interactants—affect the behavior of interactants in an immediate situation of action. An example would be the effect on the behavior of a female (or male) actor in a mixed gender group of information that females (or males) in the past have outperformed members of the opposite gender on the same task that now confronts them (Freese, 1974; Pugh & Wahrman, 1983; Wagner *et al.*, 1986).

- Research on second-order expectations. If we consider the expectations that a first actor holds for self and a second actor as *first-order expectations*, then the expectations that the first actor believes the second holds for self and the first actor are labeled as *second-order expectations*. This research investigates the conditions under which second-order expectations are transformed into or affect the structure of first-order expectations (Fisek, Berger, & Moore, 2002; Moore, 1985; Troyer & Younts, 1997; Webster *et al.*, 2004; Whitmeyer, Webster, & Rashotte, 2005).

- Research on the effects of sentiments on status behavior. This research is concerned with the conditions and processes by which positive and negative sentiments (such as patterns of like and dislikes) either accentuate or attenuate the differentiation generated by status inequalities (Bianchi, 2005; Driskell & Webster, 1997; Fisek & Berger, 1998; Lovaglia & Houser, 1996; Shelly, 1993).

- Research on status cues. Status cues are behaviors, which may be expressive and nonverbal, that communicate information on the high or low performance capacities of an actor (e.g., rates of fluency and nonfluencies of speech). This research is concerned with the processes by which status cues affect the formation of interaction hierarchies in groups, and in turn how status expectation differences affect the display of different types of status cues (Foddy & Riches, 2000; Rainwater, 1987; Riches & Foddy, 1989; Tuzlak & Moore, 1984; Webster & Rashotte, 2006).

- Research on status legitimation. A legitimated status order is one in which (1) generalized deferential relations are prescriptive, (2) these relations have become embedded in power and prestige behaviors (e.g., the high-status individual should control the group's time and attention), and (3) there is the presumption that there will be collective support in maintaining the existing status order. This research is concerned with the emergence of legitimated status hierarchies and its consequences on group behavior (Berger, Ridgeway, Fisek, & Norman, 1998; Kalkhoff, 2005; Ridgeway & Berger 1986).
- Research on multiple standards. This research is concerned with the use of different standards to evaluate the same performances of individuals who differ in status positions. The activation of multiple standards is seen as a mechanism that in general operates to maintain existing status distinctions (Foschi, 1996, 2000).
- Research on social identity and status characteristic processes. This research is concerned with the effect of social identity and status characteristic processes examined singly as well as the combined effect of these processes on behavior as studied in the same experimental situation (Kalkhoff & Barnum, 2000).

E. SES AND OTHER RESEARCH SITES

We never entertained the idea that the SES would be the only source of data relevant to theories in the expectation states program and, in fact, it has not been the only source. Data relevant to theories in the program have come from a wide variety of research settings, including the following:

- Open-interaction studies (Gallagher, Gregory, Bianchi, Hartung, & Harkness, 2005; Propp, 1995; Shelly & Munroe, 1999; Skvoretz, Webster, & Whitmeyer, 1999). Usually, these are studies of problem-solving groups whose members confront each other in face-to-face interaction.
- Studies in simulated organizational settings (Hysom & Johnson, 2006; Johnson, 2003; Lucas, 2003a).
- Actual on-site studies—for example, in classroom settings (E. G. Cohen & Lotan, 1997), research and development teams (B. P. Cohen & Zhou, 1991), and on New York City Police teams (Gerber, 2001).
- Studies in hypothetical task situations (Balkwell, Berger, Webster, Nelson-Kilger, & Cashen, 1992; Fisek & Hysom, 2004; Shelly, 2001). Typically, these are studies in which the subject is asked to "imagine" himself in a social situation and then describe how he would react. This contrasts with studies in SES where the subject finds himself in a social situation created by the researcher, who also then observes and assesses the subject's reactions.

- Studies that involve social applications and social interventions. Such studies cover a very broad range of settings from Israeli soccer teams (Yuchtman-Yaar & Semyonov, 1979) to breast cancer nursing wards (Ludwick, 1992).
- Studies in other kinds of experimental situations (Dovidio, Brown, Heltmann, Ellyson, & Keating, 1988; Driskell, Olmstead, & Salas, 1993).
- Studies involving confederates (Ridgeway & Erickson, 2000). These are studies in which confederates are used to induce subordinate or superordinate behavior patterns on the part of the subjects.

Thus, for example, there is a status legitimation study done within SES (Kalkhoff, 2005) and another status legitimation study in a markedly different, highly controlled experimental situation (Johnson, 2003). There is research on the "reverse process" (reward allocations leading to task expectations) done using the SES (Cook, 1975) and not using the SES (Bierhoff, Buck, & Klein, 1986) that yield comparable findings. In addition, there are research results from status cues studies in the SES (Riches & Foddy, 1989), as well as from the Nemeth-Wachtler jury-judging situation (Mohr, 1986), that are interrelated by theoretical research on the status cues formulation (Berger, Webster, Ridgeway, & Rosenholtz, 1986; Fisek, Berger, & Norman, 2005).

One of the most interesting examples of research involving a site other than SES that directly tests an expectation state theory is a study by Dovidio and his colleagues (1988). This study was done in a setting they developed that involves two-person open-interaction groups whose members are working on short problem-solving tasks.

A prediction from the status characteristics theory (a major branch of the expectation states program) is that if a male and a female are working on a task that is not initially relevant to gender (a neutral task), then the male will have an expectation advantage over the female. If they are working on a masculine-typed task, the male's expectation advantage will be even greater than when they are working on a neutral task; if they are working on a feminine-typed task, the expectation advantages will be reversed so as to favor the female. Furthermore, both observable power and prestige and status cues behaviors are predicted to be functions of one actor's expectation advantage over a second actor.

We had actually considered doing research in SES to test for these arguments, but to our surprise and pleasure, we found that it already had been done by Dovidio and colleagues, and that they had come up with results that were exactly what we had predicted. They found that, when working on gender-neutral tasks, males initiated more speech, spoke more, made more eye contact when speaking, and gestured more than females. These inequalities favoring males increased on masculine-type tasks and, as predicted, were actually reversed when groups were working on feminine-type tasks.

The fact that other research settings and methods have been used to test and apply the theories in the expectation program clearly adds to the credibility of these theories (B. P. Cohen, 2003).

V. SOME CONCLUDING COMMENTS

Although the major components of the standardized situation had been developed by the early 1960s, graduate students and I (including Thomas Conner, Hamit Fisek, Murray Webster, James Moore, Barbara Meeker, Gordon Lewis, Karen Cook, and Ruth Cronkhite) continued to work on refining the situation throughout the 1960s and early 1970s. We continued to make changes in the basic scenario to strengthen task and collective orientation by introducing the notion of a critical choice situation, and we shifted from using unit feedback on the individual's performances in manipulating expectations to simply handing out final scores to subjects. In fact, I began to wonder how little information we could give subjects and still manipulate their expectations, but we never pursued that problem. Eventually, a laboratory manual for doing research in SES was prepared by Cook, Cronkhite, and Wagner (1974) and was subsequently revised by Wagner and Harris (1993).

By the end of the 1960s, B. P. Cohen introduced the use of video techniques in the SES, which gave us additional resources in our research (B. P. Cohen et al., 1969). Among other things, by enabling us to present major portions of experimental procedures and information on videotape, it allows us to reduce the variability of theoretically irrelevant features in our experiments (e.g., variations in the behavior of the host experimenter). Subsequently, Foschi et al. (1996) introduced a computerized version of the SES, and still later Lisa Troyer (1999, 2000) developed a second and Webster et al. (2004) developed a third computerized version of the situation. We recognized at the time that these innovations introduced major changes in the basic standardized situation. Most recently, research by Troyer (2001, 2002) and by Kalkhoff and Thye (2006) has been concerned with examining the properties of the situation and, in particular, the variations in behavior introduced by the video and computerized versions.[11]

---

[11]On the basis of a meta-analysis of 26 expectation states experiments, Kalkhoff and Thye (2006) have documented the fact that the two situational parameters of the graph theory vary as between experiments conducted in the basic standardized situation (the standardized situation whose history is described in this chapter), the video version of the SES, and the Foschi computerized version of the situation. They offer explanations to account for these effects, as well as effects due to number of experimental trials and sample size. The analysis provides systematic information that researchers can use in planning future expectation states studies in standardized experimental settings.

I think it is fair to say that we have witnessed major advances in our theoretical and empirical knowledge of status processes since the early 1960s (see Berger & Webster, 2006; Wagner & Berger, 2002). In these advances, the SES has played an important (but certainly not an exclusive) role. It has provided us with a sizable quantity of comparable and relatively precise information that has facilitated the growth and testing of theories within the expectation states program. But we should also observe that status research during this period may have provided us with information such as that on status cues behaviors (Fisek *et al.*, 2005), that on status latency effects (Conner, 1977, 1985), or that on status relations to acoustic behaviors (Gallagher *et al.*, 2005). This information can be used to develop more elaborated as well as completely new versions of the standardized experimental situation for future research on expectation state and related social processes.

APPENDIX

The Structure of Status Characteristics Theory

I. METATHEORETICAL COMPONENTS

Working strategies that are involved in defining theoretical problems, and in formulating concepts and principles for their solution— for example, the concept of a status organizing process in the status characteristics theory (see Figure 1).

II.   THEORETICAL COMPONENTS

A. *Scope conditions.* Statements that describe in abstract terms the social conditions under which the theory is assumed to hold.
B. *Concepts and principles.* A set of abstract concepts and general principles that describe the operation (activation, evolution, deactivation) of the social processes with which the theory is concerned. For status characteristics theory, see Figure 1.
C. *Logical and mathematical structure.* A logical and mathematical structure within which the concepts and general principles are formulated in an interrelated and consistent structure (e.g., graph formulation in status characteristics theory).
D. *General and specific derivations.* Statements about the relevant social processes that derive from the concepts and principles of the theory using the embedded logical or mathematical structure. General derivations describe behaviors in classes in situations (e.g., effects of increasing the number, relevance, or

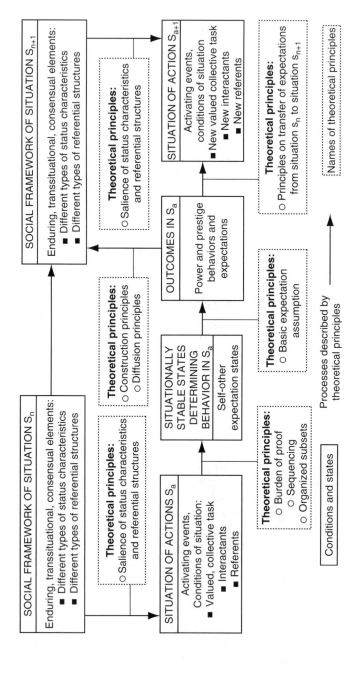

FIGURE 1   Status characteristics theory as a state organizing process. (From Wagner & Berger, 2002.) Legend: a=n

consistency of status distinctions on status behaviors). Specific derivations that describe behaviors for specific status situations.

III.   THEORY-BASED EMPIRICAL MODELS

A. *Instantiation of the theory*. This involves factual information or assumptions that relate abstract concepts in the theory to concrete conditions and phenomena—for example, information or assumptions to the effect that race or gender for a given population at a given time is a diffuse status characteristic.
B. *Specification of the theory*. This involves information on the specific initial conditions that hold true of a given situation for which predictions are to be made (e.g., number of actors involved, type and states of status characteristics they possess, etc.).
C. *Observations and the theory*. This involves assumptions and information that relate abstract theory to empirical observations. These include *coordinating* assumptions, as in those relating the abstract concept of "influence" in status theory with change in behavior in the standardized experimental situation. They also include *translation* assumptions, as in those that relate "expectation advantage" (theoretical) to the "proportion of stay responses" (observational). The latter assumption involves situational parameters whose values must be estimated to make specific behavioral predictions.
D. *Simplifying assumptions*. One or more assumptions used to simplify representation of status situations, or to simplify analysis and computations using status theory (e.g., that certain status task connections beyond a specified length are so weak that they can be ignored).

ACKNOWLEDGMENTS

I would like to acknowledge support from the Hoover Institution for work on this chapter, and to express my appreciation to Robert Shelly, Jane Sell, Murray Webster, and Morris Zelditch for their comments and suggestions on the earlier version.

REFERENCES

Bales, R. F. (1953). The equilibrium problem in small groups. In T. Parsons, R. Bales, & E. H. Shils (Eds.), *Working papers in the theory of action* (pp. 111–161). Glencoe, IL: Free Press.
Bales, R. F., & Slater, P. (1955). Role differentiation in small decision-making groups. In T. Parsons, & R. F. Bales (Eds.), *Family, socialization and interaction process*. Glencoe, IL: Free Press.
Bales, R. F., Stodbect, F. L., Mills, T. M., & Roseborough, M. E. (1951). Channels of communication in small groups. *American Sociological Review, 16*, 461–468.

Balkwell, J. W. (1991). Status characteristics and social interaction: An assessment of theoretical variants. In E. J. Lawler, B. Markovsky, C. Ridgeway, & H. Walker (Eds.), *Advances in group processes* (Vol. 8, pp. 135–176). Greenwich, CT: JAI Press.

Balkwell, J. W., Berger, J., Webster, M., Jr., Nelson-Kilger, M., & Cashen, J. (1992). Processing status information: Some tests of competing theoretical arguments. In E. J. Lawler, B. Markovsky, C. Ridgeway, & H. A. Walker (Eds.), *Advances in group processes* (Vol. 9, pp. 1–20). Stamford, CT: JAI Press.

Bavelas, A. (1950). Communication patterns in task-oriented groups. *Journal of the Acoustical Society of America, 22,* 725–730.

Berger, J. (1958). *Relations between performance, rewards, and action-opportunities in small groups.* Ph.D. dissertation, Harvard University.

Berger, J. (1960). An investigation of processes of role-specialization in small problem-solving groups. Proposal funded by The National Science Foundation (July).

Berger, J. (1974). Expectation states theory: A theoretical research program. In J. Berger, T. L. Conner, & M. H. Fisek (Eds.), *Expectation status theory: A theoretical research program* (pp. 3–22). Cambridge, MA: Winthrop Press.

Berger, J., Cohen, B. P., & Zelditch, M., Jr. (1966). Status characteristics and expectation states. In J. Berger, M. Zelditch, Jr., & B. Anderson (Eds.), *Sociological theories in progress* (Vol. 1, pp. 29–46). Boston: Houghton Mifflin.

Berger, J., Cohen, B. P., & Zelditch, M., Jr. (1972). Status characteristics and social interaction. *American Sociological Review, 37,* 241–255.

Berger, J., & Conner, T. L. (1966). Performance expectations and behavior in small groups. Technical Report No. 18. Laboratory for Social Research, Stanford University.

Berger, J., & Conner, T. L. (1969). Performance expectations and behavior in small groups. *Acta Sociologica, 12,* 186–97.

Berger, J., & Fisek, M. H. (1970). Consistent and inconsistent status characteristics and the determination of power and prestige orders. *Sociometry, 33,* 278–304.

Berger, J., & Fisek, M. H. (1974). A generalization of the theory of status characteristics and expectation states. In J. Berger, T. L. Conner, & M. H. Fisek (Eds.), *Expectation states theory: A theoretical research program* (pp. 163–205). Cambridge, MA: Winthrop.

Berger, J., Fisek, M. H., & Crosbie, P. V. (1970). Multicharacteristics status situations and the determinations of power and prestige orders. Technical Report No. 35. Laboratory for Social Research, Stanford University.

Berger, J., Fisek, M. H., & Norman, R. Z. (1989). The evolution of status expectations: A theoretical extension. In J. Berger, M. Zelditch, Jr., & B. Anderson (Eds.), *Sociological theories in progress: New formulations* (pp. 100–30). Newbury Park, CA: Sage.

Berger, J., Fisek, M. H., Norman, R. Z., & Zelditch, M., Jr. (1977). *Status characteristics and social interaction: An expectation states approach.* New York: Elsevier.

Berger, J., Norman, R. Z., Balkwell, J. W., & Smith, R. F. (1992). Status inconsistency in task situations: A test of four status processing principles. *American Sociological Review, 57,* 843–55.

Berger, J., Ridgeway, C. L., Fisek, M. H., & Norman, R. Z. (1998). The legitimation and delegitimation of power and prestige orders. *American Sociological Review, 63,* 379–405.

Berger, J., & Snell, J. L. (1961). A stochastic theory for self-other expectations. Technical Report No. 1. Laboratory for Social Research, Stanford University.

Berger, J., & Webster, M., Jr. (2006). Expectations, status, and behavior. In P. J. Burke (Ed.), *Contemporary social psychological theories* (pp. 268–300). Stanford, CA: Stanford University Press.

Berger, J., Webster, M., Jr., Ridgeway, C. L., & Rosenholtz, S. (1986). Status cues, expectations, and behavior. In E. J. Lawler (Ed.), *Advances in group processes* (Vol. 3, pp. 1–22). Greenwich, CT: JAI Press.

Berger, J., & Zelditch, M., Jr. (Eds.). (1993). *Theoretical research programs: Studies in the growth of theory*. Stanford, CA: Stanford University Press.

Bianchi, A. (2005). Rejecting others' influence: Negative sentiment and status in task groups. *Sociological Perspectives, 47*(4), 339–355.

Bierhoff, H. W., Buck, E., & Klein, R. (1986). Social context and perceived justice. In H. W. Bierhoff, R. L. Cohen, & J. Greenberg (Eds.), *Justice in social relations* (pp. 165–85). New York: Plenum Press.

Camilleri, S. F., & Berger, J. (1967). Decision-making and social influence: A model and an experimental test. *Sociometry, 30* (December), 367–78.

Camilleri, S. F., Berger, J., & Conner, T. L. (1972). A formal theory of decisionmaking. In J. Berger, M. Zelditch, Jr., & B. Anderson (Eds.), *Sociological theories in progress* (Vol. II). Boston: Houghton Mifflin Company.

Cohen, B. P. (2003). Creating, testing, and applying social psychological theories. *Social Psychology Quarterly, 66*(1), 5–16.

Cohen, B. P., Kiker, J. E., & Kruse, R. J. (1969). The use of closed circuit television in expectation experiments. Technical Report No. 29, Laboratory for Social Research, Stanford University.

Cohen, B. P., & Zhou, X. (1991). Status processes in enduring work groups. *American Sociological Review, 56*, 179–88.

Cohen, E. G. (1982). Expectation states and interracial interaction in school settings. *Annual Review of Sociology 8*, 209–235.

Cohen, E. G., & Lotan, R. A. (1995). Producing equal-status interaction in the heterogeneous classroom. *American Educational Research Journal, 32*, 99–120.

Cohen, E. G., & Lotan, R. A. (1997). *Working for equity in heterogeneous classrooms*. New York: Columbia University Teachers College Press.

Cole, S. (2001). Why sociology doesn't make progress like the natural sciences. In S. Cole (Ed.), *What's wrong with sociology?* (pp. 37–60). New Brunswick, NJ: Transaction Books.

Conner, T. L. (1964). Three tasks for use in laboratory small-group experiments. A technical report from the Laboratory for Social Research, Stanford University, Stanford, CA.

Conner, T. L. (1965). *Continual disagreement and the assignment of self-other performance expectations*. Unpublished Ph.D. dissertation, Department of Sociology, Stanford University.

Conner, T. L. (1977). Performance expectations and the initiation of problem solving attempts. *Journal of Mathematical Sociology, 1977*, 187–198.

Conner, T. L. (1985). Response latencies, performance expectations, and interaction patterns. In J. Berger & M. Zelditch, Jr. (Eds.), *Status, rewards, and influence: How expectations organize behavior*. San Francisco: Jossey–Bass Publishers.

Cook, K. S. (1975). Expectations, evaluations, and equity. *American Sociological Review, 40*, 372–388.

Cook, K. S., Cronkite, R., & Wagner, D. G. (1974). *Laboratory for Social Research manual for experiments in expectation state theory*. Laboratory for Social Research, Stanford University.

Dovidio, J. F., Brown, C. E., Heltmann, K., Ellyson, S. L., & Keating, C. F. (1988). Power displays between women and men in discussions of gender-linked tasks: A multichannel study. *Journal of Personality and Social Psychology, 55*, 580–587.

Driskell, J. E., Olmstead, B., & Salas, E. (1993). Task cues, dominance cues, and influence in task groups. *Journal of Applied Psychology, 78*, 51–60.

Driskell, J. E., & Webster, M., Jr. (1997). Status and sentiment in task groups. In J. Szmatka, J. Skvoretz, & J. Berger (Eds.), *Status, network, and organization* (pp. 179–200). Stanford, CA: Stanford University Press.

Fisek, M. H. (1991). Complex task structures and power and prestige orders. In E. J. Lawler, B. Markovsky, C. Ridgeway, & H. A. Walker (Eds.), *Advances in group processes* (Vol. 8, pp. 115–134). Greenwich, CT: JAI Press.

Fisek, M. H., & Berger, J. (1998). Sentiment and task performance expectations. In J. Skvoretz & J. Szmatza (Eds.), *Advances in group processes* (Vol. 15, pp. 23–40). Greenwich, CT: JAI Press.

Fisek. M. H., Berger, J., & Moore, J. C., Jr. (2002). Evaluations, enactment and expectations. *Social Psychology Quarterly, 65,* 329–345.

Fisek, M. H., Berger, J., & Norman, R. Z. (2005). Status cues and the formation of expectations. *Social Science Research, 34,* 80–102.

Fisek, M. H., & Hysom, S. J. (2004). *Status characteristics and reward expectations: Test of a model.* Paper presented at the annual meeting of the American Sociological Association in San Francisco, August 14–17.

Fisek, M. H., Norman, R. Z., & Nelson-Kilger, M. (1992). Status characteristics and expectation states theory: A priori model parameters and test. *Journal of Mathematical Sociology, 16,* 285–303

Foddy, M., & Riches, P. (2000). The impact of task and categorical cues on social influence: Fluency and ethnic accent as cues to competence in task groups. In S. R. Thye, E. J. Lawler, M. W. Macy, & H. A. Walker (Eds.), *Advances in group processes* (Vol. 17, pp. 103–130). Stamford, CT: JAI Press.

Foschi, M. (1996). Double standards in the evaluation of men and women. *Social Psychology Quarterly, 59,* 237–254.

Foschi, M. (2000). Double standards for competence: Theory and research. *Annual Review of Sociology, 26,* 21–42.

Freese, L. (1974). Conditions for status equality. *Sociometry, 37,* 147–188.

Gallagher, T. J., Gregory, S. W., Jr., Bianchi, A. J., Hartung, P. J., & Harkness, S. (2005). Examining medical interview asymmetry using the expectation states approach. *Social Psychology Quarterly, 68,* 187–203.

Gerber, G. L. (2001). Women and men police officers: Status, gender, and personality. Westport, CT: Praeger.

Goar, C., & Sell, J. (2005). Using task definition to modify racial inequality within task groups. *Sociological Quarterly, 46,* 525–543.

Harrod, W. (1980). Expectations from unequal rewards. *Social Psychology Quarterly, 43,* 126–130.

Harvey, O. J. (1953). An experimental approach to the study of status relations in informal groups. *American Sociological Review, 18,* 357–367.

Heinecke, C., & Bales, R. F. (1953). Developmental trends in small groups. *Sociometry, 16,* 7–38.

Hysom, S. J., & Johnson, C. (2006). Leadership structures in same-sex task groups. *Sociological Perspectives, 49,* 391–410.

Johnson, C. (2003). Consideration of legitimacy processes in teasing out two puzzles in the status literature. In S. R. Thye & J. Skvoretz (Eds.), *Advances in group processes, power and status* (pp. 251–284). Greenwich, CT: JAI Press.

Kalkhoff, W. (2003). Collective validation in multi-actor task groups: The effects of status differentiation. *Social Psychology Quarterly, 68,* 57–88.

Kalkhoff, W., & Barnum, C. (2000). The effects of status-organizing and social identity processes on patterns of social influence. *Social Psychology Quarterly, 63,* 95–115.

Kalkhoff, W., & Thye, S. (2006). Expectation states theory and research: New observations from meta-analysis. *Sociological Research and Methods.*

Leavitt, H. J. (1950). Some effects of certain communication patterns on group performance. *Journal of Abnormal and Social Psychology, 46,* 38–50.

Lockheed, M. E., & Hall, K. P. (1976). Conceptualizing sex as a status characteristic: Applications to leadership training strategies, *Journal of Social Issues, 32,* 111–124.

Lovaglia, M. J., & Houser, J. A. (1996). Emotional reactions and status in groups. *American Sociological Review, 61,* 867–883.

Lucas, J. W. (2003a). Status processes and the institutionalization of women as leaders. *American Sociological Review, 68,* 464–80.

Lucas, J. W. (2003b). Theory-testing, generalization, and the problem of external validity. *Sociological Theory, 21,* 236–253.

Ludwick, R. (1992). Registered nurses' knowledge and practices of teaching and performing breast exams among elderly women. *Cancer Nursing, 15,* 61–67.

Markovsky, B., Smith, R. F., & Berger, J. (1984). Do status interventions persist? *American Sociological Review, 49,* 373–382.

Mohr, P. B. (1986). Demeanor, status cue or performance? *Social Psychology Quarterly, 49,* 228–236.

Moore, J. C., Jr. (1965). Development of the spatial judgment experimental task. Technical Report No. 15. Laboratory for Social Research, Stanford University.

Moore, J. C., Jr. (1985). Role enactment and self-identity. In J. Berger & M. Zelditch, Jr. (Eds.), *Status, rewards, and influence: How expectations organize behavior* (pp. 262–316). San Francisco: Jossey–Bass Publishers.

Munroe, P. (2001). *Creating a legitimated power and prestige order: The impact of status consistency and performance evaluations on expectations for competence and status.* Ph.D. dissertation, Stanford University.

Norman, R. Z., Smith, R. F., & Berger, J. (1988). The processing of inconsistent status information. In M. Webster, Jr. & M. Foschi (Eds.), *Status generalization: New theory and research* (pp. 169–87). Stanford, CA: Stanford University Press.

Platt, J. R. (1964). Strong inference. *Science, 146,* 347–53.

Prescott, W. S. (1986). *Expectation states theory: When do interventions persist?* Unpublished manuscript, Dartmouth College.

Propp, K. M. (1995). An experimental examination of biological sex as a status cue in decision-making groups and its influence on information use. *Small Group Research, 26,* 451–474.

Pugh, M. D., & Wahrman, R. (1983). Neutralizing sexism in mixed-sex groups: Do women have to be better than men? *American Journal of Sociology, 88,* 736–62.

Rainwater, J. A. (1987). *Status cues: A test of an extension of status characteristics theory.* Ph.D. dissertation, Department of Sociology, Stanford University.

Rashotte, L. S. (2006). *Controlling the status effects of gender.* Paper presented at the International Society of Political Psychology Annual Meeting, Barcelona.

Riches, P., & Foddy, M. (1989). Ethnic accent as status cue. *Social Pscyhology Quarterly, 52,* 197–206.

Ridgeway, C. L., & Berger, J. (1986). Expectations, legitimation, and dominance behavior in task groups. *American Sociological Review, 51,* 603–617.

Ridgeway, C. L., & Erickson, K. G. (2000). Creating and spreading status beliefs. *American Journal of Sociology, 106,* 579–615.

Ridgeway, C. L., Johnson, C., & Diekma, D. (1994). External Status, legitimacy, and compliance in male and female groups. *Social Forces, 72,* 1051–1077.

Rosenholtz, S. J. (1977). *The multiple ability curriculum: An intervention against the self-fulfilling prophecy.* Unpublished doctoral dissertation, Department of Sociology, Stanford University.

Shelly, R. K. (1993). How sentiments organize interaction. In E. J. Lawler *et al.* (Eds.), *Advances in group processes* (Vol. 10, pp. 113–132). Greenwich, CT: JAI Press.

Shelly, R. K. (2001). How performance expectations arise from sentiments. *Social Psychology Quarterly, 64,* 72–87.

Shelly, R., & Munroe, P. (1999). Do women engage in less task behavior than men? *Sociological Perspectives, 42,* 49–67.

Sherif, M., White, B. J., Harvey, O. J. (1955). Status in experimentally produced groups. *American Journal of Sociology, 60,* 370–379.

Skvoretz, J., Webster, M., Jr., & Whitmeyer, J. (1999). Status orders in task discussion groups. In S. R. Thye, E. J. Lawler, M. W. Macy, & H. A. Walker (Eds.), *Advances in group processes* (Vol. 16, pp. 199–218). Stamford, CT: JAI Press.

Stewart, P., & Moore, J. C. (1992). Wage disparities and performance expectations. *Social Psychology Quarterly, 55*, 78–85.

Troyer, L. (1999). MacSES v. 5.0. Unpublished software manual.

Troyer, L. (2000). MacSES v. 5.0. Unpublished software manual.

Troyer, L. (2001). Effects of protocol differences on the study of status and social influence. *Current Research in Social Psychology*. Available online at *http://www.uiowa.edu/~grpproc*.

Troyer, L. (2002). The relation between experimental standardization and theoretical development in group processes research. In J. Szmatka, M. Lovaglia, & K. Wysienska (Eds.), *The growth of social knowledge: Theory, simulation, and empirical research in group processes* (pp. 131–147). Westport, CT: Praeger Publishers.

Troyer, L. (2007). Technological issues related to experiments. In M. Webster & J. Sell (Eds.), *Laboratory experiments in the social sciences* (pp. 173–191). Burlington, MA: Elsevier.

Troyer, L., & Younts, C. W. (1997). Whose expectations matter? The relative power of first-order and second-order expectations in determining social influence. *American Journal of Sociology, 103*, 692–732.

Tuzlak, A., & Moore, J. C., Jr. (1984). Status, demeanor, and influence: An empirical assessment. *Social Psychology Quarterly, 47*, 178–183.

Wagner, D. G., & Berger, J. (1985). Do sociological theories grow? *American Journal of Sociology, 90*, 697–728.

Wagner, D. G., & Berger, J. (2002). Expectation states theory: An evolving research program. In J. Berger & M. Zelditch, Jr. (Eds.), *New directions in contemporary sociological theory* (pp. 41–76). Lanham, MD: Rowman & Littlefield Publishers, Inc.

Wagner, D. G., Ford, R. S., & Ford, T. W. (1986). Can gender inequalities be reduced? *American Sociological Review, 51*, 47–61.

Wagner, D. G., & Harris, R. O. (1993). *Manual for experimenters in expectation states theory* (3rd ed.). Technical report. Group Processes Research Office, Department of Sociology, University at Albany.

Walker, H. A., & Cohen, B. P. (1985). Scope statements: Imperatives for evaluating theory. *American Sociological Review, 40*, 288–301.

Webster, M., Jr. (1969). Sources of evaluations and expectations for performance. *Sociometry, 32*, 243–258.

Webster, M., Jr. (2005). Laboratory experiments in social science. *Encyclopedia of Social Measurement, 2*, 423–433.

Webster, M., Jr., & Entwisle, D. R. (1974). Raising children's expectations for their own performance: A classroom application. In J. Berger, T. L. Conner, & M. H. Fisek (Eds.), *Expectation states theory: A theoretical research program* (pp. 211–243). Cambridge, MA: Winthrop.

Webster, M., Jr., & Rashotte, L. S. (2006). *How behavior affects performance expectations*. Paper presented at the International Society of Political Psychology Annual Meeting, Barcelona.

Webster, M. A., Jr., & Sobieszek, B. I. (1974). *Sources of self-evaluation: A formal theory of significant others and social influence*. New York: Wiley.

Webster, M., Jr., Whitmeyer, J. M., & Rashotte, L. S. (2004). Status claims, performance expectations, and inequality in groups. *Social Science Research, 33*, 724–745.

Whitmeyer, J. M., Webster, M., Jr., & Rashotte, L. (2005). When status equals make status claims. *Social Psychology Quarterly, 68*, 179–186.

Whyte, W. F. *Street Corner Society*. Chicago: Chicago University Press.

Yuchtman-Yaar, E., & Semyonov, M. (1979). Ethnic inequality in Israeli schools and sports: An expectation states approach. *American Journal of Sociology, 85*, 576–590.

Zelditch, M., Jr. (1980). How are inconsistencies between status and ability resolved? *Social Forces, 58*, 1025–1043.

Zelditch, M., Jr., Lauderdale, P., & Stublarec, S. (1975). How are inconsistencies in status and ability resolved? Technical Report No. 54. Laboratory for Social Research, Stanford University, Stanford, CA.

CHAPTER 15

Experiments on Exchange Relations and Exchange Networks in Sociology

LINDA D. MOLM
University of Arizona

ABSTRACT

The late 1970s marked the beginning, in sociology, of sustained programmatic research on exchange relations and exchange networks. Because of the emphasis on testing and building theory, nearly all of this work has used in experimental methods and standardized laboratory settings. This chapter reviews the background and development of the field, the standardized laboratory settings that were created to study different forms of exchange, and the technological developments that contributed to advances in both theory and research. Exemplary experiments from four different research programs illustrate the transformation of theoretical concepts into experimental operations.

I. INTRODUCTION

This chapter reviews experimental research programs on social exchange and exchange networks as they have developed in sociology over the last 30 years.

Many forms of interaction outside the economic sphere can be conceptualized as an exchange of benefits. Both social and economic exchanges are based on a fundamental feature of social life: much of what we need and value (for example, goods, services, companionship) can only be obtained from others. People depend on one another for these valued resources, and they provide them to each other through the process of exchange. Social exchange theories focus on this aspect of social life—the benefits that people obtain from, and contribute to, social interaction, and the patterns of interdependence that govern those exchanges. The social structures within which exchange takes place (exchange relations and networks), the different processes through which exchange occurs (such as bargaining, reciprocal gift-giving, and generalized exchange), and the behavioral and affective outcomes of exchange (including power inequalities, coalition formation, commitment, and trust) are all addressed by contemporary exchange theories and experimental programs testing these theories.

In this chapter, I review the theoretical background and historical development of the field that led to the strong tradition of experimental research in social exchange. I describe features of the major standardized settings and designs for the study of exchange, discuss the role of technological advances in their development, and describe several experiments that illustrate the use of the standardized settings to test, modify, and extend exchange theories. I conclude with an assessment of the current state and future prospects for experimental research in this area.

II. BACKGROUND AND DEVELOPMENT

A. Basic Concepts and Assumptions

Social exchange occurs between two or more actors who are dependent on one another for valued outcomes. Social exchange theories assume that actors are motivated to obtain more of the outcomes that they value and others control, that actors provide each other with these valued benefits through some form of social exchange, and that exchanges between the same actors are recurring over time (rather than "one-shot" transactions). These scope assumptions are shared by most theories of exchange and must be met in the experimental settings in which the theories are tested.

The simplest form of social exchange involves just two actors, A and B, each of whom possesses at least one resource that the other values. The actors can be either individuals or corporate groups (e.g., organizations), and the resources can include not only tangible goods and services, but also capacities to provide socially valued outcomes such as approval or status. In exchange experiments, the actors are always individual persons, but sometimes they are given roles as representatives of organizations. Exchange theories make no assumptions about *what* actors value and assume that interaction is unaffected by actors' values or the resources exchanged; this makes them broadly applicable to social relations regardless of content and means that experimental tests of exchange theories can use any resource of known value. Some exchange theories assume "rational" actors who cognitively weigh the potential benefits and costs of alternative partners and actions and make choices that maximize outcomes; others adopt a learning model that assumes actors respond to consequences of past choices, without conscious weighing of alternatives and without necessarily maximizing outcomes.

As Figure 1 illustrates, social exchange can take several distinct forms: direct exchange, generalized exchange, and productive exchange. In relations of *direct exchange* between two actors, each actor's outcomes depend directly on another actor's behaviors; that is, A provides value to B, and B to A, as in the example of two coworkers helping each other with various projects. As Figure 1a shows, such direct exchange relations can occur either in isolated dyads or within larger networks. In relations of *generalized exchange* among three or more actors, each actor gives benefits to another and eventually receives benefits from another, but not from the same actor. Consequently, the reciprocal dependence is indirect; a benefit received by B

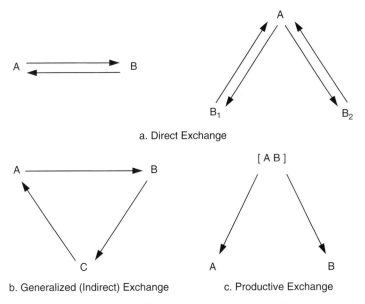

FIGURE 1   Direct, generalized, and productive exchange structures: a. direct exchange; b. generalized (indirect) exchange; c. productive exchange.

from *A* is not reciprocated directly by *B*'s giving to *A*, but, rather, indirectly, by *B*'s giving to another actor in the network. Some forms of indirect exchange (e.g., the classic Kula ring) take a specific circular form, as shown in Figure 1b. Other examples, such as donating blood and reviewing journal manuscripts, do not. Finally, in *productive exchange* (Figure 1c), *both* actors in the relation must contribute in order for either to obtain benefits (for example, coauthoring a book).

Although generalized exchange was a particular interest of early anthropological exchange theorists, the study of direct exchange relations has dominated research and theorizing in sociology since Homans (1961). Direct exchanges can be *negotiated* or *reciprocal*. Both forms of direct exchange have been the subject of long-term research programs. In negotiated exchange, actors jointly negotiate the terms of an agreement (usually binding) through a series of offers and counteroffers. Each agreement comprises a discrete transaction that provides benefits for both actors. In reciprocal exchange, actors perform individual acts that benefit another, such as giving assistance or advice, without negotiation and without knowing whether, when, or to what extent the other will reciprocate. Exchange relations develop when beneficial acts prompt reciprocal benefit.

B. Historical Development

The early theories. The development of the exchange perspective in sociology began around 1960, with the publication of theories by George Homans (1961), Peter Blau (1964), and John Thibaut and Harold Kelley (1959). The experimental research tradition, in contrast, did not begin until the late 1970s. The early works of Homans and Blau demonstrated the ubiquity of exchange processes in social life and stimulated interest in (and controversy over) the perspective, but their narrative style was not conducive to experimental test. Thibaut and Kelley (both psychologists) provided a more analytical framework for the study of social exchange and introduced a tool for describing patterns of interdependence in relations—the outcome matrix, which did stimulate research, although its usefulness is largely restricted to dyads. The outcome matrix was used in experimental research throughout the 1960s and 1970s. Much of this research on gaming, bargaining, cooperation, and the like provided the impetus for later, more sophisticated studies of social exchange.

Emerson's power-dependence theory. The contemporary tradition of social exchange and the rise of programmatic experimental research on exchange began with the publication of Richard Emerson's exchange formulation in 1972 (1972a, b). Emerson's theory, which built upon his earlier (1962) work on power-dependence relations and, consequently, came to be called power-dependence theory, influenced the development of theory and research on exchange in several important ways. First, his rigorously derived system of propositions moved the tradition towards a more formal, analytical approach to theory that was more amenable to experimental test. Second, Emerson made the structure of relations, rather than the actors themselves, the central focus of his theory, and exchange networks replaced dyadic relations as the primary structural unit. Third, the theory established power and its use as the major topics of exchange theory, topics that would dominate research on social exchange until the late 1990s. Fourth, the research program that Emerson, Karen Cook, and their students began in the late 1970s shifted the focus from the reciprocal exchanges of the classical exchange theorists to negotiated exchanges in which actors bargained over the terms of agreements.

One of Emerson's most influential contributions was the concept of *exchange networks,* defined as sets of direct exchange relations that are *connected* to one another. Connected relations are linked by a focal actor (for example, A–B–C), and exchange in one relation (for example, the A–B relation) affects the frequency or value of exchange in the other (the B–C relation). Connections are *positive* to the extent that exchange in one relation increases exchange in the other (for example, B's exchange with A might give B a resource that B can then use in exchange with C), and *negative* to the extent that exchange in one relation decreases exchange in the other.

Negatively connected relations form the structural foundation for Emerson's theory of power-dependence and for most theories of power in exchange networks that were subsequently developed.[1] Negative connections provide actors with *alternative* exchange partners who compete with one another for the opportunity to exchange with the focal actor. For example, in the negatively connected network B_1–A–B_2, A might be an employer and the Bs applicants for the same job, or the Bs might be potential tennis partners for A. Alternative partners decrease an actor's dependence on any one exchange partner, giving that actor an advantage in exchanges with more dependent partners. Alternative partners can vary in both their *value* and their *availability* as an exchange partner. The former is a function of the value to an actor of the resources controlled by the alternative, and the latter is a function of the size and shape of the exchange network. The more valuable and the more available A's alternative to B is, the less dependent A is on B.

Emerson proposed that power in the A–B relation can be described by two dimensions: the power balance or imbalance in the relation (i.e., actors' *relative* power over each other, determined by the difference in their dependencies) and the cohesion in the relation (actors' *absolute* power over each other, i.e., their total dependencies on each other). Later theories focused almost exclusively

---

[1] In the network exchange tradition developed by Barry Markovsky and David Willer, negative connections are called exclusive connections (Szmatka & Willer, 1995).

on relative power. Over time, the structure of power produces predictable effects on the frequency and distribution of exchange benefits. The more dependent actors are on each other, the more frequently they exchange with each other; the more imbalanced (unequal) their power dependencies are, the more unequal their exchange is, with the less dependent, more powerful actor receiving more benefits at lower cost from the exchange relation.

The emergence of new theories of power in networks. With the concept of exchange networks, theory and research on social exchange shifted from the study of dyadic relations to the larger opportunity structures within which those relations are embedded. How network structures affect the availability of alternatives and how the value and availability of alternatives affect the distribution of power became the central focus of experimental research on exchange networks during the 1980s and 1990s. Richard Emerson and Karen Cook conducted the first experimental tests of Emerson's power-dependence theory in the late 1970s and early 1980s, showing that power use—the asymmetry in exchange benefits between *A* and *B*—can be predicted from the structural power relations in exchange networks. As their research program developed, however, it became apparent that Emerson's formulation was inadequate for analyzing power relations in more complex networks (Cook, Emerson, Gillmore, & Yamagishi, 1983). As a result, new theories of network exchange and power emerged (including a reformulation of power-dependence theory), with varying algorithms capable of predicting the distribution of power in networks as a whole. The competition among these theories led to substantial growth in experimental research on exchange and became one of the most significant developments of the contemporary exchange tradition.

The major competitor to power-dependence theory to emerge, and the one that has produced the largest body of experimental research, is the network exchange theory (NET) of Barry Markovsky, David Willer, and associates (Markovsky, Willer, & Patton, 1988). This theory draws on Willer's elementary theory; later, it was modified to incorporate a resistance model. Other approaches include Friedkin's (1992) expected value theory and Bienenstock and Bonacich's (1992) application of core theory. All of these theories use formal mathematical models to predict the distribution of power in exchange networks, and all focus primarily on negotiated exchanges in negatively connected networks that vary in size, shape, and complexity.

Network exchange theory uses a path-counting algorithm called the graph-theoretic power index (GPI) to predict the relative power of each position in a network; positions with higher GPI scores are predicted to receive a larger share of the profit from agreements. Like power-dependence theory, NET assumes structural power is derived from the availability of alternative partners, but, unlike power-dependence theory, NET does not consider variations in the potential value of alternatives as a determinant of power. Alternatives are more available to an actor if they lie on odd- rather than even-length paths (i.e., alternatives on odd-length paths have no alternatives themselves, or their alternatives are less available because they have other alternatives). Therefore, odd-length paths increase power while even-length paths decrease it. Differences in availability affect the likelihood that some actors will be excluded from exchange on each opportunity; exclusion increases power use by driving up the offers that excluded actors make on subsequent negotiation opportunities.

Other questions, new directions. At the same time that research on power in exchange networks was developing, other research programs, particularly one headed by Edward Lawler and a second directed by Linda Molm, were studying somewhat different exchange questions: how dynamic processes of power use (and not only structure) affect exchange patterns, and how the capacity and use of both reward-based and punitive power affect exchange. Both Lawler and Molm argued that power use can be strategic as well as structurally induced, and both studied power that was based not only on control over rewards, but also on control over punishments. Lawler's (1992) work continued the focus of exchange researchers on bargaining relations (negotiated exchanges), while Molm's (1981, 1997) work introduced the study of reciprocal exchanges in which actors provide benefits to each other without bargaining or negotiation. Such exchanges, which were the focus of the classical exchange theorists, are arguably more common in social life than the negotiated exchanges that most contemporary researchers have studied.

In their work on power, both Lawler and Molm continued to emphasize structure as the dominant force on exchange. But they also brought in consideration of other factors: cognitions, affect, risk, and fairness. This work set the stage for the development of new directions in exchange theories and research.

Beginning in the 1990s and continuing to the present, exchange theorists began to pay renewed attention to some of the long neglected concerns of the classical theorists: the risk and uncertainty inherent in exchange (particularly generalized exchange and reciprocal exchange), the emergence of affective ties between exchange partners and their ability to transform the structure and nature of exchange, and the effects of different forms of exchange—negotiated, reciprocal, generalized, and productive—on the development of trust, commitment, and affective attachments. Lawler and colleagues developed theories of relational cohesion and affect in social exchange that emphasized the causal role of emotions in social exchange processes; Molm and colleagues began systematic comparisons of different forms of exchange and studied how the structure of reciprocity in exchange affects the development of trust and solidarity; and Peter Kollock, Toshio Yamagishi, and Karen Cook studied the relations among uncertainty, trust, and commitment.

This work shifted theories and research in the social exchange tradition in two important ways: first, toward consideration of forms of exchange other than direct negotiated exchange, and, second, to the study of integrative outcomes (trust, commitment, affective ties) rather than the differentiating outcomes (power and inequality) that had dominated exchange research.[2] These changes also led to the development of new experimental settings, particularly for the study of different forms of exchange.

III. STANDARD SETTINGS AND DESIGNS

Because most theories of exchange are concerned with basic causal processes linking various aspects of exchange, most of the research testing these theories uses experimental methods and standardized laboratory settings. Laboratory experiments have several advantages for studying exchange and for testing theories formulated at the abstract level of most exchange theories.

As we have seen, most contemporary theories of exchange are concerned, in some way, with the effects of exchange *structures*. In natural settings it can be very difficult to separate the effects of characteristics of structures from the effects of characteristics of actors who occupy particular positions in the structures. In laboratory experiments, random assignment of subjects to positions in exchange relations or networks accomplishes this task. It can also be difficult in natural settings to measure or compare exchange relations that are based on different resources (such as approval, expert advice, or status); many exchange resources cannot easily be quantified, and their value varies across different actors and changes over time. In laboratory experiments, a single, widely valued, quantifiable resource is used: money. Typically, subjects earn points from exchanges that are equal to money and those points are exchanged for money at the conclusion of the experiment.

In developing a laboratory setting for the study of social exchange, several considerations are important. First, the setting must meet the general scope conditions of the theory. For most exchange theories, this means that the setting must (1) give actors control over resources that provide outcomes of value to other actors, (2) structure exchange relations and networks that create mutual dependencies among actors for those outcomes, and (3) provide repeated opportunities for exchange among the same actors

---

[2]For examples of work on these topics, see Kollock (1994), Lawler (2001), Lawler and Yoon (1996), Lawler, Thye, and Yoon (2000), Molm, Collett, and Schaefer (2006a), Molm, Takahashi, and Peterson (2000, 2003), Yamagishi and Cook (1993), and Yamagishi, Cook, and Watabe (1998).

(or among interchangeable occupants of the same positions). Second, the setting must make it possible to manipulate dimensions of the structure and process of exchange and measure their effects on exchange outcomes (such as power use, behavioral commitment, and affective ties) in ways that are consistent with the theory's conceptual definitions of these concepts. Third, variables that are unrelated to the theory and extraneous to the research must be either controlled or randomly distributed across the experimental conditions.

Several standardized laboratory settings have been created that meet these requirements in various ways. The first standardized setting for the study of social exchange networks was developed by Karen Cook and Richard Emerson (1978),[3] who created a setting for the study of negotiated exchange. Basic parameters of their setting were later adopted, with some notable differences, by researchers testing other theories of network exchange and power. Samuel Bacharach and Edward Lawler (1981) also developed a standardized setting for studying power in negotiated exchange, but in dyads rather than networks. Linda Molm (1981) created a setting for the study of reciprocal exchange and later (1997) extended it to include both reward-based and coercive exchanges. More recently, several researchers have created settings for studying generalized and productive exchange, although no standardized setting—that is, a setting used in multiple experiments—has emerged yet.

All of these settings share certain features. Subjects are randomly assigned to positions in a particular network or dyadic structure (sometimes varied as one of the factors in the experiment), and their exchange behavior within that structure is studied. The network creates an opportunity structure for exchange and determines the actors in the network with whom each subject can exchange. Variations in the size, shape, types of connections, and potential value of exchange relations create differences in power and dependence. To assure that behavior is affected solely by manipulated (typically structural) characteristics of the exchange relations or networks and not by actors' personal characteristics, subjects typically interact with one another through computers, rather than face to face. They engage in repeated exchanges with other actors in the network, varying in number from 20 to several hundred in different experiments and different research programs.

The benefits that subjects receive from these exchanges are operationalized as points, equal to money; subjects earn money through exchanges and at the

---

[3]An earlier setting was developed by John Stolte and Richard Emerson (1977); however, it was never established as a standardized setting used in multiple experiments.

conclusion of the experiment are paid what they earn.[4] To meet the assumption of self-interested actors, subjects are recruited on the basis of their interest in earning money. Because no other individual characteristics are theoretically relevant, most researchers take advantage of the convenience of undergraduate students as the potential pool of subjects.

The following sections describe the specific features of the settings developed to study the different forms of exchange: negotiated, reciprocal, generalized, and productive.

A. NEGOTIATED EXCHANGE SETTING

In the original Cook and Emerson (1978) setting, subjects negotiated the terms of exchange, through a series of offers and counteroffers, to reach binding agreements on each of a series of negotiation opportunities. The agreements reached determined how many points—equal to money—each subject received. During each transaction period, subjects could send offers and counteroffers to any of the partners to whom they were connected in their network. If an agreement was reached during that time, both subjects received the benefits they had agreed upon; if no agreement was reached before the end of the transaction period, no benefits were received on that opportunity.[5] To eliminate effects of equity concerns on exchange agreements, subjects had no knowledge of the benefits their partners received from their agreements. In reality, subjects divided a fixed amount of profit between them, but they were unaware of the division, the total profit, or their partner's gain.

The Cook and Emerson setting (like most others that followed) was specifically designed for the study of power in negatively connected networks.[6] To create the negative connections, a subject's exchange with one partner precluded exchange with another partner on that opportunity. Thus, each actor

---

[4]Money is used as the valued benefit because of its advantages for experimental control: money is widely valued, it can be quantified to produce a ratio level of measurement, and it is resistant to the effects of satiation or diminishing marginal utility (which would alter the value of the resource in unknown ways). The exchange resource in the experiments is not money per se—that is, money is not transferred from one actor to another, as in economic exchanges—but rather the capacity to produce valued outcomes, operationalized as money, for another.

[5]In the Cook and Emerson setting, each transaction period was of a specified length of time; in other negotiated exchange settings, a specified number of rounds of offers and counteroffers is allowed. For example, in Lawler and Yoon's (1996) negotiated exchange setting (which Molm and colleagues adopted with minor modifications), subjects had five rounds to make an agreement.

[6]Positively connected networks have been studied by Yamagishi, Gillmore, and Cook (1988); related distinctions in the network exchange tradition between exclusive and inclusive connections have been studied by Szmatka and Willer (1995).

in the network could make only one agreement per opportunity. The setting was also designed to test Emerson's assertion that power leads to power use regardless of actors' knowledge or intentions; accordingly, subjects were not informed of the size or shape of the network beyond their immediate connections. This meant that actors were unaware of any power advantage (or disadvantage) that they or others in the network enjoyed.

As competitors to power-dependence theory emerged, including Markovsky and Willer's network exchange theory, Friedkin's expected value theory, and Bienenstock and Bonacich's core theory, these researchers adopted the basic parameters of the Cook and Emerson setting, particularly the process of having subjects bargain over the division of a fixed pool of profit points. Some important modifications were made, however. In the original experiments testing network exchange theory (e.g., Markovsky et al., 1988), subjects were given full information of the network structure, the points each actor received from agreements, and the earnings of all positions in the network.

Rather than using restricted information to control effects of subjects' equity concerns, the network exchange setting rotated subjects through all power positions in a network. The rationale for this procedure (as a control for equity effects) is that since all subjects know they will be in both high- and low-power positions at some point, all should try to maximize their earnings in each power position rather than striving for equal agreements. From an experimental standpoint, however, the rotation is a within-subject variable with potential order effects; that is, the order in which subjects occupy the different positions may affect their behavior. Later experiments in this tradition returned to the restricted information procedures of Cook and Emerson, limiting information about the network and other subjects' earnings to control equity effects rather than rotating subjects through positions (e.g., Thye, Lovaglia, & Markovsky, 1997).

The network exchange tradition also treated the Cook and Emerson procedure for operationalizing negative connections—restricting the number of exchanges that a subject could make on each negotiation opportunity—as a variable condition of exchange. Varying this number can change the network connections, depending on the size of the network relative to the number of exchanges. As long as the number of exchanges that an actor can make on an opportunity is *less* than the number of potential partners with whom the actor can exchange, the relations that connect that actor to those partners will be negatively connected; that is, the actor's exchange with one partner will decrease the probability of exchange with another partner. But if the number of exchanges that an actor can make on an opportunity *equals* or *exceeds* the number of potential exchange partners for that actor, then the actor's exchange with one partner will have no effect on his or her exchange with other partners; the actor can exchange with *all* of the partners on each opportunity.

B. Reciprocal Exchange Setting

In Molm's (1981, 1997) reciprocal exchange setting, subjects exchange by individually performing acts—adding to or subtracting from a partner's points—that have rewarding or punishing consequences for the partners. In the standard setting, the points are of fixed value, which means that variations in the equality or inequality of exchange can occur only over time, according to differences in subjects' rates of giving to one another.

On each of a series of exchange opportunities, subjects choose which partner to give points to (or, if both rewarding and punishing actions are possible, subjects choose both a partner and an action). Initiating exchange with one partner precludes initiating exchange with an alternative partner on that opportunity, thus creating negative connections between exchange relations. All subjects in the network make these choices simultaneously, without knowing in advance whether or when the target partner will reciprocate. They are then informed of their partners' behaviors: whether each of their potential exchange partners added to (or subtracted from) their earnings or did not act toward them.

Thus, on any given exchange opportunity, a subject might give to another without receiving, receive without giving, or reciprocally give and receive. Reflecting the learning model on which this exchange setting is based, subjects typically engage in exchanges for several hundred opportunities—far more than in the negotiated exchange experiments. In the absence of explicit bargaining or knowledge of others' intentions, subjects can influence one another by making their behavioral choices contingent on their partners' previous choices.

Like the Cook and Emerson setting for negotiated exchange, the Molm setting for reciprocal exchange typically gives subjects only limited information about the network structure (i.e., subjects know only their immediate connections) and no information about the value of exchange benefits to their partners or their partners' cumulative earnings. The exception is experiments that are explicitly designed to study perceptions of fairness; then, subjects necessarily are given the full information necessary for making fairness judgments. They know the size and shape of the network, the potential value of exchange in each of their relations, and the number of points that they and their partners received from their exchanges.

Other variations in the standard reciprocal exchange setting have included changing the amount given to the partner from a fixed value to a variable value, chosen by the subject from a range of points, and allowing subjects to keep any points not given to the partner (Molm, Collett, & Schaefer, 2006a). In the latter variation, points given to another triple in value, while points kept for self remain the same. These variations introduce costs other than the opportunity costs that are part of the standard setting. They also illustrate, more generally, the use of systematic variation in experimental parameters within a standardized setting. Such variations help the researcher to tease out what aspects of a variable are theoretically important for producing particular effects and also help the researcher to rule out alternative explanations of findings.

C. GENERALIZED EXCHANGE AND PRODUCTIVE EXCHANGE SETTINGS

Generalized and productive exchange settings, both of which involve collective systems of exchange rather than the dyadic relations of direct exchange, are far less developed. Few sociologists have studied these forms of exchange, and no true standardized settings (i.e., settings used across multiple experiments) have been developed. Yamagishi and Cook (1993) created a chain- or network-generalized exchange structure by giving participants a divisible resource (10¢) on each trial and allowing them to decide how much of this resource to give to the next person in the chain; points given to the other were doubled in value, while those kept for self remained the same. Molm and colleagues (Molm, Collett, & Schaefer, in press) and Lawler and colleagues (Thye, Lawler, & Yoon, 2006) have recently developed generalized exchange settings, as well. Molm operationalizes chain-generalized exchange as a variant of her reciprocal exchange setting, in which subjects chose whether or not to give a fixed number of points to their recipient in the chain (while receiving, or not receiving, points from their benefactor in the chain). The Thye, Lawler, and Yoon setting is similar to Molm's, but includes a default benefit for subjects who choose not to give to their recipient.

Productive exchange has been studied only by Lawler, Thye, and Yoon (2000), who operationalize it as a variant of Lawler and Yoon's (1996) negotiated exchange setting. Three subjects engage in a joint venture, with each negotiating for a share of the profits produced by the joint venture. As in dyadic negotiations, subjects must agree on a division of profit—in this case, a three-way division—before any can benefit.[7]

---

[7]Experiments on public goods (see Sell, Chapter 18, this volume) also involve a form of productive exchange in which actors contribute to a collective good, which then benefits all members of the group. The main distinction is that in productive exchange, free riding is not possible because (1) all members must contribute to produce the collective good, and (2) the collective good is not a "public" good whose benefits can be enjoyed by anyone; instead, its benefits are restricted to those who contributed to its creation.

D. Measurement of Dependent Variables

In the two direct-exchange settings, behavioral measures include the frequency of exchange (measured by the frequency of agreements in negotiated exchange and the frequency of reciprocated giving in reciprocal exchange), the inequality of exchange, behavioral commitment to exchange partners, and (more rarely) the formation of coalitions of actors as a power-balancing strategy. In negotiated exchange, inequality (or power use) is measured by the agreed-upon division of the profit pool—that is, by the difference in points received by the two actors from their agreements. In reciprocal exchange, inequality is measured over time, by comparing the relative frequencies of rewarding behaviors that the two partners perform for one another. Exchange researchers also study the behavioral commitments that develop in particular dyadic relations within a network; commitment measures typically compare the number of exchanges made with two or more alternative partners (Cook & Emerson, 1978; Kollock, 1994). Most exchange experiments do not allow actors to form coalitions; in the few that have, measures of which coalitions form and how often coalitions form have served as dependent variables (Cook & Gillmore, 1984).

Integrative outcomes such as trust, affective regard, perceptions of relational cohesion, and positive emotions are measured both through responses to semantic differential items asking subjects to evaluate their partners and relationships and through some behavioral tasks. Lawler and Yoon (1996), for example, have measured commitment to partners by giving subjects the opportunity to give gifts to one another or to contribute to a joint venture. Responses to multiple semantic differential items are typically combined in scales to increase the reliability of the measure.

E. Designs and Design Issues

Most exchange experiments employ mixed between- and within-subject designs, typically varying such factors as network structure, relative power imbalance, or form of exchange "between subjects" (often in factorial designs, to test interactions), while treating actor positions within networks and trial blocks within the exchange period as "within-subject" variables.[8] Changes in behaviors over trial blocks are often studied to examine specific theoretical questions (e.g., whether power use increases or decreases over time), or to determine

---

[8]Here, of course, the experimental terms "between subject" and "within subject" refer to between or within networks (or, in the case of trial blocks, the exchange period), since the experimental unit is the network rather than the individual subject.

whether or when exchange patterns stabilize. Actor positions are compared to measure the inequality of exchange benefits (as a measure of power use) or to examine how actors who are relatively advantaged or disadvantaged on power differ in trust or feelings toward their partners.

The design employed also depends on the level of theory development. Theories that make ordinal predictions (i.e., predicting that mean values of a dependent variable will be higher in one condition than in another) typically use designs and methods of analysis that compare experimental conditions with each other. Theories that make specific point predictions compare expected values with observed values within experimental conditions.

Because both the actors within a relation or network and their transactions or exchange behaviors over time are interdependent, the unit of analysis for exchange experiments must be the relation or network and, for analyses of behaviors, some single measure of interaction for the exchange period must be used (e.g., mean behavior for the entire exchange period or for the last half of the exchange period). Recognition of these interdependencies is also theoretically consistent with the assumption of recurring exchanges between the same actors and the development of *social exchange relationships*. In this respect, the network exchange tradition makes assumptions more typical of economic exchanges, often treating the transaction (rather than the exchange relation) as the unit of analysis and correcting for interdependencies through statistical means (Skvoretz & Willer, 1991).

IV. TECHNOLOGICAL DEVELOPMENTS

A. COMPUTER-MEDIATED EXCHANGE

Technological advances in computers and computer networks have shaped the development of exchange programs in several important ways. First, they facilitate the high level of control that is a hallmark of laboratory experimentation. Laboratories physically isolate subjects from the influence of extraneous stimuli that might produce either systematic bias or random "noise" in experimental data. In the study of social exchange, these stimuli include characteristics of the exchange partner. Computer-mediated interaction allows participants to interact with one another without knowing (or potentially being influenced by) characteristics of the other such as gender, race, year in school, etc. It also allows researchers to structure the exchange process in particular ways, assuring that all participants have the same kinds of information, interact for the same amount of time, make decisions and receive feedback in the same sequences, and so forth.

A 1991 study by John Skvoretz and David Willer compared a face-to-face exchange setting (used in the early research of Markovsky *et al.*, 1988) with a computer-mediated setting to study the distribution of power in networks of negotiated exchange. The face-to-face setting used partitions to separate participants who were not connected to each other in the network (and thus could not exchange with each other), but subjects could hear one another's offers and comments. Despite instructions to the contrary, these comments included suggestions of various kinds of collective strategies, claims of injustice, and communication of "symbolic sanctions." In the computer-mediated setting, subjects were seated in separate rooms and could not see or hear others; thus, influence from such comments was avoided. While the data generally fit predictions in both settings, there were some indications that comments related to justice may have suppressed power use in at least one structure.

More generally, computer-mediated interaction allows information and communication to be as highly constrained or as free as desirable, according to the research objectives. Earlier technologies were more restrictive. In early research on reciprocal exchange, for example, subjects interacted via computer-operated human test consoles that contained only push buttons, stimulus lights, and counters (Molm, 1981). In this setting, information was necessarily very minimal. When the test consoles were replaced with computer monitors, a much greater range of information was possible, and the kinds and amounts of information that subjects had about various aspects of the exchange—the size and shape of networks, the behavioral choices of other actors, the value of benefits that others receive—could be varied or controlled.

Second, advances in computer networking capabilities have made the manipulation of exchange networks and subjects' assignments to positions in those networks a relatively simple matter. Subjects can be physically situated in a particular room and yet be assigned to any network, occupy any position within a network, and be connected (or not connected) to any other positions in the network, as specified by the researcher. No cumbersome partitions are required, and the physical use of space is consequently more efficient. Computers have also made it possible to refine and control the bargaining process in negotiated exchange settings in ways that were not originally anticipated. Early experiments on negotiated exchanges simply allowed subjects to make offers and counteroffers to whomever they wanted, whenever they wanted, within a particular time frame. Later settings added more control over this process—for example, by specifying that all actors must make offers to all partners on each exchange opportunity, by limiting the number of rounds of offers and counteroffers that are possible, and by constraining the range of offers and counteroffers that can be made.

Third, computers have made the use of "programmed" or "simulated" partners easier, more sophisticated, and more realistic. While many experiments on exchange are conducted on networks of all real subjects, certain research questions are best answered with the help of simulated partners whose behavior is either held constant or manipulated as one of the experimental factors.

If, for example, the researcher is interested in comparing perceptions of justice—either distributive or procedural—in different exchange networks or different forms of exchange, it is typically desirable to manipulate or control the equality or inequality of exchange outcomes as well as some aspects of the partner's behavior. More generally, any time the research focus shifts from interaction patterns at the level of relations or networks to the responses (behavioral or affective) of individual actors, it is desirable to control the behavior of the other actors in the network by using computer-simulated actors for those positions.

B. Web-Based Experiments

A recent development with potential implications for experimental work on exchange networks is the advent of Web-based experiments, which make use of the internet to conduct experiments outside the laboratory. Rather than having subjects go to a laboratory to participate in a 2-hour experiment with other subjects (typically other students at the same university), participants from across the country—or from multiple countries—can participate in experiments in which they interact with others through the Internet. Web-based experiments potentially allow much larger and more complex exchange networks to be studied, using more diverse populations of subjects, with interaction extending over longer periods of time.

While the potential of Web-based experiments is great, the hurdles are also formidable. These include the challenges of software development and problems of ensuring experimental control, effectiveness of manipulations, and security of data. At this point in time, it remains to be seen whether Web-based experiments will become widely used and, if so, what kinds of contributions they will make to the experimental study of exchange and exchange networks.

V. SOME EXAMPLES OF EXCHANGE EXPERIMENTS

To illustrate how different researchers have tested theories of exchange and exchange networks in standardized laboratory settings, let us consider four experiments. These particular experiments represent four distinct research programs and illustrate the diversity of work conducted in the exchange tradition. They employ different settings for studying different forms of exchange; place varying emphasis on structure or process, networks or relations; and study different exchange outcomes, including power and inequality, commitment, and trust. They also span more than 20 years of research in the development of the contemporary exchange tradition.

A. COOK AND EMERSON: POWER AND EQUITY
IN EXCHANGE NETWORKS

The first experiment is the classic Cook and Emerson (1978) experiment studying how power and equity affect the use of power in exchange networks. The primary objective of this study was to test the central thesis of Emerson's power-dependence theory: that networks of power-balanced relations will produce equal benefits for actors, while networks of power-imbalanced relations will produce unequal benefits, with the distribution of benefits favoring the less dependent, more powerful actor. A second objective was to examine the constraining effects of equity concerns on the use of power, testing the prediction that power use would be lower under conditions where significant equity concerns would be expected to operate.[9]

The researchers tested these predictions in the negotiated exchange setting described earlier. Undergraduate student subjects were recruited on the basis of a desire to earn money and were randomly assigned to positions in one of two networks: a power-balanced network or a power-imbalanced network (see Figure 2a). Relations in the networks were negatively connected, operationalized by allowing subjects to make an agreement with only one of their three alternative partners on each of 40 opportunities to exchange. The balance or imbalance of power was manipulated by varying the *value* of actors' alternatives.

In the power-balanced network, all potential exchange relations were of equal value (worth 24 points, divided through negotiations); thus, all actors in the network were in equivalent positions of power. In the power-imbalanced network, actors were *not* in equivalent positions of power. One actor, *A,* had three high-value (24 points) alternative exchange relations, while the three *B*s had only one relation (with *A*) of high value and two others (with the other *B*s) of low value (8 points). Thus, *A* had a power advantage over the *B*s. The theory predicted that over time, power use would increase in *A*'s favor: the *B*s—who were competing with each other to make the more valuable agreements with *A* on each opportunity—would offer more and more of the 24 points to *A* in order to be the one *B* with whom *A* made an agreement on that opportunity.

An important assumption of power-dependence theory that Cook and Emerson were testing was the principle that power use is *structurally determined*—in this case, by the availability of high-value exchange relations—regardless of actors' awareness of power or intent to use power. To test this central tenet, subjects were told only of their immediate connections to others in the network; they were unaware of the size or shape of the network as a whole.

---

[9]The study also examined the effects of emerging commitments among exchange partners on the use of power.

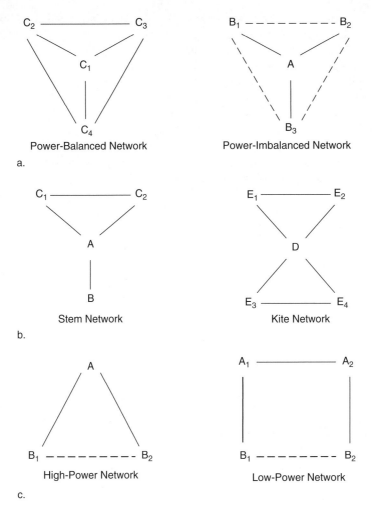

FIGURE 2 Networks in three exchange experiments. (Solid lines indicate relations with high exchange value; dashes indicate relations with low exchange value.) a. Networks studied by Cook and Emerson (Cook, K. S., & Emerson, R. M., 1978, *American Sociological Review, 43,* 721–739); b. Weak-power networks studied by Markovsky and colleagues (Markovsky, B., Skvoretz, J., Willer, D., Lovaglia, M., & Erger, J., 1993, *American Sociological Review, 58,* 197–209); c. Networks studied by Molm and colleagues (Molm, L. D., Takahashi, N., & Peterson, G., 2000, *American Journal of Sociology, 105,* 1396–1427).

The researchers manipulated equity concerns by varying the information subjects had about others' earnings. For the first 20 trials, subjects knew only their own cumulative earnings; then, they received information of the cumulative earnings of *all* participants in the network. That meant that, for the last

20 trials, subjects in the power-imbalanced networks bargained with knowledge of the inequalities in benefits between actors in the A and B positions. This knowledge should trigger equity concerns, which were predicted to reduce power use in the power-imbalanced networks.

The design of this study, then, was a mixed design, containing a between-subjects variable (power balance/imbalance) and a within-subjects variable (equity). In addition, four trial blocks of five trials each were nested within the two conditions of the equity manipulation, allowing the examination of change over time. The primary dependent variable was power use, measured as a difference score (the difference in benefits received by the two actors in an agreement). Because of the interdependencies between subjects in a network and transactions across the exchange period, the entire four-person network, interacting through 40 3-minute transaction periods, was the statistical unit for the analyses. Fourteen networks were studied in each of the two power conditions.

Findings supported the main predictions: mean power use was significantly greater in the power-imbalanced network than in the power-balanced network, and it increased over time up to the time of the equity manipulation. Then, as predicted, knowledge of everyone's earnings produced a significant and marked decrease in power use in the power-imbalanced networks.

B. Markovsky and Colleagues: Strong and Weak Power

Theoretical advances sometimes occur when experiments produce unexpected findings that challenge current theoretical formulations and lead to modifications of a theory and further experimental test. This study, conducted by Markovsky, Skvoretz, Willer, Lovaglia, and Erger in 1993, is an example of this process.

In the original formulation of network exchange theory, power in exclusionary networks (called "negatively connected" in power-dependence theory) was measured solely by the GPI. This index predicted power use quite accurately in some networks, but in others, weak power differences emerged even though the GPI predicted no differences in power. In this study, Markovsky and colleagues developed the distinction between "strong power" networks and "weak power" networks and tested a modified version of the theory designed to do a better job of predicting the small but consistent differences that occur in weak power networks.

Powerful actors in strong- and weak-power networks vary in their capacity to exclude others consistently from negotiated exchanges without suffering costs themselves. In strong power networks, one or more disadvantaged actors are excluded on every opportunity from exchanges with a powerful actor, at no cost to the powerful actor. For example, $B_1–A–B_2$ is a strong power network because one of the Bs is always excluded from exchange with A on each opportunity. According to the path-counting algorithm of the GPI, this occurs because A's two alternatives—the two Bs—lie on paths of length 1 and are highly available for A, while the Bs' single alternative—A—lies on a path of length 2 and is less available to either B.

In weak power networks, either all positions are equally subject to exclusion or no position can consistently exclude another without incurring cost to itself. For example, B_1–A_1–A_2–B_2 is a weak power network: A_1's immediate alternative to B_1—the other A—lies on a path of length 2 and has another potential partner. The two As can exclude the Bs only by exchanging with each other, but this action is costly because agreements between the equal-power As should provide each with no more than half the benefits. Nevertheless, positions in weak-power networks still differ in their *probabilities* of inclusion or exclusion, and an iterative refinement of the GPI was developed to capture these more subtle differences in power. Markovsky and colleagues then tested the predictions of the modified theory in two weak-power network structures, a four-actor network called the "stem," and a five-actor network called the "kite" (see Figure 2b).

The negotiated laboratory setting included the features described earlier: subjects' computer screens displayed full information of the network configuration, current offers, and completed exchanges, and subjects were rotated (by a software configuration) through the different network positions, with four negotiation rounds in each position. A total of eight networks were run in the stem configuration and six networks in the kite configuration. Predictions specified the relative ordering of exchange outcomes (i.e., which actor of each two-party pair would get more from his or her agreements) for structurally distinct relations within the two networks. The units of analysis were the observed exchanges and nonexchanges among pairs of subjects, and a dummy-variable, constrained-regression analysis was used to estimate positional effects (Skvoretz & Willer, 1991). The findings confirmed the predicted "weak-power" effects: in the stem network, agreements between A and either B or C favored A; in the kite network, agreements between D and E favored D.[10]

C. Lawler and Yoon: Commitment in Exchange Relations

Lawler and Yoon also investigated the effects of power structures in a series of experiments, but the exchange outcome in which they were interested was the

[10]This study also tested hypotheses about the power use of "experienced" versus "inexperienced" subjects, and about strong versus weak power networks. For the latter prediction, the results from this experiment were compared with results from earlier experiments on strong power networks (Markovsky et al., 1988).

development of cohesion and commitment between exchange partners, not power use, and their focus was on dyadic relations, not exchange networks. In a 1996 experiment, they tested a theory of relational cohesion, which predicted that Emerson's two dimensions of structural power (the total power-dependence in the relation, and the relative power-dependence—or power imbalance—in the relation) affect the frequency of agreements between two parties, and that frequent, successful exchanges produce positive emotions that lead to perceptions of the relation as a cohesive unit and various forms of commitment behaviors. They conducted three experiments, each examining a different kind of commitment behavior: giving token gifts to the partner, staying with the partner in the face of alternatives, and contributing to a joint venture. The gift-giving experiment is described here.

Undergraduate subjects were assigned to roles representing either Alpha Company or Beta Company. Alpha was trying to buy iron ore from Beta, and the task of both representatives was to negotiate the best possible deal for their companies. Subjects knew only their own profit, not their negotiation partner's profit, from any agreements reached. The subjects negotiated over 12 episodes but were told to expect 15, a procedure designed to prevent "end effects" (i.e., strategies based on the knowledge that the partner cannot respond contingently after the last episode). In each episode, if Alpha and Beta failed to reach an agreement, their profit was determined by an agreement with a hypothetical alternative supplier or buyer. Subjects knew the probabilities of different profits that they and the other actor could earn from the alternative; a random drawing determined the actual profit received.

The two dimensions of power were crossed in a factorial design (with 20 dyads per cell) and manipulated by varying the probabilities of agreements with the hypothetical alternative. In the equal-power conditions, the expected value from the alternative was 60 points for both actors when total power was high and 80 points for both actors when total power was low (i.e., low total power implied that dyad members were less dependent on each other relative to the alternative). In the unequal power conditions, the expected value of the alternative was greater for the high-power actor than for the low-power actor: 75/50 under low total power and 100/65 under high total power. Thus, although the dyad was not embedded in an exchange network in these experiments, the same notion of alternatives underlying power and dependence relations governs the manipulation of power.

All of the other variables in the analysis were measured: the proportion of negotiation episodes in which Alpha and Beta reached agreement, the emotions the subjects reported experiencing (measured after episodes 4 and 8, on dimensions of pleasure/satisfaction and interest/excitement), perceptions of the cohesiveness of the relationship (measured after episode 8), and the gift-giving measure of commitment. Giving a token gift—a voucher that subjects could exchange for pieces of candy after the experiment—was an option after each of episodes 9 through 12. Subjects could give the voucher to their partner or keep it for themselves; they were told they would not know if the other gave them gifts until the experiment was over.

The analysis tested the endogenous process predicted by the theory: that high total power and equal power increase exchange frequency, which increases positive emotions, which increase perceptions of relational cohesion, which increase the frequency of gift-giving as a behavioral indicator of commitment. A series of regression analyses, testing the predicted indirect paths of the model, generally supported the theory.

Lawler, Thye, and Yoon (2000) used a variant of this setting to extend the theory of relational cohesion to productive exchange. Rather than dyads negotiating two-party agreements, triads negotiated three-party agreements. Again, relative dependence (equal versus unequal) and total dependence (high versus low) were manipulated through the use of a hypothetical alternative, and effects on the endogenous processes examined. Comparisons with the 1996 dyadic experiment showed that agreements were less frequently reached in the triad than in the dyad. Consequently, cohesion and commitment were more problematic: positive emotions were lower, perceptions of perceived cohesion were lower, and fewer gift opportunities were used.

D. Molm, Takahashi, and Peterson: Risk and Trust

So far all of the examples we have discussed have been experiments on negotiated exchange. In a series of experiments, Molm and colleagues compared negotiated exchange with reciprocal exchange, testing the thesis that reciprocal exchange produces stronger trust, affective regard, and relational solidarity than negotiated exchange. These effects are produced through three intervening mechanisms: the greater risk of nonreciprocity in reciprocal exchange, the higher perceived relational conflict in negotiated exchange, and the stronger expressive value of reciprocity in reciprocal exchange (Molm, 2003).

The first experiment in this program tested the classical prediction that trust is more likely to develop between exchange partners when exchange occurs without explicit negotiations or formal agreements. Negotiated exchanges with binding agreements provide assurance against exploitation, but give actors little opportunity to develop trust in one another. Acts of trust and attributions of trustworthiness can only be made in situations that involve some risk and uncertainty; that is, the partner must have both the incentive and the opportunity to exploit the actor. Reciprocal exchanges provide the necessary risk and uncertainty for trust to develop; actors in these exchanges initiate exchange without knowing whether or when the partner will reciprocate. If the partner behaves in a trustworthy manner under these conditions, then the actor's trust in the partner should increase.

Molm, Takahashi, and Peterson (2000) tested this logic in a laboratory experiment that compared negotiated and reciprocal forms of exchange in equivalent, negatively connected networks of three or four actors (see Figure 2c). Within each network, a power advantage for A over B was created by giving A a high-value (16 points) alternative to exchange with B, while B had only a low-value (4 points) alternative to exchange with A. Across networks, the strength of A's power advantage was manipulated by varying the availability of A's alternatives. Note that these manipulations combine aspects of both the Cook and Emerson manipulation of power (based on the value of alternatives) and the Markovsky and associates manipulation of power (based on availability of alternatives). A factorial design crossed the form of exchange (negotiated or reciprocal) with the power advantage for the powerful actor (high or low power). Ten networks were studied in each of the four conditions.

In the negotiated exchange setting, subjects jointly negotiated the division of a fixed amount of benefit on each of a series of exchange opportunities; on each opportunity, subjects had five rounds of offers and counteroffers to make an agreement with another subject to whom they were connected in the network. In the reciprocal exchange setting, as described earlier, each actor gave points to one of his or her partners on each opportunity, without knowing whether the partner would reciprocate. The total value of exchange within relations was held constant across the two forms of exchange. As in the Cook and Emerson (1978) and Lawler and Yoon (1996) experiments, subjects did not know the value their partners received from their exchanges or their partners' cumulative earnings. Their knowledge of the network structure was also limited to their immediate exchange relations. Subjects' trust in their partners, as well as other integrative outcomes (positive affect, feelings of commitment), was measured at the end of the exchange period by a series of semantic differential scales.

The manipulation of the form of exchange affected the *risk* predicted to be necessary for the development of trust (reciprocal exchanges are riskier than negotiated exchanges); the manipulation of power imbalance affected two behaviors used as indicators of the partner's *trustworthiness*: behavioral commitment and equality of exchange, both of which were greater when power imbalance was low. Molm and colleagues predicted that average trust would be greater in the reciprocal exchange relations than in the negotiated exchange relations, that an actor's trust in the partner would increase with the partner's behavioral commitment to the actor and with the equality of their exchange, and that the effects of these behaviors on trust would be stronger in the reciprocal exchange conditions than in the negotiated exchange traditions.

These predictions were supported. Subsequent experiments in the research program demonstrate that these effects are quite broad; reciprocal exchanges produce more positive outcomes than negotiated exchanges on a wide range of relational measures—not only greater trust, affective regard, and feelings of commitment, but also stronger perceptions of fairness and of relational solidarity or social unity.

Recently, Molm, Collett, and Schaefer (2006b) tested an extension of this logic to generalized exchange. In generalized exchange, reciprocity is not only uncertain but *indirect*; that is, *A*'s giving to *B* is reciprocated, not by *B*, but by another actor in the generalized exchange system. Thus, the risk of nonreciprocity is even greater. They compared chain-generalized exchange networks of three or four actors with equivalent networks of negotiated and reciprocal exchange; as predicted, trust was greater in the chain-generalized exchange with indirect reciprocity than in either of the forms of exchange with direct reciprocity.

VI. ASSESSMENT AND FUTURE PROSPECTS

The experimental tradition of social exchange research is one of the best examples in sociology of sustained theory construction and testing through programmatic, cumulative research. The sheer quantity of research during the last 30 years stands in sharp contrast to the earlier years of exchange theory in sociology, when little empirical work was conducted. In a relatively short time, the combined efforts of a number of researchers have produced a strong empirical base that offers substantial support for the perspective.

The use of standardized laboratory settings has contributed to the cumulation of knowledge in the field by making it easier for researchers to compare results across experiments. The process of theory construction necessarily involves numerous experimental tests, with each experiment investigating only one part of a theory. Most of the research programs reviewed in this chapter also involve an interplay between deduction and induction, with experiments playing a vital role in both aspects of theory construction. Work typically begins with experiments testing theoretically derived hypotheses (deduction); results of those experiments sometimes challenge theoretical assumptions and pose empirical puzzles that must be solved theoretically (induction), and new experiments then test the revised theory (deduction).

The study of power in exchange networks dominated the field during much of this period and engaged the attention of numerous researchers working in several distinct programs. As a result, we have learned a great deal about how particular network structures affect patterns of exchange and the distribution of power. The various programs on power in exchange networks have expanded the size of networks studied, increased the precision of predictions, and reinforced the linkage of exchange theory with social network analysis. Other areas, such as the effects of different forms of exchange and the study of affect, commitment, and trust, are newer, but knowledge in these areas is also rapidly cumulating as a result of several highly productive research programs and the growth of new theories on these topics.

There are many topics in exchange that have hardly been tapped, however, and our experimental laboratories could be used in more creative ways to address some of these topics. Two of the biggest ones involve change: changes in network structures over time (either environmentally induced or actor induced) and changes in relationship dynamics. The large body of research on power in exchange networks has concentrated almost exclusively on the effects of static structures on behavior. Some early research allowed subjects to change the structure by forming coalitions (Cook & Gillmore, 1984), but in general, research on structural change is rare. As we have seen in several examples, researchers commonly assign subjects to positions in particular network structures, which comprise one of the key manipulated variables in virtually all of this work, and the effects of those structures on power use, exchange frequency, commitment, trust, and the like are then studied.

But networks are rarely static; they expand and contract, network connections change, and the value of resources attached to different positions varies. Studying what produces structural change, how change affects established patterns of interaction, and how the structural history of a network alters its current impact are all questions that should be pursued and could be pursued in laboratory experiments that change features of network structures. Such changes could occur at programmed times, they could be contingent on particular patterns of exchange, or they could be under the control of the subjects themselves.

Different forms of exchange have also been studied in isolation from each other. But as our knowledge of their effects becomes more established, researchers can begin to pursue more complex questions: How do different sequences or histories of negotiated and reciprocal exchange affect behavior? How do concurrent opportunities to engage in both forms of direct exchange affect relationships? Some preliminary work has begun on these topics, and laboratory settings can easily be designed to address these questions. Such work would be particularly valuable in connecting the experimental work on these topics with the organizational work on socially embedded relationships, which typically involve both negotiated and reciprocal forms of exchange.

Finally, some of the most neglected questions in exchange concern more collective forms of exchange, including productive exchange and various forms of generalized exchange. The sociological tradition of exchange that George Homans initiated explicitly rejected the study of generalized (indirect) exchange in favor of a focus on direct exchange. But both generalized and productive forms of exchange are highly relevant to contemporary concerns with the development of trust and solidarity in social life, and both are pervasive throughout society. By combining laboratory experiments on relatively small collective systems with computer-simulated studies of larger systems, researchers can study the role of these more collective forms of exchange in creating strong bonds in groups and networks. Such work would also reconnect the contemporary exchange tradition in sociology with its early anthropological roots, thus bringing the tradition full circle.

REFERENCES

Bacharach, S. B., & Lawler, E. J. (1981). *Bargaining: Power, tactics, and outcomes*. San Francisco: Jossey–Bass.

Bienenstock, E. J., & Bonacich, P. (1992). The core as a solution to exclusionary networks. *Social Networks, 14*, 231–243.

Blau, P. M. (1964). *Exchange and power in social life*. New York: Wiley.

Cook, K. S., & Emerson, R. M. (1978). Power, equity and commitment in exchange networks. *American Sociological Review, 43*, 721–739.

Cook, K. S., Emerson, R. M., Gillmore, M. R., & Yamagishi, T. (1983). The distribution of power in exchange networks: Theory and experimental results. *American Journal of Sociology, 89*, 275–305.

Cook, K. S., & Gillmore, M. R. (1984). Power, dependence, and coalitions. In E. J. Lawler (Ed.), *Advances in group processes* (Vol. 1, pp. 27–58). Greenwich, CT: JAI Press.

Emerson, R. M. (1962). Power-dependence relations. *American Sociological Review, 27*, 31–41.

Emerson, R. M. (1972a). Exchange theory, part I: A psychological basis for social exchange. In J. Berger, M. Zelditch, Jr., & B. Anderson (Eds.), *Sociological theories in progress* (Vol. 2, pp. 38–57). Boston: Houghton–Mifflin.

Emerson, R. M. (1972b). Exchange theory, part II: Exchange relations and networks. In J. Berger, M. Zelditch, Jr., & B. Anderson (Eds.), *Sociological theories in progress* (Vol. 2, pp. 58–87). Boston: Houghton–Mifflin.

Friedkin, N. E. (1992). An expected value model of social power: Predictions for selected exchange networks. *Social Networks, 14*, 213–229.

Homans, G. C. (1961). *Social behavior: Its elementary forms*. New York: Harcourt Brace and World.

Kollock, P. (1994). The emergence of exchange structures: An experimental study of uncertainty, commitment, and trust. *American Journal of Sociology, 100*, 313–345.

Lawler, E. J. (1992). Power processes in bargaining. *The Sociological Quarterly, 33*, 17–34.

Lawler, E. J. (2001). An affect theory of social exchange. *American Journal of Sociology, 107*, 321–352.

Lawler, E. J., Thye, S. R., & Yoon, J. (2000). Emotion and group cohesion in productive exchange. *American Journal of Sociology, 106*, 616–657.

Lawler, E. J., & Yoon, J. (1996). Commitment in exchange relations: Test of a theory of relational cohesion. *American Sociological Review, 61*, 89–108.

Markovsky, B., Skvoretz, J., Willer, D., Lovaglia, M., & Erger, J. (1993). The seeds of weak power: An extension of network exchange theory. *American Sociological Review, 58*, 197–209.

Markovsky, B., Willer, D., & Patton, T. (1988). Power relations in exchange networks. *American Sociological Review, 53,* 220–236.

Molm, L. D. (1981). The conversion of power imbalance to power use. *Social Psychology Quarterly, 16,* 153–166.

Molm, L. D. (1997). *Coercive power in social exchange.* Cambridge, U.K.: Cambridge University Press.

Molm, L. D. (2003). Theoretical comparisons of forms of exchange. *Sociological Theory, 21,* 1–17.

Molm, L. D., Collett, J. L., & Schaefer, D. R. (2006). Conflict and fairness in social exchange. *Social Forces, 84,* 2331–2352.

Molm, L. D., Collett, J. L., & Schaefer, D. R. (In press). Building solidarity through generalized exchange: A theory of reciprocity. *American Journal of Sociology.*

Molm, L. D., Takahashi, N., & Peterson, G. (2000). Risk and trust in social exchange: An experimental test of a classical proposition. *American Journal of Sociology, 105,* 1396–1427.

Molm, L. D., Takahashi, N., & Peterson, G. (2003). In the eye of the beholder: Procedural justice in social exchange. *American Sociological Review, 68,* 128–152.

Sell, J. (2007). Social dilemma experiments in sociology, psychology, political science, and economics. In M. Webster & J. Sell (Eds.), *Laboratory experiments in the social science* (pp. 459–479). Burlington, MA: Elsevier.

Skvoretz, J., & Willer, D. (1991). Power in exchange networks: Setting and structural variations. *Social Psychology Quarterly, 54,* 224–238.

Stolte, J. R., & Emerson, R. M. (1977). Structural inequality: Position and power in network structures. In R. L. Hamblin & J. Kunkel (Eds.), *Behavioral theory in sociology* (pp. 117–138). New Brunswick, NJ: Transaction.

Szmatka, J., & Willer, D. (1995). Exclusion, inclusion and compound connection in exchange networks. *Social Psychology Quarterly, 55,* 123–131.

Thibaut, J. W., & Kelley, H. H. (1959). *The social psychology of groups.* New York: Wiley.

Thye, S., Lawler, E. J., & Yoon, J. (2006). Social exchange and micro social order: Comparing four forms of exchange. Paper presented at the Annual Meeting of the American Sociological Association, Montreal.

Thye, S. R., Lovaglia, M. J., & Markovsky, B. (1997). Responses to social exchange and social exclusion in networks. *Social Forces, 75,* 1031–1047.

Yamagishi, T., & Cook, K. S. (1993). Generalized exchange and social dilemmas. *Social Psychology Quarterly, 56,* 235–248.

Yamagishi, T., Cook, K. S., & Watabe, M. (1998). Uncertainty, trust and commitment formation in the United States and Japan. *American Journal of Sociology, 104,* 165–194.

Yamagishi, T., Gillmore, M. R., & Cook, K. S. (1988). Network connections and the distribution of power in exchange networks. *American Journal of Sociology, 93,* 833–851.

FURTHER READING

Bacharach, S. B., & Lawler, E. J. (1981). *Bargaining: Power, tactics, and outcomes.* San Francisco: Jossey–Bass.

Cook, K. S., & Rice, E. (2003). Social exchange theory. In J. Delamater (Ed.), *Handbook of social psychology* (pp. 53–76). New York: Kluwer Academic.

Molm, L. D. (1997). *Coercive power in social exchange.* Especially Chapter 2, "An Experimental Setting for Studying Power in Exchange Relations." Cambridge, U.K.: Cambridge University Press.

CHAPTER 16

Solving Coordination Problems, Experimentally

Giovanna Devetag
University of Perugia

Andreas Ortmann
Charles University and Academy of Sciences of the Czech Republic

ABSTRACT

We discuss four major classes of coordination problems: pure coordination or rendezvous games; Pareto-ranked coordination games such as stag hunt and order-statistic games; mixed-motives coordination games such as "Battle of the Sexes"; and critical-mass games. We review "classic" implementations of exemplars of the four classes of coordination problems. En passant, we discuss experimental practices in economics. We conclude with an assessment of the literature: we discuss what we have learned so far from the literature as well as what we consider particularly promising avenues of future research.

I. INTRODUCTION

Coordination problems are pervasive in real life: for example, how do people coordinate on one of many meeting places, or how do firms coordinate their investment projects if they cannot communicate directly about their options?

More abstractly, how do people coordinate on one of many possible outcomes? Does it help if the possible outcomes are of different desirability? How do risk and other people's characteristics and choice sets affect such interactive decision situations?

Over the past 2 decades, coordination problems, classified in economics as coordination "games" because of their interactive nature, have become an important research topic. Many researchers have tried to answer the preceding and many related microeconomics and macroeconomics questions through laboratory studies of coordination problems (see Ochs, 1995; Camerer, 2003; Devetag & Ortmann, in press, for two surveys and a critical assessment of the literature).

In this chapter, we discuss four major classes of coordination games in order to convey to our readers the diversity of coordination problems analyzed by economists. We then discuss the design of typical coordination experiments, and the experimental protocols that economists use. Then we review "classic" implementations of exemplars of the four classes of coordination games. We conclude with an assessment of the literature and discuss what we have learned so far from the literature as well as what we consider particularly promising avenues of future research.

II. BACKGROUND AND DEVELOPMENT

We consider four major classes of coordination games.

A. PURE COORDINATION GAMES (RENDEZVOUS GAMES)

Pure coordination games differ from social dilemmas in a key aspect: in social dilemmas, players' preferences over the game outcomes conflict, while in pure coordination games, players' preferences over outcomes coincide. Players in coordination games, however, are uncertain about the behavior of the other players. This induces what is now called in economics a situation of "strategic uncertainty."

The prototypical example of a pure coordination problem, occasionally also called a matching game, is illustrated in the "payoff" table in Table 1, where the numbers represent, for example, earnings that experimental subjects might be able to earn as a function of their choices.

In the matching game in Table 1, players or subjects have to choose simultaneously between two available actions, A and B. The players receive positive, and in fact identical payoffs if they choose the same action, and they both get

TABLE 1

Your choice	Other player's choice	
	A	B
A	1,1	0,0
B	0,0	1,1

payoffs of zero if their action choices differ. (In each pair of payoffs, the first number denotes the payoff of the row player, and the second number denotes the payoff of the column player.) The game has two strict Nash equilibria (i.e., equilibria that are robust against slight trembles).[1] Although the players in the game will be indifferent about which of the equilibria they end up with, clearly they face a problem of coordinating their action choices so as to avoid a disequilibrium outcome. The matching game exemplifies the most fundamental, and by no means trivial, coordination problem.

Importantly, any preplay agreement that would allow players to coordinate on one of the equilibria is self-enforcing: once players have agreed on playing, say, action A, no one has an incentive to deviate from such an agreement. Apart from communication, which would solve the matching game in a trivial way, anything that allows players to break the symmetry between the two equilibria (and make this common knowledge) can function as a coordination device. Symmetry can also be broken when one of the two (or several) possible equilibria is a focal point.

Schelling gave more than half a dozen examples of focal points in his widely cited book, *The Strategy of Conflict* (1960). In a famous example, he asked his students to imagine they had to meet someone in New York City without having had a chance to specify a place and time, but knowing the other person was also trying to meet them. Where and when would they show up? A large majority of students indicated Grand Central Station at noon as an answer.

In other informal experiments, Schelling asked students to name either "heads" or "tails" during a coin toss, knowing that somebody whom they would be paired with was asked the same thing. In this case, if they both chose the same side they would get a positive reward; otherwise, no one would receive anything. Most people gave "heads" as an answer.

From a game-theoretic standpoint, both "Grand Central at noon" and "heads" can be seen as labels attached to the strategies of a matching game. These labels stand out as unique or more salient than others by virtue of psychological, historical, perceptual, cognitive, or possibly linguistic factors that,

---

[1]Throughout the chapter we focus on equilibria supported by pure strategies and ignore (the quite possibly numerous) mixed-strategy equilibria.

of course, beg for explanation. Focal points are the equilibria that result from choosing strategies with salient labels (Mehta, Starmer, & Sugden, 1994).

B. Coordination Games with Pareto-Ranked Equilibria

Pareto-ranked equilibria also do not entail, in principle, players' competing preference orderings over the game outcomes, and, like pure coordination games, represent a problem of "strategic uncertainty." A prototypical example of a coordination problem with Pareto-ranked equilibria goes by the name "stag-hunt game," denoted here as $g(1,0,x,x)$, where $0 < x < 1$. An instance of such a game is shown in Table 2.

Again, players have to choose between two available actions, A and B. The situation can be conceptualized as a weak-link or team effort game in which two players exert either a standard effort independently from one another (hunting hares on their own), or a joint extra effort (hunting a stag together).[2] They both receive a positive payoff if they choose the same action; further, the payoff from the joint stag hunt is higher than the one from hunting hares alone. This game, too, has two strict Nash equilibria; however, rather than being payoff equivalent, these equilibria are Pareto ranked, with one of the payoffs (resulting from the choice of A by both players) being higher for both players than the other (resulting from the choice of B by both players).

The stag-hunt game, at first glance, seems to exemplify another fundamental, yet more trivial, coordination problem. It also seems that preplay agreements should again be self-enforcing. Or should they?

Assume players have agreed on playing action A: does anyone have an incentive to deviate from the agreement? Unfortunately (and this is at the heart of some of the most intriguing results in the literature on coordination), yes: action B yields a positive payoff to the player who chooses it, *no matter what the other player does;* action A is more risky in that it can lead to a payoff of zero to the player who chooses it if he or she is the only one doing so. Hence, the secure action B can undermine the risky action A which therefore under-

TABLE 2

Your choice	Other player's choice	
	A	B
A	1,1	0,X
B	X,0	X,X

[2]The game derives its name, "stag hunt," from a parable contained in Rousseau's writings about the social contract.

mines the possibility of the Pareto-efficient outcome of both players choosing A. In fact, even if players agree on choosing A prior to playing the game, subsequently they might be tempted to pick B if they do not trust the other player (whereas in the matching game this risk is not present). Intuitively, such a change of mind is also a function of the value that x takes. If x is greater than 0.5, then the secure action becomes too attractive; if x is less than 0.5, the secure action turns out to be not attractive enough.

Note that the concept of "security" does not distinguish these two cases. If subjects have concerns for security only, then they would always select B whether x is greater or less than 0.5. This contradicts intuition captured by an alternative solution concept called "risk dominance." Risk dominance compares the product of the deviation losses of the two equilibria[3] and it formalizes our intuition that safe actions that are too attractive may induce people to select B, while safe actions that are not attractive enough might induce people to select A. Risk dominance therefore makes the prediction that, for high values of x, the inferior action profile (B,B) will be selected, while for low values of x, the superior and Pareto-efficient action profile (A,A) will be selected.

Stag-hunt games are the basic building block of "global games" (Carlsson & Van Damme, 1993; Morris & Shin, 2003), which are essentially stag-hunt games with payoff perturbances. This literature has gained some notoriety because it suggests—initially theoretically and more recently also experimentally—that risk dominance has more predictive power than payoff dominance.

Stag-hunt games are also the simplest exemplars of so-called order-statistic games. These games tend to feature payoff matrices of higher dimension and more than two Pareto-ranked equilibria. They otherwise feature the same tension between a secure action and riskier actions that have the potential for higher (but also lower) payoffs.

An example of an order-statistic game is the minimum-effort or weak-link game. In this game, the outcome is determined by the minimum effort exerted (the "weak link" of the chain); any extra effort above such minimum is costly for the player who exerts it without having any positive effect on the output. Hence, all players gain if the minimum is the highest possible, but each player individually has the incentive not to choose an effort above the minimum. This feature characterizes many team production situations. Respecting deadlines for parts of projects (e.g., chapters of books) is an appropriate example: all contributors prefer the book to be published as soon as possible, but no one has an incentive to put in an extra effort to respect the deadline if he or she expects that at least one other contributor will be late (Camerer, 2003; Camerer & Knez, 1997).

[3]A risk-dominant equilibrium has a greater Nash product of deviation losses relative to the efficient equilibrium (e.g., Harsanyi & Selten, 1988).

The payoff function of a generic order-statistic game is $\pi_i = f(OS - |e_i - OS|)$, where OS stands for the order statistic chosen (which could be the median or the minimum—the weak link—or something else), e_i denotes the effort choice, $|e_i - OS|$ denotes the (symmetric) deviation cost, and f is some scalar function of these terms. Obviously, the terms can be arbitrarily modified by setting the coefficients of the two terms on the RHS not equal to 1, by squaring the second term, or by defining the deviation costs asymmetrically. We will give more specific examples—in fact, some rather famous examples of such games—next.

C. MIXED-MOTIVES COORDINATION GAMES ("BATTLE OF THE SEXES" GAMES)

While in pure coordination games players have the same preference ordering over the game outcomes, in mixed-motives games the players differ with respect to the equilibrium they prefer (although they prefer ending up on the equilibrium they dislike to reaching a disequilibrium outcome). The prototypical mixed-motives coordination game is called the "battle of the sexes" and is represented in Table 3.

The game takes its name from the example of a couple that has to decide which activity to attend: a football game or the theater. The husband prefers the theater, while the wife prefers the football game, but they both prefer attending either activity together over going to their preferred event alone. The game has two Nash equilibria in pure strategies represented by the two action profile combinations (theater, theater) and (football, football).

Firms often play games captured by the battles of the sexes when they engage in collaboration agreements to standardize technologies: both firms are better off if a single standard develops in the industry, but at the same time each firm prefers its own standard to prevail over the rival's (Farrell & Saloner, 1986; Stango, 2004). The game also captures situations in which firms have incentives to cooperate to create new business opportunities, but in doing so, compete to get the major part of the profits that the new business generates (Brandenburger & Nalebuff, 1997). For example, hardware and software producers have an obvious interest in producing highly complementary

TABLE 3

Your choice	Other player's choice	
	Theater	Football
Theater	2,1	0,0
Football	0,0	1,2

technologies to maximize product penetration in the market; at the same time, each wants to extract the largest share of the profits generated by the collaboration.

When the game is played only once, chances are high that a disequilibrium outcome occurs; preplay communication—either one- or two-sided—might facilitate the chances of reaching equilibrium. If the positions of the two players are asymmetric, the one with higher bargaining power has an advantage. If the game is played repeatedly, a reasonable (and fair) pattern of coordination is to alternate between equilibria, so that, at time one, one player is advantaged and, at time two, the other is advantaged.

D. Critical-Mass Games (Panics, Revolutions)

Equilibria in a coordination game can be interpreted as conventions that arise within a population so that no one has an incentive to deviate from a conventional behavior if everybody else adheres to it. Sometimes the emergence of specific conventions depends on a critical number of adopters. If people who start behaving in a certain way are numerous enough, that behavior will eventually become a convention; otherwise, it will disappear or remain customary only within a minority. Schelling (1978) mentioned several examples of "critical-mass" effects in the social sciences, from fashion to panic crises to revolutions.

Critical-mass effects are also important in economics. For example, they are important in understanding the adoption of technologies or products subject to "network externalities" (where the incentive to adopt a technology increases as the number of other adopters of the technology increases).[4] The large-scale diffusion of these technologies often depends on whether the "installed base" of early adopters is large enough to generate a bandwagon effect (e.g., Farrell & Saloner, 1986).

An example of a critical-mass game with binary choices is the following: N players simultaneously must decide whether to invest a certain amount Z in a network technology that is installed if the total revenue is at least $Z K$, where $1 \leq K \leq N$. Contribution yields a return greater than Z if and only if at least K players contribute (Heinemann, Nagel, & Ockenfels, 2004). In some cases, different conventions are possible, some of which yield greater returns than others but require a higher critical number of adopters.

An example of a critical-mass game, with $N = 7$ possible choices and with $N = 7$ players, is shown in Table 4.

---

[4]All communication technologies (e.g., fax, e-mail, mobile phones) are subject to network externalities: the utility for any adopter increases with the number of other adopters.

TABLE 4

Your choice of X	Number of people who have chosen X						
	7	6	5	4	3	2	1
7	7	0	0	0	0	0	0
6	6	6	0	0	0	0	0
5	5	5	5	0	0	0	0
4	4	4	4	4	0	0	0
3	3	3	3	3	3	0	0
2	2	2	2	2	2	2	0
1	1	1	1	1	1	1	1

The rows of the table report each player's possible choices in the game (i.e., numbers from 1 to 7), the columns report the number of players in the group who have chosen that number, and the cells report the individual payoff associated with each possible combination of choices. Higher numbered actions yield greater returns but are also riskier because they require higher critical numbers of adopters in order to yield positive payoffs. There are seven equilibria corresponding to all players choosing the same action. The equilibria are Pareto-ranked, and higher payoff equilibria are riskier than lower payoff equilibria. As in stag-hunt and order-statistic games, there is a trade-off between efficiency and security. Unlike order-statistic games, however, payoffs depend on absolute frequencies of people choosing certain actions and not on "minimum" or "median" players.

Other real-world examples are processes of organizational change: low-complementarity norms and routines (regarding, e.g., division of labor among divisions, offices, or coworkers) usually provide a high level of redundancy against "breakdowns" but are relatively inefficient; more efficient norms may require the effective contribution of more (or all) organization members in order to be effective. A related example is the gradual emergence of a new norm or behavioral "standard," which may have to be adopted by a large proportion of organizational members to lead to improvements.

Yet other examples of which the game in Table 4 captures the essence are choices between, say, activities that are more or less enjoyable but some of which cannot be performed unless enough people adhere, where "enough" has a different value for each activity. I can rent a movie and watch it alone at home, but I need a companion to play chess; at the same time, we both would prefer going to a party, as long as everybody else will go. Schelling himself mentions the example of the last day of a class when some students hesitantly begin clapping as the teacher prepares to leave the room. "If enough clap, the whole class may break into applause; if a few clap indecisively, it dwindles to an embarrassed silence" (Schelling, 1978, p. 93).

III. DESIGN AND IMPLEMENTATION OF COORDINATION EXPERIMENTS IN ECONOMICS

The design and implementation of economics experiments follows, by and far, a well-established canon of methodological precepts (Hertwig & Ortmann, 2001). One of the key precepts of experimental economics is that deception should not be used. Essentially, experimental economists are concerned that deception could lead to negative reputational spillover effects that might contaminate their subject pool in future experiments. Indeed, suspected deception has been shown to provoke a range of motivational, emotional, and cognitive reactions on the part of subjects (for a review of the evidence from psychology, where deception is still often used, see Ortmann & Hertwig, 2002). There are also reasons to believe that deception and the reaction it causes lead to important changes in the readily available subject pools. Obviously, suspicion of deception could immensely affect subjects' choices in coordination games. Therefore, the de facto proscription of deception in experimental economics—top economics journals refuse to publish results of experiments in which deception was used—is a crucial step toward experiments with some external validity.

Another relevant tenet of economic experiments is that participants in experiments role-act. They are made to play, for example, the role of a buyer or seller in a transaction of a good or service, or the role of a decision maker in a coordination game situation. Until very recently, it was argued that the specific situation ought to be presented to subjects in abstract (and therefore allegedly neutral) terms: rather than being called a buyer or a seller, the two participants in an experimental transaction would be called participant A and participant B, and they would trade an unspecified good. This procedure was rationalized by the assumption that it would guarantee that subjects would not bring prior expectations (e.g., from previous buyer–seller interactions) into the lab that might contaminate their reactions to experimental stimuli.

More recently, in light of important evidence, this assumption has been questioned and it has been argued that a better experimental strategy would be to study the impact of context systematically (e.g., Ortmann & Gigerenzer, 1997; see also the important debate about field experiments in Harrison & List, 2004). To the extent that some classes of coordination games are considered to be appropriate toy models of teamwork or other forms of coordination within firms, this issue is of obvious importance. Recent attempts by coordination problem researchers to add context to their laboratory settings is therefore another welcome step toward experiments to ascertain the framing effects and how this may affect generalizabililty of the results to a particular theoretical settings.

The issue of financial incentives is probably the most prominent character-istic of economics experiments. The term "financial incentives" typically denotes monetary rewards that are the function of the actions a player chooses as well as the actions other players choose. For example, while subjects might be paid a show-up fee (or participation fee) up front, a significant part of their earnings comes from "performance-based" payments where performance is typically relative to a game or decision-theoretic benchmark.

This strictly enforced experimental requirement, by making participants' decisions to some extent costly for them (through foregone earnings, for example), was hypothesized to affect participants' decisions. The available evidence supports this conjecture. After reviewing the available evidence empirically, Hertwig and Ortmann (2001) stated:

> To conclude, concerning the controversial issue of the effects of financial incen-tives, there seems to be agreement on at least the following points: First, financial incentives matter more in some areas than in others (e.g., see Camerer & Hogarth's distinction between judgment and decision versus games and markets). Second, they matter more often than not in those areas that we explore here (in particular, research on judgment and decision making), which are relevant for both psycholo-gists and economists. Third, the obtained effects seemed to be two-fold, namely, convergence of the data toward the performance criterion and reduction of the data's variance (p. 395).[5]

A number of recent studies have confirmed these results (e.g., the study of Gneezy & Rustichini, 2000; see also the reanalysis of these data in Rydval & Ortmann, 2004). Importantly, in light of the basic structure of many coordi-nation problems—risky but potentially payoff-improving action choices versus safe action choices that, however, pay relatively little—a recent controversy reported in the *American Economic Review* is noteworthy. Holt and Laury (2002, 2005) showed that increasing the financial incentives for decisions over lotteries (i.e., risky choices with potentially high payoffs), and doing so quite dramatically, increased risk aversion.[6]

---

[5]Hertwig and Ortmann also argue, thereby questioning the current practice among experi-mental economists, that "researchers seeking maximal performance ought to make a decision about appropriate incentives. This decision should be informed by the evidence available... In cases where there is no or only mixed evidence, we propose that researchers employ a simple 'do-it-both-ways' rule. That is, we propose that the different realizations of the key variables discussed here, such as the use or nonuse of financial incentives ... be accorded the status of independent variables in the experiments" (Hertwig & Ortmann, 2001, p. 400).

[6]Harrison, Johnson, McInnes, and Rutström (2005) have made a persuasive case that order effects confounded the original results and that the effect of increased financial incentives, while an important qualitative phenomenon, was only about half of what had been reported in Holt and Laury (2002).

By and far, the design and implementation of coordination games is straightforward from a methodological point of view. Subjects are recruited (either through flyers or e-mails) and come to a laboratory where they are seated and presented with stimuli materials. The stimuli materials, in coordination games, almost always contain earnings tables of the kind presented in the preceding section (Tables 1 through 4). Subjects are then asked, typically repeatedly, to decide which of the available action choices they are willing to choose. How often they are asked turns out to be of some importance, roughly for the reason reported by Holt and Laury (2002): the less that is at stake at each decision (typically that means more rounds overall are being conducted), the less expensive is experimentation for the players (i.e., the choice of the risky but potentially very rewarding action choice), as convincingly shown by Berninghaus and Ehrhart (1998) for games with Pareto-ranked equilibria. Other implementation details that have similar effects are an increase in the number of actions available to subjects, or a reduction in the costs of deviating from equilibrium play (see more details in Section VI).

Another important design choice in coordination game experiments concerns the interaction protocol: in some studies each player always interacts with the same opponent throughout the game duration (the so-called "fixed matching" protocol), while in other studies subjects play a game repeatedly but change the person they are matched with every round, either by a random scheme or by a deterministic rotation scheme. In games with groups of players (i.e., more than two), the composition of the groups can remain fixed throughout the game, or, at some point, different groups can be "scrambled" randomly to form new groups.

Random rematching is typically introduced to induce subjects to use behavioral strategies that they would use if the game were to be played only once. This design strategy allows one to observe behavior in a series of independent "one-shot" games without implying huge numbers of subjects (of course, the random scheme does not impede learning from experience, but it does impede, for example, coordination based on precedent). The choice of the interaction structure in coordination games is obviously relevant in that some coordination devices can be used by players in the first interaction protocols but not in the second ones, with relevant differences for the possibility of observing coordination failure (more on this in the section describing the experiments).

IV. TECHNOLOGICAL DEVELOPMENTS

More than 50 years have passed since the first economics experiments were conducted. During that time, some implementation features have changed as a consequence of developments in technology. Whereas the earlier experiments were conducted with paper and pencil, the norm nowadays is to have experimental subjects interact through networked computers. The advantages of computerized experiments over the traditional paper-and-pencil methods are evident and have allowed the discipline to make enormous progress in a relatively short time. Games that are repeated for many rounds (e.g., 100 or more) in order to test alternative learning theories would, for example, not be feasible without the aid of computers. Computerized experimental sessions, furthermore, allow the experimenters to keep track of an enormous amount of data well beyond the decision itself (e.g., time responses, patterns of information search).

A downside of computerized experiments is the possibility of a decrease in transparency (and hence a slightly higher probability that participants may think they are being deceived). If, for example, a random device must be used in an experiment, a computerized random number generator is less transparent than the act of throwing a dice in front of participants. Such implementation details have been shown to matter in some circumstances (see Hertwig & Ortmann, 2001).

The path-breaking experiments on coordination games by Van Huyck, Battalio, and Beil (1990, 1991) and by Cooper, DeJong, Forsythe, and Ross (1990, 1992) were run with paper and pencil, but most subsequent studies were computer-aided experiments. One of these (Van Huyck, Battalio, & Rankin, 2001) used a graphical interface whereby players could see payoffs corresponding to alternative choices by moving a cursor along the rows and columns of a matrix instead of having an earnings table in paper format. Since the actions available to a subject in this experiment (a variant of the "weak link" game discussed before) were 100 instead of the 7 used in the early experiments, this implementation option was almost unavoidable. However, some doubts remain on whether discovering potential payoffs in this fashion—as opposed to having them displayed on a table—makes a difference in terms of behavior. No comparative study has been done so far; hence, we can only conjecture that the presentation format may have had an impact on, for example, the different salience of available actions in the first round, or on the ease of processing the payoff consequences of out-of-equilibrium choices.

Surprisingly enough, another well-established computerized methodology to record patterns of information search, MouseLab, has never been used in coordination game experiments.[7] With the MouseLab technology, participants must click on a cell of the earnings table to visualize its content, thereby making it possible to observe which pieces of information are taken into consideration by subjects in deciding which action to take, and in which sequence

---

[7]For examples of experiments on games conducted with MouseLab, see Johnson, Camerer, Sen, and Rymon (2002) and Costa-Gomes, Crawford, and Broseta (2001).

such information is acquired. In experiments on repeated coordination games, for example, using MouseLab techniques would provide insights on which payoff information drives first-round choices.

V. EXAMPLES

A. EXPERIMENTING WITH PURE COORDINATION GAMES: MEHTA ET AL. (1994)

Matching games are especially useful in studying the influence of culture, language, and shared codes on coordination. Standard game theory does not discriminate between payoff-equivalent equilibria, and the usual refinements such as payoff dominance and risk dominance do not apply to matching games. Therefore, matching games are an ideal point of departure for the empirical investigation of how people solve coordination problems. Following Schelling's original intuition on focal points, Mehta et al. (1994) decided to test experimentally the influence of salience on coordination and replicated some of Schelling's early informal experiments in a more rigorous fashion.[8]

Mehta and colleagues (1994) started by providing a definition of different forms of salience. They distinguish among *primary salience,* consisting in an individual's personal preference for a particular option among other available options; *secondary salience,* which is derived from knowing that a particular option has primary salience for the person with whom one has to coordinate; and *Schelling salience,* which is derived only from the probability that a particular option may be selected as the solution to a coordination problem. An option may not need to be characterized by all three types of salience.

For instance, if 9 is my favorite number, such a number will have primary salience for me. Number 5 has secondary salience for me if I know that it is my best friend's favorite number. Further, if my best friend and I play a matching game in which we have to pick the same number to be rewarded, we may pick number 1 which is characterized by high Schelling salience. In this example, three different options have different types of salience. If, instead, we both had 9 as our favorite number and if this information was common knowledge between us, we could pick number 9 in a matching game. In this last case, the three types of salience would select the same option.

The experimental studies reported in Mehta et al. (1994) were aimed at verifying whether coordination success in matching games is related to primary

---

[8]That is, by providing financial incentives to subjects and by preventing any form of communication in the typical tradition of experimental economics experiments (see the preceding section for a discussion).

salience (which would imply that some options simply tend to be preferred to others by most people), or to Schelling salience. For this purpose, they divided their pool of subjects into two groups. In one group (the "picking" group), subjects were simply asked to select, from a set of options, the one they preferred; subjects in the other group (the "coordinating" group) were asked to select an option in the same set knowing that they would be paired with another participant randomly and both would earn a prize if the option selected was the same. Subjects in the second group, hence, were playing a matching game. The options were taken from several sets including dates, flowers, cities, and male names.

The results clearly indicate the existence of Schelling salience, which is often uncorrelated with the other types of salience.[9] For example, when asked to specify a date, subjects in the "picking" group indicated 75 different dates; however, in the "coordinating" group, 50% of subjects indicated December 25. Analogously, when asked to specify a male name, only 9% of subjects indicated John in the "picking" condition, but 50% did so in the "coordinating" condition. These experiments show that people are generally good at solving coordination problems. Coordination success is not related to the fact that some options are more preferred than others,[10] but to the fact that people within a population tend to agree on which options are unique and distinctive regardless of individual preferences (which are often highly heterogeneous). The results, of course, beg an answer to the questions of why people are so good at solving pure coordination games in certain situations, and how Schelling salience comes into existence.

B. Experimenting with Mixed-Motives Games: Cooper, DeJong, Forsythe, and Ross (1989)

In the "battle of the sexes" game described earlier, players' conflict of preferences over the two equilibria is dominated by their common incentive to coordinate. Path-breaking experiments on mixed-motive games of coordination were run by Cooper et al. (1989), who used the payoff matrix in Table 5.

The Nash equilibria in pure strategies correspond to the two action profiles (1,2) and (2,1). (The game also has an equilibrium in mixed strategies in which players play strategy 1 with probability .25 and strategy 2 with probability .75.

---

[9]Harrison (2005) argues, and makes a persuasive case empirically, that at least some treatments were confounded by natural language: "There is a fundamental confound in experiments such as these: The fact that natural language has been used to present the task to subjects, and that the task itself uses natural language. That language itself has salient labels, which is just to say that some words are prominent or conspicuous" (p. 21).

[10]Primary salience and Schelling salience were usually unrelated in the experiment.

TABLE 5

| | Other player's choice | |
Your choice	1	2
1	0,0	200,600
2	600,200	0,0

The expected payoff of the mixed-strategy Nash equilibrium is 150.) Without a device that would allow breaking the symmetry between players and without possibility of communicating, the chances that a disequilibrium outcome occurs are high. Indeed, Cooper et al. (1989) report data of an experiment in which cohorts of 11 players were playing a series of 22 one-shot games like the one shown in Table 5, with different anonymous opponents, and with players alternating between the role of row and column player. More than half of the rounds resulted in a disequilibrium outcome, suggesting that subjects were not able or willing to capture the significant gains from coordination.

Given this baseline result, Cooper et al. (1989) then tested the effect of communication on the chances of coordination; in doing so, they distinguished between one-way and two-way communication. In the one-way communication treatment, the row player can send the opponent a nonbinding, costless message on his choice intentions prior to playing the game. The column player cannot respond. In the two-way communication treatment, both players can send each other costless and nonbinding messages declaring what they intend to play. Messages are nonbinding in the sense that a player is not constrained to choose according to what he has announced ("cheap talk"). After messages have been sent, pairs of subjects play out the game.

For one-way communication, a plausible outcome is for the player who can send the message to declare his intention to play the strategy supporting his preferred equilibrium. If the opponent considers the announcement credible enough, he will best respond to the announced strategy. This way the player who can communicate has the advantage of favoring coordination on his preferred equilibrium, and both players are better off, on average, than without the announcement. The data show that, in fact, one-way communication improves coordination dramatically and favors, in the majority of cases, the player that had the possibility to send the message.

Equilibrium play in the one-way communication treatment is observed 95% of the time, and in the large majority of cases row players announce their intention to play strategy 2 and actually play it. Column players, on their part, almost always play strategy 1. Two-way communication, in contrast, makes coordination *less* likely compared to one-way communication and almost as likely as coordination without communication. In fact, in the majority of cases, both players announce their intention to play the strategy implementing their preferred equilibrium, thus reintroducing the perfect symmetry that one-way communication had broken. As a result, although 80% of equilibrium announcements result in equilibrium play, 45% of rounds results in an outcome of disequilibrium.

C. EXPERIMENTING WITH ORDER-STATISTIC AND STAG-HUNT GAMES: VAN HUYCK ET AL. (1990, 1991), COOPER ET AL. (1990, 1992), RANKIN, VAN HUYCK, AND BATTALIO (2000)

Van Huyck et al. (1990, 1991) used the earnings tables in Table 6 and Table 7 for their two path-breaking studies. These studies are particular implementations of the order-statistic games mentioned earlier.[11]

TABLE 6   Earnings Table for the "Median Game"[a]

Your choice of X	Median value of X chosen						
	7	6	5	4	3	2	1
7	1.30	1.15	0.90	0.55	0.10	−0.45	−1.10
6	1.25	1.20	1.05	0.8	0.45	0.00	−0.55
5	1.10	1.15	1.10	0.95	0.70	0.35	−0.10
4	0.85	1.00	1.05	1.00	0.85	0.60	0.25
3	0.50	0.75	0.90	0.95	0.90	0.75	0.50
2	0.05	0.40	0.65	0.80	0.85	0.80	0.65
1	−0.5	−0.05	0.3	0.55	0.70	0.75	0.70

[a]Table Γ in Van Huyck, J. B., Battalio, R. C., & Beil, R. O. (1991).

TABLE 7   Earnings Table for the "Minimum Game"[a]

Your choice of X	Smallest value of X chosen						
	7	6	5	4	3	2	1
7	1.30	1.10	0.90	0.70	0.50	0.30	0.10
6	—	1.20	1.00	0.80	0.60	0.40	0.20
5	—	—	1.10	0.90	0.70	0.50	0.30
4	—	—	—	1.00	0.80	0.60	0.40
3	—	—	—	—	0.90	0.70	0.50
2	—	—	—	—	—	0.80	0.60
1	—	—	—	—	—	—	0.70

[a]Table A in Van Huyck, J. B., Battalio, R. C., & Beil, R. O. (1990).

[11]The results of these studies are—in our view rightly so—among the most celebrated in the literature on coordination failure (e.g., Camerer, 2003; Ochs, 1995; or scholar.google.com).

In both experiments, subjects had to choose numbers between 1 and 7. In the median game, everybody's payoff increases in the median of all numbers and decreases with the distance between the number chosen and the median. In the minimum, or "weak link" game described in Section II, everybody's payoff increases in the minimum of all numbers chosen and again decreases with the distance in the number chosen and the minimum. Note that while the Pareto-ranked equilibria on the main diagonal are the same, the off-diagonal elements differ. That is because they were generated by slightly different payoff functions.

The payoff-dominant or efficient equilibrium is in the upper left corner for both the minimum game and the median games, while the secure action induces an equilibrium (the secure equilibrium from here on) in the lower right corner for the minimum game and two rows up from the bottom in the median game. Both games feature seven (identical) Pareto-ranked strict equilibria on the main diagonal. There is a tension between the secure action—the lowest action in the minimum game, and the third lowest in the median game—and the action required for the efficient equilibrium. If payoff dominance, or efficiency, were the focal equilibrium, as intuited by almost all economists initially and also suggested by Harsanyi and Selten (1988), then subjects should have selected the strategy with the highest number.

Was efficiency psychologically salient in Van Huyck et al. (1990, 1991) or were competing concepts such as security, or risk dominance, more salient? The key result of their 1990 work is the stable and speedy unraveling of action choices to the worst of the strict equilibria. In the treatments based on Table 7, 14 to 16 participants played the stage game repeatedly (10 times in treatment A, and 5 times in treatment A′) and for money, receiving only information about their payoffs and about the minimum number chosen after each stage. The outcome was essentially the same even after payoff-efficient precedents emerged in a treatment (B) that was inserted between treatments A and A′ for four out of six sessions.

Several other researchers (with baseline treatments for various modifications reported in those papers) replicated this unraveling result with the same payoff matrix, with subject numbers varying from 6 to 14. Other experimenters (also with baseline treatments for various modifications reported in those papers) chose structurally similar payoff matrices (e.g., linear deviation costs, no negative payoffs) with slightly more or less action choices and also replicated this result. The detailed references may be found in Devetag and Ortmann (in press).

Van Huyck et al. (1991) demonstrate the influential role of the initial action choices. For the baseline treatment, neither the unique payoff-dominant equilibrium nor the unique secure equilibrium emerged when nine participants played the stage game repeatedly (again, 10 times), receiving only information about their payoffs and the median of all numbers after each stage. The initial median constituted a strong precedent from which subjects had trouble extracting themselves. This result, too, has been replicated by other authors. The detailed references may be found in Devetag and Ortmann (in press).

Not surprisingly, the remarkable coordination failure results of Van Huyck *et al.* (1990, 1991) (produced under conditions that, on the surface, seemed rather conducive to coordination successes—for example, small groups, perfect information, easy-to-understand task), drew considerable attention and a steady flow of attempts to test their robustness.[12] We shall discuss some of these attempts in Section VI.

The basic stag-hunt game explored experimentally by Cooper *et al.* (1992) was 1000 g(1,0,.8,.8) and is represented in Table 8.

As in the order-statistic games discussed earlier, the payoff-dominant equilibrium is in the upper left corner while the secure equilibrium is in the lower right corner. There is thus again a tension between A, the risky action (that might induce the efficient equilibrium), and B, the secure action (that induces the inefficient equilibrium, or an outcome even worse for the other player). The same kind of stag-hunt game was also the basic building block of Cooper *et al.* (1990). A key question that these researchers asked in both of their articles was whether the Pareto-dominant equilibrium would always be selected. The answer to this question is no.

Following up on related work published in 1989, Cooper *et al.* (1992) also explored whether the coordination failure results in their earlier work (1990) were robust to the use of both one-way and two-way communication, for this particular parameterization of the stag-hunt game. Coordination failure turned out to be endemic in the no-communication baseline conditions (and still significant with one-way communication); coordination failure was eliminated by two-way communication between players.

TABLE 8

Your choice	Other player's choice	
	A	B
A	1000,1000	0,800
B	800,0	800,800

[12]Van Huyck *et al.* (1990, 1991) themselves conducted a number of important robustness tests. Among their key insights are the importance of the number of participants, the matching protocol, the feedback conditions, and the deviation cost. In Van Huyck *et al.* (1990), for example, the authors demonstrated (in the already mentioned treatment B) that setting the coefficient on the deviation cost equal to zero lead to quick convergence to efficiency. They also demonstrated that two participants, when matched repeatedly and with the same person (but not with randomly drawn others), were able to coordinate on the efficient outcome.

It is important to mention that the coordination failure results of Cooper *et al.* (1990, 1992) came about under a matching protocol that differed sharply from the one used by Van Huyck *et al.* (1990, 1991) and other multiplayer studies afterwards. Specifically, while Van Huyck *et al.* routinely used multiplayer, finitely repeated coordination games with fixed matching, Cooper *et al.* (1989, 1990, 1992) used two-player sequences of one-shot games resulting from a random matching or rotation matching (Kamecke, 1997) protocol. As was shown later, for example, by Clark and Sefton (2001), the choice of the latter interaction patterns makes an efficiency-reducing difference.[13] It has also been shown, for example, that in multiplayer repeated games whereby group composition remains fixed over time, some players are willing to choose efficient actions in initial rounds, and by doing so forego higher initial payoffs with the purpose of "teaching" the other players to do the same and converge to the efficient equilibrium eventually. This behavior obviously is not applicable under a random rematching scheme.

Arguably the most intriguing article in this area is that produced by Rankin *et al.* (2000). The authors use a scaled-up version of $g(1,0,x,x)$ where x is, for each round, drawn randomly from the unit interval and then, ever so slightly, perturbed. Rankin and colleagues had their subjects play a sequence of 75 such games, in addition scrambling the action labels so that the payoff-dominant equilibrium and the secure equilibrium would not show up in the same corner throughout the 75 rounds. The intriguing result of this experiment, which was explicitly motivated by an attempt to increase external validity, was the high percentage of efficient play both when $x < 0.5$ (making the secure strategy less attractive and making payoff-dominant and risk-dominant equilibrium coincide) and when $x > 0.5$ (making the secure strategy more attractive and positioning the payoff-dominant and the risk-dominant equilibrium at opposite ends of the main diagonal).[14]

Rankin *et al.* point out that their setup inhibits learning from experience and focuses subjects on the exploration of deductive principles. In addition, in about half of the rounds, subjects faced a situation in which payoff dominance and risk dominance selected the same equilibrium. Obviously the results these

[13]The authors had their subjects participate in a stag-hunt game either as a sequence of one-shot games implying a random matching protocol, or as a repeated game with a fixed matching protocol. Their data show that, indeed, in the first round of play the frequencies of choice of the risky action were 0.3 in the random matching and 0.6 in the fixed matching protocol, a highly significant difference (0). Moreover, the fixed matching protocol reduced the instances of disequilibrium outcomes and   increased the overall proportion of risky choices across rounds.

[14]Specifically, for the first 10 periods, 65% (85%) of choices corresponded to the efficient action when $x > 0.5$ ($x < 0.5$). For the last 10 periods, about 90% (almost 100%) of the choices corresponded to the efficient action when $x > 0.5$ ($x < 0.5$). Thus, payoff dominance clearly carried the day.

researchers have reported (2000) seem to contradict the claim that coordination failure is common.

D. Experimenting with Critical-Mass Games

Devetag (2003) studied coordination in a repeated critical-mass game with seven strategies. The baseline version of the critical-mass game has been shown already in Table 4 and, for integers in a range [1, ..., *I*], is generated by the following payoff function:

$$\pi_i = \begin{cases} i & \text{if } k_i \geq i \\ 0 & \text{otherwise} \end{cases}$$

where *i* is the integer chosen by a player, and k_i is the total number of players in the group who have chosen that integer. A version of the critical-mass game featuring increasing returns can be generated by the following payoff function:

$$\pi_i = \begin{cases} i + k_i - 1 & \text{if } k_i \geq i \\ 0 & \text{otherwise} \end{cases}$$

In the second payoff function, given that the threshold for a number is matched, the individual payoff is higher the higher the number of players who choose it. Theoretically and anecdotally, critical-mass effects tend to be associated with increasing returns, implying that once a threshold is reached, the more individuals who engage in some activity, the more others will be inclined to do the same. If the number of players *N* is set equal to the number of available choices *I*, the attainment of the highest payoff implies that all players in the group must pick the highest integer.

Devetag (2003) investigated coordination in the critical-mass game both with and without increasing returns. Groups of seven subjects played one of the two games for 14 periods with a fixed matching protocol. The main treatment variable was the information condition. In the full information condition, players knew the distribution of all choices after each round. In this information condition, coordination on the payoff-dominant equilibrium was the most frequent outcome, and such outcome was even more frequent when groups played the version of the game with increasing returns. The path of play was often a coordinated, one-step-at-a-time movement toward the payoff-dominant equilibrium, sometimes allowing subjects to escape even from inefficient equilibria. Such "creeping up" did not occur when subjects only knew the median of all choices. Players did use this information to coordinate on the historical median, but this was an inefficient median in all groups. Finally, no convergence, even on suboptimal equilibria, was observed when players only had information about their own payoff. Hence, average efficiency increased monotonically with the increase in information, although only full feedback allowed convergence to the best equilibrium.

The individual behavior analysis reveals that the full feedback condition was actively used by some players (the "leaders") to signal to others the choice of the efficient equilibrium, in most cases inducing the remaining players eventually to pick the efficient action.

VI. ASSESSMENT

The early results of Van Huyck et al. (1990, 1991) and Cooper et al. (1990, 1992) seemed to indicate, somewhat in contrast to the results on pure coordination games (e.g., Mehta et al., 1994), critical-mass games (Devetag, 2003), and also market entry games (e.g., and famously, Kahneman, 1988),[15] that coordination failure, at least for order-statistic games and stag-hunt games, was a common phenomenon. Not surprisingly then, a number of authors followed up on these studies. For stag-hunt games we have already discussed the remarkable results of Rankin and colleagues (2000) that suggested strongly that payoff dominance had been counted out as a selection principle too early. Other related work is discussed in Camerer (2003) and Devetag and Ortmann (in press).

The order-statistic games by Van Huyck and colleagues (1990, 1991), especially the minimum game, have recently seen a renewed interest from top economics journals (e.g., Blume & Ortmann, 2007; Brandts & Cooper, 2006; Chaudhuri, Schotter, & Sopher, 2005; Weber, 2006). What has made the minimum game so attractive is the claim, not undisputed, that it is a good model of teamwork and therefore a basic building block of organizations, a theme already explored in the early work by Camerer and Knez (1996, 1997). In light of that claim, the initial evidence of coordination failure being pervasive was almost bound to lead to attempts to test experimentally the robustness of this evidence.

In Devetag and Ortmann (in press) we have discussed the myriad ways through which researchers have explored whether coordination failure is indeed a robust phenomenon. We document there that the original results

---

[15]Kahneman (1988) had N participants choose simultaneously, without communicating, whether to enter a market or not. The market had a "carrying capacity" of firms that the market could sustain. Entry in excess of the carrying ended in losses for all participants while lack of entry led to forgone welfare gains. The results of his experiment (later prominently used in Camerer & Lovallo, 1999) stunned Kahneman, who famously said, "To a psychologist, it looks like magic." A good discussion of market entry games may be found in Camerer (2003).

have indeed been shown to be robust in that they have been easy to replicate. We have, however, also documented what can be done to engineer coordination successes. For example, it is now well established that lowering the attractiveness of the secure action relative to the risky action that is required to implement the efficient equilibrium (e.g., Brandts & Cooper, 2006) is an efficiency-enhancing design choice.

Likewise, lowering the costs (expressed as foregone payoffs) that subjects suffer from trying out riskier but potentially more rewarding strategies ("experimentation") has been shown to work quite well. This has been achieved by lowering the payoffs associated to out-of-equilibrium choices (e.g., Battalio, Samuelson, & Van Huyck, 2001; Goeree & Holt, 2005; Van Huyck et al., 1990), by increasing the number of rounds while keeping the overall earnings roughly the same (e.g., Berninghaus & Ehrhart, 1998), or, finally, by refining the action grid while keeping the range of available payoffs the same (Van Huyck et al., 2001).

In addition, fixed matching protocols promote efficiency (e.g., Clark & Sefton, 2001; Schmidt, Shupp, Walker, & Ostrom, 2003; Van Huyck et al. 1990). Even random matching schemes can favor efficiency if the experimental design and implementation induce subjects to focus on the deductive principles underlying the game rather than on their own payoff history associated with different actions (e.g., Rankin et al., 2000; see also Schmidt et al., 2003). Providing full informational feedback seems efficiency enhancing in "small" groups (e.g., Berninghaus & Ehrhart, 2001; Brandts & Cooper, 2005, as well as Weber, 2006; but see Devetag, 2005), and the possibility of observing other players' expressions of intent and subsequent action choices was also shown to increase efficiency (Duffy & Feltovich, 2002, 2005). Last but not least, both costly (e.g. Cachon & Camerer, 1996; Van Huyck et al., 1993) and costless pre-play communication are efficiency-enhancing devices (e.g., Bangun, Chaudhuri, Prak, & Zhou, 2006; Blume & Ortmann, 2007; Cooper et al., 1992; Duffy & Feltovich, 2002, 2005; Van Huyck et al., 1992), as is higher quality of information, when this is made common knowledge (Chaudhuri et al., 2005; see also Bangun et al., 2006.)

In Devetag and Ortmann (in press), we have argued that many of the strategies to engineer coordination successes seem to move us away from the artificiality of the laboratory. A corollary to that statement is that, to some extent, the initial results of coordination experiments, while extraordinarily successful in generating discussion and subsequent research, were too hastily declared proof positive of the pervasiveness of coordination failure. The literature that has emerged over the past 12 to 15 years seems to suggest, in our view, a different picture. Of course, it was the initial results that prompted the follow-up work that helped us to understand the determinants of (laboratory) coordination successes and failures much better than we did then. Of note in this context is the contribution of the experimental method, which allows us to design and implement experiments that are clean in that we can study cause and effect in ways that are rarely achievable in real life. Yes, the experimental method has its problems. In principle, however, whatever objection one has to a particular experiment can be evaluated experimentally (e.g., using more realistic subject pools, or stimuli).

Future research on coordination games, in our view, would benefit from the application of MouseLab-type technologies as well as from gathering sociodemographic data (such as sex, age, income, or income proxies) that could identify correlations with behavior that have been identified for other types of games (see, e.g., Cooper, 2006, and Dufwenberg & Gneezy, 2005, for two noteworthy attempts). Some researchers in this volume, for example, argue that there should be more convergence between disciplines so that the "social" concepts focused upon by sociologists and social psychologists could be theoretically integrated into the game theoretic contexts. (See Sell, Chapter 18, and Eckel, Chapter 20, this volume.)

Research on coordination problems would also, in our view, benefit from additional efforts to move laboratory scenarios closer to the real world. It is, for example, quite desirable to understand what parameterizations of the coordination games we have described in this chapter are the most suitable to reflect coordination phenomena in real-life markets and other social contexts. In other words, it seems particularly desirable to understand what would be, for a certain problem, an appropriate parameterization ("calibration") of the game that tries to model the situation. It is predictable that clever researchers will soon conduct natural field experiments (Harrison & List, 2004), quite possibly matched with complementary laboratory experiments, on coordination phenomena (e.g., within organizations) that could get us closer to a realistic assessment of the way people solve coordination problems in real-life settings. The promise of such an approach has been, in our view, remarkably well demonstrated (List, 2006), albeit for another class of games.

ACKNOWLEDGMENTS

We thank Jane Sell for her constructive comments and Julie Ann VanDusky for improving the readability of the manuscript.

REFERENCES

Bangun, L., Chaudhuri, A., Prak, P., & Zhou, C. (2006). Common and almost common knowledge of credible assignments in a coordination game. *Economics Bulletin, 3,* 1–10

Battalio, R. C., Samuelson, L., & Van Huyck, J. (2001). Optimization incentives and coordination failure in laboratory stag hunt games. *Econometrica, 69,* 749–764.

Berninghaus, S. K., & Ehrhart, K.-M. (1998). Time horizon and equilibrium selection in tacit coordination games: Experimental results. *Journal of Economic Behavior and Organization, 37,* 231–248.

Berninghaus, S. K., & Ehrhart, K.-M. (2001). Coordination and information: Recent experimental evidence. *Economics Letters, 73,* 345–351.

Blume, A., & Ortmann, A. (2007). The effects of costless pre-play communication: Experimental evidence from games with Pareto-ranked equilibria. *Journal of Economic Theory, 132,* 274–290.

Brandenburger, A. M., & Nalebuff, J. N. (1997). *Co-opetition.* New York: Doubleday.

Brandts, J., & Cooper, D. J. (2005). Observability and overcoming coordination failure in organizations. *Experimental Economics, 9,* 407–423.

Brandts, J., & Cooper, D. J. (2006). A change would do you good ... An experimental study on how to overcome coordination failure in organizations. *American Economic Review, 96,* 669–693.

Cachon, G. P., & Camerer, C. F. (1996). Loss-avoidance and forward induction in experimental coordination games. *Quarterly Journal of Economics, 111,* 165–194.

Camerer, C. (2003). *Behavioral game theory. Experiments in strategic interaction.* Princeton, NJ: Princeton University Press.

Camerer C., & Knez M. (1996). Coordination, organizational boundaries and fads in business practice. *Industrial and Corporate Change, 5,* 89–112.

Camerer, C., & Knez, M. (1997). Coordination in organizations: A game-theoretic perspective. In Z. Shapira (Ed.), *Organizational decision making* (pp. 158–188). Cambridge: Cambridge Series on Judgment and Decision Making.

Camerer, C., & Lovallo, D. (1999). Overconfidence and excess entry: An experimental approach. *American Economic Review, 89,* 306–318.

Carlsson, H., & van Damme, E. (1993). Global games and equilibrium selection. *Econometrica, 61,* 989–1018.

Chaudhuri, A., Schotter, A., & Sopher, B. (In press). Talking ourselves to efficiency: Coordination in intergenerational minimum effort games with private, almost common and common knowledge of advice. *Experimental Economics.*

Clark, K., & Sefton, M. (2001). Repetition and signaling: Experimental evidence from games with efficient equilibria. *Economics Letters, 70,* 357–362.

Cooper, D. J. (2006). Are experienced managers experts at overcoming coordination failure? *Advances in Economic Analysis & Policy, 6.2,* Article 6 [50 pages].

Cooper, R., De Jong, D., Forsythe, R., & Ross, T. (1989). Communication in the battle of the sexes game. *Rand Journal of Economics, 20,* 568–587.

Cooper, R., De Jong, D., Forsythe, R., & Ross, T. (1990). Selection criteria in coordination games: Some experimental results. *American Economic Review, 80,* 218–233.

Cooper, R., De Jong, D., Forsythe, R., & Ross, T. (1992). Communication in coordination games. *Quarterly Journal of Economics, 107,* 739–771.

Costa-Gomes, M., Crawford, V., & Broseta, B. (2001). Cognition and behavior in normal form games: An experimental study. *Econometrica, 69,* 1193–1235.

Devetag, G. (2003). Coordination and information in critical mass games: An experimental study. *Experimental Economics, 6,* 53–73.

Devetag, G. (2005). Precedent transfer in coordination games: An experiment. *Economics Letters, 89,* 227–232.

Devetag, G., & Ortmann, A. (In press). When and why: A critical survey on coordination failure in the laboratory. *Experimental Economics.*

Duffy, J., & Feltovich, N. (2002). Do actions speak louder than words? Observation vs. cheap talk as coordination devices. *Games and Economic Behavior, 39,* 1–27.

Duffy, J., & Feltovich, N. (2006). Words, deeds and lies: Strategic behavior in games with multiple signals. *Review of Economic Studies, 73,* 669–688.

Dufwenberg, M., & Gneezy, U. (2005). Gender and coordination. In A. Rapoport & R. Zwick (Eds.), *Experimental business research* (Vol. 3, pp. 253–262). Boston: Kluwer.

Eckel, C. (2007). Economic games for social sciences. In M. Webster & J. Sell (Eds.), *Laboratory experiments in the social sciences* (pp. 497–515). Burlington, MA: Elsevier.

Farrell J., & Saloner G. (1986). Installed base and compatibility: Innovation, product preannouncements and predation. *American Economic Review, 76,* 940–955.

Gneezy, U., & Rustichini, A. (2000). Pay enough or don't pay at all. *Quarterly Journal of Economics, 115,* 791–811.

Goeree, J. K., & Holt, C. A. (2005). An experimental study of costly coordination. *Games and Economic Behavior, 51,* 349–364.

Harrison, G. (2005). Field experiments and control. In J. Carpenter, G. W. Harrison, & J. A. List (Eds.). *Field experiments in economics.* Greenwich, CT: JAI Press, *Research in experimental economics* (Vol. 10, pp. 17–50).

Harrison, G. W., Johnson, E., McInnes, M. M., & Rutström, E. E. (2005). Risk aversion and incentive effects: Comment. *American Economic Review, 95,* 897–901.

Harrison, G., & List, J. (2004). Field experiments. *Journal of Economic Literature, 42,* 4.

Harsanyi, J., & Selten, R. (1988). *A general theory of equilibrium selection in games.* Cambridge, MA: The MIT Press.

Heinemann, F., Nagel, R., & Ockenfels, P. (2004). The theory of global games on test: Experimental analysis of coordination games with public and private information. *Econometrica, 72,* 1583–1599.

Hertwig, R., & Ortmann, A. (2001). Experimental practices in economics: A challenge for psychologists? *Behavioral and Brain Sciences, 24,* 383–403. http://www.cogsci.soton.ac.uk/bbs/Archive/bbs.hertwig.html.

Holt, C., & Laury, S. (2002). Risk aversion and incentive effects in lottery choices. *American Economic Review, 92,* 1644–1655.

Holt, C., & Laury, S. (2005). Risk aversion and incentive effects: New data without order effects. *American Economic Review, 95,* 902–912.

Johnson, E. J., Camerer, C., Sen, S., & Rymon, T. (2002). Detecting failures of backward induction: Monitoring information search in sequential bargaining. *Journal of Economic Theory, 104,* 16–47.

Kahneman, D. (1988). Experiments in economics: A psychological perspective. In W. A. Tietz & R. Selten (Eds.). *Bounded rational behavior in experimental games and markets* (pp. 11–18R). New York: Springer.

Kamecke, U. (1997). Rotation: Matching schemes that efficiently preserve the best response structure of a one-shot game. *International Journal of Game Theory, 26,* 409–417.

List, J. (2006). The behavioralist meets the market. Measuring social preferences and reputation effects in actual transactions. *Journal of Political Economy, 114,* 1–37.

Mehta, J., Starmer, C., & Sugden, R. (1994). The nature of salience: An experimental investigation of pure coordination games. *American Economic Review, 84,* 658–673.

Morris, S., & Shin, H. S. (2003). Global games: Theory and applications. In M. Dewatripont, L. Hansen, & S. Turnovsky (Eds), *Advances in economics and econometrics* (Proceedings of the Eighth World Congress of the Econometric Society). Cambridge: Cambridge University Press.

Ochs, J. (1995). Coordination problems. In J. K. Kagel & A. E. Roth (Eds.), *Handbook of experimental economics* (pp. 195–252). Princeton, NJ: Princeton University Press.

Ortmann, A., & Gigerenzer, G. (1997). Reasoning in economics and psychology: Why social context matters. *Journal of Institutional and Theoretical Economics, 153,* 700–710.

Ortmann, A., & Hertwig, R. (2002). The costs of deception: Evidence from psychology. *Experimental Economics, 5,* 111–131.

Rankin, F., Van Huyck, J. B., & Battalio, R. C. (2000). Strategic similarity and emergence of conventions: Evidence from payoff perturbed stag hunt games. *Games and Economic Behavior,* 32, 315–337.

Rydval, O., & Ortmann, A. (2004). How financial incentives and cognitive abilities affect task performance in laboratory settings: An illustration. *Economics Letters, 85,* 315–320.

Schelling, T. (1960). *The strategy of conflict.* Cambridge, MA: Harvard University Press.

Schelling, T. (1978). *Micromotives and macrobehavior.* New York: E. W. Norton.

Schmidt, D., Shupp, R., Walker, J. M., & Ostrom, E. (2003). Playing safe in coordination games: The role of risk dominance, payoff dominance, social history, and reputation. *Games and Economic Behavior, 42,* 281–299.

Sell, J. (2007). Social dilemma experiments in sociology, psychology, political science, and economics. In M. Webster & J. Sell (Eds.), *Laboratory experiments in the social sciences* (pp. 459–479). Burlington, MA: Elsevier.

Stango, V. (2004). The economics of standards wars. *Review of Network Economics, 3,* 1–19.

Van Huyck, J. B., Battalio, R. C., & Beil, R. O. (1990). Tacit coordination games, strategic uncertainty, and coordination failure. *The American Economic Review, 80,* 234–248.

Van Huyck, J. B., Battalio, R. C., & Beil, R. O. (1991). Strategic uncertainty, equilibrium selection, and coordination failure in average opinion games. *The Quarterly Journal of Economics, 106,* 885–911.

Van Huyck, J. B., Battalio, R. C., & Beil, R. O. (1993). Asset markets as an equilibrium selection mechanism: Coordination failure, game form auctions, and tacit communication. *Games and Economic Behavior, 5,* 485–504.

Van Huyck, J. B., Battalio, R. C., & Rankin, F. W. (2001). Evidence on learning in coordination games. Texas A&M University laser script.

Van Huyck, J. B., Gillette, A., & Battalio, R. C. (1992). Credible assignments in coordination games. *Games and Economic Behavior, 4,* 606–626.

Weber, R. (2006). Managing growth to achieve efficient coordination in large groups. *American Economic Review, 96,* 114–126.

CHAPTER 17

Voting and Agenda Setting in Political Science and Economics

RICK K. WILSON
Rice University

I. INTRODUCTION

Political scientists and economists have a standard theoretical model for understanding group decision making that involves individual actors making a collective decision using well-defined rules for choice. An enormous body of theoretical literature is devoted to understanding how particular voting rules, agenda mechanisms, and power relationships affect the group choice. Many of those theoretical conjectures have been tested in the laboratory using spatial committee voting experiments.

Typically, political scientists are interested in formal (and informal) institutions and how those institutions produce incentives or sanctions for individual action. When it comes to group decision making, there is particular interest in the ways in which rules are adopted in order to translate individual preferences into a collective choice. How, then, does a group with diverse interests decide on a single collective pursuit? For example, how do individual members of a department jointly decide on which subfield will be allowed to hire a new faculty line?

The rules for deciding who votes, how votes are counted, and who controls the agenda are all critical. My focus is on this domain of decision making.

In the next section I detail the "standard experiment" used by political scientists and economists when understanding group decision making. This experiment and the accompanying model serve as the foundation for most studies of collective decision making. The third section presents findings for "institution-free" settings with and without equilibrium. The fourth section turns toward agenda-setting mechanisms, while the fifth looks at asymmetric power. The final section concludes with cautionary comments for theorists and experimentalists.

II. THE CANONICAL EXPERIMENT

Suppose a three-person department had to decide what kind of position to hire. The possible choices might be to hire an experimentalist, a statistician, or a theorist. How should the department decide? Of course, if all three members of the department (rightly) agree that an experimentalist is needed, then any one of the members could make the decision and everyone would be happy with what was chosen. However, what happens if there is disagreement over the type of position? In this instance, some decision rule must be invoked in order to make the decision. The department might decide on simple majority rule (in this case, the "winning" type of position determined by two of the three department members). While this seems "democratic" in that no single individual makes the decision for the group, it turns out that guaranteeing the collective (group) choice is difficult.

To illustrate the nature of this problem, consider a three-person department that has faculty with the following preferences over an experimentalist [E], a statistician [S], and a theorist [T]:

Amy: E > S > T
Bob: S > T > E
Cathy: T > E > S

Let ">" note a preference relation—that is, in Amy's case, an experimentalist is preferred to a statistician; a statistician is preferred to a theorist; and by transitivity an experimentalist is preferred to a theorist. From these preferences it is clear that the department does not agree over what type of person should be hired. So, how should the decision be made?

Appealing to democratic principles, Bob suggests that majority rule be used and proposes that the department first vote between a theorist and an experimentalist. In such a vote, the theorist would win, with Bob and Cathy voting in favor. Bob then proposes that a vote take place between the theorist and the statistician. Here, the statistician wins with Bob and Amy voting in favor (both preferring a statistician to an experimentalist). The department's decision, then, is a statistician. However, Amy recognizes that there has been no vote between an experimentalist and a statistician. If she is allowed to call a vote, then an experimentalist will defeat the statistician, with the department's choice being an experimentalist. Of course, Cathy would recognize what has happened and call for a vote on the theorist. This could continue ad infinitum with no end to the agenda.

The Marquis de Condorcet recognized this problem with simple majority voting in 1785. In the more contemporary period, Kenneth Arrow (1963) detailed how any decision rule violates basic axioms concerning collective choice, calling into question appeals to democratic fairness when making group decisions. It turns out that Arrow's findings, which were limited to discrete alternatives, are extremely general. The principal finding from 40 years of mathematical modeling is that an equilibrium is unlikely. This is worrisome because, in the absence of equilibrium, "anything can happen." Under the standard model, any outcome can be selected under any voting rule. This is problematic for those who believe that democratic systems will reveal the popular will.

A useful place to start is with the "standard model," a spatial model of committee voting that has served as a source of inspiration for theorists and laboratory experimentalists alike.[1] That model had its clearest statements in the late 1970s in a series of papers running through McKelvey (1976, 1979), Schofield (1978), Cohen (1979), and Cohen and Mathews (1980). In a simple sense, the standard model states that (1) if actors have well-defined preferences over a multidimensional policy space; (2) no actor has specialized agenda power; (3) simple majority rule is employed; and (4) a forward moving, binary-comparison, agenda mechanism is used, then equilibrium will be rare.[2] These results culminated in Riker's (1980) pessimistic conclusion that disequilibrium in collective choice should be pervasive and that political science unfortunately was proper heir to the title of "the dismal science." However, it is rare to find natural settings matching such conditions, so confirming or refuting the standard model is difficult.[3]

---

[1]For a useful overview of spatial modeling and its historical antecedents, see Enelow and Hinich (1984).

[2]The conditions required to fit such a model are more stringent than those pointed to here. In order to save space, the reader is invited to look at any of the papers cited previously. Cox (1987) develops explicit conditions for equilibrium, and Krehbiel (1988) reviews the general foundations of the model and empirical findings in legislative settings; Ostrom, Gardner, and Walker (1994) and Walker and Willer (Chapter 2, this volume) offer a brief discussion of what would be minimally necessary to describe this setting.

[3]Jillson and Wilson (1994) claim that the Continental Congress in the later 1770s resembled just such a "McKelvey world." I find that study compelling.

Of course, as Amy might say, "That's fine for theory, but we constantly see groups making decisions, so how do these models stack up?" Experimentalists were interested in this question and developed a canonical experimental design in order to test a variety of models. The first of these experiments appeared in Berl, McKelvey, Ordeshook, and Winer (1976), Fiorina and Plott (1978), and McKelvey, Ordeshook, and Winer (1978). I will use an experiment I developed based on these early designs. This experiment is computerized and is representative of all spatial committee experiments; I will demonstrate sample results in later sections. The experiments are based on spatial committee games that require knowing about the actors, the choice space, preferences, voting rules, and the agenda. I go through each separately.

A. Actors

In the experimental design I detail, there are five committee members. In other committee experiments the number of subjects ranges from three members to eleven. An odd number of committee members are typically used in order to avoid tied votes under simple majority rule. Given that each committee decision usually yields a single observation, increasing the size of the committee is costly and normally not worth it. Moreover, results from five-member committees generalize to all of the standard theoretical outcomes.

Subjects in my computerized experiments were separated by partitions and could not see one another's computer screens. Computerizing these experiments was deliberate and aimed at eliminating discussion between subjects. An early experiment by Laing and Olmsted (1978) pointed out that groups often made odd choices that could be due to subject communication. Eavey (1991) suggests that when subjects talk with one another, they key in on a concept of fairness within the group (but see Grelak & Koford, 1997, for another view on this matter). To eliminate collusion, subjects were identified in the experiment by a randomly assigned letter. Because the research questions are concerned with the ways in which different institutional mechanisms affect group choices, it is important to rule out prearranged strategies between subjects. Subjects were randomly assigned to conditions and were told not to talk during the course of the experiment.

B. The Policy Space

What did subjects make decisions about? In the economics and political science literature, the decision space is usually referred to as a "policy space." In the usual committee experiment, the policy space is nothing more than a plane made up of two orthogonal dimensions labeled X and Y. The intuitive notion involves trade-offs between spending on the military and education (or some other pair of competing policies). An outcome (a decision) is simply a point in the space given by a pair of (x,y) coordinates. This can be thought of as selecting an amount to spend on the military and an amount to spend on education. The policy space used in the computerized experiment, like many of the committee experiments, is quite dense. It was made of 300 points on the X dimension and 300 points on the Y dimension, representing 90,000 distinct outcomes that could be chosen. Such a policy space closely approximates theoretical models that make strong assumptions about the mathematical topology of the policy space (see McKelvey, 1976)

Why were there two dimensions and not one or three? For much of the theoretical literature one dimension is not of much interest. It (almost) always yields equilibrium and is well understood. Three dimensions, obviously, are difficult to represent to subjects. It turns out that a two-dimensional policy space has many of the same characteristics as an m-dimensional policy space.[4] As such it has become the standard design.

The policy space is kept abstract in order to minimize subjects' bringing their own values into the experiment. If the policy space instead were represented concretely as some mixture of military and education expenditures (or any other pairing of policies), then subjects might think about the space in ways that the experimenter does not control. For example, I might be predisposed to education and ill disposed to military spending. Even if the experimenter tries to get me to think otherwise about the space (by telling me that I am a "hawk" in this particular committee decision), I might resist because of my own personal preferences. To keep it abstract, subjects are told they are making decisions over the X and Y grid.

C. PREFERENCES

A crucial feature of the experimental design involves motivating subjects and manipulating what they prefer in the policy space. This is accomplished by assigning each subject an "ideal point" in the policy space and defining a payoff function that details exactly how much every alternative is worth in that space. The ideal point is worth the most to a subject; representative payoff functions and ideal points are given in Table 1 for several different experimental designs. Typically, subject payoffs are represented as circles (indifference curves) with the value of any alternative decreasing as a function of distance from the ideal point.

[4]The interested reader could take a look at Schofield (1985) or Saari (1994).

TABLE 1   Ideal Points and Payoffs Used in Experiments

| Member | Core preferences | | |
	Ideal points	Max. value	Loss rate (γ)
1	(120,125)	$15.00	−.018
2	(34,168)	$19.00	−.013
3	(242,247)	$25.00	−.011
4	(222,74)	$19.00	−.013
5	(30,35)	$19.00	−.013
Status quo = (175,265)			

| Member | Star preferences | | |
	Ideal points	Max. value	Loss rate (γ)
1	(22,214)	$25.00	−.013
2	(171,290)	$25.00	−.013
3	(279,180)	$25.00	−.013
4	(225,43)	$25.00	−.013
5	(43,75)	$25.00	−.013
Status quo = (280,280)			

| Member | Skew preferences | | |
	Ideal points	Max. value	Loss rate (γ)
1	(75,290)	$30.00	−.0129
2	(270,118)	$20.00	−.0129
3	(240,43)	$20.00	−.0170
4	(195,21)	$20.00	−.0129
5	(30,64)	$25.00	−.0129
Status quo = (280,280)			

Utility for any X and for the ith's member's ideal point, X_i, is given by:

Nonlinear payoff: $U_i = $ (max. value) $* \exp\left(\gamma * \left(||X - X_i|| \right) \right)$

Subjects are paid in the experiment for the value of the outcome that the group selects. By assigning subjects an ideal point and an associated payoff function, such a design adheres to principles of "induced valuation" promulgated by Smith (1982). This aims to control the motivation of subjects. In order to do so, three points are of primary concern:

- The payoff medium ought to be monotonic in the sense that a subject values more of the reward medium to less. Here, money, rather than some other reward, is useful. If subjects were rewarded with candy, rather than dollars, they might tire of too much candy. However, with something as fungible as money, each additional increment allows more of something to the subject.

- The reward medium ought to be salient. Again, the fungibility of money makes it more salient to a subject than many other forms of reward (e.g., additional grade points).
- The payoff medium ought to be dominant. That is, the amount of the reward in the experiment should overcome boredom by the subjects, should trump experimenter demand, and exceed the subject's own opportunity costs.

One nice feature of spatial committee experiments is that subject ideal points and payoffs can easily be manipulated. If the payoffs meet conditions for "induced valuation," then subject preferences are a part of the experimental control and variation across outcomes should be due to experimental manipulations. Figure 1 gives a three-dimensional visual representation of the ideal points and payoff functions for the "star" preference configuration given in Table 1. The policy dimensions X and Y are labeled on the figure. The third axis presents the dollar values for subjects across the policy space. This figure illustrates the steepness of the payoff functions for each subject. The maximal amount that a subject earns is the "peak" of the plotted function. The downward "slopes" indicate how fast payoffs decline as a function of distance from each ideal point. Because of the way in which the plot is generated,

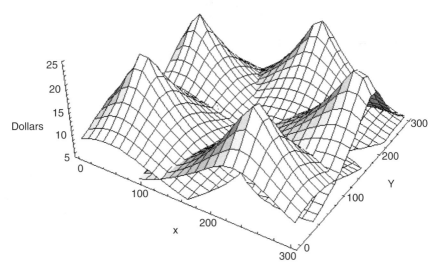

FIGURE 1   Plot of utility functions for five committee members in a spatial committee game. The X and Y axes constitute the policy space. The "dollars" axis indicates the amount a committee member would receive for each specific (x,y) policy.

the decreasing slope for each subject is obscured. The point to the figure is that each subject has a unique payoff function that is mapped onto the policy space.

D. Voting and Agenda Rules

The outcome of a committee decision is a single point in the space. In most of the committee experiments discussed here, a simple majority rule is used in which three of the five committee members must agree on the outcome. The experiments used a forward moving, open-agenda procedure in which proposing alternatives, voting, and adjourning are governed under a modified version of *Robert's Rules of Order*. At the outset of the decision, a status quo is presented to all subjects. That status quo is usually far removed in the policy space from all of the subjects, and as such is not worth much to anyone. Any subject can place a proposal on the floor. A proposal is a coordinate pair different from the status quo and could appear anywhere in the policy space (including a committee member's own ideal point).

Proposals remain "on the floor" (and on the screen) until seconded by a different person. Once seconded, the proposal is treated as an amendment and a vote is called between it and the status quo. All amendments are treated as an amendment in the nature of a substitute. If a majority votes in favor of keeping the status quo, the experiment continues and the floor is opened to further amendments. If a majority votes for the amendment, it becomes the (amended) status quo, and the floor is open to further amendments. The experiment continues in this fashion until a motion is made to adjourn. Anyone, at any time, can call for adjournment. If a majority votes to adjourn, the decision period ends and subjects are paid in cash the value of the current status quo. If a majority votes against adjournment, the experiment continues, with the floor open to further amendments. It is up to a majority of the committee to decide when to end the decision period.

Everyone in the experiment sees the location of the current status quo, all of the proposals, and the idea points of the other players; if an amendment is seconded, they see that as well. To make a proposal, a subject uses a pull-down menu and clicks on the point in the screen. The proposal is posted on everyone's screen within 200 milliseconds. Likewise, in order to second a proposal, a subject uses the pull-down menu and then clicks on an existing proposal. When a vote is called, the status quo and the amendment flash for 15 seconds, with a warning that a vote is impending. During that time additional proposals can be put on the floor, although no one can second another proposal. Once the vote is called, the screen changes and subjects see the location and value of the status quo and amendment. Subjects are instructed to vote for one or the other. When everyone finishes voting, the number voting in favor of the status quo and the amendment (but not who voted for each) is reported and the screen

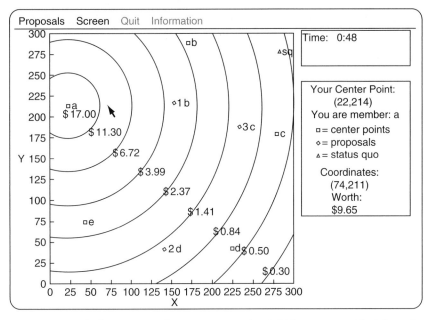

FIGURE 2 Sample decision screen for committee member "a." The screen provides information concerning payoffs, proposals on the floor, and the ideal points of other members.

switches back to the policy space. A vote to adjourn has a similar screen, except subjects are told the value of the current status quo and then they are asked whether they wish to quit or continue.

E. AN EXAMPLE

Figure 2 provides a picture of the main screen used in one of the experiments. The large square box on the left is the policy space for the experiment. This particular screen belonged to member "a" whose ideal point (the most valuable point to member a) was located at (22,214). A subject's ideal point is referred to as his or her "center point." Also plotted for subjects are representative indifference contours (which were called "value circles" in the instructions) and values associated with those contours. In this particular instance, member a's ideal point was worth $19, and payoffs declined rapidly as a function of distance away from that ideal point.[5] The smaller box to the right provides

[5]See Grelak and Koford (1997) for a useful discussion about the steepness of payoff functions and what it means for stability in the distribution of outcomes.

a legend for various pieces of information contained in the policy space. To be consistent with models of complete information, the ideal points of the other players are displayed as are any proposals and the current status quo.

Also displayed is the location of the cursor. On the figure, an arrow represents the cursor's position with its tip pointing to (74,211), which is worth $9.65. The computer automatically calculated the value of that point for the subject. In this way, subjects could obtain much finer readings of value than by trying to interpolate value from the indifference contours. All that was necessary for each subject was to move the cursor to the appropriate spot on the screen.

In this example, the status quo is located in the upper right corner at the point (280,280). Also on the screen are four different proposals. A proposal is represented as a diamond and is given a number based on when it was put on the floor. The letter to the right of the number indicates who proposed that alternative. The first proposal, located at (152,218), was made by committee member "b." That member's ideal point is given by the square and letter located at (171,290). This particular configuration of preferences is equivalent to the "star" configuration given in Table 1.

III. EQUILIBRIUM AND DISEQUILIBRIUM

The first spatial committee experiments turned to the question of equilibrium and disequilibrium. Spatial theories of group decisions make three very clear predictions:

P1. Equilibrium will be rare.
P2. The equilibrium will be chosen when it exists.
P3. In the absence of equilibrium, anything can happen.

The first prediction makes a general point about the distribution of citizen preferences. Plott (1967) and later Cox (1987) axiomatically prove the difficulty of ensuring an equilibrium in a multidimensional space. For a five-person committee in a two-dimensional space, it means that four of the committee members must be pairwise symmetrically dispersed around the fifth member.[6] While there are many ways in which preferences can be arranged to yield equilibrium, there are far more ways in which no equilibrium will occur.

---

[6]There are additional, very strong, assumptions that must be made about the shape of the indifference curves of all of the committee members, among other features of the model. For a readable introduction to spatial models, see Enelow and Hinich (1984).

The second prediction is a point prediction. For spatial models when equilibrium exists, it falls at a single point in the policy space. This means that if an experimenter manipulates the ideal points of subjects so as to guarantee equilibrium, it provides an explicit prediction.

The third prediction is the most vexing and has driven most of the theoretical and experimental literature on voting. If there is no equilibrium, then there should be no pattern to the group choice. In effect, the decision process will be unpatterned, leading to different outcomes even when beginning from the same initial point and using the same distribution of preferences. While these predictions are difficult to test in a natural setting because preferences cannot be controlled or the agenda cannot be fixed, spatial committee experiments allow for direct tests. Next, I report on results from my own experiments.

A. Equilibrium Outcomes

Results from an experiment with equilibrium are troubling. Seven trials were run manipulating preferences so that a unique equilibrium exists. The group decisions are plotted in Figure 3. Also on the figure are the ideal points of the five actors and the initial status quo.[7] The unique equilibrium is located at committee member 1's ideal point. The lines connecting members 2 and 4 and 3 and 5 illustrate that these subjects are pairwise symmetric around 1. From visual inspection it is easy to see that no outcome is located at the equilibrium, although two of the final committee choices are quite close. The equilibirum does not do well as a point prediction, although this finding is true for a number of spatial committee experiments. Berl *et al.* (1976), Fiorina and Plott (1978), Wilson (1986), Eavey (1991), and Grelak and Koford (1997) all find that outcomes consistently deviate from the equilibrium. At best, these outcomes are "close."

Fortunately, spatial theories predict more than an outcome. An additional prediction holds that the equilibrium is attractive. In other words, successful amendments will converge to it. In the experiment, data were collected concerning the proposals that were made, the time at which those proposals were made, and all of the votes on amending and adjourning. Consequently, the agenda can be reconstructed to see whether successive amendments were closer to the equilibrium than to the status quo they replaced. In five of the seven trials, *every* successful amendment was closer to the equilibrium than its predecessor. This strictly satisfies the "attractiveness" component of the

---

[7]It is important to note that in various trials, players' ideal points were rotated so that the equilibrium was located in different parts of the alternative space. All of the outcomes have been normalized to the same distribution of ideal points.

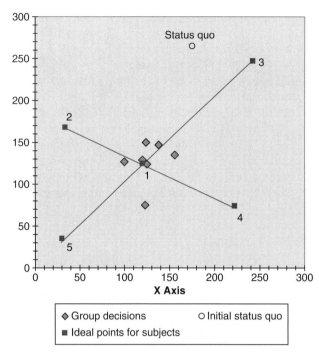

FIGURE 3   Group decisions under experiments with equilibrium. The equilibrium is located at the ideal point for committee member 1.

model, since the agenda converged toward the equilibrium. Moreover, the process converged quickly.

As can be seen from Table 2, subjects typically took under 10 minutes of floor discussion to end the period.[8] On average, the final outcome was selected within 41.6 seconds of beginning the trial. For the five trials converging on the equilibrium, subjects averaged 1.4 amendments before adjourning. This meant that the final outcome was quickly selected, was chosen from a handful of proposals on the floor, and was reached via a short agenda. However, this did not mean that subjects were quick to end the trial. The number of unsuccessful amendments outpaced successful amendments by almost 4 to 1 and subjects averaged 7.3 unsuccessful adjournment votes before ending the trial.

The findings seem a bit odd. On the one hand, the equilibrium is never chosen, but on the other hand it is attractive. Why this discrepancy? First, only a small number of proposals made it to the floor (on average, 27). Interestingly

[8]The time that elapsed while taking a vote was excluded.

TABLE 2   Descriptive Data for the Baseline Core Trials

Trial	Total time (sec)	Final outcome	Number of proposals	Number of amendment votes	Number of adjournment votes
BCORE1	610	(125,124)	39	14	12
BCORE2	156	(123,75)	17	3	3
BCORE3	548	(156,135)	31	8	8
BCORE4	84	(138,147	6	2	1
BCORE5	166	(124,150)	17	1	5
BCORE6	1972	(100,127)	40	62	25
BCORE7	398	(120,129)	39	9	4

the equilibrium was *never* proposed in *any* of the experimental trials. Second, subjects tended to make proposals near their own ideal points, but they had to depend on someone else bringing them to a vote. Proposals located at some distance from one's own ideal point were not financially worthwhile and this led to a very limited set of proposals being brought to a vote. The standard spatial model with an equilibrium enjoys more support than it might seem at first glance. This is the conclusion that two generations of axiomatic theorists and experimentalists have reached.

B. No-Equilibrium Outcomes

The more interesting case involves group decisions when there is no equilibrium. This setting is expected to be more common and is all the more troubling because there is no useful prediction.[9] Anything can happen in this case.

I examine two different manipulations of ideal points: a star and a "skew star" distribution. In both instances there is no equilibrium. Under the star distribution, subjects' ideal points are scattered about the policy space and no one occupies a central position. Given the relative symmetry of the ideal points, there is no obvious minimum winning coalition and no obvious focal point (see Figure 4). The skew star manipulates preferences so that three committee members are near one another and form a natural winning coalition.

The trials under the star manipulation are conducted in the same manner as those with the equilibrium manipulation. Outcomes from 18 trials are plotted on Figure 4. Once again, the ideal points of the five committee members are dis-

---

[9]A number of different solution concepts have been proposed and tested. See, for example, Fiorina and Plott (1978), Feld, Grofman, Hartlet, Kilgour, and Miller (1987), and Bianco, Lynch, Miller, and Sened (2006) for different approaches.

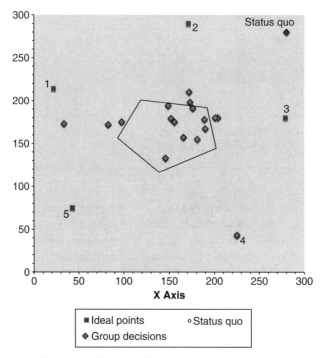

FIGURE 4   Group decisions under disequilibrium "star" preferences.

played, as is the status quo from which each agenda begins. The axiomatic models provide little insight into the distribution of outcomes. The usual view is that, since anything can happen, outcomes should be broadly distributed across the policy space. Instead, it is easy to see that outcomes are concentrated in the middle of the space. Half (9 of 18) of the outcomes are located in the (small) interior pentagon made up by finding the convex hulls of all possible winning coalitions. That pentagon defines a central portion for the alternative space. While three outcomes are located well outside that central space (one at the status quo where a majority voted to end the experiment unusually quickly), most are centrally distributed. The same finding is reported by Fiorina and Plott (1978) and McKelvey and Ordeshook (1990) in their survey of spatial committee experiments.

There is an additional empirical implication for disequilibrium settings: these agendas should be longer and more extensive than agendas in equilibrium settings. This implication is strongly supported by the data reported in Table 3. First, I find that subjects spend a good deal of time in these trials; subjects averaged just over 15 minutes per trial. By comparison, subjects in tri-

TABLE 3   Descriptive Data for Baseline Star Trials

Trial	Total time (sec)	Final outcome	Number of proposals	Number of amendment votes	Number of adjournment votes
BSTAR1	951	(176,191)	76	24	3
BSTAR2	77	(225,43)	6	1	1
BSTAR3	712	(204,180)	71	24	9
BSTAR4	782	(172,210)	48	15	6
BSTAR5	1901	(190,167)	174	55	18
BSTAR6	422	(34,173)	31	10	2
BSTAR7	1474	(146,133)	115	38	26
BSTAR8	1377	(173,198)	72	30	20
BSTAR9	39	(280,280)	5	0	1
BSTAR10	1905	(83,172)	96	37	14
BSTAR11	1845	(152,179)	72	26	14
BSTAR12	921	(201,180)	41	14	9
BSTAR13	477	(149,194)	39	9	3
BSTAR14	292	(156,175)	19	4	2
BSTAR15	1376	(98,175)	92	24	1
BSTAR16	1302	(166,157)	70	21	4
BSTAR17	961	(189,178)	34	20	14
BSTAR18	1097	(181,155)	61	20	6

als with equilibrium averaged just under 10 minutes of proposal making.[10] Given that subjects spent 50% more time in the experiment, I would expect them to make more proposals. They do, but at a rate that is two and one-half times their counterparts' in the equilibrium manipulation. In the most extreme case, subjects in the trial BSTAR5 had 174 proposals on the floor. By comparison, no more than 40 were placed on the floor in any equilibrium trial. Subjects under the star configuration also called a significantly larger number of amendment votes than did their counterparts under the equilibrium configuration (on average, just under 21 for the former compared with slightly over 14 for the latter).

These findings for the star baseline are concentrated in the central portion of the space. Is there something about that region, or is it simply an artifact of the preference configuration used? In order to address this question, the ideal

---

[10]Several of the experiments under this star baseline condition were conducted requiring that subjects use a minimum of 15 minutes of proposal-making time. This was done to make these experiments consistent with other manipulations not reported here. If those trials, BSTAR10 through BSTAR18, are removed from consideration, subjects still spent almost 15 minutes of proposal time in these experiments (on average, 859 seconds). The same point is true with the other statistics mentioned: subjects in trials using star preferences behave differently from those in the equilibrium configuration.

points for subjects were manipulated while keeping the structure of the institution the same.

The skew preference configuration, like the star configuration, has no majority rule equilibrium. Three players have ideal points located in the same quadrant of the alternative space, with two other players located some distance away (see the distribution in Figure 5). This preference configuration allows us to determine whether the central portion of the policy space has an independent effect on outcomes. Also plotted on this figure are outcomes from the trials.

As with the star configuration, these outcomes are clustered in a specific region of the alternative space, but removed from the central part of the space. Consequently, it appears there is nothing special about the central region. Two points are important when eyeballing the distribution of these outcomes. First, it is remarkable that outcomes shift so markedly with a change in the distribution of preferences. Clearly, subjects are responding to the manipulation. Second, and more critically, these outcomes are not widely distributed. Once again, I find that outcomes are tightly clustered and are not widely distributed across the policy space. This echoes findings by Laing and Olmsted (1978) and

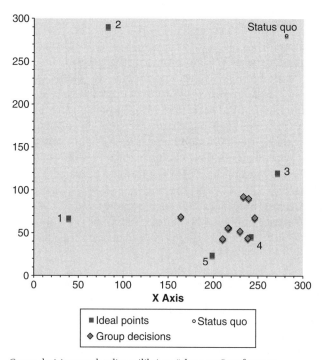

FIGURE 5   Group decisions under disequilibrium "skew star" preferences.

TABLE 4   Descriptive Data for Baseline Skew Star Trials

Trial	Total time (sec)	Final outcome	Number of proposals	Number of amendment votes	Number of adjournment votes
BSKEW1	317	(227,49)	32	8	6
BSKEW2	358	(237,88)	32	6	6
BSKEW3	478	(207,40)	36	12	6
BSKEW4	147	(244,65)	32	3	3
BSKEW5	105	(214,53)	8	2	1
BSKEW6	2346	(213,53)	137	76	21
BSKEW7	415	(231,90)	30	10	6
BSKEW8	474	(236,41)	27	9	8
BSKEW9	1998	(159,66)	123	46	29

McKelvey *et al.* (1978), who examined a variety of distributions of ideal points and observed a similar phenomenon.

Even though outcomes are clustered, this did not mean that subjects had an easy time selecting them. Again I find that the agenda process is lengthy and extensive. First, under the skew star configuration, subjects took a good deal of time to settle on an outcome (see Table 4). On average they spent about 12.5 minutes in proposal making—less than under the star but greater than under the core configuration. While subjects in the skew star manipulation cast almost as many amendment votes as subjects in the star configuration, they were less successful in finding amendments to the status quo. On average, they replaced the status quo only 4.0 times versus 6.7 times in the star replications. In part this was due to the fact that fewer proposals were made that could overturn the status quo. Only 13.7% of the proposals on the floor could have defeated the final status quo, compared with 20.8% for the star replications.

The skew star trials ended more quickly than the star trials. Agendas were typically shorter and a specific coalition dominated when choosing an outcome. Outcomes were not scattered throughout the alternative space, but neither were they concentrated in the center of that space. Both of the disequilibrium manipulations raise interesting questions about standard spatial models. On the one hand, the experiments show that outcomes converge.[11] On the other hand, the agendas for many of these trials demonstrate the kind of incoherence that theorists expect.

---

[11]Of course, a large number of solutions have been proposed. Among the most promising is the "uncovered set." For a thorough discussion, see Bianco *et al.* (2006).

IV. AGENDAS

The absence of equilibrium poses a serious challenge for scholars. A good deal of theoretical work focuses on institutional mechanisms that impose equilibrium (see Shepsle, 1989; Diermeier & Krehbiel, 2003). The most obvious theoretical work turns to the structure of the agenda. Romer and Rosenthal (1978) and Shepsle (1979) provide the earliest statements of the ways in which agenda mechanisms directly affect outcomes in settings in which there are disequilibrium preferences. The former shows the advantage held by those close to the status quo, while the latter points to germaneness rules in legislative settings that yield equilibrium.

Plott and Levine (1978) illustrate agenda power both in a natural setting (an airplane club) and in an experimental setting. When the preferences of others are well known and when control over the agenda is ceded to an agenda setter, predicting the group decision is easy. Of course, as Eavey and Miller (1995) show, agenda setting may not always matter if agenda setters have extreme preferences. Eavey and Miller (1984) show the power of the status quo when a status quo is the reversion point (the fall-back position if there is no agreement by the committee). Krehbiel (1986) makes a similar point in a legislative context. Kormendi and Plott (1982) and Endersby (1993) point to a variety of procedural rules affecting the agenda and their impact on outcomes. In short, there are a numerous studies showing the effect of basic institutional rules that govern the order in which alternatives are considered.

To provide an example of an agenda rule, I detail experience from my own research. In many deliberative bodies a "backward moving" agenda is used. Unlike the forward moving agenda described in the previous section, in this setting the status quo is voted last. The agenda is built so that the final seconded amendment is voted on first and is treated as a perfecting amendment to the prior amendment. The last amendment standing is voted against the status quo in the final vote. With an agenda that looks like {sq, a, b, c, d}, where the letters represent amendments and their order is the order in which the amendments were proposed, the first vote is between "c" and "d." The winner of that vote is then paired with "b," and so forth. The winning amendment is then paired in the final vote with "sq," the status quo. Shepsle and Weingast (1984) develop the equilibrium for this procedural rule and the experiment I designed tests it directly.

In this experiment, a star preference configuration is used and the policy space is 350 × 350 points. Rather than beginning with the status quo in the upper right corner of the policy space, a status quo was chosen that minimized the size of the equilibrium under a backward moving agenda (no equilibrium exists under a forward moving agenda). The status quo was the same for both the backward and forward moving agendas. Figure 6 plots the committee outcomes for both manipulations. While it may appear there is no difference between the two manipulations because of the distribution of outcomes, the differences are dramatic. Outcomes under the forward moving agenda range across the policy space in much the same manner as noted in the prior section. By comparison, all of the outcomes under the backward moving agenda are in equilibrium. Eight of the twelve committee decisions are located at the status quo, indicating its advantage under this institutional mechanism. None of the outcomes from the forward moving agenda returned to the initial status quo. The conclusion from this experiment is that agendas matter.

The earliest findings from spatial committee experiments turned on questions of the agenda procedures that induce equilibrium. For many, it was refreshing to discover that institutional design could ensure equilibrium and that the equilibrium could be predicted. The question next raised was what other types of institutional rules affect group choices.

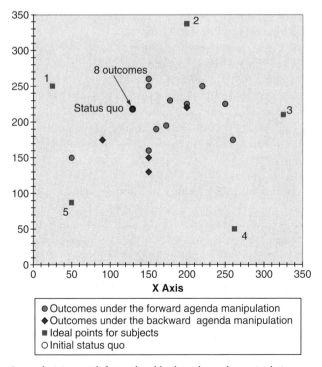

FIGURE 6   Group decisions with forward and backward agenda manipulations.

V. ASYMMETRIC RELATIONS

Political scientists and economists naturally understood that the symmetric power relations usually assumed in spatial committee theoretical models were unrealistic. After all, political institutions typically cede power to some at the expense of others. An early discussion by Buchanan and Tullock (1962) pointed to the trade-offs when differently sized aggregation rules are used. For example, the advantage that a person holds under unanimity is very different than when a simple majority is needed.

Theorists like Schofield (1985) spent a good deal of time demonstrating the circumstances under which equilibrium emerges as the voting rule and the dimensionality of the policy space changes. The possibility that a single player could be given the power to veto was discussed by Slutsky (1979); later, Tsebelis (2002) pointed to how institutions can be designed to effect veto powers. Others pointed to equilibrium appearing through multiple institutions, noting that a balance between the U.S. House and Senate effectively provides vetoes by each branch (Miller, 1987). Finally, Baron and Ferejohn (1989) developed a model illustrating the power of the initial proposer in legislative settings. This is only a sampling of the work that sorts out what happens when some actors are given special powers.

There is also a rich set of experiments illustrating what happens as special rules are granted to some players. Experiments by Laing and Slotznick (1983) and King (1994) are important for demonstrating the effect of changing aggregation rules. Unsurprisingly, building an agenda gets more difficult as the size of the majority increases. When unanimity is required, committee decisions are virtually deadlocked. Wilson and Herzberg (1987) and Haney, Herzberg, and Wilson (1992) looked at different versions of vetoing and blocking decisions. Any actor assigned veto power is greatly advantaged and committee decisions reflect this fact.

Miller, Hammond, and Kile (1996) and Bottom, Eavey, Miller, and Victor (2000) provided evidence noting the balance provided by institutions when committee members must decide first within a group and then across groups. This simple change in the decision setting yields an equilibrium and subjects converge on it. Bottom, Eavey, and Miller (1996) illustrated the power of the initial proposer in spatial committee games, though they found that such power is not absolute. By and large, the experimental data support the theoretical models, noting that when some are given power over others, the group decisions reflect that advantage.

What does it mean to give a committee member institutional advantage? To illustrate this, I again use my own research. In this experiment, one subject was randomly chosen and granted the sole power to bring a proposal to a vote and to call a vote to adjourn. Other committee members kept the power to place proposals on the policy space, but they could not second any proposals and could not call for a vote to adjourn. In effect, monopoly power was assigned to a single individual. Other aspects of a standard committee experiment remained the same. A vote to amend the status quo still required a simple majority and so too did adjournment.

Figure 7 illustrates the outcomes from a standard committee experiment using a forward moving agenda and outcomes from the monopoly agenda-setter experiment. Under the agenda-setting manipulation, the subject at member 5's position was *always* assigned to be the agenda setter and this point was also the equilibrium for the manipulation. These outcomes are represented as diamonds and outcomes under the standard manipulation are circles. Two points are clear from the figure. First, even though the structure of preference is the same for both manipulations, the pattern of outcomes is quite different. Under the standard manipulation, outcomes are more scattered in the policy space. Outcomes under the agenda-setter manipulation are more compact and they are anchored to the agenda setter's ideal point.

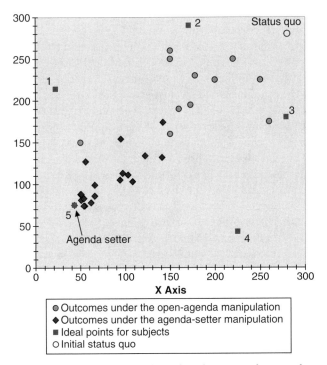

FIGURE 7　Group decisions with specialized agenda and open-agenda manipulations.

Second, group decisions under the agenda-setting manipulation do not fall at the equilibrium. Instead, those outcomes range from the agenda setter's ideal point to the central portion of the alternative space. This points out that the power of the agenda setter is not total. However, the agenda setter is clearly advantaged by having the right to call votes. What is valuable about this experiment is that it points to simple ways in which rules can be changed, how it is easy to instantiate those changes, and how theory and experiment can work together and it makes clear that there remains a gap between theory and empirical outcomes.

VI. EQUILIBRIUM AND DISEQUILIBRIUM, REDUX

I have sampled only a few of the chief results from several decades of work on group decision making in political science and economics. This work is rather stylized and focuses on institutional mechanisms that yield equilibrium. Much of this work was in response to basic theoretical models that predicted disequilibrium. Moreover, results on agenda setting and asymmetric power reoriented political scientists and economists to understanding basic institutional mechanisms.

But the standard axiomatic model predicting disequilibrium remains an intriguing empirical problem. As I briefly noted in Section III, group decisions do not necessarily converge on the equilibrium. As well, group decisions do not wander throughout the policy space when there is no equilibrium; in fact, in these settings the outcomes tend to cluster in distinct regions and respond to a shift in committee members' ideal points. As such, it seems there is room for continued theorizing and making sense of the empirical findings.

Several directions have been taken when confronting these anomalous findings. Among the axiomatic theorists, Miller (1980) shows that, with finite alternatives, the collective choice will not cycle among all alternatives when there is no unique equilibrium. McKelvey (1986) extends Miller's conjectures to multidimensional spaces and proves that outcomes will be limited to a distinct subset of the alternative space that has been coined the "uncovered" set. Bianco et al. (2006) have gone back through many of the spatial committee experiments and calculated the uncovered set for each distribution of ideal points. This has been no mean feat since, although the theoretical properties of this set are well known, it is only with very fast computers that it has been possible to identify this set. Bianco and colleagues find that, while disequilibrium remains the rule of the day, at least outcomes are limited to a subset of the policy space.

Other efforts to understanding deviations from equilibrium or the patterning of outcomes under disequilibrium have focused on decisions about fairness made by the subjects. Eavey and Miller (1984) and Eavey (1991) have argued that fairness considerations predominate in many committee experiments. Certainly in face-to-face discussion this is a possibility. Modeling fairness and its predictions, however, is difficult (see Grelak & Koford, 1997, for some of the concerns). On a different tack, Roberta Herzberg and I have pointed out that the transaction costs for decision making are never zero. Even though subjects may increase their earnings by continuing to vote, for many it is not worth the effort. Herzberg and Wilson (1991) detail some of the work on this front.

I think the more worthwhile path to take is to join with others who call for behavioral models that inform theoretical models. Camerer (2003) clearly makes this point when examining bargaining and negotiation experiments in economics. Behavioral biases, simple heuristics, and cognitive constraints may all play a part informing our basic models. It seems obvious that people do not finely search a complex policy space, that people are unwilling to invest considerable time pondering a cognitively demanding task, and that patience may wear thin. Fairness concerns or even envy may intervene when making choices. The point is that by understanding behavior in the laboratory, we can better inform our axiomatic models. Rational choice models have made a lot of headway in political science and economics. However, it is time to pay attention to what sociologists and psychologists can tell us about behavior.

ACKNOWLEDGMENTS

I gratefully acknowledge the support of the Workshop in Political Theory and Policy Analysis and the National Science Foundation (SES 8721250), neither of which bears any responsibility for the conclusions reached in this paper. Thanks is also due to Dean Dudley for his assistance in conducting these experiments.

REFERENCES

Arrow, K. J. (1963). *Social choice and individual values.* New Haven: Yale University Press.

Baron, D. P., & Ferejohn, J. A. (1989). Bargaining in legislatures. *American Political Science Review,* 83(4), 1181–1206.

Berl, J. E., McKelvey, R. D., Ordeshook, P. C., & Winer, M. D. (1976). An experimental test of the core in a simple N-person cooperative nonsidepayment game. *Journal of Conflict Resolution,* 20(3), 453–476.

Bianco, W. T., Lynch, M. S., Miller, G. J., & Sened, I. (2006). "A theory waiting to be discovered and used": A reanalysis of canonical experiments on majority rule decision making. *Journal of Politics,* 68(4), 837–850.

Bottom, W. P., Eavey, C. L., & Miller, G. J. (1996). Getting to the core—Coalitional integrity as a constraint on the power of agenda setters. *Journal of Conflict Resolution,* 40(2), 298–319.

Bottom, W., Eavey, C. L., Miller, G., & Victor, J. N. (2000). The institutional effect on majority rule instability: Bicameralism in spatial policy decisions. *American Journal of Political Science, 44*(3), 523–540.

Buchanan, J., & Tullock, G. (1962). *Calculus of consent.* Ann Arbor: University of Michigan Press.

Camerer, C. F. (2003). *Behavioral game theory: Experiments in strategic interaction.* New York: Russell Sage Foundation, Princeton University Press.

Cohen, L. (1979). Cyclic sets in multidimensional voting models. *Journal of Economic Theory, 20*(1), 1–12.

Cohen, L., & Mathews, S. (1980). Constrained Plott equilibrium, directional equilibria and global cycling sets. *Review of Economic Studies, 97,* 975–986.

Cox, G. (1987). The uncovered set and the core. *American Journal of Political Science, 31,* 408–22.

Diermeier, D., & Krehbiel, K. (2003). Institutionalism as a methodology. *Journal of Theoretical Politics, 15*(2), 123–144.

Eavey, C. L. (1991). Patterns of distribution in spatial games. *Rationality and Society, 3,* 450–474.

Eavey, C. L., & Miller, G. J. (1984). Bureaucratic agenda control: Imposition or bargaining. *American Political Science Review, 78*(4), 719–733.

Eavey, C. L., & Miller, G. J. (1995). Subcommittee agenda control. *Journal of Theoretical Politics, 7*(2), 125–156.

Endersby, J. W. (1993). Rules of method and rules of conduct: An experimental study on two types of procedure and committee behavior. *Journal of Politics, 55*(1), 218–236.

Enelow, J. M., & Hinich, M. J. (1984). *The spatial theory of voting.* Cambridge: Cambridge University Press.

Feld, S., Grofman, L. B., Hartlet, R., Kilgour, M., & Miller, N. (1987). The uncovered set in spatial voting games. *Theory and Decision, 23,* 129–155.

Fiorina, M. P., & Plott, C. R. (1978). Committee decisions under majority rule: An experimental study. *American Political Science Review, 72*(2), 575–598.

Grelak, E., & Koford, K. (1997). A re-examination of the Fiorina-Plott and Eavey voting experiments: How much do cardinal payoffs influence outcomes? *Journal of Economic Behavior & Organization, 32*(4), 571–589.

Haney, P., Herzberg, R., & Wilson, R. K. (1992). Advice and consent: Unitary actors, advisory models and experimental tests. *Journal of Conflict Resolution, 36*(4), 603–633.

Herzberg, R. Q., & Wilson, R. K. (1991). Costly agendas and spatial voting games: Theory and experiments on agenda access costs. In T. Palfrey (Ed.), *Experimentation in political science.* Ann Arbor: University of Michigan Press.

Jillson, C. C., & Wilson, R. K. (1994). *Congressional dynamics: Structure, coordination and choice in the first American Congress, 1774–1789.* Palo Alto, CA: Stanford University Press.

King, R. R. (1994). An experimental investigation of super majority voting rules—Implications for the Financial Accounting Standards Board. *Journal of Economic Behavior & Organization, 25*(2), 197–217.

Kormendi, R. C., & Plott, C. R. (1982). Committee decisions under alternative procedural rules. *Journal of Economic Behavior & Organization, 3*(3), 175–195.

Krehbiel, K. (1986). Sophisticated and myopic behavior in legislative committees: An experimental study. *American Journal of Political Science, 30*(3), 542–561.

Krehbiel, K. (1988). Spatial models of legislative choice. *Legislative Studies Quarterly, 13*(3), 259–319.

Laing, J. D., & Olmsted, S. (1978). An experimental and game theoretic study of committees. In P. C. Ordeshook (Ed.), *Game theory and political science.* New York: New York University Press.

Laing, J. D. S. N., & Slotznick, B. (1983). Winners, blockers, and the status quo: Simple collective decision games and the core. *Public Choice, 40*(3), 263–279.

McKelvey, R. D. (1976). Intransitivities in multidimensional voting models and some implications for agenda control. *Journal of Economic Theory, 12*(3), 472–482.

McKelvey, R. D. (1979). General conditions for global intransitivities in formal voting models. *Econometrica, 47*(5), 1085–1111.

McKelvey, R. D. (1986). Covering, dominance, and institution free properties of social choice. *American Journal of Political Science, 30*(2), 283–314.

McKelvey, R. D., & Ordeshook, P. C. (1990). A decade of experimental research on spatial models of elections and committees. In J. M. Enelow & M. J. Hinich (Eds.), *Advances in the spatial theory of voting.* Cambridge: Cambridge University Press.

McKelvey, R. D., Ordeshook, P. C., & Winer, M. D. (1978). The competitive solution for N-person games without transferable utility with an application to competitive games. *American Political Science Review, 72*(2), 599–615.

Miller, G. J. (1987). Core of the Constitution. *American Political Science Review, 81*(4), 1155–1174.

Miller, G. J., Hammond, T. H., & Kile, C. (1996). Bicameralism and the core: An experimental test. *Legislative Studies Quarterly, 21*(1), 83–103.

Miller, N. R. (1980). A new solution set for tournaments and majority voting: Further graph-theoretical approaches to the theory of voting. *American Journal of Political Science, 24*(1), 68–96.

Ostrom, E., Gardner, R., & Walker, J. (1994). *Rules, games and common-pool resources.* Ann Arbor: University of Michigan Press.

Plott, C. R. (1967). A notion of equilibrium and its possibility under majority rule. *American Economic Review, 57*(3), 787–806.

Plott, C. R., & Levine, M. E. (1978). A model of agenda influence on committee decisions. *American Economic Review, 68*(1), 146–160.

Riker, W. H. (1980). Implications from the disequilibrium of majority rule for the study of institutions. *American Political Science Review, 74*(2), 432–446.

Romer, T., & Rosenthal, H. (1978). Political resource allocation, controlled agendas, and the status quo. *Public Choice, 33*(1), 27–43.

Saari, D. (1994). *Geometry of voting.* Berlin: Springer–Verlag.

Schofield, N. (1978). Instability of simple dynamic games. *Review of Economic Studies, 45*(141), 575–594.

Schofield, N. (1985). *Social choice and democracy.* Heidelberg: Springer–Verlag.

Shepsle, K. A. (1979). Institutional arrangements and equilibrium in multidimensional voting models. *American Journal of Political Science, 23*(1), 27–59.

Shepsle, K. A. (1989). Studying institutions: Some lessons from the rational choice approach. *Journal of Theoretical Politics, 1*(2), 131–149.

Shepsle, K. A., & Weingast, B. R. (1984). Uncovered sets and sophisticated voting outcomes with implications for agenda institutions. *American Journal of Political Science, 28*(1), 49–74.

Slutsky, S. (1979). Equilibrium under a majority voting. *Econometrica, 46*(5), 1113–1126.

Smith, V. L. (1982). Macroeconmic systems as an experimental science. *American Economic Review, 72*(5), 923–955.

Tsebelis, G. (2002). *Veto players: How political institutions work.* New York: Russell Sage Foundation.

Walker, H. A., & Willer, D. (2007). Experiments and the science of sociology. In M. Webster & J. Sell (Eds.), *Laboratory experiments in the social sciences* (pp. 25–55). Burlington, MA: Elsevier.

Wilson, R. K. (1986). Results on the Condorcet winner: A committee experiment on time constraints. *Simulations and Games, 17*(2), 217–243.

Wilson, R. K., & Herzberg, R. Q. (1987). Negative decision powers and institutional equilibrium: Experiments on blocking coalitions. *Western Political Quarterly, 40*(4), 593–609.

Social Dilemma Experiments in Sociology, Psychology, Political Science, and Economics

JANE SELL

Texas A&M University

ABSTRACT

Social dilemmas are so named because their structure creates an incentive problem: what is in the individual's benefit at any given point in time is contrary to the group benefit over time. Social dilemmas are pervasive and can be found at all levels of interaction from dyads to nation-states. As a consequence, all of the social sciences have both a theoretical and applied interest in them. Much of the research across disciplines has been experimental. This chapter details some specific theoretical and methodological approaches that have demonstrated both problems and solutions to issues in social dilemmas.

I. INTRODUCTION

Social dilemmas occur in settings in which there is a conflict between individual short-term incentives and overall group incentives (see Dawes, 1980). There are two types of social dilemmas: public goods in which people must decide whether or not to contribute and resource goods in which people must decide whether to consume or refrain from consuming. Public goods are things that are available for all group members to use; examples include public radio and national defense. Resource goods are maintained for group members' use at some future time; examples include the maintenance of fragile rainforests and the protection of endangered species.

Public and resource goods are quite different from private goods. For private goods, an individual pays for a good (and thus reveals her preference) and then can consume the good individually. With resource and public goods, there is no direct relationship between contribution (or restraint from taking) and private acquisition. So, for example, regardless of whether you contribute to public radio, you can still consume or listen to public radio. This property, called nonexcludability, is at the heart of the incentive issue of social dilemmas. A person can refuse to contribute but still consume the good; that is, the person can "free ride" on the contributions of others. As this is the case, from the point of view of the individual, it is tempting not to contribute and to hope that others will contribute. But if all share that point of view, nobody contributes and the public or resource good is not provided. This is the dilemma.

We are faced with these dilemmas everywhere. Sometimes they involve our relationships with intimates—for example, while everyone in a family might prefer it if everyone else cleaned, if everyone waited for others, the house would never be clean. Sometimes these dilemmas involve institutions or countries. On an international level, if some countries voluntarily reduce their pollution rates, other countries benefit, even if they themselves do not reduce rates.

Since social dilemmas are so pervasive, they have been the subject of attention for sociology, psychology, political science, and economics. There is a different emphasis in these different fields and there are many examples of interdisciplinary research. The interdisciplinarity has been aided by the acceptance and use of experimental research. I will examine some central issues in the study of social dilemmas that cut across disciplines. In particular, I will emphasize structural aspects of social dilemmas.

I first consider the basic structure of the social dilemma. Over the past 20 years, there has been a movement toward more precision in specifying the type of social dilemma. For the most part, this has been in response to unexpected, sometimes surprising results from both laboratory and field experiments.

II. DIFFERENT KINDS OF SOCIAL DILEMMAS

A. Two-Person Dilemmas

There is a long tradition of two-person social dilemmas. In fact, the most famous is termed the "prisoner's dilemma" (PD), named for a story about two

TABLE 1 Illustration of Choices Associated with Two
Actors in a Prisoner's Dilemma

	Actor 1	
	Cooperate	Defect
Cooperate		
Defect		

(Actor 2 labels the left side: Cooperate, Defect)

prisoners, their separation, and the manner in which police try to structure the incentive for each to inform on the other. These two-person dilemmas are often characterized by a 2 × 2 table typology in which two choices for each person are characterized: cooperate or defect (see Table 1). The payoffs associated with cooperating with or defecting from an other (or partner) determine the incentive structure. Listing Person 1 first and Person 2 second, the four possible options then are cooperate/cooperate, cooperate/defect, defect/cooperate, and defect/defect. If the payoffs associated with cooperate/cooperate are the highest, then the incentive dilemma (the defining characteristic of social dilemmas) is absent, but there may still be a coordination problem. That is, how do the participants make sure that everyone knows the best choice?

The two-person dilemma has structured an enormous amount of research. Much of this research has centered around the issue of trust (and this will be discussed later in the chapter.) These kinds of scenarios, while tremendously important for two-party interactions, are usually considered apart from larger group analysis of social dilemmas. There are some sound theoretical reasons for a division based on size of the group. As Dawes (1980) indicated, when only two people are involved, the source of the defection or cooperation is completely known by both parties when they know the outcome. Each person, of course, knows his own behavior, and when he learns his outcome, he can tell immediately whether the partner cooperated or defected. Because this is the case, monitoring is complete, and sanctions can be specifically targeted to a single person. In cases in which multiple actors are involved (usually termed N-PDs), responsibility for actions is more diffuse and consequently monitoring is more problematic. Kollock and Smith (1996) correctly point out that it is not necessarily the case that two-person and N-person dilemmas are completely different. So, at times, N-person dilemmas could have these same characteristics of the two-person case.

Kollock and Smith's observation is important and relates to the direct tie between the theoretical question being asked and the development of an experimental test of those questions. For example, designing the experiment is a matter of what theoretical question is being asked—not necessarily the same thing as the particular number of participants. It is possible to design an experiment with many participants that involves the complete monitoring that is the default case for a two-person dilemma.

The intimate link between the theoretical question asked and the design is, of course, the biggest strength of the experiment. Since social dilemmas are so pervasive in everyday situations, experience based on actual settings has sometimes been taken for granted. The power of the experimental method to separate out factors has been instrumental sometimes in demonstrating that investigators were wrong about what factors lead to what behaviors.

B. N-Person Social Dilemmas

As mentioned, social dilemmas can be categorized into two groups: public goods and resource goods. Each of these has a classic statement or set of statements. For public goods, there is the theoretical explication by Paul Samuelson (1954, 1955) and Mancur Olson's (1965) *The Logic of Collective Action*. The goods and collective movements described in these pieces survive through contributions of members. Garrett Hardin's (1968) "The Tragedy of the Commons" is certainly the most mentioned article depicting resource goods and the temptation to take from the public domain. The tragedy mentioned by Hardin was the temptation for herders to overgraze their livestock on the commons and cause the degradation of the very resource that was necessary for them as well as the community.

For the most part, these two kinds of goods were viewed as equivalent, so the theoretical principles developed in one were generally applied to the other. Such equivalence seemed reasonable, especially from an economics point of view since the payoffs created the same set of incentives (Ledyard, 1995). In all cases, the social dilemma incentives favored individual-level defection (or non-cooperation) at each individual point in time. However, some in psychology argued that the two types of settings were different from a social psychological point of view (see Brewer & Kramer, 1986). Sell (1988) argued that the distinction between them could be gauged on the basis of the setting and then actors' responses to those structures. In particular, she argued that while public goods could be characterized by the production function, which linked the contributions or resources of group members to the public good, resource goods (often created experimentally to "mirror" natural resources) are characterized by a replenishment function that creates a different type of good.

Brewer and Kramer were the first to investigate the difference in 1986; this was followed by several other studies (see, for example, McCusker & Carnevale, 1995; Messick, Allison, & Samuleson, 1988; Rutte, Wilke, & Messick, 1987; Schwartz-Shea & Simmons, 1995). Results were mixed. Although all this research was experimental, there were differences in the good being examined. Son and Sell (1995) and Sell and Son (1997) argued that some of the differences might be due to the nature of the public good, but that the static versus dynamic nature of the setting also could affect outcomes dramatically, partly in terms of uncertainty of outcome. Theoretically, there is a large difference between a one-time event or decision and multiple series of decisions, and also a difference between a static series in which each decision is like every other and a series of decisions in which a replenishment function transforms the resource.

In a series of studies designed to determine how the type of good affected levels of cooperation (either restraint from taking or contributing), Sell and Son (1997) and Sell, Chen, Hunter-Homes, and Johansson (2002) tried to compare giving versus restraint directly in exactly the same settings. One-time decision making was compared to one-time decision making; repeated decision-making was compared to repeated decision-making and dynamic decision making was compared to dynamic decision making. To make comparisons, the simplest version of the public good setting and the resource good setting was utilized. The experimental scenario was adapted from the standard linear public goods setting developed and modified by Gerald Marwell and Ruth Ames in their research beginning in 1979 and then Mark Isaac, James Walker, and Susan Thomas in 1984.

For public goods, subjects are provided an initial *endowment,* an amount of money individually allocated to the subjects. This endowment is usually specified in terms of tokens so that the tokens can take on different monetary values. The subject then is given the choice to keep his or her endowment in the private fund or invest in a group fund, a fund that has a payoff different from that of the private fund and is shared with group members regardless of whether those members have contributed or not. This standard linear public goods setting has a number of advantages. It is simple and it provides a simple way to quantify the incentive structure. The different values for the different funds allow a method of calculating what Isaac and Walker (1988) termed the marginal per-capita return (MPCR) on the public good. For example, in a particular experiment, if there are four members of a group and if every token in that experiment is worth 1¢ in the private fund, but worth 3¢ in the public fund (the public good), the marginal per-capita return on the public good is 0.75. To ensure the defining property of a public good, the marginal per-capita return on the public good (at a given point in time) is always less than 1. Such measurement enables comparison across different settings and more precise estimates of the nature of the incentives.

In comparing public goods and resource goods, it was important that the framing or language aspects were kept minimal, since language can affect subjects' decisions. For example, Andreoni (1995) found that framing choices as focusing on others' benefits or losses rather than just making decisions created increased contributions to the public good. Consequently, no descriptions of actual settings were used and wording differences between the resource and public goods were minimized. In computerized versions of the experiments, two simple blue boxes appeared on a subject's screen. In the public good setting, an individual's tokens appeared in the blue box marked "private fund," and subjects could choose whether to keep tokens in the private fund, which paid a certain amount and was not shared with others. Their other choice was to contribute to a group fund that paid another amount and was divided (equally) among all group members. In the resource good setting, the same blue boxes appeared, but the tokens appeared first in the group fund and subjects could choose whether to keep the tokens in the group fund (which again was divided equally among group members) or withdraw tokens to invest in the private fund (see Figure 1). For both these cases, the payoffs were exactly the same for the two funds and these payoffs instantiated the property of the social dilemma such that, for any given period of time, the individual's payoff for the private fund was more than his or her payoff from the group fund.

When the experiments were not computerized (when, for example, this was not possible in cultures where computers were not available), the blue boxes were replaced by two different columns. Those subjects randomly assigned to the resource good condition found their token amount listed in the group fund

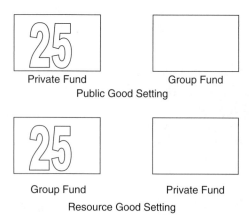

FIGURE 1   Comparison of screens for static public goods and resource goods.

column initially, while those randomly assigned to the public good condition found their token amount listed in the individual fund initially.

Record sheets indicated other aspects of the good setting such as replenishment rates and results from previous trials. In this manner, each subject could see what behaviors had occurred in the past, and these were made salient. (This was particularly important because some studies had been criticized for being confusing or unclear about the payoffs and the information.)

Results from these studies indicated that, indeed, resource and public goods did generate different levels of cooperation for static settings. In particular, in violation of traditional expected utility formulations, but explained through alternative theoretical perspectives, resource goods settings generated more cooperation than did public goods settings. When tokens started out in the group fund, subjects were more likely to cooperate than when the tokens started out in their own individual fund. Further, this process builds up over time; cooperation builds more cooperation.

While these results are important, there is a central mystery in them as in many results: why do people cooperate *at all* in any setting that creates a social dilemma? The incentive structure seems to dictate that people should NOT cooperate because it is not in their best interest. Much research was directed at answering this question, and both theoretical and empirical advancement resulted.

III. REJECTION OF THE STRONG FREE-RIDING HYPOTHESIS

From the viewpoint of traditional economic models, public goods must entail coercion to sustain them. This is the case because, if an actor is rational and seeking only to maximize her own utility, she will not contribute. Furthermore, with such goods, there must be some way to develop measures of willingness to pay (or preferences) to optimize the welfare of those in the particular setting. This is problematic because, unlike the private good situation where there is no incentive to misrepresent preferences, there *is* incentive for people to understate preferences since the good can be consumed by all. This theoretical perspective was explicated by Paul Samuelson in 1954, 1955, and 1958.

Much of welfare economics takes these two issues as a given, but experimentalists sought to explore this, initially to develop an important baseline. But, there were problems in the empirical analyses of such settings. Bohm (1972) designed a survey to study individuals' willingness to pay for a new television show. One group was told that their personal cost would be tied to their stated willingness to pay while another group was told that their cost was irrespective of their stated willingness to pay. The primary prediction was that those who were asked their true willingness to pay would understate their willingness (and therefore would free ride) while those whose cost was independent of their statements would overstate preferences to ensure that the show would be provided. However, the predictions were not supported. There was no difference in the willingness to pay among groups; the free-rider hypothesis was not supported.

This prompted a series of further tests. Many contended that there were interpretation problems with the Bohm experiment. Perhaps subjects were confused about the good; preferences were not induced and not known. Other researchers conducted studies to try to eliminate different alternative interpretations for the results.

Sociologists Gerald Marwell and Ruth Ames were interested in these general phenomena as well and conducted a series of experiments in natural settings. In a 1979 study, the researchers contacted a fairly large number of high school students (256). Students were contacted several times by phone and also received directions by mail. They were told that they could invest in two types of exchanges. The individual exchange was a sure investment, so, for every token invested, the subject received a set amount. The group exchange, however, was shared by all group members irrespective of their individual contributions. This group exchange constituted the public good and contained a provision point—a point beyond which the public good payoff was greatly increased. As predicted by economic theory, if there was an individual in the group whose interest exceeded the cost of its provision, that group invested far more in the public good than did other types of groups.

However, other effects predicted by basic economic theory were not supported. Important among those predictions, far less free riding occurred than what would be predicted. Following up on the 1979 research and responding especially to critics who claimed that their experimental design could not rule out effects that might result from misunderstanding, Marwell and Ames (1980, 1981) systematically replicated parts of their 1979 research to eliminate several plausible alternative reasons for their results. For instance, to address critics who claimed that the stakes were not high enough, in one study they increased the stakes fivefold and found that it did not affect subjects' behavior.

In essence, these first groups of studies demonstrated that the strong free-riding hypotheses did not hold. This is an interesting and important contribution of experimental methodology because the control possible through the experimental design in terms of separation of effects of reputation, anticipation of future interaction, and knowledge of the characteristics of others in the group could be held constant. Even when possible interpretations, such as participants somehow trying to please the experimenters, were controlled for, contributions never completely disappeared. Although free riding could be increased by making it more materially lucrative, free riding was never complete. For sociologists and psychologists, the reason seems clear: there was a normative aspect to the phenomenon even when all usual normative trappings and language were stripped away. Economists have given this residual level of cooperation various terms, including "warm glow" (see Andreoni, 1995). Most recently, some economists have argued that there may be a biological basis for cooperation that may need to be incorporated into economic theories of exchange. Kevin McCabe and Vernon Smith (2001), for example, argue that humans engage in a "goodwill accounting" that is part of humans' evolutionary heritage.

IV. TESTING PAYOFF PROPERTIES

While people may not completely free ride, it still is common, and in the experiments it displayed predictable patterns according to the incentive structure; for example, the MPCR on a public good does make a difference. When the MPCR is lower, then contributions are lower also (Isaac & Walker, 1988). Using the MPCR also allows developing precise comparisons for cases in which some players receive high returns while others receive low returns.

This scenario can be modified in many ways, including varying the amount or return on the public good both to the group and to the individual. For example, as cited by Goeree, Holt, and Laury (2002), research by Carter, Drainville, and Poulin (1992) investigated the possible of differential returns—for example, they set an "internal return" for the public good (say, 5¢) and an external return for the public good (say, 2¢). In this way, the marginal value of the public good can also be contrasted to the net costs of contributing versus the benefit to others. This experimental strategy was also followed by Goeree, Holt, and Laury (2002) in an analysis of altruism in one-shot or one-time public goods settings.

A. REPEATED DECISIONS VERSUS ONE-SHOT OR ONE-TIME DECISIONS

One of the most important theoretical differences in resource settings concerns whether or not the actors anticipate future interaction with their partner in two-person settings, or with other members of their group in larger settings. In social psychological terms, anticipation of interaction carries with it many different factors. The interaction carries with it the ability of the other to respond, of course. So, if actor A contributes at time 1 while actor B does not contribute at time 1, actor A can retaliate against actor B at time 2; conversely, actors can also reinforce each other.

One-shot or one-time decisions do not carry the anticipation of future interaction and so also do not admit the possibility to reward or punish other actors. They are restrictive interactions and cooperation should be, from all theoretical perspectives, difficult to maintain. Cooperation levels should be at the lowest in one-shot settings. Still, as noted earlier, cooperation to the public or resource good is not zero.

In some experimental studies, one-shot decisions have been repeated. But when the composition of the group is continuously shuffled so that the group members in decision 1 are completely different from the group members at decision 2, the claim is usually made that these are equivalent to one-shot decisions.

When groups are interacting over time, there are several theoretical conditions important to the establishment of the theoretical predictions and consequently the experimental design. If actors know the exact number of times that they will be interacting, they can engage in calculations associated with the endpoint. If actors know the endpoint, they realize that, at that endpoint, there is minimal incentive to cooperate, since other members will no longer have any sanctioning ability. This can lead to an "unraveling" of cooperation—or at least that has been the theory of "backward induction." Although experimental research does not support the lack of cooperation or the extent of free riding posited though a mechanism of backward induction (see a review of resource dilemmas by Rapoport, 1997, for example), there are still problems with subjects knowing exactly how many trials will obtain.

Experimentalists have dealt with endpoint issues in several ways. Sometimes the number of trials is announced to the subjects, but the data for the endpoint are analyzed separately from the other data. There is some controversy about doing this because subjects do know the endpoint and this knowledge may create different behavior as the trials proceed. Another way this has been handled is for the experimenter to announce a range of trials that will occur, and then randomly select the number for a given experiment. This technique has the advantage of not having a specific endpoint, but practically speaking, gives subjects an idea of about how long they will be interacting.

At times the amount of interaction is a very specific and important parameter of the study. According to what is usually called the "folk theorem," contributing to a public good can be a "purely rational" individual strategy for an individual actor if the time horizon (or discount parameter) is sufficiently large. Basically, if an actor knows that the interaction is likely to occur for a long time, it is better to cooperate (as long as others are cooperating) because his or her share of the resource or public good could be greater. The statement "as long as others are cooperating" is an important theoretical proviso. The folk theorem is critical for theoretical development because it states that while cooperating at any one point in time is not individually rational, cooperating over longer periods of time can be justified without sacrificing any assumptions about rationality. However, while the folk theorem allows for the possibility of cooperation, it does not rule out very many possible outcomes, so additional assumptions are necessary for predicting likely outcomes. (For discussion, see Sell & Wilson, 1999.)

If the discount parameter is either a scope condition of the theory or a manipulated portion of the study, it can be specifically implemented. For example, subjects can be "given" a projected number of trials. To eliminate the endpoint effects, a random component is added. In Sell and Wilson (1999), treatments involved several different discount parameters. It was explained to subjects that there was a likelihood or probability that each trial would be the last (this remained constant throughout the experiment). Subjects were told that this number represented the likelihood that the study would end and that another way to think about this was the likely number of trials. For example, if the likelihood was .95, that meant that the probability of this decision period being the last was .05 or 5 out of 100. Subjects were told, "Another way to think about this is as the average number of trials that you can expect to play. You figure this out by simply considering the fractions. If the probability was .95, then you could expect, on the average, to play 20 trials (5/100 or 1/20)." This probability was then concretely symbolized by chips in a basket. In the case of .95, for example, there might be 19 red chips and 1 white chip. After each decision, a chip was drawn to see if the decision periods would continue.

In many resource goods experiments, the time dimension is controlled by the replenishment rate and, of course, the rate at which subjects might harvest the good. This replenishment rate can mirror the way in which biological resources such as fish might reproduce or, alternatively, die off from causes other than overfishing.

V. WHAT GROUP MEMBERS KNOW ABOUT EACH OTHER

One of the most important components for social dilemmas is what information actors have about each other. This can take the form of what characteristics of group members are revealed and what information about payoffs or motivations is available.

A. SYMMETRICAL VERSUS ASYMMETRICAL INFORMATION

Information is symmetric if each actor knows that his or her information is the same as others'. Asymmetric information is the instance in which some actors possess private information, or information that others do not have. This information can relate to the endowments that group members possess, the payoff structure, or characteristics of the actors themselves.

Yamagishi and his colleagues have developed a manipulation to examine how a common identity might affect actors' cooperation (or contributions) in a one-shot prisoner's dilemma setting. This asymmetric information is used to determine the mechanism through which group identity might affect cooperation.

Group identity has been an important variable in the cooperation and trust literature. As discussed in Yamagishi, Jin, and Miller (1998) and Yamagishi *et al.* (2005), much research has been interpreted as supporting the general idea that actors give preference to those who share a group membership characteristic. A series of studies beginning with Billig and Tajfel (1973) and Tajfel, Billig, Bundy, and Flamen (1971) developed the concept of group differentiation based upon a trivial label (such as preference between two artists (Klee or Kandinski) or whether an individual was likely to overestimate or underestimate the number of dots when exposed to slides containing vast numbers of dots). Surprisingly, actors used this label to give preference to those with whom they shared the label and consequently gave them greater resources.

Yamagishi and his colleagues conducted a series of experiments to examine the nature of this preference. To separate out the effects of favoritism built upon the commonality versus the effects of expectations that other in-group members will acknowledge (and potentially reciprocate), the researchers developed a condition they called "unilateral knowledge" and compared its outcomes to those from mutual knowledge. Unilateral knowledge (a particular kind of asymmetrical knowledge) meant that a subject knew that the partner was in the same group as he or she was; however, the subject also knew that the partner *did not* know. (See Jin & Yamagishi, 1997).

What was found was that more cooperation was generated in settings in which there was mutual rather than unilateral knowledge. Such results support the idea that in-group favoritism is not just based upon a shared membership; rather, it seems to be based upon what Jin and Yamagishi called "group heuristics," an expectation of generalized reciprocity from members of the same group. This general result was supported in the instance of a trivial group membership as well as the nationality (specifically, subjects knowing that their partner was either Australian or Japanese).

B. Simultaneous or Sequential Decisions

The implementation of particular decision-making rules can also affect the information that group members have about each other. One type of decision rule pertains to *when* group members make decisions relative to each other. In one-time decisions, if decisions are simultaneous, group members have no information about the others' decisions when they make their own decision. In sequential decisions, there is an order to the decision making and, consequently, those who follow always have more information about others' decisions than the actors who preceded them did.

Yamagishi and Kiyonari (2000) used the difference between simultaneous and sequential decision making to further examine the impact of group identity. They manipulated group identification in terms of a trivial category—a given group member was classified according to his or her preference for either Klee or Kandinski paintings. Subjects were then paired with a partner and told that they would make the first decision while their partner would make the second decision. In this particular manipulation of a dilemma, subjects received an endowment and had to decide how much of the endowment they would give to the partner. The money provided to the partner was doubled; then, subjects were told that the partner was faced with the same decision: how much of their endowment should they provide to the partner? (This is a very common experimental paradigm called the "trust game.")

Yamagishi and Kiyonari reasoned that if group identity was the only factor affecting decisions, there would be no difference between allocations made when decisions were simultaneous and when they were sequential. Their prediction, in line with the formulations discussed before, was that cooperation would *not* be equivalent. That is, regardless of whether the group member was or was not a member of the same group, cooperation should be high in the sequential game. This, indeed, was the case and supported the idea that the hope of reciprocity prompted cooperation in the sequential game.

C. Common Knowledge

In economic investigations, in particular, the requirement of common knowledge is important for theoretical derivations and empirical application. This relates to some of the assumptions posited in game theory, the basis of much of the economic theory of social dilemmas. Basically, the common knowledge requirement is that everyone knows the parameters of the setting and everyone knows that everyone knows. This terminology is a little different from that

used about information symmetry and asymmetry. Whereas information usually refers to particular characteristics about actors or their past behavior,
common knowledge usually refers to understandings about how others
process information. Generally speaking, this means that actors assume others
have the information to act in a consistent (rational) manner. For example,
a setting could be one in which there was asymmetric information but still be
one in which there was common knowledge.

To ensure that this requirement is met, it is usual to illustrate to all subjects
that everyone is reading and seeing the same information. (This information
could include the idea that some subjects have information that others do not.)
While it is important that all subjects have their own sets of instructions and
their own records of payoffs and earnings for interactions over time, for many
studies it is also important for a researcher to read or demonstrate general
study parameters. Along this line, it is a good idea to interview subjects over
the relevant parameters of the studies and to answer any questions before
beginning data collection. In this way, a researcher can assess that subjects
believe that those with whom they will interact have the same information and
understand the information in the same way as they do. (See discussion in
Holt & Davis, 1995.)

D. INFORMATION ABOUT CHARACTERISTICS OF GROUP MEMBERS

As mentioned earlier, characteristics of individuals can be used to examine
group identity issues. Sometimes the group identity relates to contrived differences such as a particular personal style or ability that relates to overestimating
or underestimating dots. Sometimes, the group identity is an actual characteristic such as nationality (Bahry, Kosolopov, Polina, & Wilson, 2005; Yamagishi
et al., 2005).

There has also been some investigation of other factors that might serve
as indicators of how people might or might not cooperate. Some research, for
example, has considered how knowledge of the sex category of those in the
group affects cooperation (Sell, Griffith, & Wilson, 1993; Sell, 1997;
Simpson, 2003). Some research has considered how ethnic differences in the
United States may be related to willingness to cooperate with a partner
(Eckel & Wilson, 2004). Research also suggests that cues provided by group
members, separate from individual characteristics, can affect the degree to
which group members trust them or cooperate with them. Such cues
include, for example, smiling (Scharlemann, Eckel, Kacelnik, & Wilson,
2001).

Methodologically important to these studies is the separation of the information alone from other content. Thus, information about sex category as carried through the gender-specific names of participants (e.g., "Michelle" versus "Michael" as an indicator of sex category) is different from showing group members pictures of each other. Pictures carry other content as well: attractiveness, ethnicity, and general facial expression, for example. It is not that one manipulation is correct while the other is not, but simply that the researcher must be aware of exactly what information is being conveyed.

E. Communication

Related to the nature of what information is available and how it is delivered is communication. A very consistent finding in the social dilemma literature is that face-to-face communication among group members tends to increase cooperation. (See extended discussions of this consistency in Sally, 1995, and Ostrom, 1998.) The idea that communication alone helps solve dilemmas is contrary to much of noncooperative game theory because communication that is not binding (in the sense of contractual) is consequently considered "cheap talk." But, despite the inability to enforce commitments, group communication still has a strong effect.

What is it about communication that increases cooperation? Researchers argue that communication creates commitments and often groups will try to encourage public commitment to a particular cooperative strategy (Kerr & Kaufmann-Gilleland, 1994; Orbell, van de Kragt, & Dawes, 1988; Ostrom, Walker, & Gardner, 1992), and that group discussion leads to in-group identity (Brewer & Kramer, 1986). Further, there is evidence that, after discussion, participants are better able to predict whether others will cooperate (see Frank, Gilovich, & Regan, 1993). It is also the case that, in repeated interactions, verbal monitoring and punishment can be invoked (Ostrom et al., 1994).

In Elinor Ostrom's presidential address to the American Political Science Association (1998), she argued that communication was one of the most important methods for potential solutions to the social dilemma or collective action problem. In her analysis, she argues that face-to-face communication does many things to help provide solutions for social dilemmas. In particular, it creates an opportunity to develop "contingent agreements." Essentially, these agreements say that one person or group is willing to contribute or cooperate at a particular level *if* others will also cooperate. Communication then enables each participant to estimate his or her degree of trust in those contingent agreements. Contingent agreements may be very simple ("Let's all contribute equally") or complex ("Let's contribute according to how much our endowments are"). They may also deal with punishment for defection. After an initial decision, reputation and reciprocity become factors: reputation of group member is the result of how their communicated commitment relates to actual behavior, and reciprocity is the behavior that reflects matching prior behavior.

An important point about that face-to-face communication made clear in Ostrom's discussion is that it contains an incredible variety of factors. Because this is the case, it must also be the case that under certain conditions face-to-face communication may NOT lead to solution of social dilemmas. For example, if a group member refuses to cooperate initially, this refusal can create problems for the group and recovery may be difficult. All this is to say that "face-to-face" communication is an incredibly complicated set of factors. In fact, this was a point made in Chapter 13 by Driskell and King in their applied experiment about communication and the importance of gestures. "Simple communication" is hardly ever really simple.

But, indeed it is important. Some evidence indicates that just sending computerized signals as a form of communication does *not* increase cooperation; in fact, it really does seem to be cheap talk, not creating credible commitment (Wilson & Sell, 1997). So while the results of experimental research in this area seem consistent, this also seems an important arena for trying to tease out what factors of communication are most important for determining trust under specific conditions. This is precisely where the experimental method has its strength.

VI. NEW DIRECTIONS AND NEW STRATEGIES

There are some issues that serve to keep research on social dilemmas separated by discipline. In terms of experimental methodology, the use of deception dramatically divides the disciplines. Economists are militantly opposed to the use of deception, while there are varying degrees of acceptance in psychology, political science, and sociology. The economists' argument against deception relates to how subjects are changed by the deception and, consequently, what this means for the research results. This argument sometimes carries objections on ethical grounds, but for the most part is based upon problems that deception may pose for ensuring that participants believe the experimenter. For example, if part of the research study is related to different endowments and the subjects did not believe the information in the study, the results could not be interpreted.

Further, since uncertainty is such an important concern in many economic investigations, subjects' suspicions about the truth of the information could dramatically affect the processes under investigation. Charles Holt (2007), an experimental economist, develops this argument using public goods language:

> Even if deception is successfully hidden during the experiment, subjects may learn
> the truth afterward by sharing their experiences with friends so the careful adherence
> to non-deceptive practices provides a public good for those who run experiments in
> the same lab at a later time. Ironically, the perverse incentive effects of deception in
> social psychology experiments may be aggravated by an ex post confession or debrief-
> ing, which is sometimes required by human subjects committees (p. 12).

Ortmann and Hertwig (1997, 1998, 2002) and Hertwig and Ortmann (2001, 2002) specifically discuss such issues, and offer some creative suggestions for when and how to use deception in experimental designs. In particular, they argue against the routine use of deception or the use of deception for conven- ience. On the other hand, sociologists, psychologists, and political scientists often argue that deception, in the sense of providing false information to subjects, is necessary for certain types of research questions. In particular, questions that involve assessing the effects of false labels or false information, by definition, involve some deception.

While both views have merits, the discussion of deception can be quite volatile. In practical terms, it is important to understand the views from the different research groups. Reviewers who are economists will not view experi- ments with deception favorably. It is especially important to discuss deception and design issues if any collaboration across disciplines is being considered.

There are also different theoretical approaches that can separate the disciplines studying social dilemmas. Although there is much overlap in the research issues investigated, differences in theoretical language and assump- tions sometimes prevent productive accumulation. Game theory, so much a part of economics and political science, is not even taught in many graduate programs in sociology and psychology, for example.

On the other hand, in economics and political science, there has been a focused effort to delineate conditions predicted on the basis of rational choice theories, usually specified in terms of game theory. For example, as discussed, there was quite a bit of research oriented to developing a baseline to demonstrate the prediction of noncooperation or free riding. When demonstrations of these predictions failed, there was a flurry of activity, some of which generated alterna- tive game theoretic models (see, for example, Benoit & Krishna, 1985). Sociologists and psychologists, on the other hand, tended to emphasize the importance of norms, reciprocity, and group identity to examine the conditions under which participants were more or less likely to cooperate for a group. They were less affected than economists and political scientists by the lack of "rational action" evidence because their formulations tended to focus upon empirical reg- ularities rather than the purely mathematical foundations of rational choice.

More recently, there is a movement that seems to have the potential to bring together these two sets of disciplines. In particular, behavioral economics and behavioral political science are research movements that focus on a more full integration of mathematical theory (in particular, game theory) and insights from psychology and sociology. The goal here is to take the findings of research into account and to incorporate them into the traditional rational choice theories. Sociology and psychology were always "behavioral" in the sense that the theories of the disciplines had to withstand empirical test and incorporated understandings, and the advantage of teaming up with economics and political science comes from the strength of the mathematics in these fields. For example, expected utility theory provides an important foundation for many kinds of decisions. It enables predictions of future choices by observation of past choices between various alternatives to determine the relative value or utility of a choice. Further, the mathematics implied by expected utility allow researchers to represent uncertainty as the likelihood of the event. Consequently, the probability of a particular choice for an actor is based upon the likelihood or probability of the event multiplied by the value of the choice.

Another potential source of integration across these fields relates to examination of institutional rules. (For more discussion of this suggestion, see Lovaglia, Mannix, Samuelson, Sell, & Wilson, 2005, and Sell, Lovaglia, Mannix, Samuelson, & Wilson, 2004.) Institutional rules are general laws or principles that specify who may do what particular actions and when they are allowed to do them. For example, voting rules specify what constituency has the right to vote and what form that vote takes, such as majority rule. Institutional rules may be formal or informal, stable or unstable. (See Crawford & Ostrom, 1995, for a discussion of the grammar of institutional rules.) Using this perspective and language enables comparison of groups who have vastly different purposes and composition. For example, institutional boundary rules delineate who is a member of the group and how a person might acquire that membership. In the case of social dilemmas, the defining property of nonexcludability—that nobody may exclude anyone else from the group—for group members is an indication of the importance of boundary.

Very little experimental research has specifically considered the rules by which individuals or groups become members, but in-group literature suggests the importance of "common fate" in definition of who is and is not a member. Sometimes boundaries are permeable (members may leave and outsiders may become members), while under other conditions boundaries are not permeable. Conceptualizing the degree of permeability might be a way to examine the structure of strength of a social dilemma as well as its solution. For example, it may be that when boundaries are permeable, group identity is lower and consequently social dilemmas become more difficult to solve.

Other institutional rules that have particular importance in the study of social dilemmas are information rules. (Some of these rules are discussed in the previous section on information.) *Information rules* describe how information is shared and what each actor can know. The information linkages determine the extent to which each member knows what the others have done or what they are planning to do. Information conditions can vary from complete information (everyone sees what everyone else has done to the point) to incomplete information or information characterized by uncertainty. The point here is that the language of institutional rules provides a method to connect across disciplines. It also provides a way to examine relations among nested groups or to highlight how institutions might reach from group to group.

The study of social dilemmas already reaches across discipline boundaries. In part this is because there has been an acceptance of the experimental method. Future prospects for further interdisciplinary research seem even more promising if more common language develops to enable routine interaction and theoretical coordination.

REFERENCES

Andreoni, J. (1995). Warm-glow versus cold prickle: The effects of positive and negative framing on cooperation in experiments. *The Quarterly Journal of Economics, 110*, 1–21.

Bagnoli, M., & McKee, M. (1991). Voluntary contribution games: Efficient private provision of public goods. *Economic Inquiry, 29*, 351–366.

Bahry, D., Kosolopov, M., Kozyreva, P., & Wilson, R. K. (2005). Ethnicity and trust: Evidence from Russia. *American Political Science Review, 99*(4), 521–532.

Benoit, J. P., & Krishna, V. (1985). Finitely repeated fames. *Econometrica, 53*, 905–922.

Billig, M., & Tajfel, H. (1973). Social categorization a similarity in intergroup behavior. *European Journal of Social Psychology, 3*, 27–52.

Bohm, P. (1972). Estimating demand for public goods: An experiment. *European Economic Review, 3*, 111–130.

Brewer, M., & Kramer, R. M. (1986). Choice behavior in social dilemmas: Effects of social identity, group size and decision framing. *Journal of Personality and Social Psychology, 50*, 543–549.

Carter, J. R., Drainville, B. J., & Poulin, R. P. (1992). A test for rational altruism in a public goods experiment. Working paper. College of Holy Cross.

Crawford, S. E. S., & Ostrom, E. (1995). A grammar of institutions. *American Political Science Review, 89*, 582–600.

Dawes, R. M. (1980). Social dilemmas. *Annual review of psychology* (pp. 169–193). Annual Reviews, Inc.

Eckel, C., & Wilson, R. K. (2004). Is trust a risky decision? *Journal of Economic Behavior and Organization, 55*, 447–466.

Fisher, J., Isaac R. M., Schatzber, J., & Walker, J. (1995). Heterogeneous demand for public goods: Effects on the voluntary contributions mechanism. *Public Choice, 85*, 249–266.

Frank, R. H., Gilovich, T., and Regan, D. T. (1993). The evolution of one-shot cooperation: An experiment. *Ethology and Sociobiology, 14*, 247–256.

Goeree, J. K., Holt, C. A., & Laury, S. K. (2002) Private costs and public benefits: Unraveling the effects of altruism and noisy behavior. *Journal of Public Economics, 83*, 257–278

Hardin, G. (1968). The tragedy of the commons. *Science, 162*, 1243–1248.

Hertwig, R., & Ortmann, A. (2001). Experimental practices in economics: A methodological challenge for psychologists? *Behavioral and Brain Sciences, 24*, 383–451.

Hertwig, R., & Ortmann, A. (2002). Economists' and psychologists' experimental practices: How they differ, why they differ and how they could converge. In I. Brocas & J. D. Carillo (Eds.), *The psychology of economic decisions* (pp. 253–272). New York: Oxford University Press

Holt, C. A. (2007). *Markets, games, & strategic behavior.* Boston: Pearson Education, Inc.

Holt, C. A., & Davis, D. (1993). *Experimental economics.* Princeton, NJ: Princeton University Press.

Isaac, R. M., & Walker, J. (1988). Group size effects in public goods provision: The voluntary contributions mechanism. *Quarterly Journal of Economics, 103,* 179–199.

Isaac, R. M., Walker, J., & Thomas, S. (1984). Divergent evidence on free riding: An experimental examination of possible explanations. *Public Choice, 43,* 113–149.

Kerr, N., & Kaufman-Gilleland, C. (1994). Communication, commitment, and cooperation in social dilemmas. *Journal of Personality and Social Psychology, 66*(3), 513–529.

Kollock P., & Smith, M. (1996). Managing the virtual commons: Cooperation and conflict in computer communities. In S. Herring (Ed.), *Computer-mediated communication: Linguistic, social, and cross-cultural perspectives* (pp. 109–128). John Benjamins: Amsterdam.

Ledyard, J. (1995) Public goods: A survey of experimental research. In J. H. Kagel & A. E. Roth (Eds.), *Handbook of experimental economics* (pp. 111–194). Princeton, NJ: Princeton University Press.

Lovaglia, M., Mannix, E. A., Samuelson, C. D., Sell, J., & Wilson, R. K. (2005). Conflict, power and status in groups. In M. S. Poole & A. B. Hollingshead (Eds.), *Theories of small groups: Interdisciplinary perspectives* (pp. 139–184). Thousand Oaks, CA: Sage.

Marwell. G., & Ames, R. E. (1979). Experiments on the provision of public goods, I. Resources, interest, group size and the free-rider problems. *American Journal of Sociology, 84,* 1335–1360.

Marwell, G., & Ames, R. E. (1980). Experiments on the provision of public goods II: Provision points, stakes, experience and the free rider problem. *American Journal of Sociology, 85,* 926–937.

Marwell, G., & Ames, R. E. (1981). Economists free ride, does anyone else? Experiments on the provision of public goods IV. *Journal of Public Economics, 15,* 295–310.

McCabe, K. A., & Smith, V. L. (2001). Goodwill accounting and the process of exchange. In G. Gigerenzer & R. Selten (Eds.), *Bounded rationality: The adaptive toolbox* (pp. 319–340). Boston: MIT Press.

McCusker, C., & Carnevale, P. J. (1995). Framing in resource dilemmas: Loss averiosn and the moderating effects of sanctions. *Organizational Behavior and Human Decision Processes, 61,* 190–201.

Messick, D. M., Allison, S. T., & Samuelson, C. D. (1988). Framing and communication effects on group members' responses to environmental and social uncertainty. In S. Maital (Ed.), *Applied behavioral economics* (Vol. 2, pp. 677–700). New York: New York University Press.

Olson, M., Jr. (1965) *The logic of collective action: Public goods and the theory of groups.* Cambridge, MA: Harvard University Press.

Orbell, J. M., van de Kragt, A. J. C., & Dawes, R. M. (1988). Explaining discussion-induced cooperation. *Journal of Personality and Social Psychology, 54,* 811–819.

Ortmann, A., & Hertwig, R. (1997). Is deception acceptable? *American Psychologist, 52,* 746–747.

Ortmann, A., & Hertwig, R. (1998). The question remains: Is deception acceptable? *American Psychologist, 53,* 806–807.

Ortmann, A., & Hertwig, R. (2002). The empirical costs of deception: Evidence from psychology. *Experimental Economics, 5,* 111–131.

Ostrom, E. (1998). Rational choice theory of collective action. *American Political Science, 92,* 1–22.

Ostrom, E., Walker, J., & Gardner, R. (1992). Covenants with and without a sword: Self-governance is possible. *The American Political Science Review, 86,* 404–417.

Rapoport, A. (1997). Order of play in strategically equivalent games in extensive form. *International Journal of Game Theory, 26,* 113–136.

Rutte, C. G., Wilke, H. A. M., & Messick. D. M. (1987) The effects of framing social dilemmas as give-some or take-some games. *British Journal of Social Psychology, 26,* 103–108.

Sally, D. (1995). Convention and cooperation in social dilemmas: A meta-analysis of experiments from 1958 to 1992. *Rationality and Society, 7*, 58–92.

Samuelson, P. A. (1954). The pure theory of public expenditure. *Review of Economics and Statistics, 36*, 387–389.

Samuelson, P.A. (1955). Diagrammatic exposition of a theory of public expenditure. *Review of Economics and Statistics, 37*, 350–356.

Samuelson, P. A. (1958). Aspects of public expenditure theories. Review. *Economics and Statistics, 40*, 332–338.

Scharleman, J. P., Eckel, C. C., Kacelnik, A., & Wilson, R. K. (2001). The value of a smile: Game theory with a human face. *Journal of Economic Psychology, 22*, 617–640.

Schwartz-Shea, P., & Simmons, R. T. (1995). Social dilemmas and perceptions: Experiments on framing and inconsequentiality. In D. A. Schroeder (Ed.), *Social dilemmas: Perspectives on individuals and groups* (pp. 87–103). Westport, CT: Praeger.

Sell, J. (1988). Types of public goods and free-riding. In E. J. Lawler & B. Markovsky (Eds.), *Advances in group processes* (5: 119–140). Greenwich, CT: JAI Press.

Sell, J. (1997). Gender, strategies, and contributions to public goods. *Social Psychology Quarterly, 60*, 252–265.

Sell, J., Lovaglia, J. J., Mannix, E. A., Samuelson, C. A., & Wilson, R. K. (2004). Investigating conflict, power and status within and among groups. *Small Group Research, 35*, 44–72.

Sell, J., & Wilson, R. K. (1999). The maintenance of cooperation: Expectations of future interaction and the trigger of group punishment. *Social Forces, 77*, 1551–1570.

Simpson, B. (2003). Sex, fear, and greed: A social dilemma analysis of gender and cooperation. *Social Forces, 82*, 35–52.

Tajfel, H., Billig, M. G., Bundy, R. P., & Famen, T. C. (1971). Social categorization and integroup behavior. *European Journal of Social Psychology, 11*, 439–443.

Wilson, R. K., & Sell, J. (1997). Cheap talk and reputation in repeated public goods settings. *Journal of Conflict Resolution, 41*, 695–717.

Yamagishi, T., Jin, N., & Miller, A. S. (1998). In-group bias and culture of collectivism. *Asian Journal of Social Psychology, 1*, 315–328.

Yamagishi, T., & Kiyonari, T. (2000). The group as the container of generalized reciprocity. *Social Psychology Quarterly, 63*, 116–132.

Yamagishi, T., Makimura, Y., Foddy, M., Matsuda, M., Kiyonari, T., & Platow, M. (2005). Comparisons of Australians and Japanese on group-based cooperation. *Asian Journal of Social Psychology, 8*, 173–190.

Experiments in the Twenty-First Century

Part 4 of the book includes three chapters in which authors review development of experimental research in their disciplines. They provide a way to understand the historical growth of experiments in their fields, and, by extension, in other social sciences.

Chapter 19, by Rose McDermott, traces experimental studies in political science going back close to a century in the United States to 1923. She identifies the difficult problems of attributing cause to factors in the highly complex situations of interest to political scientists, and shows how early experiments were guided by a desire to reduce the number of interactions so that researchers could study a few of them at one time. Political scientists, like others, have shown considerable ingenuity developing situations and technology to study processes of interest. The chapter concludes with an assessment of contemporary studies in political science, identifying strands of new investigations for the twenty-first century.

Chapter 20, by Catherine Eckel, provides a historical examination of experimental economics. Experiments have been relatively recent in economics and were initially met with skepticism because often the results did not match the predictions. As Eckel describes, however, anomalies discovered in experiments played important roles in theoretical modification, and in some cases, entirely new lines of theory development. She examines four canonical games or experimental areas: the double auction, the public goods game, the ultimatum game, and the trust game. She concludes by integrating this history with assessments of future developments.

Chapter 21, by Morris Zelditch, Jr., identifies different eras in sociological experiments, defined by the primary goals of practitioners. Early experiments were "effect experiments"—studies whose main aim was to show that some process, such as conformity or norm creation, could be created in a laboratory. The growth of sociological theory led to theory-testing experiments, and theoretical research programs spawned sequential, theory-driven experimental designs of the kind described in the chapters of this book. Zelditch describes research questions and goals that led to the contemporary kinds of experimental studies. Like McDermott and Eckel, Zelditch concludes his chapter (and this book) by identifying strands of new and promising research directions in sociology and in social science generally.

CHAPTER 19

Experimental Political Science

Rose McDermott

University of California–Santa Barbara

ABSTRACT

This chapter describes some of the historical developments in experimental political science. Some innovative new techniques provide contemporary illustrations of how to address old questions with new methods. An assessment of the field and a discussion of future challenges is provided.

I. INTRODUCTION

In February 2006, *The New York Times* published an article entitled "Study Finds Low Fat Diet Won't Stop Cancer or Heart Disease," reporting results of an 8-year study of almost 50,000 women (Kolata, 2006). The popular press had a field day with this report, and certainly many women felt relief that the extra chocolate they crave may not prove as diabolical for their future health as they had feared. On the other hand, health experts appeared exasperated, knowing that these reports did not fully characterize their findings, thus distorting the results. Why? Because, in fact, following the low-fat diet for close to a decade proved exceedingly difficult for even the most extremely motivated women in the study. As a result, hoped-for differences in health out-comes were obscured by the low variance in the independent variable between groups. Yet, in addition to possibly affecting the preferred diets of millions of women, these results will certainly have an impact on future health policy and public health funding priorities in significant ways. This confound nicely illus-trates the reason why experiments provide an extremely useful method of social inquiry: sufficiently large populations can display precisely the kind of variance in behavior and outcome necessary to disentangle the impact of var-ious causes on particular outcomes of interest.

Although the use of experimental methodology by political scientists has increased greatly in the last decade or so, some scholars still do not embrace its viability for the nondiscursive and normative questions that preoccupy the discipline (Smith, 2002). However, Elinor Ostrom (2002), among others, has called for a greater integration of experimentation, along with other methodolo-gies, to address more successfully enduring questions in political science such as collective action, voting and multicultural group behavior. Although most exper-imental applications in political science have focused on American politics and voting behavior, increasingly scholars in international relations and comparative politics have become interested in incorporating experiments as one potentially useful method to employ in investigating questions and problems of interest.

While topics related to voting continue to dominate the use of experiments in political science up until the present, experiments do not represent the most common method by which to examine such topics. In other words, while experiments are increasing in number, sophistication, and breadth of applica-tion, they still have not begun to enter common use in the discipline at large. Indeed, in a recent survey of over 1,000 international relations scholars conducted by the editors of *Security Studies* (Peterson & Tierney, 2005), less than 4% reported using experimental methodology at all in their work.

This section begins by briefly examining some of the historical developments in experimental political science. Next, some contemporary examples illustrate the ways in which innovative new techniques can throw light on old problems in new ways. Finally, an assessment of the field, along with a discussion of future challenges confronting the effective incorporation of experimental methodology in political science, will be provided.

II. HISTORICAL DEVELOPMENT

The use of experiments in political science dates back at least to the work of Harold Gosnell in 1926. Starting in 1923, he conducted a field experiment to examine the reasons why people did not vote in the mayor's election in Chicago. In his first study, he reported on the effect of direct mail on voter turnout.

Following up on an earlier study involving 6,000 personal interviews, Gosnell identified an additional 6,000 voters in Chicago residing in 12 districts. In each district, he randomly assigned half the voters to receive postcard reminders to register to vote, while the other half of citizens received no intervention.

The first postcard, printed in several languages, resulted in 42% of 3,000 potential voters registering to vote, while only 33% of the 2,700 citizens who did not receive mail did so. A second follow-up postcard resulted in 56% of the 1,700 subjects being stimulated to register, while only 47% of the 1,770 unstimulated counterparts registered. This card had a note and a cartoon portraying nonvoters as slackers; the cartoon proved slightly more effective with women. Overall, 75% of those who received a card registered while only 65% of those who did not receive mail registered. The impact of direct mail on actual vote varied across district, and it was related to the strength of the local party organizations. This experiment provided a model for the development of future experimental work in political science for decades to come, in at least three ways.

First, it focused on issues relating to American voting behavior, still the most common topic for experimental investigation by political scientists (for a review, see McDermott, 2002). Second, it examined an applied issue of concern to academic as well as real-world policymakers and, sometimes, even the interested public. This approach continues today through work on topics such as voting; campaigns and elections; committee and jury decision making; and problems relating to coordination, cooperation, bargaining, negotiation, and conflict resolution. Last, Gosnell's (1926) experiment examined a topic that had generated a great deal of interest among political scientists and been studied widely from the perspective of other methodologies, but without any causal consensus emerging. This tradition continues today, as political scientists often invoke experimental methodology to investigate topics that have received widespread attention while yielding either inconsistent or confusing results. The most influential modern trend in such voting research embeds experimental manipulations in large survey instruments to examine the impact of both cognitive and emotional processes on attitudes and actions concerning topics such as racial prejudice and affirmative action (Kuklinski *et al.*, 1997).

Starting relatively early in political science, many scholars believed that their experimental work was being unfairly rejected from established political science journals. In response to this concern, a new journal, *Experimental Study of Politics*, was founded in 1971. However, the journal lasted only 4 years, at least partly because a great deal of the experimental work published in this venue was not as sophisticated as work on similar topics coming out in psychology journals. Indeed, many of the best experimental political scientists of the time published their work on voting in psychology journals, or in edited volumes devoted to specific topics, such as political psychology or race. In general, while the number of experimental articles published in political science journals has increased over time, they continue to be concentrated in a group of specific journals, most notably those focused on American voting behavior, and *Political Psychology*. Predominant topics of interest continue to include voting, games such as prisoner's dilemma and game behavior, bargaining, committee rules and work, race, and media effects.

As the use of experimental methodology has broadened beyond voting in recent years, scholars in international relations and even comparative politics have begun to undertake such work as well. In international relations, such research has tended to focus on simulated arms races, bargaining and negotiations (Deutsch, Epstein, Canavan, & Gumpert, 1967), war games and crisis simulations (Beer, Healy, Sinclair, & Bourne, 1987; Beer, Sinclair, Healy, & Bourne, 1995), and foreign policy analysis and decision making (Geva, Mayhar, & Skorick, 2000; Mintz & Geva, 1993; Mintz, Geva, Redd, & Carnes, 1997).

One of the most productive areas of experimental research has involved the study of public goods and free-rider problems. John Orbell and colleagues (Dawes, Orbell, Simmons, & van De Kragt, 1986; van de Kragt, Orbell, & Dawes, 1983) have conducted a number of experiments in this area. In one set of small-group experiments, van de Kragt *et al.* examined the effect of communication on contributions to public goods. They found that communication resulted in an efficient production of the public good, while lack of communication resulted in failure to create the good over one-third of the time. In further work, Dawes *et al.* (1986) explored the effectiveness of various incentives on contributions to public goods. In three experiments, they discovered that enforcing contributions in a "fair share" agreement does increase contributions. Ostrom, Walker, and Gardner (1992) have investigated common-good problems experimentally as well, to assess the effect of various strategies on developing credible commitments among contributors. These techniques, which included "covenants" allowing communication, "swords" offering opportunities for sanctions, and combinations of both, did result in self-governance without the presence of external enforcement.

Interestingly, some scholars in comparative politics have begun to undertake field experiments as well. While some of these studies examine voting behavior in other countries (Wantchekon, 2003), others have adapted the methodology for investigating topics including cultural differences between ethnic groups in Africa (Posner, 2004) and nation building and public goods problems there as well (Miguel, 2004).

Rick Wilson and Donna Bahry achieved a remarkable accomplishment when they conducted a series of experimental laboratory economic games, including the ultimatum game, between and among several different ethnic groups in Russia at various locations (Bahry, Kosolopov, Kozyreva, & Wilson, 2005; Bahry & Wilson, 2006). These studies examined the nature of trust between ethnic groups and found, perhaps surprisingly, higher levels of interethnic trust than previous observers appeared to expect.

III. CONTEMPORARY EXAMPLES

Contemporary examples of the use of experiments in political science constitute two types. First, experimental examination of particular topics has accumulated and aggregated in certain areas, most notably voting behavior, but also in areas such as collective action. Second, experiments have recently provided some innovative ways to explore socially sensitive or previously inaccessible topics using new methods. Each of these will be discussed briefly in turn.

First, several topics in American voting behavior have been subjected to extended and repeated experimental tests and manipulations. Notable in the tradition of field experiments exploring abstention, originally spawned by Gosnell, lies the work of Donald Green and Alan Gerber on voter turnout. This work includes examinations of the effect of leafleting (Gerber & Green, 2000a), canvassing, direct mail and phone banks (Gerber & Green, 2000b; Green, 2004), and habit (Gerber, Green, & Shachar, 2003). Other scholars have expanded on the use of field experiments to examine other aspects of voting behavior, including race. In a 2002 study of Asian American voters in Los Angeles County, Wong (2005) found that both telephone calls and postcards increased voter turnout. In a similar but much larger study of Latino voters in 2002, Ramirez (2005) randomly assigned more than 465,000 Latino voters to receive direct mail, live phone calls, or robotic calls. In this study, contrary to the results Wong reported among Asians, Ramirez found that only live calls increased voter turnout among Latinos. More recent extensions have incorporated the use of cluster analysis into field experiments to study voting behavior (Arceneaux, 2005).

Turnout does not represent the only aspect of experimental investigation into voting. In fact, not all voting research takes place in the context of field experiments. Indeed, the majority of experimental work in voting takes place in laboratory settings. Such topics include campaigns and elections, most notably the impact of television news (Iyengar, Peters, & Kinder, 1982) and negative advertising (Ansolabehere, Iyengar, & Simon, 1999; Ansolabehere, Iyengar, Simon, & Valentino, 1994) on voter preference. Interesting work has attempted to use experimental manipulations to examine the way in which political advertising can cue and prime socially sensitive topics like race to influence attitudes during campaigns (Valentino, Hutchings, & White, 2002).

Indeed, other areas of experimental voting research engage explicitly psychological models of human decision making as well. Such work includes experimental investigations into the dynamics underlying framing effects

(Druckman, 2001a, b), as well as the affective forces that support political party identification (Burden & Klofstad, 2005). Lau and Redlawsk (2001) have used a computer-based decision platform to examine experimentally the influence of cognitive heuristics on political decision making; they find a widespread use of such strategies that interact with an individual's degree of sophistication. Specifically, experts prove more adept and effective in their use of mental shortcuts than novices. Additional experimental research examining psychological processes has manipulated the impact of threat on individuals with varying levels of authoritarianism to examine how individuals seek out and assimilate information under various conditions, which holds implications for voting as well as other political topics in areas such as foreign policy (Lavine, Lodge, & Freitas, 2005).

Experimental work that incorporates psychological processes has delved explicitly into realms outside voting behavior as well. Work that examines the impact of cognitive and affective processing on decision making has preoccupied experimental work in the area of international relations in particular. Geva and colleagues (2000) have used an experimental decision board similar to that employed by Lau and Redlawsk (2001) in their domestic context to investigate the impact of various cognitive processes on foreign policy decision making.

Some interesting work has examined the impact of process on individuals' attributions of outcomes. In one study, procedural justice appears to affect individuals' acceptance of their particular distribution of resources, leading the authors to posit underlying processes favoring cooperation (Hibbing & Alford, 2004). This study demonstrated that individuals do not like those who take a disproportionate share of payoffs, or set themselves above others. Such abuse of hierarchy and authority appears to predispose individuals against such leaders. Investigations of the effects of procedural justice, rules, and norms also inform experimental work on jury decision making as well (Guarnaschelli, McKelvey, & Palfrey, 2000).

A second way in which contemporary work in political science can be examined is through the lens of new technologies, which can be used to examine previously inaccessible or sensitive topics in new ways. Two research prospects deserve particular mention in this regard. First, Shanto Iyengar and colleagues have developed a very interesting and sophisticated design to manipulate similarity or familiarity or both. Using morphing technology, these authors integrate the face of the subject with that of noteworthy politicians or celebrities to examine the impact of familiarity and similarity on voter preference. In one study of similarity, Bailenson, Garland, Iyengar, and Yee (2006) found intriguing gender differences, such that men proved more likely to vote for a candidate who resembled them than women did. Although a Democratic candidate provided the only stimulus in this particular study, political party identification did not appear to influence voter preference in this experiment.

Another version of this study examining familiarity used large, if somewhat biased, convenience samples derived from readership of the *Washington Post*. In fact, Iyengar agreed to provide one such online survey instrument for the newspaper per month for a year. In one study, of over 2,000 participants in both political parties demonstrated a decided preference for an unknown face morphed with that of Senator Hilary Clinton over the same face morphed with Senator John McCain. In a second study, which adapted the popular "whack-a-mole" game into "whack-a-pol," Iyengar demonstrated partisan whacking. When subjects had to hit famous celebrities (Michael Jackson, Brad Pitt, Angelina Jolie) or famous dictators (Joseph Stalin, Adolph Hitler, Saddam Hussein), whacking appeared rapid and indiscriminate in nature. But when subjects had to attack politicians (John Kerry, George Bush), people took much more time and care in whom they hit, demonstrating clear partisan bias in comparison to the other conditions, where subjects were able to hit many more targets much faster because they did not care whom they whacked (Iyengar, 2002).

In another particularly telling demonstration, readers were asked how much federal assistance a victim of Hurricane Katrina should receive. In each case, the story remained the same, but the picture of the victim provided to readers varied by race. Those readers who saw a white victim advocated for more aid than those who received the black or Hispanic picture of the ostensible victim (for a review, see *Stanford Magazine*, 2006). Such innovative techniques offer unique and unprecedented ways of examining socially sensitive topics such as race in less obtrusive ways. For example, recent work by Jennifer Eberhardt, a psychologist, demonstrated that subjects were more likely to give harsher sentences, including the death penalty, to perpetrators whose skin was experimentally manipulated to be darker in tone in the pictures presented to subjects (Eberhardt, Davies, Purdie-Vaughns, & Johnson, 2006).

A second area of innovative research involves the use of new technologies in brain mapping to investigate the former "black box" of decision making. Two of these advances deserve particular mention. The first involves the use of functional magnetic resonance imaging (fMRI) technology to better understand the parts of the brain that are activated during particular judgments, feelings, or behaviors. Darren Schreiber and colleagues (Lieberman, Schreiber, & Ochsner, 2003) have used such techniques to explore the differences in political thinking and processing between experts and novices. The second uses electroencephalogram (EEG) technology to examine the timing of particular processes in the brain. Rick Wilson and colleagues (Wilson, Stevenson, & Potts, 2006) have used this strategy to explore the mental differences in subjects playing both dominant and mixed strategy games. Note that MRIs provide better information regarding the geography of brain activity, while EEGs offer more precise data on the timing of such events. Additional research in the biological bases of human decision making might explore specific genetic and hormonal markers to examine their impact on various behaviors of interest, including aggression and cooperation.

IV. ASSESSMENT AND CHALLENGES

There are several ways in which these new techniques might be effectively exploited to examine topics of interest and importance to political scientists across subfields. Methodological innovations alone cannot drive creativity, of course. Innovative technology should be harnessed in service of cohesive and comprehensive theoretical models of the problems and topics under investigation. In addition, institutional and structural challenges confront any attempt to make the use of experiments more widespread in political science.

V. POTENTIALS FOR FUTURE WORK

There are at least three areas in which innovative technologies hold promise for creative experimental explorations in political science. First, the use of brain mapping technologies such as fMRI, EEG, and positron emission topography (PET) offer previously unprecedented access into the workings of the human mind. Although these technologies remain relatively new, increasingly widespread use in the cognitive neurosciences has led to an explosion of new discoveries in the operation and psychological dynamics underlying human thoughts, feelings, and behaviors.

For applications in the political realm, several possibilities in addition to the obvious ones relating to judgments of candidates appear particularly appealing. Since such technology can now be used in a synchronous nature, studies of strategic interaction seem an obvious first step for exploration. Such research might investigate structural and psychological factors, which can affect bargaining, negotiation, reputation, status, power, and other topics related to achieving cooperative outcomes.

Three specific areas of application in this arena seem particularly promising. First, studies that examine the difference in process and outcome when monetary versus nonmonetary payoffs are under consideration could prove particularly illuminating for understanding political motivation in noneconomic realms. Essentially, all experimental research in behavioral economics uses monetary incentives to motivate subject participation and action. However, as psychologists, sociologists, and political scientists recognize, many people remain additionally, or even solely, motivated by nonmonetary considerations, especially social forces. For example, Colin Camerer (personal communication) finds that subjects playing experimental games related to

saving money for retirement learn at a different rate, and adopt different strategies, if they are playing for juice after having been made thirsty than when
playing for money.

Comparing strategies when monetary versus social incentives are at play
offers a particularly interesting way to investigate other forces that may decisively
motivate human action in the political and social realms. Sell, Griffith, and
Wilson (1993) found gender differences by resources as well. In two experiments, these authors examined the effect of the subject's gender, the gender
composition of his or her group, and the resources involved within the context
of social dilemma to explain gender differences in contributions. In this study,
gender did not affect individual contributions when money was the relevant
resource. However, when other resources were involved, gender differences in
fact emerged.

Second, examining the differences in decision making when people play
against other people as opposed to playing against a computer may help provide
a model for individual action and apathy in modeling such processes as citizen
engagement, civic action, and disobedience, and processes surrounding and
underlying rebellion and revolution. Do individuals respond to bureaucracies
in ways that resemble their response to computers? If individuals can person-
alize their experience of government, does it change the strategies and actions
they take with regard to processes of civic engagement? Such questions might
be illuminated through a comparison of such processes.

Third, the impact of rhetoric and communication styles and strategies
receives different treatment in various areas of academic and public political
discourse. In American politics, such attention to "spin" appears ubiquitous, if
often undertheorized. In international relations, the impact of such factors is
often ignored, or taken for granted but assumed not to influence outcomes in
any material fashion. But MRI technology might provide an avenue by which
to examine the effect of communication on actions such as aggression and
cooperation. It may be that certain types of individuals are more affected by
such discourse than others. If such impact were predictable and discernible—
based, for example, on genetic differences—it could influence strategies of
public education in important arenas such as health policy, as well as have an
impact on campaign strategies.

A second area of investigation that offers promise for advances in political
science lies in the domain of genetic and biological influences on behavior.
This arena encompasses at least three separate kinds of studies. The classic
way to investigate such differences involves twin studies. Indeed, Alford,
Funk, and Hibbing (2004) used twin studies to investigate the potential
genetic basis of political attitudes in one work. Such studies offer the definitive
way to determine the extent to which behavioral outcomes are relatively influenced by genetic and environmental factors. Twin studies might be employed

to compare temperamental characteristics of interest, such as hostility, trustworthiness, aggression, and competitiveness. They might also be used to examine differences in the preferred strategic strategies given types of individuals employ when confronting particular challenges, threats, or puzzles. For example, twins might play certain types of economic games where various options for action are manipulated in systematic ways to allow for either aggressive, prosocial, or avoidance responses to examine the interaction between genetic type and particular environmental challenges.

Another way to investigate the genetic determinants of behavior involves DNA analysis, using cheek swabs or hair samples to divide populations into those who differ on some genetic variation of interest. For example, some recent discussion suggests that those with genetic mutations in arginine vasopressin (AVP) will exhibit different patterns of behavioral aggression. Similarly, other important genetic markers can now be extracted from saliva, including dopamine and monoamine oxidase (MAO) inhibitors. Such differences may help illuminate important behavioral tendencies in behaviors such as aggression and impulsivity. Clearly, many behaviors remain far too complex to be driven by simple mutations on single genes. But such challenges should not prevent the basic research needed to understand the impact that such basic genetic variations might have on behaviors or attitudes of interest, and how such predispositions might interact with specific environmental cues and triggers either to mute or exacerbate the expression of particular behavioral tendencies.

A final way in which to examine the impact of genetic and biological influences on political behavior of interest is through the use of hormonal markers, such as testosterone and cortisol. Such hormones may not decisively influence behavior on their own, but can potentiate certain behaviors under specific conditions. For example, high levels of testosterone appear to increase the likelihood of responding to threats or challenges in a hostile manner.

A third way in which experimental work might help illuminate processes of interest concerns the way in which emotion affects decisions and behaviors. Mood manipulation offers one way in which specific emotions, such as fear or anger, can be induced to examine their impact on specific behaviors, such as cooperation, willingness to compromise, or proclivity to respond aggressively to threat. Such studies seem important to learn the ways in which emotion serves as a signal of commitment or intention between interested parties. Both verbal and nonverbal cues may help understand and moderate such important factors as audience effects and compromise in situations involving bargaining and negotiation. Further, emotion appears critical in forming the basis and maintenance of processes of social and political identity. Bonds of affiliation promote the formation of social and political groups, just as feelings of enmity can dissolve such associations; understanding such processes clearly would prove a key piece of understanding the nature of political action and involvement.

Emotion may also prove essential in establishing attempts at reconciliation following ethnic strife, civil war, or genocide in particular. Experimental manipulations of specific emotions can go far in helping to establish their influence on behaviors and outcomes of interest.

VI. CHALLENGES

Despite the promise offered by experiments to help investigate processes of interest to political scientists, several challenges remain.

The main intellectual challenge involves the difficulty of designing simple and compelling demonstrations of realistic and important political processes either at the individual or small-group level. Coherent and cohesive theoretical hypotheses must motivate and inform empirical investigation. In particular, studies should be designed to test competing theoretical propositions against each other, not simply to engage in "dust bowl" empiricism to see what emerges from the data after the study is run. Deriving testable hypotheses from dominant models, especially models that may predict divergent outcomes, needs to take place prior to the design of particular experimental protocol materials. Such specification allows investigators to know what variables to measure and how to do so.

In addition, clear expectations about which independent variables might produce particular outcomes provides experimenters with direction concerning the design of dependent variables. For effective demonstrations, such variables need to encompass sufficient variation for effects to occur, as well as sufficient magnitude for differences to be witnessed. Nothing is more frustrating than designing a careful experiment that results in ceiling or floor effects because too much attention was paid to operationalizing the independent variable, and not enough to constructing a viable dependent measure. Carefully operationalizing variables to suit the issue at hand can prove daunting; making tasks that involve without distracting requires special skill. Findings need to be replicated, extended, and expanded to determine the limits of applicability. Aggregation provides the key to discovering and understanding larger macrolevel political processes, particularly in the realm of cross-cultural or broad-based population studies. Such pursuits are not trivial in nature. Often technology can improve and change faster than scholarly ability to assimilate its meaning and importance. Yet, the topic must always drive the inquiry.

Two important intertwined institutional and structural concerns emerge as central challenges as well. First, to accomplish much of this experimental work successfully in an optimal and efficient manner requires truly interdisciplinary research. An MRI study, for example, will require collaboration with someone who has access to a machine; similarly, hormonal or genetic studies need laboratories in which to analyze samples and assays. At a deeper level, understanding the impact of these microlevel variables on macrofactors of political interest often requires collaboration with biologists, geneticists, evolutionary psychologists, or biological anthropologists. Such collaboration not only can enrich and enlighten studies in both theoretical and empirical ways, but also challenges individual researchers to go outside their home departments to locate those who share their interest to find colleagues willing to work with them on particular projects of mutual concern.

Second and related, experiments such as those discussed previously demand effective collaboration. Both interdisciplinarity and collaboration often run contrary to a discipline that historically prizes and rewards the lone scholar doing mainstream work. This proves particularly problematic in the arena of new technologies because young scholars are most likely to prove interested and adept at learning how to use these techniques, and yet they are the ones most likely to be punished at tenure time when the majority of their publications are coauthored. Yet, it remains almost wholly unrealistic for a political scientist to undertake an MRI or hormonal assay study entirely on his or her own. But the professional costs associated with collaboration mean that a great deal of important work may be neglected in favor of paths of less resistance, if less interest as well.

VII. CONCLUSIONS

Experiments do offer unparalleled ability to ascertain causal inferences from a complex, confusing, and often chaotic world. The ability to isolate factors of interest to determine their influence on outcomes of concern provides one of the most compelling reasons to undertake experimental study. However, in many scholars' views, the messy environment of real-world politics—rife with unintended consequences, blowback effects, and self-conscious actors—renders the applicability of such sterile experimental environments limited.

However, the past use of experimental methodology has led to a number of important insights on topics of widespread interest to political scientists, including the impact of certain tactics like direct mail on voter turnout, the influence of interethnic identity on trust, and the effect of negative advertising on candidate evaluation. No doubt further experimental work remains warranted as new technologies offer unprecedented opportunities to access previously inaccessible aspects of human decision-making processes.

REFERENCES

Alford, J., Funk, C., & Hibbing, J. (2005). Are political orientations genetically transmitted? *American Political Science Review*, 99(2), 153–167.

Ansolabehere, S., Iyengar, S., & Simon, A. (1999). Replicating experiments using aggregate and survey data: The case of negative advertising and turnout. *American Political Science Review,* 93(4), 901–909.

Ansolabehere, S., Iyengar, S., Simon, A., & Valentino, N. (1994). Does attack advertising demobilize the electorate? *American Political Science Review, 88*(4), 829–838.

Arceneaux, K. (2005). Using cluster randomized field experiments to study voting behavior. *Annals of the American Academy of Political and Social Sciences, 601,* 169–179.

Bahry, D., Kosolopov, M., Kozyreva, P., & Wilson, R. (2005). Ethnicity and trust: Evidence from Russia. *American Political Science Review, 99*(4), 521–532.

Bahry, D., & Wilson, R. (2006). Confusion or fairness in the field? Rejection in the ultimatum game under the strategy method. *Journal of Economic Behavior and Organization, 60*(1), 37–54.

Bailenson, J., Garland, P., Iyengar, S., & Yee, N. (2006). Transformed facial similarity as a political cue: A preliminary investigation. *Political Psychology, 27*(3), 373–385.

Beer, F., Healy, A., Sinclair, G., & Bourne, L. (1987). War cues and foreign policy acts. *American Political Science Review, 81*(3), 701–716.

Beer, F., Sinclair, G., Healy, A., & Bourne, L. (1995). Peace agreement, intractable conflict, escalation trajectory: A psychological laboratory experiment. *International Studies Quarterly, 39*(3), 297–312.

Burden, B., & Klofstad, C. (2005). Affect and cognition in party identification. *Political Psychology,* 26(6), 869–886.

Dawes, R., Orbell, J., Simmons, R., & van De Kragt, A. (1986). Organizing groups for collective action. *American Political Science Review, 80*(4), 1171–1185.

Deutsch, M., Epstein, Y., Canavan, D., & Gumpert, P. (1967). Strategies of inducing cooperation: An experimental study. *Journal of Conflict Resolution, 11*(3), 345–360.

Druckman, J. (2001a). The implications of framing effects for citizen competence. *Political Behavior, 23*(3), 225–256.

Druckman, J. (2001b). On the limits of framing effects: Who can frame? *Journal of Politics, 63*(4), 1041–1066.

Eberhardt, J., Davies, P., Purdie-Vaughns, V., & Johnson, S. L. (2006). Looking deathworthy: Perceived stereotypicality of black defendants predicts capital-sentencing outcomes. *Psychological Science, 17*(5), 383–386.

Gerber, A., & Green, D. (2000a). The effect of a nonpartisan get-out-the-vote drive: An experimental study of leafletting. *Journal of Politics, 62*(3), 846–857.

Gerber, A., & Green, D. (2000b). The effects of canvassing, telephone calls and direct mail on voter turnout: A field experiment. *American Political Science Review, 94*(3), 653–663.

Gerber, A., Green, D., & Shachar. (2003). Voting may be habit-forming: Evidence from a randomized field experiment. *American Journal of Political Science, 47*(3), 540–550.

Geva, N., Mayhar, J., & Skorick, J. M. (2000). The cognitive calculus of foreign policy decision making: An experimental assessment. *Journal of Conflict Resolution, 44*(4), 447–471.

Gosnell, H. F. (1926). An experiment in the stimulation of voting. *American Political Science Review, 20*(4, November), 869–874.

Green, D. (2004). Mobilizing African-American voters using direct mail and commercial phone banks: A field experiment. *Political Research Quarterly, 57*(2), 245–255.

Guarnaschelli, S., McKelvey, R., & Palfrey, T. (2000). An experimental study of jury decision making. *American Political Science Review, 94*(2), 407–423.

Hibbing, J., & Alford, J. (2004). Accepting authoritative decisions: Humans as wary cooperators. *American Journal of Political Science, 48*(1), 62–76.

Iyengar, S. (2002). *Experimental designs for political communication research: From shopping malls to the Internet.* Presented to the Workshop in Mass Media Economics, Department of Political Science, London School of Economics. Accessed 8/9/2006 at: http://pcl.stanford.edu/common/docs/research/iyengar/2002/expdes2002.pdf.

Iyengar, S., Peters, M., & Kinder, D. (1982). Experimental demonstrations of the "not-so-minimal" consequences of television news advertising. *American Political Science Review, 76*(4), 848–858.

Kolata, G. (February 7, 2006). Study finds low fat diet won't stop cancer or heart disease. *The New York Times*, A1.

Kuklinski, J., Sniderman, P., Knight, T. P., Tetlock, P., Mellers, G., & Mellers, B. (1997). Racial prejudice and attitudes toward affirmative action. *American Journal of Political Science, 41*, 402–419.

Lau, R., & Redlawsk, D. (2001). Advantages and disadvantages of cognitive heuristics in political decision making. *American Journal of Political Science, 45*(4), 951–971.

Lavine, H., Lodge, M., & Freitas, K. (2005). Threat, authoritarianism, and selective exposure to information. *Political Psychology, 26*(2), 219–244.

Lieberman, M., Schreiber, D., & Ochsner, K. (2003). Is political cognition like riding a bicycle? How cognitive neuroscience can inform thinking on political thinking. *Political Psychology, 24*(4), 681–704

McDermott, R. (2002). Experimental methods in political science. *Annual Review of Political Science, 5*, 31–61.

Miguel, E. (2004). Tribe or nation? Nation building and public goods in Kenya and Tanzania. *World Politics, 56*(3), 327–362.

Mintz, A., & Geva, N. (1993). Why don't democracies fight each other? An experimental study. *Journal of Conflict Resolution, 37*(3), 484–503.

Mintz, A., Geva, N., Redd, S., & Carnes, A. (1997). The effect of dynamic and static choice sets on political decision making: An analysis using the decision board platform. *American Political Science Review, 91*, 553–566.

Ostrom, E. (2002). Some thoughts about shaking things up: Future directions in political science. *PS: Political Science and Politics, 35*(2), 191–192.

Ostrom, E., Walker, J., & Gardner, R. (1992). Covenants with and without a sword: Self-governance is possible. *American Political Science Review, 86*(2), 404–417.

Peterson, S., & Tierney, M., with Maliniak, D. (2005). Teaching and research practices, views in the discipline, and policy attitudes of international relations faculty at U.S. colleges and universities, at http://mjtier.people.wm.edu/intlpolitics/teaching/papers.php.

Posner, D. (2004). The political salience of cultural difference: Why Chewas and Tumbukas are allies in Zambia and Adversaries in Malawi. *American Political Science Review, 98*(4), 529–545.

Ramirez, R. (2005). Giving voice to Latino voters: A field experiment on the effectiveness of a national nonpartisan mobilization effort. *Annals of the American Academy of Political and Social Science, 601*, 66–84.

Sell, J., Griffith, E., & Wilson, R. (1993). Are women more cooperative than men in social dilemmas? *Social Psychology Quarterly, 56*(3), 211–222.

Smith, R. (2002). Should we make political science more of a science or more about politics? *PS: Political Science and Politics, 35*(2), 199–201.

Stanford Magazine. July/August. (2006). Morph, whack, choose.

Valentino, N., Hutchings, V., & White, I. (2002). Cues that matter: How political ads prime racial attitudes during campaigns. *American Political Science Review, 96*(1), 75–90.

Van de Kragt, A., Orbell, J., & Dawes, R. (1983). The minimal contributing set as a solution to public goods problems. *American Political Science Review, 77*(1), 112–122.

Wantchekon, L. (2003). Clientalism and voting behavior: Evidence from a field experiment in Benin. *World Politics, 55*(3), 399–422.

Wilson, R., Stevenson, R., & Potts, G. (2006). Brain activity in dominant and mixed strategy games. *Political Psychology, 27*(3), 459–478.

Wong, J. (2005). Mobilizing Asian American voters: A field experiment. *Annals of the American Academy of Political and Social Science, 601*, 102–114.

CHAPTER 20

Economic Games for Social Scientists

CATHERINE ECKEL

University of Texas at Dallas

ABSTRACT

Experimental economics has developed since the 1960s, establishing a methodology, a professional organization, and finally a journal. Economics experiments differ methodologically from other social science experiments in two ways: subjects are paid their earnings in the experiments, and no deception of subjects is allowed. They differ in content as well, focusing on tests of economic theory and careful examinations of behavior that deviate from the "rational actor model" dominating the field of economics. Four canonical games are examined (double auction market, public goods game, ultimatum game, and trust game), and contributions of experiments in economics are summarized.

I. INTRODUCTION

Experimental economics is a relatively new field, with the first experimental games published in the 1960s. Only in the last few years has experimental economics become well enough established that every university economics department wants at least one experimental economist on its faculty. In this section, I outline some of the contributions of experimental economics as it has matured, and sketch the development of the professional infrastructure in the field.

The earliest experiments in economics journals focused almost exclusively on markets (e.g., Smith, 1962), with particular emphasis on finding the conditions under which a market would converge to the competitive price and quantity. Meanwhile, on parallel tracks, psychologists and game theorists began to investigate simple games like the prisoner's dilemma, and decision theorists tackled individual decision making under risk and uncertainty, including (but not limited to) expected utility theory. (See Roth, 1993, for a discussion of experiments before 1960.)

More than 20 years later, in 1986, as the field really began to take off and gain widespread readership, the first U.S. professional organization of experimental economists, the Economic Science Association (ESA), was formalized by a group of pioneer experimentalists who had been meeting at the Westward Look Resort in Tucson, Arizona, for several years. (The association continues to hold annual meetings at the same resort, more than two decades later.) It was a small group in 1986; everyone could fit into a single photograph taken on the hotel's stairs. Vernon Smith was the first president, followed by Charlie Plott.

When I first attended the ESA meetings in 1992, the association was somewhat larger, with around 80 papers presented over the 2 days. The program was dominated by two types of experiments: markets and public goods. In the market sessions, auctions and asset markets figured prominently, while in the public goods sessions the word was out: people apparently were much more cooperative than a straightforward extrapolation from simple game theory would predict, and everyone was asking "why" in one way or another. There were also quite a few sessions on collective choice and committee decision making. Today, this work seems nearly to have died out, suggesting it may be time for a resurgence. The bargaining games that have come to dominate recent meetings had only a small place—two sessions, as I remember— including my own paper on ultimatum bargaining.[1]

Since the early 1990s a long series of increasingly clever experiments has established regular patterns of violations of certain implications of game the-

---

[1]Personal history: my first public act as an academic economist was to discuss an experimental paper presented by Tom Palfrey at the Southern Economic Association meetings the fall of my third year in graduate school. I commented favorably on the paper (it was published as Forsythe, Palfrey, and Plott, 1982, and has become a classic asset market experiment). My first experimental research paper, published 10 years later, was written with Charlie Holt, whom I had met when he was visiting the University of Virginia (Eckel & Holt, 1989). The one I presented at ESA in 1992 was my first foray into what became the "social preferences" or "fairness" agenda. (It was eventually published as Eckel & Grossman, 2001). Since then I have wandered pretty far from standard economics; at times I have imagined both Charlies shaking their heads in dismay.

ory and expected utility theory. This has led to a remarkable growth of both experimental and theoretical research concerned with how people make real decisions when rationality and information processing are less than perfect, and when social considerations such as fairness and cooperation play an important role. I do not mean to suggest that worrying about fairness and bounded rationality is all that experimental economists have done (though for me it is the most engaging element of the experimental agenda).

Experimentalists contributed to the design of the FCC microwave spectrum auction for cellphone frequencies, which raised more revenue than has been spent on economic research by all government agencies combined. Experimentalists have developed "designer markets" for research slots on the space shuttle (Ledyard, Porter, & Wessen, 2000), pollution permits (e.g., Cason & Plott, 1996), and wholesale electricity (e.g., Smith, Rassenti, & Wilson, 2002), just to name a few. See Plott (1994), who is responsible for the term "designer markets," for more examples. I believe these are the two main contributions of experimental economics to date: the challenge to the rational actor model and the resulting development of theory that incorporates social preferences and bounded rationality, and the study and design of economic institutions.[2]

In the early years, National Science Foundation (NSF) support was very important for experimental research and contributed greatly to its growth. (The requirement that subjects be paid—see later discussion—makes experimental economics more dependent on funding than are most other fields of economics.) The first experimental grant was to Tom Schelling at Harvard in 1960, and Vernon Smith received the second, titled "Behavior in Competitive Markets," in 1962. It was some years before Martin Shubik received the third in 1969. Charlie Plott was first funded in 1972, for a proposal titled "Political Economic Decision Processes," and John Kagel, Ray Battalio, and Robert Basmann at Texas A&M were funded in 1972 for a proposal that included the first Kagel/Battalio experiments with animals (which later won Wisconsin Senator William Proxmire's Golden Fleece Award [see Battalio et al., 1985]).[3] Shortly thereafter, in 1974, Dan Newlon became program director for the economics program at NSF. Newlon appreciated experimental research, and under his guidance the review panel has included at least one experimentalist since then. I do not think Newlon realizes how important his support has been for the development of the field.[4]

---

[2]Experimental economists are, of course, not alone in their emphasis on the importance of institutions. Eleanor Ostrom is probably best known for her work on institutions in the field that helps address public goods and commons problems (e.g., Ostrom, 1998; Ostrom, Walker, & Gardner, 1994) and Douglas North for his work on historical institutions (e.g., North, 1990).

[3]Personal communication with Dan Newlon, director of the economics program, NSF, October 29, 2006.

[4]Personal communication with Charles Plott, October 31, 2006.

It was more than 10 years after the founding of ESA that the association finally established its own journal, *Experimental Economics*, and that journey was a rocky one. There was much opposition from members of the ESA to starting a journal, primarily because of a concern that a dedicated journal would provide editors of top economics journals with an excuse to reject experimental papers. Charlie Holt, ESA president from 1991 to 1993 and one of the founding editors of the journal, remembers bringing a motion before the ESA board during his presidency and having it rejected.

Finally, in 1997, Holt and his coeditor Arthur Schram decided to go ahead with the journal with or without ESA support, but still hoping the association would go along. Under the leadership of Tom Palfrey, the president at the time, ESA adopted the journal, and the association became fully international, with a structure that included a president, a North American vice president, and a European vice president.[5] I served two terms as vice president of ESA (2000–2004), largely because no one remembered to have elections the year my term was up, so it was agreed we would just continue. John Hey as European vice president was in the same boat. In the middle of our double term, economist Vernon Smith, together with psychologist Daniel Kahneman, won the Nobel Prize in economics. We were proud and happy.[6]

II. METHODOLOGY

As is true of any academic field, experimentalists are fussy about their methodology. We have two unbreakable rules: subjects must be paid, and there can be no deception of any kind. The rules are enforced by reviewers and editors who refuse to publish exceptions. This rigid approach may seem a bit extreme to experimentalists in other social sciences, but the rules are there for a reason.

Economists are a tough and skeptical audience. As theorists Paul Milgrom and John Roberts (1987) note: "No mere fact ever was a match in economics for a consistent theory." Presenting an experimental paper in a mainstream economics department used to be hazardous duty. Both market experiments and games came under fire. The criticism of market experiments was always, "How can you learn anything about how markets work from what undergraduates do in a lab for 10 bucks?" For games like public goods or bargaining games where results nearly always deviated from game theory predictions, the

---

[5]Personal communication with Charlie Holt, October 28, 2006.

[6]Not everyone knows that Vernon Smith is a dedicated (and wonderful) country-Western dancer. Charlie Holt recalls being in a cowboy bar one night, with Vernon, as usual, dancing with all the ladies. One woman Charlie danced with said to him, "Isn't he that guy who's going to win the Nobel Peace Prize?"

question was whether this would "hold up with high stakes," followed by, "How can you learn anything about bargaining from what undergraduates do in a lab for 10 bucks?"

To add credibility to my experimental results, I often began my talks by conducting the game in my experiment with the audience. In one such pretalk game, an ultimatum game (see later discussion), a well known theorist stood up and shouted to everyone in the room that he would reject anything below 50%, perfectly illustrating by example the result I was about to present: that small offers are rejected in this game, in contradiction to predictions of game theory. All this is by way of making a case that our methods had to be strong enough to convince the rest of the profession that they should pay attention to experimental research. Naturally, economists required substantial monetary stakes in experimental games; they wanted to be sure that enough money was at stake that subjects would be truly motivated to pursue their best interests. This practice continues, even though it has been demonstrated that stakes are not always important. (Camerer & Hogarth, 1999, provide a survey of when and how much stakes affect behavior in experiments.)

A prohibition against deception arose as part of the same project to convince the profession that experiments constituted serious and important research. To those outside the field, the term "experimental research" evokes extreme examples of abusive deception, such as those employed in the Tuskegee medical experiments or the study conducted by Stanley Milgram (1963) in psychology.[7] The popular perception of experiments in the other social sciences is that deception is commonplace, indeed more the rule than the exception. Whether or not that is true, most economists strongly reject introducing any deception in experiments.

From an economist's perspective, the most serious deception involves payment, such as when an experimenter says he will pay subjects their earnings in a negotiation, then in the end pays everyone a fixed participation fee instead. Particular care is taken over the design of the economic incentives in an experiment, as this is the way the details of the theory are implemented in the lab (as explained later). Experimentalists feared that subjects would not believe they were playing the game that was set up in the lab, and that they (the experimenters) would lose control over subjects' motivation. Control over motivation through payoffs is probably the most critical element of an

---

[7]In the infamous Tuskegee experiments of the 1940s and 1950s, some prisoners who had syphilis were left untreated to study the natural course of the disease, long after penicillin, which could have cured all of them, became available. In Milgram's experiments, subjects were led to believe they had inflicted serious harm on another subject, and some of them reported long-lasting harmful psychological consequences.

economics experiment. Thus, the prohibition against deception in economic experiments was born, and it persists today.

To test an economic theory in the lab requires a kind of translation of the theory into a set of instructions and protocols—and payments. For example, suppose you want to test the theory of supply and demand, the bedrock of Economics 101. When we teach supply and demand, we talk about where demand comes from (buyers' valuations of a good, taking income and all other goods' prices as given) and where supply comes from (sellers' costs), and we ask students to imagine those factors coming together. Equilibrium in the market is something we explain to students: we ask them to reason out what will happen when demanders with different values and sellers with different costs come together, and conclude that the lower cost sellers and higher value buyers will end up with the good, all buying or selling at the equilibrium price—the one where the quantity sellers want to sell agrees with the quantity the buyers want to buy (i.e., where the curves cross—perhaps with the aid of a Walrasian auctioneer). But to test supply and demand in a lab requires that buyers have real values and sellers have real costs and that there be some real mechanism for them to come together and trade. Enter two of the experimentalist's building blocks: *induced values* and *institutions*.

Induced value theory (Smith, 1982) was developed to drive home the point that as long as subjects care about money and do not care too much about other things, we can use the structure of payoffs to induce values—in this case, buyer values and seller costs. The institution is the set of rules that determine who can do what in the experiment. This includes rules about the message space (what information can flow—bids and offers, in this case) and who can send what messages (buyers bid, sellers ask), as well as rules about prices and trades (in this case, if a buyer and seller agree to trade at a price, the trade occurs). Induced values and carefully specified institutions go a long way toward accomplishing the translation of an economic theory into the lab.

In the case of demand and supply, an experimenter has to decide how many buyers and sellers there are, how many "units" each can buy or sell, and their values or costs. Buyers' values are induced by paying subjects the difference in cash between their induced value and the price at which they purchase a unit; sellers earn a more intuitive profit of selling price less (induced) unit cost. How to implement the institution—the rules by which information is exchanged and trade takes place—is not obvious, however. In the supply and demand world of Economics 101 there are no institutions; demanders and suppliers come together as if by magic. But there is no magic in the lab (that would be deception). Instead, an experimenter must choose and implement an institution. Sometimes the institution is part of the theory, but very often it is not, as in the case of supply and demand. This brings me to the first canonical game in economics experiments: the double auction.

A. Canonical Games I: The Double Auction

In 1962 Vernon Smith reported the results of a classroom experiment testing supply and demand. According to published histories (see Eckel, 2004), while a graduate student at Harvard, Smith participated in a classroom experiment conducted by Edward Chamberlin. Chamberlin (1948) conducted bilateral trading experiments with his graduate students at Harvard to "prove" the failure of the competitive model. The "institution" he used was to give buyers values and sellers costs and let them wander around the room and negotiate trades. The transactions were then summarized. Price dispersion disproved the competitive model. He concluded that "economists may have been led unconsciously to share their unique knowledge of the equilibrium point with their theoretical creatures, the buyers and sellers, who, of course, in real life have no knowledge of it whatever" (p. 102).

Vernon Smith, then in his first faculty appointment at Purdue, decided to modify the institution a little, to make it more like stock, bond, and commodities markets, by having buyers and sellers call out prices, and providing a pit boss to coordinate things. A second innovation was to repeat the market, in the sense of the movie *Groundhog Day;* every trading "day" started fresh, with buyers and sellers holding the same set of values and costs as the previous day. In his experiment, the supply and demand model did a great job of predicting the outcome in the market. When Vernon tells about this experiment, he maintains he was astonished at the rapid convergence to competitive equilibrium that he observed, and says he had to replicate it a few times, with differently shaped demand and supply arrangements, to convince himself it was correct.

Experimentalists now know that the institution he implemented, the oral double auction, is about the most powerful institution there could be for inducing convergence to competitive equilibrium. It works like a charm, with groups as small as three buyers and sellers; even with low or hypothetical stakes, it reliably converges to equilibrium. Every experimentalist I know conducts this experiment in classrooms in almost every class he or she teaches. It is an unforgettable experience for students. They come away believing that markets (can) work, and that the theory of supply and demand has teeth (or jaws, as Plott would say).[8]

The second most popular market institution is probably the posted offer market. Demand and supply are induced as before. But in this institution,

---

[8]Plott has a wonderful presentation where he illustrates the dynamics of price movement using an animated, visual representation of the book (outstanding bids and offers) in an auction market. I do not believe this has been published, but his distinguished guest lecture for the Southern Economic Association contains related material (Plott, 2000).

prices are posted by one side of the market (typically sellers), and then buyers are chosen randomly to "go shopping." Since most of the action is on the seller side of the market, buyers are often simulated in posted offer markets. A comparison of these institutions has produced a remarkable result. In the double auction, market structure has little effect on price. Even a monopoly seems unable fully to exercise market power in this setting. However, in a posted offer market, a small number of sellers can much more easily enforce a price above competitive equilibrium. This very well established result is something that is not taught in Economics 101. Economists teach that market power is important, but, as the experiment shows, we should also teach that *institutions matter*. The rules of trade have a lot to do with whether and how much market power can be exercised.

The double auction has been used in hundreds of experiments, and the results comparing double auction and posted offer experiments have led to a large body of research comparing theoretically equivalent institutions. A small example: In theory, the English increasing-price auction and the Vickrey second-price auction are equivalent: both should lead people to bid up to their values. However, in practice this is not what happens. The latter leads to higher prices, as agents systematically overbid their values. Again and again we learn that institutions matter. It is this close attention to institutions that gave experimentalists the tools to design markets, such as the microwave auction mentioned in the introduction. This may prove to be the greatest contribution of experimental economics, and it is still very much work in progress.[9]

B. CANONICAL GAMES II: THE PUBLIC GOODS GAME

The first public goods experiment was conducted by Peter Bohm in 1972. His objective was to elicit willingness to pay for a television show, and he cautiously concluded that the public goods problem—the incentive to conceal one's valuation for public goods—was overstated. The experiment had flaws, but it was the first to find that subjects would contribute to a public good. The canonical implementation of the public goods situation in economics came a bit later; it is the linear voluntary contribution mechanism, first studied by Isaac, Walker, and Thomas (1984). The game is essentially a multiperson continuous prisoner's dilemma, a game that is well known to every social scientist. Ledyard (1995) provides the definitive survey up to that point, and Croson (2007) contains an overview of main issues in this research.

---

[9]For surveys of experimental research on market institutions see Davis and Holt (1993, Chapters 3 and 4) and Kagel and Roth (1995, Chapters 5–7).

Several features of Isaac *et al.* (1984) have become standard practice in public goods research. In this game, subjects are brought into the lab and formed into groups, typically of four participants, though experiments with group sizes of two to ten participants are common, and larger groups have also been studied. Each subject is given an endowment, usually of "tokens," that can be exchanged for dollars at some known exchange rate. The endowment can be invested in one of two accounts: the individual account or the group account. The individual account pays off one token per token invested; the group account pays each person in the experiment a fraction of a token less than 1 but greater than $1/n$ where n is the group size ($n = .3$ in this study). This fraction is called the marginal per capita return (MPCR). The game is usually repeated, and 10 rounds is a typical length, though longer and shorter versions have been conducted.

This game mimics the incentive structure of a public good: consumption is nonexcludable and nonrival. Each token invested in the group account produces one unit of public good worth MPCR to each person. Since most of the benefit of the public good accrues to others, there will be underprovision of the good relative to the social optimum, which is for all tokens to be contributed to the group account. The tension, of course, is the same as in a prisoner's dilemma game. Since MPCR > $1/n$, cooperation (investing all one's tokens in the group account) maximizes efficiency and leads to higher total payoffs for the group. But since MPCR < 1, each person has an incentive to free ride off others' contributions and invest all tokens in his or her individual account.

The Nash equilibrium of the game that results from such "rational" (selfish) play is for all players to contribute zero to the group account, and this is what economists expected people to do. But the results of the experiments differ from this prediction and are very stable: contributions average 40 to 60% of the endowment in early rounds and deteriorate close to zero by the 10th round. There is also considerable heterogeneity across individuals and for a given individual over time. This is the first experimental game to show so much variation across individuals. In the market games, the only variation in play comes from the differences in induced values provided by the payoff structure of the experiment. Here, there is clear variation in what people do, with some contributing all their tokens to the group, some keeping all their tokens in the individual account, and some contributing part and keeping part.

The provocative results of early public goods game experiments were followed by publication of many variations on the game. Many studies were designed to figure out why subjects behaved in this way, contributing to the production of public goods in contradiction to the rational actor model. While some may term it irrational, it is hard to sustain that opinion when subjects succeed this way in extracting considerably higher earnings from the experimenter than they would from uniform free riding. Hypotheses include: mistakes, confusion, altruism, reciprocity, social norms, and conditional cooperation, among others. This game illustrated some obvious holes in the rational actor model, narrowly drawn, that experimentalists have continued to explore, using this game and others.

I do not intend to debate the value of rational choice analysis. As Schotter (2006) argues, the main value of building models based on the assumption of rational actors is that it allows us to separate logical "wheat" from intuitive "chaff" by carefully proving theorems about the implications of assumptions. He says, "[P]eople get confused by the rational choice methodology when they believe that the results of the theorems proven in this fashion are correct predictors of human behavior. Mature thinkers understand that this is not true" (p. 500). When I claim that this game's results revealed holes in the rational actor model, I do not mean to criticize the modeling exercise; like Schotter, I think we learn a lot from theory that is "strong and wrong."

However, many economists do seem to swallow their theory whole, if not as a predictor of human behavior, then as a normative description of how behavior *should* be conducted. I have so often seen an economics professor teach innocent students in principles of economics that it is "irrational" not to free ride in a prisoner's or social dilemma situation. (Such situations are described by Sell, Chapter 18, this volume.) Indeed, a fictional account of just such an event appears in Jane Smiley's novel of a fictional Midwestern university, *Moo*. Though merely baffling to Smiley's (female) character, to me this exercise is not only irresponsible, but a missed opportunity.

Instead, we should be teaching students that situations like this signal caution: awareness of the free-rider problem should be used to solve it, not worsen it by encouraging free riding. Some students free ride by nature; in experiments something like 25% are natural rational actors. But most pursue different strategies in these games. A few are principled cooperators and always contribute everything. More than half of subjects exhibit what looks like conditional cooperation: they cooperate when others do so, and respond to free riding with more of the same. Indeed, these conditional strategies earn the most money for their practitioners when played across lots of different groups of players. When prompted to specify a multiperiod strategy for playing the public goods game, Keser (2000) found that the most successful strategies were contingent, depending on the past actions of others in the group.

C. CANONICAL GAMES III: THE ULTIMATUM GAME

When three German economists conducted the first ultimatum bargaining game experiment (Guth, Schmittberger, & Schwarze, 1982), their purpose was to strip bargaining down to its essentials by creating the simplest possible bargaining situation. In this game there are two players, usually called the proposer and the responder. An amount of money is provisionally made available to the pair. The proposer's task is to determine a division of the money. The responder is then presented with this proposal, and his only choice is to accept or reject the proposal. If accepted, the money is divided as proposed; if rejected both players receive zero earnings, and the money reverts to the experimenter. The subgame perfect Nash equilibrium of the game is for a payoff-maximizing responder to accept any positive offer, and the proposer, knowing that, to offer the smallest positive increment available. For example, in a $10 ultimatum game with decisions restricted to whole dollar amounts, Nash equilibrium would involve payoffs of $9 to the proposer, and $1 to the responder.

This game might seem overly artificial, with little relevance to bargaining behavior in the field. On its own, that is a valid criticism. But the ultimatum game plays an important part in theory: this game makes up the final stage of many multistage games, and is relied upon by most of principal–agent theory (in which the principal pays the agent his reservation wage, and the agent accepts). So, behavior in the ultimatum game has implications for a broad set of important economic problems involving worker motivation, contracting, and the notion of a fair price or wage.

Guth et al. (1982) found that very few proposers proposed such an unequal division, and that such divisions were routinely rejected. The modal offer was 50% of the pie for inexperienced subjects, and 40% for experienced. In a second experiment, subjects played both sides of the game, stating an offer as proposer, and a minimum acceptable offer as responder. The results were essentially unchanged. When Guth and colleagues' results were published, the reaction of many economists was to assume that he had done the experiment wrong. It was treated as a mere anomaly, not a true finding. There followed dozens of studies attempting to do it "correctly"—that is, in a way that produced the results predicted by theories based on rational choice. (See reviews by Thaler, 1988, and Camerer & Thaler, 1995, and Camerer, 2003.) The real issue was not the behavior of proposers, because they might be maximizing their expected payoffs conditional on expectations about the patter of rejections. But how to explain the behavior of responders? An income-maximizing responder would accept any positive offer, and yet offers of 30% or less of the pie were regularly rejected.

Were the subjects confused? One study (Kahneman, Ketch, & Thaler, 1986) explained the game, gave subjects a quiz, and kicked out all the ones who did not give the right answer. Offers and minimum acceptable offers were about the same. Later studies attempted to "teach" second movers to accept low offers by repeating the game. Ochs and Roth (1989) had subjects repeat the game 10 times, randomly rematching subjects each round. The results of these experiments were very similar to those of the one-shot games. (See also the studies cited in Chapter 4 in Kagel & Roth, 1995.)

Methodological detour: random rematching is one of two main methodological innovations resulting from the ultimatum game agenda. It is a technique experimental economists use to give subjects an opportunity to learn about a game without turning the game into a repeated game. Repeated games are problematic theoretically because they generate so many possible equilibria. Each round, each subject is matched with a different counterpart to replay the game. In the most careful designs, subjects never meet a counterpart twice or a counterpart who has played with one of their counterparts. In practice, this is tricky and probably in vain, as some evidence suggests that subjects respond about the same to information whether it comes from a one-time or repeated counterpart (e.g., Ashley, Ball, & Eckel, 2005).

Was this anomaly the result of low stakes? You will recall that as a very common seminar challenge. Hoffman, McCabe, and Smith (1996), weary of the challenge, spent $5,000 to find out. They played the game with $100 stakes. The proportional distribution of offers was not significantly different from the $10 version of the game and, more surprisingly, three of four offers of $10 and two out of three offers of $30 were rejected! Lisa Cameron (1999) conducted very high stakes experiments in Indonesia, and found essentially the same result. Slonim and Roth (1998) combine learning and stakes. Subjects play low- and high-stakes games, with payoffs between 60 and 1,500 Slovak crowns[10] repeated for 10 rounds. In their study, the rejection rate falls slightly with higher stakes and with repetition, and offers tend to converge towards about 40% of the pie.

Were the subjects punishing unfair behavior? Kahneman *et al.* (1982) report another study that focused on punishment. In the first stage of the game, subjects could divide a $20 pie only one of two ways: by keeping $10 and giving $10 to their counterpart, or by keeping $18 and giving $2. The counterpart was passive, and did not have an opportunity to reject the offer. Seventy-six percent split it equally. (I believe this is the first dictator game, in which the respondent is a passive recipient of the proposer's offer, though the researchers did not refer to it as such.) In the second stage of the game, subjects were matched with two recipients, and could divide $10 (evenly) with someone who had previously kept $10, or $12 with someone who had previously kept $18. In this game, 74% chose the smaller pie, sacrificing $1 in payoffs to reward the fair counterpart. Clearly, fairness plays a role, and subjects are willing to sacrifice small amounts of money to punish unfair behavior.

---

[10]The exchange rate at the time was about 30 crowns per U.S. dollar, so the stakes were substantial.

My favorite illustration of the importance of fairness comes from Sally Blount (1995). Her study compares offers randomly generated by computer with those generated by the proposer. When the computer generates the offer, unfair offers are readily accepted by responders. However, the same offer generated by the proposer is rejected as usual.

Did proposers fear rejection, or were they being deliberately fair? Forsythe, Horowitz, Savin, and Sefton (1994) explicitly compare the ultimatum game with a dictator game, where the proposer can make the same set of offers as in the ultimatum game, but the responder must passively accept. In the dictator game, there is no risk of rejection, so fairness is the only possible motivation for positive offers. They find that with hypothetical decisions there is little difference between the two games, but when stakes are positive ($5), the dictator game results in substantially more selfish behavior, with about 40% of subjects keeping the full endowment. Increasing the stakes to $10 has little additional effect.

Is anonymity a factor? In game theory, agents are essentially anonymous. They have no individual or social identities, in the way psychologists or sociologists think of identity, and their counterparts do not either. There is no expectation of having to deal with one's experimental partner after the experiment concludes. An implication of this is that agents probably do not care if they are observed, and do not care about the identity of the person they are playing with. But in the experiments we conduct, subjects are not anonymous. They know who else is in the room, and they know the experimenter is going to see what they do. One hypothesis, then, is that subjects are fair because they are observed.

Hoffman et al. (1994) introduce a second methodological innovation resulting from the ultimatum game agenda: the double-blind or double-anonymous protocol. In this protocol subjects know that their actions are not observed, and they are anonymous to each other. The experiment is designed in such a way that the experimenter ends up with a set of proposer decisions that cannot be attributed to any particular person, and the responders cannot tell whom their offers came from. Their experiment is a dictator game, and the result was a notable increase in selfish play, with about 70% of subjects keeping the full endowment.

A prominent property of ultimatum game play is, like the public goods game, a great deal of heterogeneity in what people do. This game and its descendent, the dictator game, both show a great deal of variation in individual play. In the early 1990s, researchers began to realize that this property made them useful as instruments for measuring preferences. Eckel and Grossman (2001) and Solnick (2001) both investigate differences in the behavior of women and men in the ultimatum game, and Eckel and Grossman (1998) do the same for the dictator game. The ultimatum game was adopted as a tool for measuring social norms in a study comparing behavior across 15 small societies (Henrich *et al.*, 2004). The use of experiments as measures of preferences has developed on its own path since. I return to this discussion in the conclusion of the chapter.

D. CANONICAL GAMES IV: THE TRUST GAME

The final game I would like to highlight is the "investment game," first presented by Berg, Dickhaut, and McCabe (1995), later referred to primarily as the trust game. In this game, two players are each endowed with $10. The first mover can send any part of her endowment between $0 and $10 to the second mover. On the way, any amount sent is tripled, so that if the first mover sends $1 it becomes $3, and $10 becomes $30. The second mover can then decide to return any amount between $0 and the full amount he received. The returned amount is not tripled. This game is interesting because of its sequential nature. Like the public goods game, there are gains to cooperation, but to achieve those gains the first mover must first trust by putting his payoffs in the hands of the second mover, with no promise of return. The amount sent by the first mover is "trust," and the amount returned is "reciprocity."

Similar to the previous games, the Nash equilibrium of the game is for the second mover to keep any money sent to him and for the first mover, knowing that, to send zero. But this is not what happens in the game. Berg *et al.* (1995) implemented a double-anonymous protocol, similar to the one invented by Hoffman *et al.* (1994), which should have enhanced self-interested behavior. To their surprise, subjects sent on average about half of their $10 endowments, and trust nearly paid, with just under $5 returned on average.

This game was soon adapted as a way to measure trust at the individual level. Since the notion of generalized trust had recently been popularized by Francis Fukuyama (1995), and levels of trust measured around the world using several questions from the World Values Survey (WVS), it seemed natural to examine this game along with the survey questions typically used to measure it. Glaeser, Laibson, Scheinkman, and Soutter (2000) conducted a variation on this experiment with Harvard undergraduates, and found a very high level of trust and reciprocity, a great deal of individual variation, and essentially no correlation with the survey questions from the WVS. However, behavior in the game was correlated with self-reported trusting and trustworthy behavior in several domains.

This game has been replicated dozens of times with different populations around the world, and as with the ultimatum game, many of the studies focus on the question of why people behave as they do in the game, particularly when typical behavior is not rational according to economic theories. However, the tone of these studies is very different from the early public goods and ultimatum games. Economists are not asking, "Why are people behaving so stupidly?" Instead, attention is paid to how behavior in this game is related to altruism, as measured by the dictator game (e.g., Cox, 2004, or Burns, 2003), or risk aversion, as measured by behavior in risky-choice experiments (e.g., Eckel & Wilson, 2004). The focus is now on developing a better understanding of the behavior rather than "fixing" the experiment to eliminate it.

III. DISCUSSION AND CONCLUSION

Experimental economics developed as a way to test theory under controlled conditions—especially theories that could not be easily tested with available field data. While experimental research was met initially with skepticism, it has over time become accepted as a legitimate, indeed important, element of the economist's toolkit. With the awarding of the Nobel Prize in economics to Vernon Smith and Daniel Kahneman in 2003, the field officially came of age. (Their Nobel lectures, as published in the *American Economic Review,* can be found in Smith, 2003, and Kahneman, 2003.)

My particular interest in experimental research is twofold, and both interests are different from the primary impetus for experimental research, which is testing theory. The first is the use of experimental games as tools for measuring preferences; the second is the study of heterogeneity in behavior.

Measuring preferences has historically been done in the domain of survey research. Yet, experiments may be superior to surveys. Here, as in other fields, survey responses do not always predict actual behavior. Self-reported survey measures of preferences have shortcomings. From an economist's point of view, the trouble is that a person has no particular incentive to reveal his or her preferences truthfully, and may have an incentive to misrepresent them. Suppose, for example, you are faced with a survey question that asks you to rate your altruism on a scale of 1 to 5. If you are in an economics class, you may want to show your professor how rational you are by choosing 1 as an answer. If you think your answer will be observed by an experimenter whose primary interest is in showing how nice people are, you might be tempted to give her the answer you believe she wants, and mark a 5. There is no cost to misrepresenting your true preferences on a survey question. In addition to impressing others, you may choose a decision that is consistent with how you wish you behaved, consciously or unconsciously using the survey answer to affirm your self-image as a good or smart person.

Economists' view of preferences, as represented by a utility function with certain properties, suggests an alternative measurement strategy. Altruism can be measured as the trade-off between own and others' payoffs. Suppose instead that you play a dictator game where the recipient is an anonymous person, or better yet, a reputable charity. In order to show your altruism, you must forego income. Showing altruism is not free; you must actually behave altruistically.

This approach may yield a measure with a higher degree of external validity, though the jury is still out on that. Very few studies compare survey and experimental measures, but those that do tend to find low levels of correlation between survey and experimental measures, and greater predictive power for the experiments (e.g., *Glaeser et al.*, 2001).[11]

Using experiments to study individual differences is another focus of my own work. Heterogeneity arises from two sources. First, people behave differently from each other. For example, women might be more altruistic and tall people less cooperative on average than others. Second, a given person may behave differently depending on whom she interacts with. Experiments have an advantage in studying how people interact, which may be different from how they think they interact or how they say they interact. Consider racial discrimination, for example. In an experiment, we can put subjects into a situation where they are matched with another person and have to make a decision. This experiment can be set up without mentioning race or discrimination, and decisions made without knowing what the study is about.

In a study using the trust game, for example, Rick Wilson and I found evidence that the skin tone of a counterpart affects expectations, trust, and reciprocity (Eckel & Wilson, 2006a, b). In this protocol we matched people over the Internet to play a trust game. We revealed skin tone (and gender, attractiveness, etc.) by showing each player his or her counterpart's photograph. The game happens in real time (no deception, remember). This way we can reveal critical information without compromising anonymity or introducing the possibility of postgame interaction. Since discrimination is not an obvious element of the experiment, subjects may reveal latent or unconscious discrimination by their behavior.

In addition to the measurement issues discussed previously, a list of contributions of experimental economics, or what we think we know, would have to include at least the following:

- Noncooperative game theory is a powerful tool for predicting behavior in competitive environments. Competition seems to enforce self-interest, producing agents who look very much like economic man. In these settings, which include a wide range of relevant environments, the rational actor model works pretty well at predicting outcomes.

---

[11] I have a study under way with a group of colleagues that will link experimental measures of preferences to a large longitudinal household survey in Mexico. This study should be conclusive. (Eckel, Johnson, & Thomas, 2006, present an early set of results.)

- Institutions matter. That is, the exact rules about what and how information can be exchanged, and how trades take place, can have a very strong impact on economic outcomes. In particular, the double oral auction is a very efficient institution, probably because it is clearly competitive and a great deal of information is exchanged in the bidding process. The posted offer auction is less obviously competitive and less information comes out in the bidding process, making market power more easily exercised in this setting.
- People do not always behave "rationally." While some institutions appear to enforce rationality, behavior that is other-regarding—such as altruism, cooperation, spite, etc.—is observed in less "disciplining" institutions. Games where others' payoffs are salient, and especially where there are potential gains to cooperation, are especially likely to elicit other-regarding behavior. The public goods and trust games are examples.
- When game theory is predictive, it is sometimes for the wrong reason. Nash assumed that people knew each others' preferences and tastes (utilities), not merely their payoffs. Yet, we are more likely to see behavior leading to the Nash equilibrium when people do not know others' payoffs. More information improves decisions in some settings, but if it removes anonymity, it seems to change the game as subjects weigh others' payoffs in their decision-making calculus.

ACKNOWLEDGMENTS

Thanks to Judy Du and Angela Milano for help in preparing the manuscript; Charles Holt, Dan Newlon, and Charles Plott for their memories; and Jane Sell, Murray Webster, and Rick Wilson for guidance.

REFERENCES

Ashley, R., Ball, S.B., & Eckel, C.C. (2005). *Motives for giving: A reanalysis of two classic public goods experiments*. Unpublished manuscript, Virginia Tech.

Battalio, R., Kagel, J., *et al.* (1985). Animals' choices over uncertain outcomes. *American Economic Review, 75*, 597–613.

Berg, J. E., Dickhaut, J. W., and McCabe, K. (1995). Trust, reciprocity, and social history. *Games and Economic Behavior, 10*, 122–142.

Blount, S. (1995). When social outcomes aren't fair: The effect of causal attributions on preferences organizational behavior and human decision processes. *Organizational Behavior and Human Decision Processes, 63*, 131–144.

Bohm, P. (1972). Estimating demand for public goods. *European Economic Review, 3*, 111–130.

Burns, J. (2003). *Insider–outsider distinctions in South Africa: The impact of race on the behavior of high school students.* Paper presented at the Conference on Field Experiments in Economics, Middlebury College, Middlebury, VT, pp. 25–27.

Camerer, C. (2003). *Behavioral game theory: Experiments in strategic interaction.* New York: Russell Sage Foundation, Princeton University Press.

Camerer, C., & Hogarth, R. 1999. The effects of financial incentives in experiments: A review and capital–labor–production framework. *Journal of Risk and Uncertainty, 19,* 7–42

Camerer, C., & Thaler, R. H. (1995). Anomalies: Ultimatums, dictators, and manners. *Journal of Economic Perspectives, 9,* 209–219.

Cameron, L. A. (1999). Raising the stakes in the ultimatum game: Experimental evidence from Indonesia. *Economic Inquiry, 37,* 47–59.

Cason, T. N., & Plott, C. R. (1996). EPA's new emissions trading mechanism: A laboratory evaluation. *Journal of Environmental Economics and Management, 30,* 133–160.

Chamberlin, E. H. (1948). An experimental imperfect market. *Journal of Political Economy, 56,* 95–108.

Cox, J. C. (2004). How to identify trust and reciprocity. *Games and Economic Behavior, 46,* 260–281.

Croson, R. (2007). Public goods experiments. In S. Durlauf & L. Blume (Eds.), *New Palgrave dictionary of economics* (2nd ed.). New York: Palgrave Macmillan.

Davis, D. D., & Holt, C. A. (1993). *Experimental economics.* Princeton, NJ: Princeton University Press.

Eckel, C. C. (2004). Vernon Smith, Nobel laureate: Economics as a laboratory science. *Journal of Socio-Economics, 33,* 15–28.

Eckel, C. C., & Grossman, P. (1998). Are women less selfish than men?: Evidence from dictator games. *The Economic Journal, 108(448),* 726–735.

Eckel, C. C., & Grossman, P. (2001). Chivalry and solidarity in ultimatum games. *Economic Inquiry, 39,* 171–188.

Eckel, C. C., & Holt, C. A. (1989). Strategic voting in agenda-controlled committee experiments. *American Economic Review, 79,* 763–773.

Eckel, C. C., Johnson, C., & Thomas, D. (2006). *Altruism and resource sharing in Mexico.* Presented at the Economic Science Association Conference, September.

Eckel, C. C., & Wilson, R. K. (2004). Is trust a risky decision? *Journal of Economic Behavior and Organization, 55,* 447–465.

Eckel, C. C., & Wilson, R. K. (2006a). Internet cautions. *Experimental Economics, 9,* 53–66.

Eckel, C. C., and Wilson, R. K. (2006b). *Initiating trust: The conditional effect of skin color among strangers.* Unpublished manuscript.

Forsythe, R., Horowitz, J. L., Savin, N. E., & Sefton, M. (1994). Fairness in simple bargaining experiments. *Games and Economic Behavior, 6,* 347–369.

Forsythe, R., Palfrey, T. R., & Plott, C. R. (1982). Asset valuation in an experimental market. *Econometrica, 50,* 537–568.

Fukuyama, F. (1995). *Trust: The social virtues and the creation of prosperity.* Glencoe, IL: Free Press.

Glaeser, E. L., Laibson, D., Scheinkman, J. A., & Soutter, C. (2000). Measuring trust. *Quarterly Journal of Economics, 115,* 811–846.

Guth, W., Schmittberger, R., & Schwarze, B. (1982). An experimental analysis of ultimatum bargaining. *Journal of Economic Behavior and Organization, 3,* 367–388.

Henrich, J., Boyd, R., et al. (Eds.). (2004). *Foundations of human sociality: Economic experiments and ethnographic evidence from fifteen small-scale societies.* Oxford, England: Oxford University Press.

Hoffman, E., McCabe, K., Shachat, K., & Smith, V. L. (1994). Preference, property rights, and anonymity in bargaining games. *Games and Economic Behavior, 7(3),* 346–380.

Hoffman, E., McCabe, K., & Smith, V. L. (1996). On expectations and monetary stakes in the ultimatum game. *International Journal of Game Theory, 25,* 289–301.

Isaac, R. M., Walker, J. M., & Thomas, S. H. (1984). Divergent evidence on free riding: An experimental examination of possible explanations. *Public Choice, 43,* 113–149.

Kagel, J. H., & Roth, A. E. (1995). *Handbook of experimental economics.* Princeton, NJ: Princeton University Press.

Kahneman, D. Knetsch, J., (2003). Maps of bounded rationality: Psychology for behavioral economics. *American Economic Review, 93,* 1449–1475.

Kahneman, D., Knetsch, J., & Thaler, R. (1986). Fairness and the assumptions of economics. *Journal of Business, 59,* S285–S300.

Keser, C. 2000. *Strategically planned behavior in public goods experiments.* Unpublished manuscript, CIRANO, Montreal, Canada.

Ledyard, J. (1995). Public goods: A survey of experimental research. In J. Kagel & A. Roth (Eds.), *Handbook of experimental economics* (pp. 111–194). Princeton, NJ: Princeton University Press.

Ledyard, J., Porter, D., & Wessen, R. (2000). A market-based mechanism for allocating space Shuttle secondary payload priority. *Experimental Economics, 2,* 73–195.

Milgram, S. (1963). Behavioral study of obedience. *Journal of Abnormal and Social Psychology, 67,* 371–378.

Milgrom, P., & Roberts, J. (1987). Informational asymmetries, strategic behavior, and industrial organization. *American Economic Review, 77,* 184–193.

North, D. C. (1990). *Institutions, institutional change, and economic performance.* New York: Cambridge University Press.

Ochs, J., & Roth, A. E. (1989). An experimental study of sequential bargaining. *American Economic Review, 79,* 355–384.

Ostrom, E. (1998). A behavioral approach to the rational choice theory of collective action (1997 APSA presidential address). *American Political Science Review, 92,* 1–22.

Ostrom, E., Walker, J., & Gardner, R. (1994). *Rules, games, and common-pool resources.* Ann Arbor: University of Michigan Press.

Plott, C. (1994). Market architectures, institutional landscapes and testbed experiments. *Economic Theory, 4,* 3–10.

Plott, C. R. (2000). Markets as information gathering tools. *Southern Economic Journal, 67,* 1–15.

Ramis, H. (prod.) & Albert, T. (dir.) (2002). *Groundhog day.* Sony Pictures.

Rassenti, S. J., Smith, V. L., & Wilson, B. (2002). Using experiments to inform the privatization/deregulation movement in electricity. *The Cato Journal, 21*(3), 515–544.

Roth, A. E. (1993). On the early history of experimental economics. *Journal of the History of Economic Thought, 15,* 184–209.

Schotter, A. (2006). Strong and wrong: The use of rational choice theory in experimental economics. *Journal of Theoretical Politics, 18,* 498–511.

Sell, J. (2007). Social dilemma experiments in sociology, psychology, political science, and economics. In M. Webster & J. Sell (Eds.), *Laboratory experiments in the social sciences* (pp. 459–479). Burlington, MA: Elsevier.

Slonim, R., & Roth, A. E. (1998). Learning in high-stakes ultimatum games: An experiment in the Slovak Republic. *Econometrica, 66,* 569–596.

Smiley, J. (1995). *Moo.* New York: Knopf.

Smith, V. L. (1962). An experimental study of competitive markets. *Journal of Political Economy, 70,* 111–137.

Smith, V. L. (1982). Microeconomic systems as an experimental science. *American Economic Review, 72,* 923–955.

Smith, V. L. (2003). Constructivist and ecological rationality in economics. *American Economic Review, 93,* 465–508.

Solnick, S. J. (2001). Gender differences in the ultimatum game. *Economic Inquiry, 39,* 189–200.

Thaler, R. H. (1988). Anomalies: The ultimatum game. *Journal of Economic Perspectives, 2,* 195–206.

CHAPTER 21

Laboratory Experiments in Sociology

MORRIS ZELDITCH, JR.
Stanford University

ABSTRACT

Laboratory experiments in sociology have become increasingly oriented to testing, refining, and extending theories and testing their applications. The increasing number of theoretically-oriented experiments has led to an increasing number of experimentally-oriented theoretical research programs and, with them, to an increasing number of standardized experimental settings capable of comparing and contrasting conditions between experiments with the same confidence as within experiments. The increase in the number of theoretical research programs has led to more theory growth; to growth, of both theory and research, that is more cumulative; and to a greater impact of experiments on the application of theory. The chapter ends with a brief discussion of the challenge of generalizing from experiments.

I. INTRODUCTION

The 1950s saw rapid and prolific development of theory and research on small groups. The 1980s and 1990s saw even more rapid, more prolific development of theory and research on group processes. The difference in how the field was described is trivia, perhaps, but in some ways an instructive comment on a sea of change in the nature and function of experiments between the two periods. Although they were a high-water mark of research on "small groups," the 1950s were also a watershed in the history of experiments in sociology, marked by a considerable climb up the ladder of abstraction, and, with it, a considerable change in the nature of the programs that grew out of them—a change that, in turn, meant not only more programs but also more growth in them and more applications of them.

II. EFFECT EXPERIMENTS

The Asch experiment was one of the classic experiments of the 1950s. In this experiment, a naïve subject (S) made a sequence of choices between one unambiguously correct and two unambiguously incorrect stimuli in the face of unanimous and incorrect responses by seven peers, all of whom were confederates of the experimenter (E). The experiment consisted of a sequence of trials, each of which required S to match a standard with a line, among three comparison lines, of the same length as the standard. S was instructed to respond orally and the oral response of six of the seven confederates preceded S's response. The instructions led S to believe that he or she was participating in an experiment on visual perception, so there was motivation to respond correctly. On the other hand, the unanimity of the confederates exerted pressure on S to conform to an incorrect response. Individuals in a control group where there were no confederates made their choices with almost complete accuracy. A unanimously incorrect majority deflected a third of S's choices in its direction and 75% of Ss made at least one error in the presence of the majority (Asch, 1951).

Contemporary experiments in sociology mostly test, refine, or extend a theory or test its application. But the Asch experiment did not so much test a theory as demonstrate an effect. The effect it demonstrated was important and powerful. But it was also complex and underanalyzed: Was it due to group pressures for conformity, which would imply that Ss knew they were making incorrect responses but went along because they were sensitive to the attitudes of the others in the group? Or was it due to cumulative evidence that they could not trust their own senses and therefore had to trust the senses of others to make a correct choice, which would imply that Ss believed they were making correct responses (Deutsch & Gerard, 1955)? The experiment demonstrated an effect but did nothing to sort out its causes.

Many other classics, either published in or still read in the 1950s, were effect experiments. For example, in Lewin's group decision experiment (1947), a persuasive communication in either a lecture or a group discussion attempted to change customary habits such as eating practices. The persuasive communication concluded with a request for a decision, made publicly; a specific period to act on it; and information that a follow-up would be made, after which the experiment's effect was measured. Group decision was more effective than a lecture in changing habits. But was the effect due to group discussion, making a decision (hence a commitment), making the commitment public, or the fact that the decisions made in the course of group discussion were consensual (Bennett, 1955)?

It may be difficult, even contentious, to construct canons, but surely Sherif's autokinetic experiment (1935) and Bavelas's communication networks (1950) were also among the signature experiments of the period. Surely, if there was a canon, it included Bales's laboratory observations of evolving group structures (Bales, 1953; Bales & Slater, 1955; Bales, Strodtbeck, Mills, & Rosebourough, 1951). All of these, like the Asch experiment, demonstrated important and powerful but complex, underanalyzed effects.

III. EFFECT RESEARCH PROGRAMS

A single experiment is neither precise enough, on the one hand, nor rich enough, on the other to leave no questions unanswered. Answering its unanswered questions gives rise to a research program—a series, typically a sequence, of interrelated experiments. Because the effect it demonstrates is complex and underanalyzed, the research program to which an effect experiment gives rise is often concerned, first of all, with explicating it (i.e., with analyzing what exactly it was that had the effect).

What "complex" means is that more than one process was involved in one concrete effect. The idea of a process may be philosophically difficult, but, roughly speaking, it means regularity of an effect—given the same conditions, the same causes have the same effects. What "underanalyzed" means is that conceptualization of the effect has not distinguished two or more processes from each other. Hence, experiments demonstrating underanalyzed, complex effects confound them; the same causes, under the same conditions, therefore have different effects.

Whether an effect is in fact complex is an empirical question, open to experiment. Thus, the Asch experiment led to experiments like Deutsch and Gerard's (1955), which argued that Asch confounded the normative effect of group pressures for conformity with the informational effect of others as sense evidence. Distinguishing the two mattered because the former but not the latter depended on the existence of a group, while the latter but not the former would be found whether judgments were anonymous or not. These differences were confirmed by replicating the experiment in the same setting but redesigning it, distinguishing the effect of a group versus an aggregate; of public versus private, anonymous judgments; and of privately versus publicly committing oneself to them.

In the same way, Lewin's group-decision experiment was followed by experiments like Bennett's (1955). The latter replicated the experiment but redesigned it to distinguish the effect of group discussion from the effects of having to make a decision, having to make it publicly, and consensus. Results showed that, even without group discussion, commitment and consensus were capable of generating the effect Lewin had found.[1]

But effect programs are also concerned with the causes, conditions, and mechanisms of the effect. Because the effect is complex and underanalyzed, the causes, conditions, and mechanisms they investigate are frequently ad hoc and unconnected, the search for them open-ended, the outcome, incoherent. For example, an experiment by Emerson (1964; see his first experiment) sought an explanation of the Asch effect in three unrelated causes, two of them more or less ad hoc: in the motivation to participate in the group, derived from Festinger's theory of informal pressures towards uniformity (Festinger, 1950); in group expectations, derived from commonplace common sense; and in status insecurity, more or less on a hunch. The incoherence of the causes, conditions, and mechanisms sought is simply a by-product of the effect's complexity. In consequence, though the growth of the Asch program has been voluminous, it has, like the effect, been an incoherent mix of unconnected causes, conditions, and mechanisms (Allen, 1965, 1977).[2]

IV. THEORETICALLY ORIENTED EXPERIMENTS

Not all of the experiments of the 1950s were effect experiments. A few, even a few classics, like Back (1951) and Schachter (1951)—both of whom tested Festinger's (1950) theory of social pressures towards uniformity in informal groups—tested theories. The 1950s were, in fact, a time of change, a time in which experiments were increasingly becoming oriented to testing, refining, and extending theories or testing their applications.

Some of these experiments emerged out of the explication of an effect, as Berger and Conner (1969) did from Bales. Their experiment was a test of a theory, not a demonstration of an effect. But the theory it tested was itself an

---

[1]Among many other examples, see also Burke (1967) and Lewis (1972), following Bales; Mulder (1960) and Faucheux and McKenzie (1966), following Bavelas.

[2]For another example, see Glanzer and Glaser (1961) on the evolution of Bavelas's program.

analytic deconstruction of an effect: it took an effect demonstrated by Bales (Bales, 1953; Bales & Slater, 1955; Bales et al., 1951) as its starting point, but was both more abstract and simpler than the effect Bales had demonstrated. What Bales had found was that initially undifferentiated groups evolved inequalities in rates of participation: who spoke to whom; who liked whom; who was asked for orientation, opinions, and suggestions; who offered them; who agreed with them; and who made overtures to others, expressed antagonism, showed tension, or released it. For the most part, these inequalities were highly intercorrelated and, once they had emerged, tended to be stable.

Berger (1958) conceptualized them more abstractly as action opportunities, performance outputs, unit evaluations, and influence. Because they were highly intercorrelated, they were regarded as one hierarchy of power and prestige. No attempt was made to encompass Bales's social emotional categories, because they were not highly correlated with the power–prestige order, and no attempt was made to explain them.[3] Analytical simplification of Bales's effect, the purpose of which was to isolate a process, simply left out the elements of other processes.

About this more abstract, analytically simplified process, Berger (1958) reasoned that the emergence of a power–prestige order reflected an underlying structure of expectations for performance of a collective task. Differentiated expectations emerged out of differential evaluations of performance when there were disagreements, because disagreements had to be resolved in order to reach a group decision. Once they emerged, these expectations probabilistically determined the observed power and prestige order, the elements of which were highly intercorrelated because they were all functions of the same underlying expectation states. Because the expectations both determined and were determined by the observed power and prestige order, any change in it was itself a function of the order; hence, it was likely to maintain itself unless or until disturbed by some change in the conditions of the process.

Experiments that test theories are, like the theories they test, analytic simplifications of a more concrete phenomenon. Thus, Berger and Conner's (1969) work is essentially an analytic simplification not only of Bales's effect, but also of his methods of observing it. For example, Bales observed open interaction; Phase 2 of Berger and Conner's experiment controlled it. Their experiment had two phases, in the first of which pairs of university students

---

[3]The fact that they did not correlate with his task categories had led Bales to infer that role differentiation was a part of the process. Later, Lewis (1972) found that role differentiation was true only of homogeneous groups, and Burke (1967) found that it had to do with the legitimacy problems of emergent hierarchies.

were publicly given scores on a test of a fictitious ability. The test consisted of repeated trials presenting sets of three words, one in English and two in a fictitious language. Ss were told that one of the two non-English words had the same meaning as the word in English and that, by comparing the sounds of the non-English words, they would be able to decide which meant the same as the English word. The ability to do this was called "meaning insight ability." The scores, as fictitious as the test and rigged by E, were interpreted to subjects as either "exceptionally superior" or "exceptionally poor."

Because feedback of the scores in Phase 1 was public, S knew both self and other's scores. This created four experimental conditions: (1) an S whose meaning insight ability was exceptionally superior but whose partner's was exceptionally poor, or, more simply, an S whose performance-expectation state was high self–low other; (2) high self–high other; (3) low self–low other; and (4) low self–high other.

The task in Phase 2 also consisted of repeated trials, each of which presented sets of three words, except that only one was in the same fictitious language as in Phase 1 and two were in English. The task was to decide which of the two English words had the same meaning as the non-English word. But in Phase 2, selection of a correct answer required three stages. Every time Ss were presented with a set of alternatives, Ss first made a preliminary selection, exchanged their initial choices with their partner, then made a final choice. The Ss could not verbally communicate or even see each other; they indicated their choices to E and to each other using a system of lights and push-button switches.

Except for 3 of a total of 25 trials, Ss were led to believe that their initial choices disagreed. The purpose of exchanging information was defined as seeing how well they worked together as a team. The final choice was completely private. Ss were told, moreover, that evaluation would be in terms only of a team score, which was simply the sum of the number of correct final choices each made and would not record, hence not reveal, their relative contributions to the score. Because the choices were binary, the final choice in the 22 disagreement trials indicated either acceptance of or resistance to the influence of the other. The result was that the probability of acceptance of the influence of the other was greatest in the low self–high other condition, least in the high self–low other condition, and about equal in the high–high and low–low conditions.

This experiment exemplifies a growing number of contemporary experiments that increasingly test a theory, and, like the theory they test, analytically simplify a phenomenon by focusing on only one of its processes. But, like this particular example, any theory-oriented experiment tests only one, or only a few, of the implications of a larger theoretical structure that has other implications. A theoretically oriented experiment is no more likely than an effect experiment to leave no unanswered questions. But the questions it leaves unanswered are questions about a theory, not an effect. This makes a significant difference in the kinds of research programs that emerge out of the answers to them. The increasingly theoretical orientation of contemporary experiments has therefore led to an increasingly theoretical orientation of contemporary research programs.

V. THEORETICAL RESEARCH PROGRAMS

Answering a question left unanswered by a theoretically oriented experiment modifies a theory, not an effect. The outcome is in fact a theory. For example, a question left unanswered by Berger and Conner (1969) was how the process would behave if a group were initially differentiated (e.g., by race, gender, education, or occupation). Answering it led to the theory status characteristics (Berger, Cohen, & Zelditch, 1966). In the standard model of theory growth, the new theory subsumes, hence, displaces, the old theory—an account in which theory growth climbs a neat, simple, linear path. But that is not always what actually happens. There are a number of different ways in which one theory can be related to another (Berger & Zelditch, 1997) and whether new theory displaces old theory—and therefore how a program grows—depends on how they are related.

Status characteristics theory did not displace the power–prestige theory. The difference in the conditions under which the process occurred led to a difference in its effect, accounting for which led to an auxiliary theory—of status generalization—not found in the power–prestige theory. The two accounted for different effects under different conditions. On the other hand, the theory of status characteristics shared much in common with the theory of power and prestige, for example the same basic ideas about expectation states. Both were expectation states theories, part of the same program. Thus, what had been *a* theory became a *family* of theories. The power–prestige theory simply proliferated, differentiating into two distinct but related theories.

Some theories, on the other hand, *do* displace earlier theories. For example, the initial theory of status characteristics was a theory of the effect of a single characteristic (Berger et al., 1966). Extension to multiple characteristics (Berger, Fisek, Norman, & Zelditch, 1977) displaced it. That is, the multi-characteristic theory was capable of explaining anything explained by it and more; hence, it was superfluous. Thus, in the course of its growth, the program gradually came to be made up not only of two theories, but also of two continually evolving branches, each elaborating a different but related theory.

Other branches emerged as other unanswered questions were asked, such as: What was the effect on the formation of expectation states of legitimate sources of evaluations, like teachers (source theory)? Or of expectations held by other interactants (second-order expectation states)? What was the effect on both the formation of expectations and the power–prestige order of relations between performance and reward expectations (the theory of distributive justice)? Or the effects of nonverbal interaction (e.g., status cues) on the formation of expectation states? Or of legitimation on the power–prestige order? New branches also emerged out of integrating theories in already existing branches, such as the status characteristics and source theories. Finally, they also emerged out of applications of the theory—for example, of status characteristics theory and source theory to schools, and status characteristics theory to gender.[4]

This is obviously a complex structure, and I have not even touched on the theoretical and methodological strategies that guided the construction and application of its theories. Not only is it complex, but other examples also often look very different because different theory–theory relations give rise to different patterns of growth, hence, to programs with quite different structures. What they all seem to have in common is that they are all made up of a set of theoretical and methodological strategies, a network of interrelated theories embodying them, and empirical models interpreting its theories, together with a body of theoretical and applied research testing, refining, and extending the programs' theories and their applications (Berger & Zelditch, 1997).

This is true no matter what the program's methods of observation and inference, but if its method is experimental it also typically standardizes its basic experimental setting, which facilitates the cumulative impact of any one experiment on the growth of the program as a whole.[5]

Such programs are also networks of interrelated theorists and researchers, whole communities addressed to solving the unsolved problems of the program's network of theories and applications. The network of theories, applications, and research is interrelated by a core of strategies, concepts, propositions, and methods common to all of them. The network of theorists and researchers is interrelated not only by ties, but also by shared goals, standards, and a common background of theory and research.

Although I have illustrated them by describing only one example, expectation states theory is merely one of an increasing number of theoretical

---

[4]For a more complete account, see Wagner and Berger (2002).

[5]Effect programs, in which replications are frequent, also standardize experimental settings. The advantage of standardization, for either type of program, is that it makes it possible to compare and contrast the effects of different conditions between experiments with the same confidence as within experiments. The impact of experiments in a theoretical research program is more cumulative only because theoretical research programs standardize an analytic simplification of a process, comparing oranges to oranges rather than to apples.

research programs in sociology. I have not made a census of them, but have been teaching them, providing students with examples of them, and coediting anthologies of them for most of the period I am describing and there are considerably more of them now than there were in the 1950s. Proper documentation of them would overbalance what is really just a research note, but would include at the least:

- affect control theory
- affect exchange theory
- power and bargaining theory
- behavioral exchange theory
- critical mass theory
- elementary theory
- E-state structuralism
- the game theory of power
- at least two programs of theory and research on social dilemmas
- identity control theory
- justice theory
- social identity theory
- social influence network theory
- the theory of collective action
- the theory of legitimate authority

The proper documentation of at least these examples can be found in one or more of Berger and Zelditch (1993, 2002), Berger, Willer, and Zelditch (2005), Burke (2006), and Sell, Chapter 18, this volume.

A theoretical research program can grow to be quite a complex structure, but it is complex in a different way than an effect program. Its structure is more coherent than an effect program's, because, like expectation states theory, theoretical research programs have a common core, like the concept of an expectation state, which interrelates their parts. This makes a considerable difference to how they grow, how much they grow, and the impact of how they grow.

VI. ASSESSMENT

Not only were there more theoretical research programs, but also there was more growth of theory in them,[6] growth of both theory and research was more

---

[6]Not all theoretical research programs achieve the growth their nature and functions make possible. But whether or not they succeed, they all have a greater capacity for theoretical growth than an effect program.

cumulative in them, and the growth of them had more of an impact on the application of theory than in effect programs.

There was more growth of theory in them in the first instance because testing, refining, and/or extending theory is what they are all about. But more than that, like an effect program, a theoretical research program defines some problems as solved, others as unsolved. But, unlike an effect program, a theoretical research program focuses only on unsolved problems that bear on its theory. Furthermore, it defines some of them as more theoretically significant, as mattering more to the growth of its theory than others. Because the growth of theory is what they are about, they focus on those unsolved problems that are most theoretically significant.

They also define some problems as more solvable than others: they provide the conceptual and methodological resources required to solve some problems but not others; their theories guarantee the existence of some solutions but not others. They focus on the problems that are most solvable. Not only are these the problems that the program's conceptual and methodological resources most enable it to solve (making it capable of solving more problems), but also the promise of a solution mobilizes more resources, more theorists, and more researchers in solving them, and it sustains their commitment to the search for a solution for longer periods of time.

But the fact that they are capable of agreeing on a solution is also important to their growth. A theoretical research program not only defines which unsolved problems are theoretically significant or not and which are solvable or not, it also defines which solutions of them are acceptable or not. The standards by which it assesses which solutions are acceptable are shared by participants in the program. Therefore, although many competing solutions are likely to be offered, theoretical research programs are capable of reaching agreement on them. They are all likely to be variants of the same basic idea but, in any case, because the standards by which they are assessed are widely shared, at least among participants in the program, disagreements over them are reconcilable by appeal to reason and evidence.[7]

Finally, despite the increasing complexity that derives from the many ways in which it grows, the entire structure of the theoretical research program remains coherent. No matter how much it proliferates, no matter how much its theories are elaborated, no matter how many variants of it compete with each other, the entire structure remains a network of interrelated theories and models organized around a common core of basic concepts, principles, and methods. Nor does it

---

[7]For more on the functions of theoretical research programs—definition of theoretical significance, solvability, and the acceptability of solutions—with examples, see Berger *et al.*, 2005.

matter how much, or in how many directions, the body of its theoretical and applied research proliferates. Because of the way a theoretically-oriented program focuses the choice of its problems, it remains relevant to the network of theories and models of the program.

Thus, theoretical research programs have a greater capacity for growth and sustain it for a longer period of time than effect programs, which tend either to peter out, as the group decision program did, or become incoherent, as the Asch program did, or morph into a theory program, as the Bales program did. The entire career of a theoretical research program is one sustained, continually evolving, yet continually coherent, body of theory and research.

The growth of both theory and research in a theoretical research program is also more cumulative than growth in an effect program. The growth of theory is more cumulative because, although there are several different ways that its theories grow, and therefore several different types of relations between theories, each type of growth in a theoretical research program builds on and enlarges its existing network of theories. Not only is each theory in its existing network interrelated with each other theory in it, but also each theory in it is interrelated with what went before and what came after it. Each elaboration of it, each proliferation of it, each integration of the theories in it with other theories in it, builds on and enlarges on what went before. A theory cannot guarantee its fruitfulness, but if it is fruitful, a theory program can guarantee that its growth is cumulative.

If the experimental setting of the program is standardized, the growth of its research is also cumulative. The impact of each of its experiments, because the methods of each of them are comparable with those of the experiments that went before them, is cumulative because each is capable, on the one hand, of building on the existing state of the program's art, and, on the other hand, of enlarging it.

They also have a greater capacity for the growth of applied research than effect programs. One feature of effect experiments and programs that I have not dwelled on up to this point is that, because their effects are underanalyzed, their results are not generalizable. The problem of generalizing from one concrete instance to another depends on a theory; it requires definition of what constitutes an instance of the phenomenon and of its scope (i.e., the conditions under which a result is or is not applicable). Effect experiments and programs, because they do not define the instantiation and scope conditions of their effects, leave the question of their generalizability unanswered. Theoretically oriented experiments, because their instantiation and scope conditions are defined by the theories they test, make such generalization possible. Hence, they successfully bridge the gap between experiment that tests the theory and use of the theory to predict or explain the phenomena to which the theory is applicable in natural situations. The applied research that is integrated into the program is every bit as important as its theoretical research, and gives to theoretical research programs an explanatory power of which no effect program is capable.

VII. CHALLENGE

Because they randomize the allocation of subjects to treatments, experiments solve the problem of internal validity, the validity of causal inference, better than any other method. But common wisdom has it that there is a trade-off between internal and external validity—the generalizability of the inference.

An experiment has a problem of external validity, in the first instance, because the population, period, and setting of an experiment are fixed initial conditions. If any of them interact with its treatments, external validity is severely impaired. Because there is no way to randomize a constant, randomization does nothing to solve the problem. In addition, a laboratory experiment is only an analytic simplification of a more complex natural phenomenon. It is concretely very different from any natural setting. It is externally valid if its analytic simplification of a natural setting reproduces all of its theoretically relevant features with the same meaning for the subjects in the laboratory that they have in the natural setting. An experimental setting that succeeds in meaningfully creating all those properties of a more complex natural setting relevant to a theory is "experimentally real" (Berkowitz & Donerstein, 1982). No experiment reproduces the whole of any concrete natural situation. But every experiment aims for experimental realism of the phenomenon it does create, because the experiment is externally valid only if it is experimentally real. But experimental realism is easier said than done.

Nevertheless, the challenge that faces a theory-oriented experiment is not that it trades external validity for internal validity or analytic simplification, but that in fact no trade-off is acceptable: if it is theoretically oriented, then either it is generalizable or it is fatally flawed. The challenge is to achieve internal validity *and* analytic simplification *and* external validity.

But the increasing number of theoretical research programs is evidence that the challenge is often met. Application is the fundamental test of the external validity of a theory-oriented experiment: it is externally valid if the theory supported by it predicts and explains the behavior of the process it describes in any natural situation to which the theory is applicable. It is a test built into the definition of theoretical research programs and met by many of them. Not all have actually made applications and no doubt the growth of some of them has been arrested by the failure of their applications. But that any of them succeed at all implies that some experiments have met the challenge.

But those that fail present a challenge of their own. An experiment that fails the test may be externally invalid, but it also may have failed because of bad theory, bad measurement, or bad application. The explanation is a matter of empirical fact. The problem is how to assess it empirically. Assessment of the failure to meet the primary challenge is itself a challenge. In the first place, a strategic choice—which explanation to pursue and when to pursue it—is called for.

It can be an easy choice to make, if what fails is an experiment in the standardized experimental setting of a program with a history. Discrediting the external validity of its setting is likely to have too many undesirable consequences for too much else. Attention is more likely to be given to theory or application than external invalidity—a useful by-product of standardizing an experimental setting because detecting, testing, and remedying external invalidity may be a necessary condition of it but is not itself theory growth, and theory growth is the whole purpose of the experiment.

But the secondary challenge is more difficult to meet *ab initio*. Few guidelines are available that help make the strategic choice that has to be made. If the choice is to pursue external invalidity, the guidelines that help detect, test, and remedy external invalidities are, for the most part, confused and confusing. The muddled literature on external validity is not much help (Zelditch, Chapter 4, this volume). Much of it equivocates between abstract and concrete approaches to the problem. More concrete approaches are especially confusing because the more concrete the approach is, the more indeterminate the external validity of an experiment is.

On the other hand, the most unequivocally analytic approach overshoots the mark, conflating external invalidity with bad theory, bad measurement, and bad application (Zelditch, Chapter 4, this volume). A straightforward solution is to be as analytic as the theory-oriented experiment itself, but define its invalidity in purely methodological terms (Zelditch). The solution allows a distinction between external invalidity and other explanations of failed theoretical generalization. Furthermore, it refines and extends what we know about threats to external validity and methods of detecting, testing, and remedying them (Zelditch). But whether in fact it proves to be much help remains to be seen.

More helpful is the literature on the social psychology of the experiment. It is largely in psychology but is presumably familiar to anyone who experiments with human subjects. A by-product of this literature has been the evolution of a sustained body of theory and research on the use of experiments to detect, test, and remedy flaws in experiments. This has been profoundly important in meeting the challenge that faces the theory-oriented experiment. There is a problem with this literature, but in this case it is not that existing guidelines are muddled. The problem is that many of the important questions at stake are sociological, but few sociologists seem to be actively addressing them.

But, whether or not helpful, much of both literatures is after the fact. Either would be more helpful before the fact. It takes time, even years, to create a fruitful experimental setting. (For example, see Berger, Chapter 14, this volume.) Probably few researchers want to invest much time, energy, and money in a setting the theoretical payoff of which is uncertain until a very much deferred final test. It would be helpful to have guidelines that helped to anticipate the outcome before investing too much in it. But even less is available that helps before than after the fact. The problem is not that the literature is muddled. There just is not much literature. The emergence of so many continually growing theoretical research programs implies that there is in fact a lot known that would help a lot. But it is largely local, in unwritten laboratory cultures or unpublished working papers, scattered, unsystematic, and unshared.

The attentive reader has by now objected (I hope) that, at least with the secondary challenge, help is in plain sight. *Laboratory Experiments in the Social Sciences* provides the kind of guidance I have claimed is often difficult to find, both systematic guidelines, in Parts 1 and 2, and diffusing experience often available only in local laboratory lore in Parts 3 and 4. It is valuable in many ways, but one of its values is directly proportional to the value of meeting the primary challenge that faces the theory-oriented experiment—achieving analytic simplification and internal validity while also achieving experimental realism and external validity.

REFERENCES

Allen, V. L. (1965). Situational factors in conformity. *Advances in Experimental Social Psychology, 2*, 133–175.

Allen, V. L. (1977). Social support for nonconformity. *Advances in Experimental Social Psychology, 8*, 1–43.

Asch, S. E. (1951). Effects of group pressure upon the modification and distortion of judgments. In H. Guetzkow (Ed.), *Groups, leadership, and men* (pp. 177–190). Pittsburgh: Carnegie Press.

Back, K. (1951). Influence through social communication. *Journal of Abnormal and Social Psychology, 46*, 9–23.

Bales, R. F. (1953). The equilibrium problem in small groups. In T. Parsons, R. F. Bales, & E. H. Shils (Eds.), *Working papers in the theory of action* (pp. 111–161). Glencoe, IL: Free Press.

Bales, R. F., & Slater, P. (1955). Role differentiation in small decision making groups. In T. Parsons & R. F. Bales (Eds.), *Family, socialization and interaction process* (pp. 259–306). Glencoe, IL: Free Press.

Bales, R. F., Strodtbeck, F. L., Mills, T. M., & Rosebourough, M. E. (1951). Channels of communication in small groups. *ASR, 16*, 461–468.

Bavelas, A. (1950). Communication patterns in task-oriented groups. *Journal of the Acoustical Society of America, 22*, 725–730.

Bennett, E. B. (1955). Discussion, decision, commitment, and consensus in group decision. *Human Relations, 8*, 251–274.

Berger, J. (1958). *Relations between performance, reward, and action opportunities in small groups.* PhD. diss. Harvard University.

Berger, J. (2007). The external validity of experiments that test theories. In M. Webster & J. Sell (Eds.), *Laboratory experiments in the social sciences* (pp. 87–112). Burlington, MA: Elsevier.

Berger, J., Cohen, B. P., & Zelditch, M. (1966). Status characteristics and expectation states. In J. Berger, M. Zelditch, & B. Anderson (Eds.), *Sociological theories in progress* (Vol. 1, pp. 29–46). Boston: Houghton Mifflin.

Berger, J., & Conner, T. L. (1969). Performance expectations and behavior in small groups. *Acta Sociologica, 12,* 186–198.

Berger, J., Fisek, M. H., Norman, R. Z., & Zelditch, M. (1977). *Status characteristics and social interaction: An expectation states approach.* New York: Elsevier Scientific.

Berger, J. D., Willer, & Zelditch, M. (2005). Theory programs and theoretical problems. *Sociological Theory, 23,* 127–155.

Berger, J., & Zelditch, M. (Eds.). (1993). *Theoretical research programs: Studies in the growth of theory.* Stanford, CA: Stanford University Press.

Berger, J., & Zelditch, M. (1997). Theoretical research programs: A reformulation. In J. Szmatka, J. Skvoretz, & J. Berger (Eds.), *Status, network, and structure: Theory development in group processes* (pp. 29–46). Stanford, CA: Stanford University Press.

Berger, J., & Zelditch, M. (Eds.). (2002). *New directions in contemporary sociological theory.* New York: Rowman and Littlefield.

Berkowitz, L., & Donnerstein, E. (1982). External validity is more than skin deep: Some answers to criticisms of laboratory experiments. *American Psychologist, 37,* 245–257.

Burke, P. (1967). The development of task and social-emotional role differentiation. *Sociometry, 30,* 379–392.

Burke, P. (Ed.). (2006). *Contemporary social psychological theories.* Stanford, CA: Stanford University Press.

Deutsch, M., & Gerard, H. B. (1955). A study of normative and informational social influences upon individual judgment. *Journal of Abnormal and Social Psychology, 51,* 629–636.

Emerson, R. M. (1964). Power-dependence relations: Two experiments. *Sociometry, 27,* 282–298.

Faucheux, C., & Mackenzie, K. D. (1966). Task dependency of organizational centrality: Its behavioral consequences. *Journal of Experimental Social Psychology, 2,* 361–375.

Festinger, L. (1950). Informal social communication. *Psychological Review, 57,* 271–282.

Glanzer, M., & Glaser, R. (1961). Techniques for the study of group structure and behavior: II. Empirical studies of the effects of structure in small groups. *Psychological Bulletin, 58,* 1–27.

Lewin, K. (1947). Group decision and social change. In T. M. Newcomb & E. L. Hartley (Eds.), *Readings in social psychology* (pp. 330–344). New York: Henry Holt.

Lewis, G. H. (1972). Role differentiation. *American Sociological Review, 37,* 424–434.

Mulder, M. (1960). Communication structure, decision structure, and group performance. *Sociometry, 23,* 1–14.

Schachter, S. (1951). Deviation, rejection, and communication. *Journal of Abnormal and Social Psychology, 46,* 190–207.

Sell, J. (2007). Social dilemma experiments in political science, economics, and sociology. In M. Webster & J. Sell (Eds.), *Laboratory experiments in the social sciences* (pp. 459–479). Burlington, MA: Elsevier.

Sherif, M. (1935). A study of some social factors in perception. *Archives of Psychology, 27.* N. 187.

Wagner, D. G., & Berger, J. (2002). Expectation states theory: An evolving research program. In J. Berger & M. Zelditch (Eds.). *New directions in contemporary sociological theory* (pp. 41–76). New York: Rowman and Littlefield.

Zelditch, M. (2007). The external validity of experiments that test theories. In M. Webster & J. Sell (Eds.), *Laboratory experiments in the social sciences* (pp. 87–112). Burlington, MA: Elsevier.

INDEX